RSMeans®

Site Work & Landscape Cost Data

18th Annual Edition

1999

First Printing

Foreword

R.S. Means Co., Inc. is owned by CMD Group, a leading worldwide provider of proprietary construction information. CMD Group is comprised of three synergistic product groups crafted to be the complete resource for reliable, timely and actionable construction market data. In North America, CMD Group encompasses: Architects' First Source, an innovative product selection and specification solution in print and on the Internet; Construction Market Data (CMD), the source for construction activity information, as well as early planning reports for the design community; Associated Construction Publications, with 14 magazines, one of the largest editorial networks dedicated to U.S. highway and heavy construction coverage; Manufacturer's Survey Associates (MSA), a leading estimating and quantity survey firm in the U.S.; R.S. Means, the authority on construction cost data in North America; CMD Canada, the leading supplier of project information, industry news and forecasting data products for the Canadian construction industry; and BIMSA/Mexico, the dominant distributor of information on building projects and construction throughout Mexico. Worldwide, CMD Group includes Byggfakta Scandinavia, providing construction market data to Denmark, Estonia, Finland, Norway and Sweden; and Cordell Building Information Services, the market leader for construction and cost information in Australia.

Our Mission

Since 1942, R.S. Means Company, Inc. has been actively engaged in construction cost publishing and consulting throughout North America.

Today, over fifty years after the company began, our primary objective remains the same: to provide you, the construction and facilities professional, with the most current and comprehensive construction cost data possible.

Whether you are a contractor, an owner, an architect, an engineer, a facilities manager, or anyone else who needs a quick construction cost estimate, you'll find this publication to be a highly useful and necessary tool.

Today, with the constant flow of new construction methods and materials, it's difficult to find the time to look at and evaluate all the different construction cost possibilities. In addition, because labor and material costs keep changing, last year's cost information is not a reliable basis for today's estimate or budget.

That's why so many construction professionals turn to R.S. Means. We keep track of the costs for you, along with a wide range of other key information, from city cost indexes . . . to productivity rates . . . to crew composition . . . to contractor's overhead and profit rates.

R.S. Means performs these functions by collecting data from all facets of the industry, and organizing it in a format that is instantly accessible to you. From the preliminary budget to the detailed unit price estimate, you'll find the data in this book useful for all phases of construction cost determination.

The Staff, the Organization, and Our Services

When you purchase one of R.S. Means' publications, you are in effect hiring the services of a full-time staff of construction and engineering professionals.

Our thoroughly experienced and highly qualified staff works daily at collecting, analyzing, and disseminating comprehensive cost information for your needs. These staff members have years of practical construction experience and engineering training prior to joining the firm. As a result, you can count on them not only for the cost figures, but also for additional background reference information that will help you create a realistic estimate.

The Means organization is always prepared to help you solve construction problems through its five major divisions: Construction and Cost Data Publishing, Electronic Products and Services, Consulting Services, Insurance Division, and Educational Services.

Besides a full array of construction cost estimating books, Means also publishes a number of other reference works for the construction industry. Subjects include construction estimating and project and business management; special topics such as HVAC, roofing, plumbing, and hazardous waste remediation; and a library of facility management references.

In addition, you can access all of our construction cost data through your computer with Means CostWorks '99 CD-ROM, an electronic tool that offers over 50,000 lines of Means construction cost data.

What's more, you can increase your knowledge and improve your construction estimating and management performance with a Means Construction Seminar or In-House Training Program. These two-day seminar programs offer unparalleled opportunities for everyone in your organization to get updated on a wide variety of construction-related issues.

Means also is a worldwide provider of construction cost management and analysis services for commercial and government owners and of claims and valuation services for insurers.

In short, R.S. Means can provide you with the tools and expertise for constructing accurate and dependable construction estimates and budgets in a variety of ways.

Robert Snow Means Established a Tradition of Quality That Continues Today

Robert Snow Means spent years building his company, making certain he always delivered a quality product.

Today, at R.S. Means, we do more than talk about the quality of our data and the usefulness of our books. We stand behind all of our data, from historical cost indexes... to construction materials and techniques... to current costs.

If you have any questions about our products or services, please call us toll-free at 1-800-334-3509. Our customer service representatives will be happy to assist you.

Table of Contents

How the Book Is Built: An Overview

A Powerful Construction Tool

You have in your hands one of the most powerful construction tools available today. A successful project is built on the foundation of an accurate and dependable estimate. This book will enable you to construct just such an estimate.

For the casual user the book is designed to be:

- quickly and easily understood so you can get right to your estimate
- filled with valuable information so you can understand the necessary factors that go into the cost estimate

For the regular user, the book is designed to be:

- a handy desk reference that can be quickly referred to for key costs
- a comprehensive, fully reliable source of current construction costs and productivity rates, so you'll be prepared to estimate any project
- a source book for preliminary project cost, product selections, and alternate materials and methods

To meet all of these requirements we have organized the book into the following clearly defined sections.

Quick Start

This one-page section (see following page) can quickly get you started on your estimate.

How To Use the Book: The Details

This section contains an in-depth explanation of how the book is arranged . . . and how you can use it to determine a reliable construction cost estimate. It includes information about how we develop our cost figures and how to completely prepare your estimate.

Unit Price Section

All cost data has been divided into the 16 divisions according to the MasterFormat system of classification and numbering as developed by the Construction Specifications Institute (CSI) and Construction Specifications Canada (CSC). For a listing of these divisions and an outline of their subdivisions, see the Unit Price Section Table of Contents.

Estimating tips are included at the beginning of each division.

Assemblies Section

The cost data in this section has been organized in an "Assemblies" format. These assemblies combine components of construction into functional systems and arrange them according to the 12 divisions of the UniFormat classification system. For a complete explanation of a typical "Assemblies" page, see "How To Use the Assemblies Cost Tables."

Reference Section

This section includes information on Reference Numbers, Crew Listings, Historical Cost Indexes, City Cost Indexes, Location Factors and a listing of Abbreviations. It is visually identified by a vertical gray bar on the edge of pages.

Reference Numbers: At the beginning of selected major classifications in the Unit Price Section are "reference numbers" shown in bold squares. These numbers refer you to related information in the Reference Section.

In this section, you'll find reference tables, explanations, and estimating information that support how we develop the unit price data. Also included are alternate pricing methods, technical data, and estimating procedures, along with information on design and economy in construction. You'll also find helpful tips on what to expect and what to avoid when estimating and constructing your project.

It is recommended that you refer to this Reference Section if a "reference number" appears within the major classification you are estimating.

Crew Listings: This section lists all the crews referenced in the book. For the purposes of this book, a crew is composed of more than one trade classification and/or the addition of power equipment to any trade classification. Power equipment is included in the cost of the crew. Costs are shown both with bare labor rates and with the installing contractor's overhead and profit added. For each, the total crew cost per eight-hour day and the composite cost per labor-hour are listed.

Historical Cost Indexes: These indexes provide you with data to adjust construction costs over time. If you know costs for a project completed in the past, you can use these indexes to calculate a rough estimate of what it would cost to construct the same project today.

City Cost Indexes: Obviously, costs vary depending on the regional economy. You can adjust the "national average" costs in this book to 305 major cities throughout the U.S. and Canada by using the data in this section. How to use information is included.

Location Factors, to quickly adjust the data to over 930 zip code areas, are included.

Abbreviations: A listing of the abbreviations used throughout this book, along with the terms they represent, is included.

Index

A comprehensive listing of all terms and subjects in this book to help you find what you need quickly.

The Scope of This Book

This book is designed to be as comprehensive and as easy to use as possible. To that end we have made certain assumptions:

1. We have established material prices based on a "national average."
2. We have computed labor costs based on a 30-city "national average" of union wage rates.

For a more detailed explanation of how the cost data is developed, see "How To Use the Book: The Details."

Quick Start

If you feel you are ready to use this book and don't think you need the detailed instructions that begin on the following page, this Quick Start section is for you.

These steps will allow you to get started estimating in a matter of minutes.

1 First, decide whether you require a Unit Price or Assemblies type estimate. Unit price estimating requires a breakdown of the work to individual items. Assemblies estimates combine individual items or components into building systems.

If you need to estimate each line item separately, follow the instructions for **Unit Prices.**

If you can use an estimate for the entire assembly or system, follow the instructions for **Assemblies.**

2 Find each cost data section you need in the Table of Contents (either for Unit Prices or Assemblies).

Unit Prices: The cost data for Unit Prices has been divided into 16 divisions according to the CSI MasterFormat.

Assemblies: The cost data for Assemblies has been divided into 12 divisions according to the UniFormat.

3 Turn to the indicated section and locate the line item or assemblies table you need for your estimate. Portions of a sample page layout from both the Unit Price Listings and the Assemblies Cost Tables appear below.

Unit Prices: If there is a reference number listed at the beginning of the section, it refers to additional information you may find useful. See the referenced section for additional information.

- Note the crew code designation. You'll find full descriptions of crews in the Crew Listings including labor-hour and equipment costs.

Assemblies: The Assemblies (*not* shown in full here) are generally separated into three parts: 1) an illustration of the system to be estimated; 2) the components and related costs of a typical system; and 3) the costs for similar systems with dimensional and/or size variations.

4 Determine the total number of units your job will require.

Unit Prices: Note the unit of measure for the material you're using is listed under "Unit."

- Bare Costs: These figures show unit costs for materials and installation. Labor and equipment costs are calculated according to crew costs and average daily output. Bare costs do not contain allowances for overhead, profit or taxes.
- "Labor-hours" allows you to calculate the total labor-hours to complete that task. Just multiply the quantity of work by this figure for an estimate of activity duration.

Assemblies: Note the unit of measure for the assembly or system you're estimating is listed in the System Components section.

5 Then multiply the total units by . . .

Unit Prices: "Total Incl. O&P" which stands for the total cost including the installing contractor's overhead and profit. (See the "How To Use the Unit Price Pages" for a complete explanation.)

Assemblies: The "Total" in the right-hand column, which is the total cost including the installing contractor's overhead and profit. (See the "How To Use the Assemblies Cost Tables" section for a complete explanation.)

Material and equipment cost figures include a 10% markup. For labor markups, see the inside back cover of this book. If the work is to be subcontracted, add the general contractor's markup, approximately 10%.

6 The price you calculate will be an estimate for either an individual item of work, or for a completed assembly or *system.*

7 Compile a list of all items or assemblies included in the total project. Summarize cost information, and add project overhead.

Localize costs by using the City Cost Indexes or Location Factors found in the Reference Section.

For a more complete explanation of the way costs are derived, please see the following sections.

Editors' Note: We urge you to spend time reading and understanding all of the supporting material and to take into consideration the reference material such as Crews Listing and the "reference numbers."

Unit Price Pages

025 | Paving & Surfacing

104	025 100 | Walk/Rd/Parkng Paving	CREW	DAILY OUTPUT	LABOR-HOURS	UNIT	1999 BARE COSTS MAT.	LABOR	EQUIP.	TOTAL	TOTAL INCL O&P
0010	ASPHALTIC CONCRETE PAVEMENT for highways R025-110									
0020	and large paved areas									
0810	Binder course, 1-1/2" thick	B-25	630	.140	Ton	26.50	3.29	2.84	32.63	37.50
0811	2" thick		690	.128		26.50	3	2.59	32.09	37
0812	3" thick		800	.110		26.50	2.59	2.24	31.33	36
0813	4" thick		850	.104		26.50	2.44	2.10	31.04	35.50
0850	Wearing course, 1" thick	B-25B	575	.167		29	4	3.54	36.54	42
0851	1-1/2" thick		630	.152		29	3.65	3.23	35.88	41

Assemblies Pages

SITE WORK — A12.7-310 | Concrete Retaining Walls

There are four basic types of Concrete Retaining Wall Systems: reinforced concrete with level backfill; reinforced concrete with sloped backfill or surcharge; unreinforced with level backfill; and unreinforced with sloped backfill or surcharge. System elements include: all necessary forms (4 uses); 3,000 p.s.i. concrete with an 8" chute; all necessary reinforcing steel; and underdrain. Exposed concrete is patched and rubbed.

The Expanded System Listing shows walls that range in thickness from 10" to 18" for reinforced concrete walls with level backfill and 12" to 24" for reinforced walls with sloped backfill. Walls range from a height of 4' to 20'. Unreinforced level and sloped backfill walls range from a height of 3' to 10'.

System Components	QUANTITY	UNIT	COST PER L.F. MAT.	INST.	TOTAL
SYSTEM 12.7-310-1000					
CONC.RETAIN. WALL REINFORCED, LEVEL BACKFILL 4' HIGH					
Forms in place, cont. wall footing & keyway, 4 uses	2.000	S.F.	1.38	5.36	6.74
Forms in place, retaining wall forms, battered to 8' high, 4 uses	8.000	SFCA	4.80	41.20	46
Reinforcing in place, walls, #3 to #7	.004	Ton	2.24	2.24	4.48
Concrete ready mix, regular weight, 3000 psi	.204	C.Y.	13.67		13.67
Placing concrete and vibrating footing con., shallow direct chute	.074	C.Y.		1.09	1.09
Placing concrete and vibrating walls, 8' thick, direct chute	.130	C.Y.		2.56	2.56
Gravel drain and pipe installed, bank gravel, perf. pipe	1.000	L.F.	1.42	1.27	2.69
Finish walls and break ties, patch walls	4.000	S.F.	.12	2.32	2.44
TOTAL			23.63	56.04	79.67

12.7-310	Concrete Retaining Walls	COST PER L.F. MAT.	INST.	TOTAL
1000	Conc.retain.wal, reinforced, level backfill, 4' high x 2'-2" base,10"thick	23.50	56	79.50
1200	6' high x 3'-3" base, 10" thick	34.50	81	115.50
1400	8' high x 4'-3" base, 10" thick	45	107	152

How to Use the Book: The Details

What's Behind the Numbers? The Development of Cost Data

The staff at R.S. Means continuously monitors developments in the construction industry in order to ensure reliable, thorough and up-to-date cost information.

While *overall* construction costs may vary relative to general economic conditions, price fluctuations within the industry are dependent upon many factors. Individual price variations may, in fact, be opposite to overall economic trends. Therefore, costs are continually monitored and complete updates are published yearly. Also, new items are frequently added in response to changes in materials and methods.

Costs—$ (U.S.)

All costs represent U.S. national averages and are given in U.S. dollars. The Means City Cost Indexes can be used to adjust costs to a particular location. The City Cost Indexes for Canada can be used to adjust U.S. national averages to local costs in Canadian dollars.

Material Costs

The R.S. Means staff contacts manufacturers, dealers, distributors, and contractors all across the U.S. and Canada to determine national average material costs. If you have access to current material costs for your specific location, you may wish to make adjustments to reflect differences from the national average. Included within material costs are fasteners for a normal installation. R.S. Means engineers use manufacturers' recommendations, written specifications and/or standard construction practice for size and spacing of fasteners. Adjustments to material costs may be required for your specific application or location. Material costs do not include sales tax.

Labor Costs

Labor costs are based on the average of wage rates from 30 major U.S. cities. Rates are determined from labor union agreements or prevailing wages for construction trades for the current year. Rates along with overhead and profit markups are listed on the inside back cover of this book.

- If wage rates in your area vary from those used in this book, or if rate increases are expected within a given year, labor costs should be adjusted accordingly.

Labor costs reflect productivity based on actual working conditions. These figures include time spent during a normal workday on tasks other than actual installation, such as material receiving and handling, mobilization at site, site movement, breaks, and cleanup.

Productivity data is developed over an extended period so as not to be influenced by abnormal variations, and reflects a typical average.

Equipment Costs

Equipment costs include not only rental, but also operating costs for equipment under normal use. The operating costs include parts and labor for routine servicing such as repair and replacement of pumps, filters and worn lines. Normal operating expendables such as fuel, lubricants, tires and electricity (where applicable) are also included. Extraordinary operating expendables with highly variable wear patterns such as diamond bits and blades are excluded. These costs are included under materials. Equipment rental rates are obtained from industry sources throughout North America—contractors, suppliers, dealers, manufacturers, and distributors.

Crew Equipment Cost/Day—The power equipment required for each crew is included in the crew cost. The daily cost for crew equipment is based on dividing the weekly bare rental rate by 5 (number of working days per week), and then adding the hourly operating cost times 8 (hours per day). This "Crew Equipment Cost/Day" is listed in Subdivision 016.

General Conditions

Cost data in this book is presented in two ways: Bare Costs and Total Cost including O&P (Overhead and Profit). General Conditions, when applicable, should also be added to the Total Cost including O&P. The costs for General Conditions are listed in Division 1 of the Unit Price Section and the Reference Section of this book. General Conditions for the *Installing Contractor* may range from 0% to 10% of the Total Cost including O&P. For the *General* or *Prime Contractor*, costs for General Conditions may range from 5% to 15% of the Total Cost including O&P, with a figure of 10% as the most typical allowance.

Overhead and Profit

Total Cost including O&P for the *Installing Contractor* is shown in the last column on both the Unit Price and the Assemblies pages of this book. This figure is the sum of the bare material cost plus 10% for profit, the base labor cost plus total overhead and profit, and the bare equipment cost plus 10% for profit. Details for the calculation of Overhead and Profit on labor are shown on the inside back cover and in the Reference Section of this book. (See the "How to Use the Unit Price Pages" for an example of this calculation.)

Factors Affecting Costs

Costs can vary depending upon a number of variables. Here's how we have handled the main factors affecting costs.

Quality—The prices for materials and the workmanship upon which productivity is based represent sound construction work. They are also in line with U.S. government specifications.

Overtime—We have made no allowance for overtime. If you anticipate premium time or work beyond normal working hours, be sure to make an appropriate adjustment to your labor costs.

Productivity—The productivity, daily output, and labor-hour figures for each line item are based on working an eight-hour day in daylight hours in moderate temperatures. For work that extends beyond normal work hours or is performed under adverse conditions, productivity may decrease. (See the section in "How To Use the Unit Price Pages" for more on productivity.)

Size of Project—The size, scope of work, and type of construction project will have a significant impact on cost. Economies of scale can reduce costs for large projects. Unit costs can often run higher for small projects.

Location—Material prices in this book are for metropolitan areas. However, in dense urban areas, traffic and site storage limitations may increase costs. Beyond a 20-mile radius of large cities, extra trucking or transportation charges may also increase the material costs slightly. On the other hand, lower wage rates may be in effect. Be sure to consider both these factors when preparing an estimate, particularly if the job site is located in a central city or remote rural location.

In addition, highly specialized subcontract items may require travel and per diem expenses for mechanics.

Other factors—

- season of year
- contractor management
- weather conditions
- local union restrictions
- building code requirements
- availability of:
  - adequate energy
  - skilled labor
  - building materials
- owner's special requirements/restrictions
- safety requirements
- environmental considerations
- traffic control

Unpredictable Factors—General business conditions influence "in-place" costs of all items. Substitute materials and construction methods may have to be employed. These may affect the installed cost and/or life cycle costs. Such factors may be difficult to evaluate and cannot necessarily be predicted on the basis of the job's location in a particular section of the country. Thus, where these factors apply, you may find significant, but unavoidable cost variations for which you will have to apply a measure of judgment to your estimate.

Rounding of Costs

In general, all unit prices in excess of $5.00 have been rounded to make them easier to use and still maintain adequate precision of the results. The rounding rules we have chosen are in the following table.

Prices from . . .	Rounded to the nearest . . .
$.01 to $5.00	$.01
$5.01 to $20.00	$.05
$20.01 to $100.00	$.50
$100.01 to $300.00	$1.00
$300.01 to $1,000.00	$5.00
$1,000.01 to $10,000.00	$25.00
$10,000.01 to $50,000.00	$100.00
$50,000.01 and above	$500.00

How Subcontracted Items Affect Costs

A considerable portion of all large construction jobs is usually subcontracted. In fact, the percentage done by subcontractors is constantly increasing and may run over 90%. Since the workmen employed by these companies do nothing else but install their particular product, they soon become expert in that line. The result is, installation by these firms is accomplished so efficiently that the total in-place cost, even adding the general contractor's overhead and profit, is no more and often less than if the principal contractor had handled the installation himself. There is, moreover, the big advantage of having the work done right. Companies that deal with construction specialties are anxious to have their product perform well and consequently the installation will be the best possible.

Contingencies

The contractor should consider inflationary price trends and possible material shortages during the course of the job. These escalation factors are dependent upon both economic conditions and the anticipated time between the estimate and actual construction. If drawings are not complete or approved or a budget cost is wanted, it is wise to add 5% to 10%. Contingencies then are a matter of judgment. Additional allowances for contingencies are shown in Division 1.

Estimating Precision

When making a construction cost estimate, ignore the cents column. Only use the total per unit cost to the nearest dollar. The cents will average up in a column of figures. A construction cost estimate of $257,323.37 is cumbersome. A figure of $257,325 is certainly more sensible and $257,000 is better and just as likely to be right.

If you follow this simple instruction, the time saved is tremendous with an added important advantage. Using round figures leaves the professional estimator free to exercise judgment and common sense rather than being overcome and befuddled by a mass of computations.

When the estimate is done, make a rough check of the big items for correct location of the decimal point. That is important. A large error can creep in if you write down $300 when it should be $3,000. Also check the list to be sure you have not omitted any large item. A common error is to overlook, for example, compaction or to forget seeding or mulching, or to otherwise commit a gross omission. No amount of accuracy in prices can compensate for such an oversight.

It is important to keep bare costs and costs that already include the subcontractor's overhead and profit separate since different markups will have to be applied to each category. Organize your estimating procedures to minimize confusion and simplify checking to ensure against omissions and/or duplications.

The Quantity Take-Off

Here are a few simplified rules for handling numbers and measurements when "taking off" quantities from a set of plans. A correct quantity take-off is critical to the success of an estimate, since no estimate will be reliable if a mistake is made in the quantity take-off, no matter how accurate the unit price information.

1. Use preprinted forms for orderly sequence of dimensions and locations. (R.S. Means provides various forms for this purpose, including the R.S. Means *Quantity Sheet* or the *Consolidated Cost Analysis Sheet.)*
2. Be consistent when listing dimensions, for example:
   Length × Width × Height.
3. Use printed dimensions where given.
4. When possible, add up printed dimensions for a single entry.
5. Measure all other dimensions carefully.
6. Use each set of dimensions to calculate multiple quantities where possible.
7. Convert feet and inch measurements to decimal feet when listing. Memorize any decimal equivalents to .01 parts of a foot.
8. Do not "round off" until the final summary of quantities.
9. Mark drawings as quantities are assembled.
10. Keep similar items together, different items separate.
11. Identify sections and drawing numbers to aid in future checking for completeness.
12. Measure or list everything that drawings show.
13. It may be necessary to list items not called for because the quantity surveyor/ estimator is able to identify additional materials required to make the job complete.
14. Be alert for: (a) notes on plans such as N.T.S. (Not To Scale); (b) Changes in the scale used throughout the drawings; (c) Drawings reduced to 1/2 or 1/4 the original size; (d) Discrepancies between the specifications and the plans.
15. Use only the front side of each piece of paper or form, except for certain preprinted forms.

Working with Plans

When working with plans for an estimate, try to approach each job in the same manner.

Keep a uniform and consistent system. This will greatly reduce the chance of omission.

The following short cuts can be used successfully when making a quantity take-off:

1. Abbreviate when possible.
2. List all gross dimensions that can be either used again for different quantities or used as a rough check of other quantities for approximate verifications.
3. Convert to decimals when working with feet and inches.
4. Multiply the large numbers first to reduce rounding errors.
5. Do not convert units until the final answer is obtained. When estimating concrete work, keep all volumes to the nearest cubic foot, then summarize and convert to cubic yards.
6. Take advantage of design symmetry or repetition:
   Repetitive designs
   Symmetrical design around a center line
   Similar layouts
7. When figuring alternates, it is best to total all items that are involved in the basic system, then total all items that are involved in the alternates. Thus you work with positive numbers in all cases. When adds and deducts are used, it is often confusing to know whether to add or subtract a given item, especially on a complicated or involved alternate.

Final Checklist

Estimating can be a straightforward process provided you remember the basics. Here's a checklist of some of the items you should remember to do before completing your estimate.

Did you remember to . . .

- factor in the City Cost Index for your locale
- take into consideration which items have been marked up and by how much
- mark up the entire estimate sufficiently for your purposes
- read the background information on techniques and technical matters that could impact your project time span and cost
- include all components of your project in the final estimate
- double check your figures to be sure of your accuracy
- call R.S. Means if you have any questions about your estimate or the data you've found in our publications

Remember, R.S. Means stands behind its publications. If you have any questions about your estimate . . . about the costs you've used from our books . . . or even about the technical aspects of the job that may affect your estimate, feel free to call the R.S. Means editors at 1-781-585-7898 or 1-800-448-8182.

Unit Price Section

Table of Contents

How to Use the Unit Price Pages

The following is a detailed explanation of a sample entry in the Unit Price Section. Next to each bold number below is the item being described with appropriate component of the sample entry following in parenthesis. Some prices are listed as bare costs, others as costs that include overhead and profit of the installing contractor. In most cases, if the work is to be subcontracted, the general contractor will need to add an additional markup (R.S. Means suggests using 10%) to the figures in the column "Total Incl. O&P."

1 Division Number/Title (025/Paving & Surfacing)

Use the Unit Price Section Table of Contents to locate specific items. The sections are classified according to the CSI MasterFormat.

2 Line Numbers (025 104 0851)

Each unit price line item has been assigned a unique 10-digit code based on the 5-digit CSI MasterFormat classification.

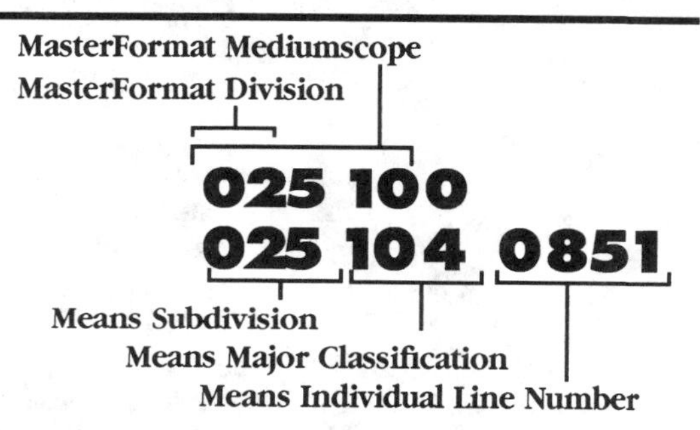

3 Description (Asphaltic Concrete Pavement, Wearing Course, 1-1/2" thick)

Each line item is described in detail. Sub-items and additional sizes are indented beneath the appropriate line items. The first line or two after the main item (in boldface) may contain descriptive information that pertains to all line items beneath this boldface listing.

4 Reference Number Information

R025 -110 You'll see reference numbers shown in bold squares at the beginning of some major classifications. These refer to related items in the Reference Section, visually identified by a vertical gray bar on the edge of pages.

The relation may be: (1) an estimating procedure that should be read before estimating, (2) an alternate pricing method, or (3) technical information.

The "R" designates the Reference Section. The numbers refer to the MasterFormat classification system.

It is strongly recommended that you review all reference numbers that appear within the major classification you are estimating.

Example: The square number above is directing you to refer to the reference number R025-110. This particular reference number shows how the unit price lines for asphaltic concrete pavement were formulated and costs derived.

025 | Paving & Surfacing

025 100 | Walk/Rd/Parkng Paving			CREW	DAILY OUTPUT	LABOR-HOURS	UNIT	1999 BARE COSTS MAT.	LABOR	EQUIP.	TOTAL	TOTAL INCL O&P
104	0010	AS[illegible]ONCR[illegible]TE PAVEMENT fo[illegible]highway[illegible] R025-110									
	0020	and [illegible]paved [illegible]eas									
	0810	[illegible]cours[illegible], 1-1/2" thick	B-25	630	.140	Ton	26.50	3.29	2.84	32.63	37.50
	0811	2" thick		[illegible]	.128		26.50	[illegible]	2.59	32.09	37
	0812	3" thick		[illegible]	[illegible]		26.50	[illegible]	2.24	31.33	36
	0813	4" thick		[illegible]	[illegible]		[illegible]	[illegible]	2.10	31.04	35.50
	0850	Wearing course, 1" thick	B-25	575	.167		29	4	3.54	36.54	42
	0851	1-1/2" thick		630	.152		29	3.65	3.23	35.88	41
	0852	2" thick		690	.139		[illegible]	[illegible]	[illegible]	[illegible]	40.50
	0853	2-1/2" thick		[illegible]	[illegible]		29	3.08	2.73	34.81	[illegible]
	0854	3" thick		800	.120		29	2.87	2.54	34.41	[illegible]
	1000	Pavement replacement over trench, 2" thick	[illegible]37	90	.533	S[illegible]	3.2[illegible]	12.15	1.70	17.10	[illegible]
	1050	4" thick		70	.686		6.4[illegible]	15.60	2.18	24.23	[illegible]

Crew (B-25B)

The "Crew" column designates the typical trade or crew used to install the item. If an installation can be accomplished by one trade and requires no power equipment, that trade and the number of workers are listed (for example, "1 Clab"). If an installation requires a composite crew, a crew code designation is listed (for example, "B-25B"). You'll find full details on all composite crews in the Crew Listings.

- For a complete list of all trades utilized in this book and their abbreviations, see the inside back cover.

Crews

Crew No.	Bare Costs		Incl. Subs O & P		Cost Per Labor-Hour	
Crew B-25B	Hr.	Daily	Hr.	Daily	Bare Costs	Incl. O&P
1 Labor Foreman	$23.45	$187.60	$36.90	$295.20	$23.93	$37.16
7 Laborers	21.45	1201.20	33.75	1890.00		
4 Equip. Oper. (medium)	28.40	908.80	43.20	1382.40		
1 Asphalt Paver, 130 H.P.		1299.00		1428.90		
2 Rollers, Steel Wheel		490.20		539.20		
1 Roller, Pneumatic Wheel		244.20		268.60	21.18	23.30
96 L.H., Daily Totals		$4331.00		$5804.30	$45.11	$60.46

Productivity: Daily Output (630)/Labor-Hours (.152)

The "Daily Output" represents the typical number of units the designated crew will install in a normal 8-hour day. To find out the number of days the given crew would require to complete the installation, divide your quantity by the daily output. For example:

Quantity	÷	Daily Output	=	Duration
12,000 Ton	÷	630 Ton/ Crew Day	=	19 Crew Days

The "Labor-Hours" figure represents the number of labor-hours required to install one unit of work. To find out the number of labor-hours required for your particular task, multiply the quantity of the item times the number of labor-hours shown. For example:

Quantity	x	Productivity Rate	=	Duration
12,000 Ton	x	.152 Labor-Hours/ Ton	=	1824 Labor-Hours

Unit (Ton)

The abbreviated designation indicates the unit of measure upon which the price, production, and crew are based. For a complete listing of abbreviations refer to the Abbreviations listing in this book.

Bare Costs:

Mat. (Bare Material Cost) (29)

The unit material cost is the "bare" material cost with no overhead and profit included. *Costs shown reflect national average material prices for January of the current year and include delivery to the job site. No sales taxes are included.*

Labor (3.65)

The unit labor cost is derived by multiplying bare labor-hour costs for Crew B-25B by labor-hour units. The bare labor-hour cost is found in the Crew Section under B-25B. (If a trade is listed, the hourly labor cost—the wage rate—is found on the inside back cover.)

Labor-Hour Cost Crew B-25B	x	Labor-Hour Units	=	Labor
$23.93	x	.152	=	$3.65

Equip. (Equipment) (3.23)

Equipment costs for each crew are listed in the description of each crew. Tools or equipment whose value justifies purchase or ownership by a contractor are considered overhead as shown on the inside back cover. The unit equipment cost is derived by multiplying the bare equipment hourly cost by the labor-hour units.

Equipment Cost Crew B-25B	x	Labor-Hour Units	=	Equip.
$21.18	x	.152	=	$3.23

Total (35.88)

The total of the bare costs is the arithmetic total of the three previous columns: mat., labor, and equip.

Material	+	Labor	+	Equip.	=	Total
$29.00	+	$3.65	+	$3.23	=	$35.88

Total Costs Including O&P

This figure is the sum of the bare material cost plus 10% for profit; the bare labor cost plus total overhead and profit (per the inside back cover or, if a crew is listed, from the crew listings); and the bare equipment cost plus 10% for profit.

Material is Bare Material Cost + 10% = $29.00 + $2.90	=	$31.90
Labor for Crew B-25B = Labor-Hour Cost ($37.16) x Labor-Hour Units (.152)	=	$ 5.65
Equip. is Bare Equip. Cost + 10% = $3.23 + $.32	=	$ 3.55
Total (Rounded)	=	$41.00

Division 1
General Requirements

Estimating Tips

The General Requirements of any contract are very important to both the bidder and the owner. These lay the ground rules under which the contract will be executed and have a significant influence on the cost of operations. Therefore, it is extremely important to thoroughly read and understand the General Requirements both before preparing an estimate and when the estimate is complete, to ascertain that nothing in the contract is overlooked. Caution should be exercised when applying items listed in Division 1 to an estimate. Many of the items are included in the unit prices listed in the other divisions such as mark-ups on labor and company overhead.

010 Overhead & Miscellaneous Data

- Before determining a final cost estimate, it is a good practice to review all the items listed in subdivision 010 to make final adjustments for items that may need customizing to specific job conditions.

013 Submittals

- Requirements for initial and periodic submittals can represent a significant cost to the General Requirements of a job. Thoroughly check the submittal specifications when estimating a project to determine any costs that should be included.

014 Quality Control

- All projects will require some degree of Quality Control. This cost is not included in the unit cost of construction listed in each division. Depending upon the terms of the contract, the various costs of inspection and testing can be the responsibility of either the owner or the contractor. Be sure to include the required costs in your estimate.

015 Construction Facilities & Temporary Controls

- Barricades, access roads, safety nets, scaffolding, security and many more requirements for the execution of a safe project are elements of direct cost. These costs can easily be overlooked when preparing an estimate. When looking through the major classifications of this subdivision, determine which items apply to each division in your estimate.

016 Material & Equipment

- This subdivision contains transportation, handling, storage, protection and product options and substitutions. Listed in this cost manual are average equipment rental rates for all types of equipment. This is useful information when estimating the time and materials requirement of any particular operation in order to establish a unit or total cost.
- A good rule of thumb is that weekly rental is 3 times daily rental and that monthly rental is 3 times weekly rental.
- The figures in the column for Crew Equipment Cost represent the rental rate used in determining the daily cost of equipment in a crew. It is calculated by dividing the weekly rate by 5 days and adding the hourly operating cost times 8 hours.

017 Contract Closeout

- When preparing an estimate, read the specifications to determine the requirements for Contract Closeout thoroughly. Final cleaning, record documentation, operation and maintenance data, warranties and bonds, and spare parts and maintenance materials can all be elements of cost for the completion of a contract. Do not overlook these in your estimate.

018 Maintenance

- If maintenance and repair are included in your contract, they require special attention. To estimate the cost to remove and replace any unit usually requires a site visit to determine the accessibility and the specific difficulty at that location. Obstructions, dust control, safety, and often overtime hours must be considered when preparing your estimate.

Reference Numbers

Reference numbers are shown in bold squares at the beginning of some major classifications. These numbers refer to related items in the Reference Section. The reference information may be an estimating procedure, an alternate pricing method or technical information.

Note: Not all subdivisions listed here necessarily appear in this publication.

010 | Overhead & Miscellaneous Data

		010 000 | Overhead	CREW	DAILY OUTPUT	LABOR-HOURS	UNIT	1999 BARE COSTS MAT.	LABOR	EQUIP.	TOTAL	TOTAL INCL O&P	
004	0011	**ARCHITECTURAL FEES** R010-010										004
	0020	For new construction										
	0060	Minimum				Project					4.90%	
	0090	Maximum				"					16%	
008	0011	**BOND, PERFORMANCE** See 010-068										008
012	0010	**CONSTRUCTION COST INDEX** (Reference) over 930 zip code locations in										012
	0020	The U.S. and Canada, total bldg cost, min. (Dalton, GA)				%					66.90%	
	0050	Average									100%	
	0100	Maximum (New York, NY)									133.90%	
016	0010	**CONSTRUCTION MANAGEMENT FEES** $1,000,000 job, minimum				Project					4.50%	016
	0050	Maximum									7.50%	
	0300	$5,000,000 job, minimum									2.50%	
	0350	Maximum									4%	
020	0010	**CONTINGENCIES** Allowance to add at conceptual stage				Project					15%	020
	0050	Schematic stage									10%	
	0100	Preliminary working drawing stage									7%	
	0150	Final working drawing stage									2%	
022	0010	**CONTRACTOR EQUIPMENT** See division 016 R016-410										022
024	0010	**CREWS** For building construction, see How To Use This Book										024
028	0010	**ENGINEERING FEES** R010-030										028
	0800	Landscaping & site development, minimum				Contrct					2.50%	
	0900	Maximum				"					6%	
032	0010	**FACTORS** Cost adjustments										032
	0100	Add to construction costs for particular job requirements										
	1100	Equipment usage curtailment, add, minimum				Costs	1%	1%				
	1150	Maximum					3%	10%				
	1400	Material handling & storage limitation, add, minimum					1%	1%				
	1450	Maximum					6%	7%				
	2300	Temporary shoring and bracing, add, minimum					2%	5%				
	2350	Maximum					5%	12%				
	2400	Work inside prisons, add, minimum						30%				
	2450	Maximum						50%				
034	0010	**FIELD OFFICE EXPENSE**										034
	0100	Field office expense, office equipment rental average				Month	133			133	146	
	0120	Office supplies, average				"	85			85	93.50	
	0125	Office trailer rental, see division 015-904										
	0140	Telephone bill; avg. bill/month incl. long dist.				Month	235			235	259	
	0160	Field office lights & HVAC				"	88			88	97	
036	0010	**FIELD PERSONNEL** Clerk average				Week		260		260	410	036
	0100	Field engineer, minimum						625		625	985	
	0120	Average						815		815	1,285	
	0140	Maximum						935		935	1,475	
	0160	General purpose laborer, average						860		860	1,355	
	0180	Project manager, minimum						1,180		1,180	1,860	
	0200	Average						1,320		1,320	2,085	
	0220	Maximum						1,490		1,490	2,350	
	0240	Superintendent, minimum						1,125		1,125	1,775	
	0260	Average						1,245		1,245	1,965	
	0280	Maximum						1,405		1,405	2,215	
	0290	Timekeeper, average						725		725	1,145	
040	0010	**INSURANCE** Builders risk, standard, minimum R010-040				Job					.22%	040
	0050	Maximum									.59%	

010 | Overhead & Miscellaneous Data

010 000 | Overhead

		010 000 Overhead	CREW	DAILY OUTPUT	LABOR-HOURS	UNIT	1999 BARE COSTS MAT.	LABOR	EQUIP.	TOTAL	TOTAL INCL O&P	
040	0200	All-risk type, minimum (R010-060)				Job					.25%	040
	0250	Maximum									.62%	
	0400	Contractor's equipment floater, minimum				Value					.50%	
	0450	Maximum				″					1.50%	
	0600	Public liability, average				Job					1.55%	
	0800	Workers' compensation & employer's liability, average										
	0850	by trade, carpentry, general				Payroll		19.92%				
	0900	Clerical						.54%				
	0950	Concrete						19.12%				
	1000	Electrical						7.04%				
	1050	Excavation						11.59%				
	1250	Masonry						18.01%				
	1300	Painting & decorating						15.62%				
	1350	Pile driving						30.30%				
	1450	Plumbing						8.99%				
	1600	Steel erection, structural						42.77%				
	1650	Tile work, interior ceramic						10.33%				
	1700	Waterproofing, brush or hand caulking						8.46%				
	1800	Wrecking						42.49%				
	2000	Range of 35 trades in 50 states, excl. wrecking, minimum						2%				
	2100	Average						18.30%				
	2200	Maximum						132.92%				
046	0010	**LABOR INDEX** (Reference) For over 930 zip code locations in										046
	0020	the U.S. and Canada, minimum (Dalton, GA)				%		35%				
	0050	Average						100%				
	0100	Maximum (New York, NY)						160.60%				
048	0010	**MAIN OFFICE EXPENSE** Average for General Contractors (R010-050)										048
	0020	As a percentage of their annual volume										
	0125	Annual volume under 1 million dollars				% Vol.				13.60%		
	0145	Up to 2.5 million dollars								8%		
	0150	Up to 4.0 million dollars								6.80%		
	0200	Up to 7.0 million dollars								5.60%		
	0250	Up to 10 million dollars								5.10%		
	0300	Over 10 million dollars								3.90%		
052	0010	**MARK-UP** For General Contractors for change										052
	0100	of scope of job as bid										
	0200	Extra work, by subcontractors, add				%					10%	
	0250	By General Contractor, add									15%	
	0400	Omitted work, by subcontractors, deduct									5%	
	0450	By General Contractor, deduct									7.50%	
	0600	Overtime work, by subcontractors, add									15%	
	0650	By General Contractor, add									10%	
	1000	Installing contractors, on his own labor, minimum						48.80%				
	1100	Maximum						102%				
054	0011	**MATERIAL INDEX** For over 930 zip code locations in										054
	0020	the U.S. and Canada, minimum (Elizabethtown, KY)				%	91.80%					
	0040	Average					100%					
	0060	Maximum (Ketchikan, AK)					140.60%					
058	0010	**OVERHEAD** As percent of direct costs, minimum (R010-050)				%				5%		058
	0050	Average								12%		
	0100	Maximum (R010-070)								30%		
062	0010	**OVERHEAD & PROFIT** Allowance to add to items in this (R010-050)										062
	0020	book that do not include Subs O&P, average				%				25%		
	0100	Allowance to add to items in this book that										
	0110	do include Subs O&P, minimum				%					5%	

010 | Overhead & Miscellaneous Data

010 000 | Overhead

		010 000 Overhead	CREW	DAILY OUTPUT	LABOR-HOURS	UNIT	1999 BARE COSTS MAT.	LABOR	EQUIP.	TOTAL	TOTAL INCL O&P	
062	0150	Average R010-050				%					10%	062
	0200	Maximum									15%	
	0300	Typical, by size of project, under $100,000								30%		
	0350	$500,000 project								25%		
	0400	$2,000,000 project								20%		
	0450	Over $10,000,000 project								15%		
064	0010	**OVERTIME** For early completion of projects or where R010-110										064
	0020	labor shortages exist, add to usual labor, up to				Costs		100%				
068	0010	**PERFORMANCE BOND** For buildings, minimum				Job					.60%	068
	0200	Maximum									1.50%	
	0300	Highways & Bridges, resurfacing, minimum									.40%	
	0350	Maximum									.94%	
070	0010	**PERMITS** Rule of thumb, most cities, minimum				Job					.50%	070
	0100	Maximum				"					2%	
082	0010	**SMALL TOOLS** As % of contractor's work, minimum				Total					.50%	082
	0100	Maximum				"					2%	
086	0010	**TAXES** Sales tax, State, average R010-090				%	4.71%					086
	0050	Maximum					7.25%					
	0200	Social Security, on first $68,400 of wages R010-100						7.65%				
	0300	Unemployment, MA, combined Federal and State, minimum						2.60%				
	0350	Average						7%				
	0400	Maximum						8.50%				

013 | Submittals

		013 100 Progress Schedules	CREW	DAILY OUTPUT	LABOR-HOURS	UNIT	1999 BARE COSTS MAT.	LABOR	EQUIP.	TOTAL	TOTAL INCL O&P	
104	0010	**SCHEDULING** Critical path, as % of architectural fee, minimum				%					2%	104
	0100	Maximum				"					4%	
	0300	Computer-update, micro, no plots, minimum				Ea.					220	
	0400	Including plots, maximum				"					2,300	
	0600	Rule of thumb, CPM scheduling, small job				Job					.10%	
	0650	Large job									.05%	
	0700	Cost control, small job									.15%	
	0750	Large job									.04%	
		013 300 Survey Data										
306	0010	**SURVEYING** Conventional, topographical, minimum	A-7	3.30	7.273	Acre	16	175		191	287	306
	0100	Maximum	A-8	.60	53.333		48	1,250		1,298	1,975	
	0300	Lot location and lines, minimum, for large quantities	A-7	2	12		25	288		313	475	
	0320	Average	"	1.25	19.200		45	460		505	760	
	0400	Maximum, for small quantities	A-8	1	32		72	745		817	1,225	
	0600	Monuments, 3′ long	A-7	10	2.400	Ea.	19	57.50		76.50	110	
	0800	Property lines, perimeter, cleared land	"	1,000	.024	L.F.	.03	.58		.61	.92	
	0900	Wooded land	A-8	875	.037	"	.05	.85		.90	1.37	
	1100	Crew for building layout, 2 person crew	A-6	1	16	Day		410		410	625	
	1200	3 person crew	A-7	1	24			575		575	885	
	1300	4 person crew	A-8	1	32			745		745	1,150	
	1500	Aerial surveying, including ground control, minimum fee, 10 acres				Total					5,000	
	1510	100 acres									8,300	
	1550	From existing photography, deduct									1,220	

Important: See the Reference Section for critical supporting data - Reference Nos., Crews, & City Cost Indexes

013 | Submittals

013 300 | Survey Data

		013 300 | Survey Data	CREW	DAILY OUTPUT	LABOR-HOURS	UNIT	1999 BARE COSTS MAT.	LABOR	EQUIP.	TOTAL	TOTAL INCL O&P	
306	1600	2′ contours, 10 acres				Acre					400	306
	1650	20 acres									270	
	1800	50 acres									85	
	1850	100 acres									72	
	2000	1000 acres									16.20	
	2050	10,000 acres				↓					10.35	
	2150	For 1′ contours and										
	2160	dense urban areas, add to above				Acre					40%	
	3000	Inertial guidance system for										
	3010	locating coordinates, rent per day				Ea.					4,000	

013 400 | Shop Drawings

		013 400 | Shop Drawings	CREW	DAILY OUTPUT	LABOR-HOURS	UNIT	MAT.	LABOR	EQUIP.	TOTAL	TOTAL INCL O&P	
406	0010	**MODELS** Cardboard & paper, 1 building, minimum				Ea.	550			550	605	406
	0050	Maximum					1,200			1,200	1,325	
	0300	Site plan layout, minimum					1,000			1,000	1,100	
	0350	Maximum				↓	1,750			1,750	1,925	
408	0010	**RENDERINGS** Color, matted, 20″ x 30″, eye level,										408
	0020	1 building, minimum				Ea.	1,500			1,500	1,650	
	0050	Average					2,500			2,500	2,750	
	0100	Maximum					3,500			3,500	3,850	
	1000	5 buildings, minimum					3,000			3,000	3,300	
	1100	Maximum					6,000			6,000	6,600	
	2000	Aerial perspective, color, 1 building, minimum					2,500			2,500	2,750	
	2100	Maximum					6,000			6,000	6,600	
	3000	5 buildings, minimum					3,000			3,000	3,300	
	3100	Maximum				↓	10,000			10,000	11,000	

013 800 | Construction Photos

		013 800 | Construction Photos	CREW	DAILY OUTPUT	LABOR-HOURS	UNIT	MAT.	LABOR	EQUIP.	TOTAL	TOTAL INCL O&P	
804	0010	**PHOTOGRAPHS** 8″ x 10″, 4 shots, 2 prints ea., std. mounting				Set	95			95	105	804
	0100	Hinged linen mounts					110			110	121	
	0200	8″ x 10″, 4 shots, 2 prints each, in color					190			190	209	
	0300	For I.D. slugs, add to all above					2.50			2.50	2.75	
	0500	Aerial photos, initial fly-over, 6 shots, 1 print ea., 8″ x 10″					815			815	895	
	0550	11″ x 14″ prints					850			850	935	
	0600	16″ x 20″ prints					1,000			1,000	1,100	
	0700	For full color prints, add					40%				40%	
	0750	Add for traffic control area				↓	260			260	286	
	0900	For over 30 miles from airport, add per				Mile	4.80			4.80	5.30	
	1000	Vertical photography, 4 to 6 shots with										
	1010	different scales, 1 print each				Set	990			990	1,100	
	1500	Time lapse equipment, camera and projector, buy					3,500			3,500	3,850	
	1550	Rent per month				↓	520			520	570	
	1700	Cameraman and film, including processing, B.&W.				Day	580			580	640	
	1720	Color				″	650			650	715	

014 | Quality Control

014 100 | Testing Services

		014 100 | Testing Services	CREW	DAILY OUTPUT	LABOR-HOURS	UNIT	1999 BARE COSTS MAT.	LABOR	EQUIP.	TOTAL	TOTAL INCL O&P	
108	0010	**FIELD TESTING**										108
	0200	Asphalt testing, compressive strength Marshall stability, set of 3				Ea.					165	

014 100 | Testing Services			CREW	DAILY OUTPUT	LABOR-HOURS	UNIT	1999 BARE COSTS MAT.	LABOR	EQUIP.	TOTAL	TOTAL INCL O&P
108	0220	Density, set of 3 (R022-220)				Ea.					87
	0250	Extraction, individual tests on sample									120
	0300	Penetration									35
	0350	Mix design, 5 specimens									200
	0360	Additional specimen									40
	0400	Specific gravity									45
	0420	Swell test									70
	0450	Water effect and cohesion, set of 6									200
	0470	Water effect and plastic flow									70
	0600	Concrete testing, aggregates, abrasion									120
	0650	Absorption									46
	0800	Petrographic analysis									850
	0900	Specific gravity									45
	1000	Sieve analysis, washed									65
	1050	Unwashed									65
	1200	Sulfate soundness									125
	1300	Weight per cubic foot									40
	1500	Cement, physical tests									300
	1600	Chemical tests									270
	1800	Compressive strength, cylinders, delivered to lab									13
	1900	Picked up by lab, minimum									15
	1950	Average									20
	2000	Maximum									30
	2200	Compressive strength, cores (not incl. drilling)									40
	2250	Core drilling, 4" diameter (plus technician)				Inch					25
	2260	Technician for core drilling				Hr.					50
	2300	Patching core holes				Ea.					24
	2400	Drying shrinkage at 28 days									260
	2500	Flexural test beams									65
	2600	Mix design, one batch mix									285
	2650	Added trial batches									132
	2800	Modulus of elasticity									150
	2900	Tensile test, cylinders									50
	3000	Water-Cement ratio curve, 3 batches									155
	3100	4 batches									205
	3300	Masonry testing, absorption, per 5 brick									50
	3350	Chemical resistance, per 2 brick									55
	3400	Compressive strength, per 5 brick									70
	3420	Efflorescence, per 5 brick									50
	3440	Imperviousness, per 5 brick									96
	3470	Modulus of rupture, per 5 brick									55
	3500	Moisture, block only									35
	3550	Mortar, compressive strength, set of 3									25
	4100	Reinforcing steel, bend test									61
	4200	Tensile test, up to #8 bar									35
	4220	#9 to #11 bar									40
	4240	#14 bar and larger									50
	4300	Seed testing, complete test									50
	4310	Purity, noxious and germination									35
	4320	Purity and noxious									20
	4330	Germination									25
	4340	Tetrazolium									25
	4350	Vigor test									10
	4360	Electrophoresis, purity analysis									150
	4365	Genetic variation									350
108	4370	Breeder evaluation				Ea.					410

014 | Quality Control

014 100 | Testing Services

108		CREW	DAILY OUTPUT	LABOR-HOURS	UNIT	1999 BARE COSTS MAT.	LABOR	EQUIP.	TOTAL	TOTAL INCL O&P
4375	Self counts (R022-220)				Ea.					130
4380	Offtype counts									130
4385	Seed type analysis									100
4400	Soil testing, Atterberg limits, liquid and plastic limits									61
4510	Hydrometer analysis									120
4530	Specific gravity									48
4600	Sieve analysis, washed									56.67
4700	Unwashed									65
4710	Consolidation test (ASTM D2435), minimum									320
4715	Maximum									475
4720	Density and classification of undisturbed sample									75
4735	Soil density, nuclear method, ASTM D2922-71									38.67
4740	Sand cone method ASTM D1556064									30.17
4750	Moisture content									5.33
4780	Permeability test, double ring infiltrometer									550
4800	Permeability, variable or constant head, undisturbed									205
4850	Recompacted									233.33
4900	Proctor compaction, 4" standard mold									125
4950	6" modified mold									75
5100	Shear tests, triaxial, minimum									450
5150	Maximum									600
5300	Direct shear, minimum									250
5350	Maximum									300
5550	Technician for inspection, per day, earthwork									215
5650	Bolting									270
5750	Roofing									250
5790	Welding									260
5820	Non-destructive testing, dye penetrant				Day					320
5840	Magnetic particle									320
5860	Radiography									480
5880	Ultrasonic									330
6000	Welding certification, minimum				Ea.					100
6100	Maximum				"					275
7000	Underground storage tank									
7500	Hydrostatic tank tightness test per tank, min.				Ea.					500
7510	Maximum				"					1,000
7600	Vadose zone (soil gas) sampling, 10-40 samples, min.				Day					1,500
7610	Maximum				"					2,500
7700	Ground water monitoring incl. drilling 3 wells, min.				Total					5,000
7710	Maximum				"					7,000
8000	X-ray concrete slabs				Ea.					200

015 | Construction Facilities & Temporary Controls

015 100 | Temporary Utilities

104		CREW	DAILY OUTPUT	LABOR-HOURS	UNIT	1999 BARE COSTS MAT.	LABOR	EQUIP.	TOTAL	TOTAL INCL O&P
0010	TEMPORARY UTILITIES									
0100	Heat, incl. fuel and operation, per week, 12 hrs. per day	1 Skwk	100	.080	CSF Flr	5	2.24		7.24	9.10
0200	24 hrs. per day	"	60	.133		7.55	3.74		11.29	14.20
0350	Lighting, incl. service lamps, wiring & outlets, minimum	1 Elec	34	.235		2.03	7.50		9.53	13.45

015 | Construction Facilities & Temporary Controls

		015 100 | Temporary Utilities	CREW	DAILY OUTPUT	LABOR-HOURS	UNIT	1999 BARE COSTS MAT.	LABOR	EQUIP.	TOTAL	TOTAL INCL O&P	
104	0360	Maximum	1 Elec	17	.471	CSF Flr	4.42	15		19.42	27.50	104
	0400	Power for temp lighting only, per month, min/month 6.6 KWH								.75	1.18	
	0450	Maximum/month 23.6 KWH								2.85	2.85	
	0600	Power for job duration incl. elevator, etc., minimum								47	51.70	
	0650	Maximum								110	121	
	0700	Temporary construction water bill per mo. average				Month	55			55	60.50	
		015 200 | Temporary Construction										
204	0010	**PROTECTION** Stair tread, 2″ x 12″ planks, 1 use	1 Carp	75	.107	Tread	3.87	2.91		6.78	8.85	204
	0100	Exterior plywood, 1/2″ thick, 1 use		65	.123		1.23	3.36		4.59	6.65	
	0200	3/4″ thick, 1 use		60	.133		1.75	3.64		5.39	7.65	
		015 250 | Construction Aids										
252	0010	**SAFETY NETS** No supports, stock sizes, nylon, 4″ mesh				S.F.	1.10			1.10	1.21	252
	0100	Polypropylene, 6″ mesh					1.55			1.55	1.70	
	0200	Small mesh debris nets, 1/4″ & 3/4″ mesh, stock sizes					.72			.72	.79	
	0220	Combined 4″ mesh and 1/4″ mesh, stock sizes					1.80			1.80	1.98	
	0300	Monthly rental, 4″ mesh, stock sizes, 1st month					.19			.19	.21	
	0320	2nd month rental					.15			.15	.17	
	0340	Subsequent months rental					.15			.15	.17	
	6220	Safety supplies and first aid kits				Month	23			23	25.50	
254	0014	**SCAFFOLDING, STEEL TUBULAR** Rent, 1 use per mo., no plank R015-100										254
	0015	Set up and take down										
	0090	Building exterior, wall face, 1 to 5 stories	3 Carp	24	1	C.S.F.	24.50	27.50		52	69.50	
	0460	Building interior, wall face area, up to 16′ high	″	25	.960	″	24.50	26		50.50	67.50	
	0910	Steel tubular, heavy duty shoring, buy										
	0920	Frames 5′ high 2′ wide				Ea.	75			75	82.50	
	0925	5′ high 4′ wide					85			85	93.50	
	0930	6′ high 2′ wide					86			86	94.50	
	0935	6′ high 4′ wide					101			101	111	
	0940	Accessories										
	0945	Cross braces				Ea.	16			16	17.60	
	0950	U-head, 8″ x 8″					16			16	17.60	
	0955	J-head, 4″ x 8″					12			12	13.20	
	0960	Base plate, 8″ x 8″					13			13	14.30	
	0965	Leveling jack					30.50			30.50	33.50	
	1000	Steel tubular, regular, buy										
	1100	Frames 3′ high 5′ wide				Ea.	58			58	64	
	1150	5′ high 5′ wide					67			67	73.50	
	1200	6′-4″ high 5′ wide					84			84	92.50	
	1350	7′-6″ high 6′ wide					145			145	160	
	1500	Accessories cross braces					15			15	16.50	
	1550	Guardrail post					15			15	16.50	
	1600	Guardrail 7′ section					7.25			7.25	8	
	1650	Screw jacks & plates					24			24	26.50	
	1700	Sidearm brackets					28			28	31	
	1750	8″ casters					33			33	36.50	
	1800	Plank 2″ x 10″ x 16′-0″					42.50			42.50	47	
	1900	Stairway section					235			235	259	
	1910	Stairway starter bar					21			21	23.50	
	1920	Stairway inside handrail					53			53	58.50	
	1930	Stairway outside handrail					73			73	80.50	
	1940	Walk-thru frame guardrail					28			28	31	
	2000	Steel tubular, regular, rent/mo.										
	2100	Frames 3′ high 5′ wide				Ea.	3.75			3.75	4.13	

015 250 | Construction Aids

R015-100

				DAILY	LABOR-		1999 BARE COSTS				TOTAL
			CREW	OUTPUT	HOURS	UNIT	MAT.	LABOR	EQUIP.	TOTAL	INCL O&P
254	2150	5′ high 5′ wide				Ea.	3.75			3.75	4.13
	2200	6′-4″ high 5′ wide					3.75			3.75	4.13
	2250	7′-6″ high 6′ wide					7			7	7.70
	2500	Accessories, cross braces					.60			.60	.66
	2550	Guardrail post					1			1	1.10
	2600	Guardrail 7′ section					.75			.75	.83
	2650	Screw jacks & plates					1.50			1.50	1.65
	2700	Sidearm brackets					1.50			1.50	1.65
	2750	8″ casters					6			6	6.60
	2800	Outrigger for rolling tower					3			3	3.30
	2850	Plank 2″ x 10″ x 16′-0″					5			5	5.50
	2900	Stairway section					10			10	11
	2910	Stairway starter bar					.10			.10	.11
	2920	Stairway inside handrail					5			5	5.50
	2930	Stairway outside handrail					5			5	5.50
	2940	Walk-thru frame guardrail					2			2	2.20
	3000	Steel tubular, heavy duty shoring, rent/mo.									
	3250	5′ high 2′ & 4′ wide				Ea.	5			5	5.50
	3300	6′ high 2′ & 4′ wide					5			5	5.50
	3500	Accessories, cross braces					1			1	1.10
	3600	U - head, 8″ x 8″					1			1	1.10
	3650	J - head, 4″ x 8″					1			1	1.10
	3700	Base plate, 8″ x 8″					1			1	1.10
	3750	Leveling jack					2			2	2.20
	6000	Heavy duty shoring for suspended slab forms to 8′-2″ high, floor area									
	6010	Set up and take down									
	6100	1 use/month	4 Carp	36	.889	C.S.F.	29.50	24.50		54	70.50
	6150	2 uses/month	″	36	.889	″	14.80	24.50		39.30	54.50
	6500	To 14′-8″ high									
	6600	1 use/month	4 Carp	18	1.778	C.S.F.	43	48.50		91.50	124
	6650	2 uses/month	″	18	1.778	″	21.50	48.50		70	101
255	0011	**SCAFFOLDING SPECIALTIES**									
	1200	Sidewalk bridge, heavy duty steel posts & beams, including									
	1210	parapet protection & waterproofing									
	1220	8′ to 10′ wide, 2 posts	3 Carp	15	1.600	L.F.	42	43.50		85.50	115
	1230	3 posts	″	10	2.400	″	63	65.50		128.50	173
	1500	Sidewalk bridge using tubular steel									
	1510	scaffold frames, including planking	3 Carp	45	.533	L.F.	4.72	14.55		19.27	28
	1600	For 2 uses per month, deduct from all above					50%				
	1700	For 1 use every 2 months, add to all above					100%				
	1900	Catwalks, 32″ wide, no guardrails, 6′ span, buy				Ea.	120			120	132
	2000	10′ span, buy				″	190			190	209
	2800	Hand winch-operated masons									
	2810	scaffolding, no plank moving required									
	2900	98′ long, 10′-6″ high, buy				Ea.	19,700			19,700	21,700
	3000	Rent per month					790			790	865
	3100	28′-6″ high, buy					25,700			25,700	28,200
	3200	Rent per month					1,025			1,025	1,125
	3400	196′ long, 28′-6″ high, buy					49,700			49,700	54,500
	3500	Rent per month					2,000			2,000	2,175
	3600	64′-6″ high, buy					73,000			73,000	80,000
	3700	Rent per month					2,900			2,900	3,200
	3800	Rolling ladders with handrails, 30″ wide, buy, 2 step					139			139	153
	4000	7 step					405			405	450
	4050	10 step					655			655	720

		015 250 | Construction Aids	CREW	DAILY OUTPUT	LABOR-HOURS	UNIT	1999 BARE COSTS MAT.	LABOR	EQUIP.	TOTAL	TOTAL INCL O&P	
255	4100	Rolling towers, buy, 5′ wide, 7′ long, 10′ high				Ea.	1,150			1,150	1,250	255
	4200	For 5′ high added sections, to buy, add				↓	188			188	207	
	4300	Complete incl. wheels, railings, outriggers,										
	4350	21′ high, to buy				Ea.	1,925			1,925	2,125	
	4400	21′ high, rent per month				″	138			138	152	
258	0010	**SURVEYOR STAKES** Hardwood, 1″ x 1″ x 48″ long				C	41.50			41.50	46	258
	0100	2″ x 2″ x 18″ long					50			50	55	
	0150	2″ x 2″ x 24″ long					62			62	68	
	0200	2″ x 2″ x 30″ long				↓	71			71	78	
259	0010	**WEATHER STATION** Remote recording, minimum				Ea.	3,000			3,000	3,300	259
	0100	Maximum				″	7,100			7,100	7,800	
		015 300 | Barriers & Enclosures										
302	0010	**BARRICADES** 5′ high, 3 rail @ 2″ x 8″, fixed	2 Carp	30	.533	L.F.	10.80	14.55		25.35	35	302
	0150	Movable	″	20	.800	″	10.80	22		32.80	46.50	
	0300	Stock units, 6′ high, 8′ wide, plain, buy				Ea.	430			430	475	
	0350	With reflective tape, buy				″	525			525	580	
	0400	Break-a-way 3″ PVC pipe barricade										
	0410	with 3 ea. 1′ x 4′ reflectorized panels, buy				Ea.	305			305	335	
	0500	Plywood with steel legs, 32″ wide					70			70	77	
	0600	Telescoping Christmas tree, 9′ high, 5 flags, buy					121			121	133	
	0800	Traffic cones, PVC, 18″ high					8.70			8.70	9.55	
	0850	28″ high				↓	12.35			12.35	13.55	
	1000	Guardrail, wooden, 3′ high, 1″ x 6″, on 2″ x 4″ posts	2 Carp	200	.080	L.F.	.99	2.18		3.17	4.53	
	1100	2″ x 6″, on 4″ x 4″ posts	″	165	.097		1.79	2.65		4.44	6.15	
	1200	Portable metal with base pads, buy					15			15	16.50	
	1250	Typical installation, assume 10 reuses	2 Carp	600	.027	↓	1.58	.73		2.31	2.89	
	5000	Barricades, see also division 016-420-1600										
304	0010	**FENCING** Chain link, 11 ga, 5′ high	2 Clab	100	.160	L.F.	4.04	3.43		7.47	9.85	304
	0100	6′ high		75	.213		3.37	4.58		7.95	10.90	
	0200	Rented chain link, 6′ high, to 500′ (up to 12 mo.)		100	.160		2.19	3.43		5.62	7.80	
	0250	Over 1000′ (up to 12 mo.)	↓	110	.145		1.59	3.12		4.71	6.65	
	0350	Plywood, painted, 2″ x 4″ frame, 4′ high	A-4	135	.178		4.06	4.71		8.77	11.80	
	0400	4″ x 4″ frame, 8′ high	″	110	.218		7.05	5.80		12.85	16.75	
	0500	Wire mesh on 4″ x 4″ posts, 4′ high	2 Carp	100	.160		5.55	4.37		9.92	13	
	0550	8′ high	″	80	.200	↓	8.35	5.45		13.80	17.80	
306	0010	**WINTER PROTECTION** Reinforced plastic on wood										306
	0100	framing to close openings	2 Clab	750	.021	S.F.	.35	.46		.81	1.11	
	0200	Tarpaulins hung over scaffolding, 8 uses, not incl. scaffolding		1,500	.011		.16	.23		.39	.54	
	0250	Tarpaulin polyester reinf. w/ integral fastening system 11 mils thick		1,600	.010		.73	.21		.94	1.14	
	0300	Prefab fiberglass panels, steel frame, 8 uses	↓	1,200	.013	↓	.70	.29		.99	1.22	
		015 400 | Security										
480	0010	**WATCHMAN** Service, monthly basis, uniformed person, minimum				Hr.					7.80	480
	0100	Maximum									14.15	
	0200	Person and command dog, minimum									10.30	
	0300	Maximum				↓					15.30	
	0500	Sentry dog, leased, with job patrol (yard dog), 1 dog				Week					195	
	0600	2 dogs				″					275	
	0800	Purchase, trained sentry dog, minimum				Ea.					800	
	0900	Maximum				″					2,000	

015 | Construction Facilities & Temporary Controls

015 500 | Access Roads

	Line	Description	Crew	Daily Output	Labor-Hours	Unit	1999 Bare Costs Mat.	Labor	Equip.	Total	Total Incl O&P
552	0010	**ROADS AND SIDEWALKS** Temporary									
	0050	Roads, gravel fill, no surfacing, 4" gravel depth	B-14	715	.067	S.Y.	1.10	1.53	.31	2.94	3.94
	0100	8" gravel depth	"	615	.078	"	2.20	1.77	.36	4.33	5.60
	1000	Ramp, 3/4" plywood on 2" x 6" joists, 16" O.C.	2 Carp	300	.053	S.F.	1.45	1.46		2.91	3.89
	1100	On 2" x 10" joists, 16" O.C.	"	275	.058		1.75	1.59		3.34	4.43
	2200	Sidewalks, 2" x 12" planks, 2 uses	1 Carp	350	.023		.55	.62		1.17	1.59
	2300	Exterior plywood, 2 uses, 1/2" thick		750	.011		.30	.29		.59	.79
	2400	5/8" thick		650	.012		.35	.34		.69	.92
	2500	3/4" thick		600	.013		.42	.36		.78	1.03

015 600 | Temporary Controls

	Line	Description	Crew	Daily Output	Labor-Hours	Unit	Mat.	Labor	Equip.	Total	Total Incl O&P
602	0010	**TARPAULINS** Cotton duck, 10 oz. to 13.13 oz. per S.Y., minimum				S.F.	.38			.38	.42
	0050	Maximum					.55			.55	.61
	0100	Polyvinyl coated nylon, 14 oz. to 18 oz., minimum					.45			.45	.50
	0150	Maximum					.65			.65	.71
	0200	Reinforced polyethylene 3 mils thick, white					.10			.10	.11
	0300	4 mils thick, white, clear or black					.12			.12	.13
	0400	5.5 mils thick, clear					.09			.09	.10
	0500	White, fire retardant					.16			.16	.18
	0600	7.5 mils, oil resistant, fire retardant					.17			.17	.19
	0700	8.5 mils, black					.22			.22	.24
	0710	Woven polyethylene, 6 mils thick					.45			.45	.50
	0730	Polyester reinforced w/ integral fastening system 11 mils thick					1			1	1.10
	0740	Mylar polyester, non-reinforced, 7 mils thick					1.10			1.10	1.21

015 800 | Project Signs

	Line	Description	Crew	Daily Output	Labor-Hours	Unit	Mat.	Labor	Equip.	Total	Total Incl O&P
804	0010	**SIGNS** Hi-intensity reflectorized, no posts, buy				S.F.	12.05			12.05	13.25

015 900 | Field Offices & Sheds

	Line	Description	Crew	Daily Output	Labor-Hours	Unit	Mat.	Labor	Equip.	Total	Total Incl O&P
904	0010	**OFFICE** Trailer, furnished, no hookups, 20' x 8', buy	2 Skwk	1	16	Ea.	5,075	450		5,525	6,275
	0250	Rent per month					134			134	148
	0300	32' x 8', buy	2 Skwk	.70	22.857		7,825	640		8,465	9,625
	0350	Rent per month					158			158	174
	0400	50' x 10', buy	2 Skwk	.60	26.667		13,400	750		14,150	15,900
	0450	Rent per month					273			273	300
	0500	50' x 12', buy	2 Skwk	.50	32		15,800	900		16,700	18,800
	0550	Rent per month					315			315	345
	0700	For air conditioning, rent per month, add					35.50			35.50	39
	0800	For delivery, add per mile				Mile	1.50			1.50	1.65
	1000	Portable buildings, prefab, on skids, economy, 8' x 8'	2 Carp	265	.060	S.F.	80	1.65		81.65	90.50
	1100	Deluxe, 8' x 12'	"	150	.107	"	87	2.91		89.91	100
	1200	Storage boxes, 20' x 8', buy	2 Skwk	1.80	8.889	Ea.	3,200	249		3,449	3,925
	1250	Rent per month					66.50			66.50	73.50
	1300	40' x 8', buy	2 Skwk	1.40	11.429		3,325	320		3,645	4,175
	1350	Rent per month					85			85	93.50

016 400 | Equipment Rental

			UNIT	HOURLY OPER. COST	RENT PER DAY	RENT PER WEEK	RENT PER MONTH	CREW EQUIPMENT COST/DAY
406	0010	CONCRETE EQUIPMENT RENTAL R033-090						
	0100	without operators						
	0200	Bucket, concrete lightweight, 1/2 C.Y.	Ea.	.14	25	75	225	16.10
	0300	1 C.Y.		.18	38.50	115	345	24.45
	0400	1-1/2 C.Y.		.25	40	120	360	26
	0500	2 C.Y.		.26	50	150	450	32.10
	0600	Cart, concrete, self propelled, operator walking, 10 C.F.		1.24	66.50	200	600	49.90
	0700	Operator riding, 18 C.F.		2.42	107	320	960	83.35
	0800	Conveyer for concrete, portable, gas, 16" wide, 26' long		3.61	167	500	1,500	128.90
	0900	46' long		3.71	227	680	2,050	165.70
	1000	56' long		3.86	250	750	2,250	180.90
	1100	Core drill, electric, 2-1/2 H.P., 1" to 8" bit diameter		.41	70	210	630	45.30
	1150	11 H.P., 8" to 18" cores		.67	93.50	280	840	61.35
	1200	Finisher, concrete floor, gas, riding trowel, 48" diameter		2.68	100	300	900	81.45
	1300	Gas, manual, 3 blade, 36" trowel		2.27	56.50	170	510	52.15
	1400	4 blade, 48" trowel		2.37	50	150	450	48.95
	1500	Float, hand-operated (Bull float) 48" wide		.10	6.65	20	60	4.80
	1570	Curb builder, 14 H.P., gas, single screw		1.60	58.50	175	525	47.80
	1590	Double screw		2.32	61.50	185	555	55.55
	1600	Grinder, concrete and terrazzo, electric, floor		1.34	58.50	175	525	45.70
	1700	Wall grinder		.62	35	105	315	25.95
	1800	Mixer, powered, mortar and concrete, gas, 6 C.F., 18 H.P.		.93	58.50	175	525	42.45
	1900	10 C.F., 25 H.P.		1.34	50	150	450	40.70
	2000	16 C.F.		1.60	66.50	200	600	52.80
	2100	Concrete, stationary, tilt drum, 2 C.Y.		7.93	325	980	2,950	259.45
	2120	Pump, concrete, truck mounted, 4" line, 80' boom		11.95	915	2,750	8,250	645.60
	2140	5" line, 110' boom		13.29	1,100	3,275	9,825	761.30
	2160	Mud jack, 50 C.F. per hr.		3.17	70	210	630	67.35
	2180	225 C.F. per hr.		6.28	310	925	2,775	235.25
	2600	Saw, concrete, manual, gas, 18 H.P.		2.83	76.50	230	690	68.65
	2650	Self-propelled, gas, 30 H.P.		5.58	110	330	990	110.65
	2700	Vibrators, concrete, electric, 60 cycle, 2 H.P.		.31	23.50	70	210	16.50
	2800	3 H.P.		.36	33.50	100	300	22.90
	2900	Gas engine, 5 H.P.		.67	33.50	100	300	25.35
	3000	8 H.P.		.93	50	150	450	37.45
	3100	Concrete transit mixer, hydraulic drive						
	3120	6 x 4, 250 H.P., 8 C.Y., rear discharge		30.08	925	2,775	8,325	795.65
	3200	Front discharge		30.18	990	2,975	8,925	836.45
	3300	6 x 6, 285 H.P., 12 C.Y., rear discharge		32.47	1,025	3,075	9,225	874.75
	3400	Front discharge		32.65	1,075	3,250	9,750	911.20
408	0010	EARTHWORK EQUIPMENT RENTAL Without operators R016-410						
	0040	Aggregate spreader, push type 8' to 12' wide	Ea.	1.13	103	310	930	71.05
	0050	Augers for truck or trailer mounting, vertical drilling R022-240						
	0055	Fence post auger, truck mounted	Ea.	10	465	1,400	4,200	360
	0060	4" to 36" diam., 54 H.P., gas, 10' spindle travel R022-250		15.91	555	1,660	4,975	459.30
	0070	14' spindle travel		16.38	650	1,955	5,875	522.05
	0075	Auger, truck mounted, vertical drilling, to 25' depth		49.54	2,300	6,900	20,700	1,776
	0080	Auger, horizontal boring machine, 12" to 36" diameter, 45 H.P.		7.31	325	980	2,950	254.50
	0090	12" to 48" diameter, 65 H.P.		12.46	650	1,955	5,875	490.70
	0100	Backhoe, diesel hydraulic, crawler mounted, 1/2 C.Y. cap.		11.43	410	1,225	3,675	336.45
	0120	5/8 C.Y. capacity		15.53	450	1,350	4,050	394.25
	0140	3/4 C.Y. capacity		17.39	515	1,550	4,650	449.10
	0150	1 C.Y. capacity		22.20	615	1,850	5,550	547.60
	0200	1-1/2 C.Y. capacity		26.59	835	2,500	7,500	712.70
	0300	2 C.Y. capacity		42.85	1,125	3,400	10,200	1,023
	0320	2-1/2 C.Y. capacity		59.38	2,025	6,100	18,300	1,695
	0340	3-1/2 C.Y. capacity		79.44	2,525	7,600	22,800	2,156
	0341	Attachments						

Important: See the Reference Section for critical supporting data - Reference Nos., Crews, & City Cost Indexes

016 400 | Equipment Rental

408	Description	Ref.	Unit	Hourly Oper. Cost	Rent Per Day	Rent Per Week	Rent Per Month	Crew Equipment Cost/Day
0342	Bucket thumbs		Ea.	.46	293	880	2,650	179.70
0345	Grapples	R016-410		.31	300	900	2,700	182.50
0350	Gradall type, truck mounted, 3 ton @ 15' radius, 5/8 C.Y.			26.47	710	2,125	6,375	636.75
0370	1 C.Y. capacity	R022-240		28.22	965	2,900	8,700	805.75
0400	Backhoe-loader, wheel type, 40 to 45 H.P., 5/8 C.Y. capacity			7.31	237	710	2,125	200.50
0450	45 H.P. to 60 H.P., 3/4 C.Y. capacity	R022-250		8.60	250	750	2,250	218.80
0460	80 H.P., 1-1/4 C.Y. capacity			12.51	315	950	2,850	290.10
0470	112 H.P.,1-3/4 C.Y. loader, 1/2 C.Y. backhoe			15.55	385	1,150	3,450	354.40
0480	Attachments							
0482	Compactor, 20,000 lb			1.29	188	565	1,700	123.30
0485	Hydraulic hammer, 750 ft-lbs			.41	305	915	2,750	186.30
0486	Hydraulic hammer, 1000 ft-lbs			.46	345	1,030	3,100	209.70
0500	Brush chipper, gas engine, 6" cutter head, 35 H.P.			4.12	137	410	1,225	114.95
0550	12" cutter head, 130 H.P.			12.51	190	570	1,700	214.10
0600	15" cutter head, 165 H.P.			16.17	255	765	2,300	282.35
0750	Bucket, clamshell, general purpose, 3/8 C.Y.			.80	60	180	540	42.40
0800	1/2 C.Y.			1.11	68.50	205	615	49.90
0850	3/4 C.Y.			1.12	76.50	230	690	54.95
0900	1 C.Y.			1.39	93.50	280	840	67.10
0950	1-1/2 C.Y.			1.65	128	385	1,150	90.20
1000	2 C.Y.			1.85	145	435	1,300	101.80
1010	Bucket, dragline, medium duty, 1/2 C.Y.			.36	38.50	115	345	25.90
1020	3/4 C.Y.			.57	48.50	145	435	33.55
1030	1 C.Y.			.60	51.50	155	465	35.80
1040	1-1/2 C.Y.			.77	70	210	630	48.15
1050	2 C.Y.			.84	86.50	260	780	58.70
1070	3 C.Y.			1.34	120	360	1,075	82.70
1200	Compactor, roller, 2 drum, 2000 lb., operator walking			1.88	130	390	1,175	93.05
1250	Rammer compactor, gas, 1000 lb. blow			.57	45	135	405	31.55
1300	Vibratory plate, gas, 13" plate, 1000 lb. blow			.71	41.50	125	375	30.70
1350	24" plate, 5000 lb. blow			1.75	63.50	190	570	52
1370	Curb builder/extruder, 14 H.P., gas, single screw			1.44	48.50	145	435	40.50
1390	Double screw			2.16	88.50	265	795	70.30
1500	Disc harrow attachment, for tractor		↓	.77	53.50	160	480	38.15
1750	Extractor, piling, see lines 2500 to 2750							
1810	Feller buncher, shearing & accumulating trees, 100 H.P.		Ea.	9.99	515	1,550	4,650	389.90
1860	Grader, self-propelled, 25,000 lb.			14.83	580	1,740	5,225	466.65
1910	30,000 lb.			17.82	720	2,165	6,500	575.55
1920	40,000 lb.			20.50	860	2,575	7,725	679
1930	55,000 lb.			33.48	1,300	3,865	11,600	1,041
1950	Hammer, pavement demo., hyd., gas, self-prop., 1000 to 1250 lb.			18.59	410	1,225	3,675	393.70
2000	Diesel 1300 to 1500 lb.			9.01	455	1,360	4,075	344.10
2050	Pile driving hammer, steam or air, 4150 ft.-lb. @ 225 BPM			1.80	288	865	2,600	187.40
2100	8750 ft.-lb. @ 145 BPM			3.24	410	1,235	3,700	272.90
2150	15,000 ft.-lb. @ 60 BPM			4.22	480	1,440	4,325	321.75
2200	24,450 ft.-lb. @ 111 BPM		↓	4.65	585	1,760	5,275	389.20
2250	Leads, 15,000 ft.-lb. hammers		L.F.	.02	2.67	8	24	1.75
2300	24,450 ft.-lb. hammers and heavier		"	.04	3	9	27	2.10
2350	Diesel type hammer, 22,400 ft.-lb.		Ea.	6.17	480	1,440	4,325	337.35
2400	41,300 ft.-lb.			9.57	755	2,265	6,800	529.55
2450	141,000 ft.-lb.			19.68	1,450	4,325	13,000	1,022
2500	Vib. elec. hammer/extractor, 200 KW diesel generator, 34 H.P.			9.33	610	1,835	5,500	441.65
2550	80 H.P.			18.23	925	2,780	8,350	701.85
2600	150 H.P.			31.36	1,625	4,840	14,500	1,219
2700	Extractor, steam or air, 700 ft.-lb.			1.44	172	515	1,550	114.50
2750	1000 ft.-lb.			1.85	240	720	2,150	158.80
2800	Log chipper, up to 22" diam, 600 H.P.			52.74	2,700	8,100	24,300	2,042
2850	Logger, for skidding & stacking logs, 150 H.P.		↓	12.77	450	1,350	4,050	372.15

016 400 | Equipment Rental

408		UNIT	HOURLY OPER. COST	RENT PER DAY	RENT PER WEEK	RENT PER MONTH	CREW EQUIPMENT COST/DAY
2900	Rake, spring tooth, with tractor (R016-410)	Ea.	7.36	267	800	2,400	218.90
3000	Roller, tandem, gas, 3 to 5 ton		6.23	172	515	1,550	152.85
3050	Diesel, 8 to 12 ton (R022-240)		9.01	288	865	2,600	245.10
3100	Towed type, vibratory, gas 12.5 H.P., 2 ton		3.86	137	410	1,225	112.90
3150	Sheepsfoot, double 60" x 60" (R022-250)		4.74	145	435	1,300	124.90
3200	Pneumatic tire diesel roller, 12 ton		8.65	292	875	2,625	244.20
3250	21 to 25 ton		15.97	345	1,030	3,100	333.75
3300	Sheepsfoot roller, self-propelled, 4 wheel, 130 H.P.		20.96	650	1,955	5,875	558.70
3320	300 H.P.		26.68	765	2,300	6,900	673.45
3350	Vibratory steel drum & pneumatic tire, diesel, 18,000 lb.		11.48	435	1,300	3,900	351.85
3400	29,000 lb.		12.72	535	1,600	4,800	421.75
3410	Rotary mower, brush, 60", with tractor		8.56	233	700	2,100	208.50
3450	Scrapers, towed type, 9 to 12 C.Y. capacity		3.19	86.50	260	780	77.50
3500	12 to 17 C.Y. capacity		5.93	258	775	2,325	202.45
3550	Scrapers, self-propelled, 4 x 4 drive, 2 engine, 14 C.Y. capacity		60.87	1,875	5,650	17,000	1,617
3600	2 engine, 24 C.Y. capacity		73.95	2,200	6,600	19,800	1,912
3650	Self-loading, 11 C.Y. capacity		27.32	785	2,350	7,050	688.55
3700	22 C.Y. capacity		33.07	1,200	3,600	10,800	984.55
3710	Screening plant 110 hp. w / 5' x 10'screen		15.97	435	1,300	3,900	387.75
3720	5' x 16' screen	↓	17.61	515	1,550	4,650	450.90
3850	Shovels, see Cranes division 016-460						
3860	Shovel/backhoe bucket, 1/2 C.Y.	Ea.	1.13	76.50	230	690	55.05
3870	3/4 C.Y.		3.19	128	385	1,150	102.50
3880	1 C.Y.		3.50	172	515	1,550	131
3890	1-1/2 C.Y.		4.21	210	630	1,900	159.70
3910	3 C.Y.		7.52	400	1,200	3,600	300.15
3950	Stump chipper, 18" deep, 30 H.P.		1.80	247	740	2,225	162.40
4110	Tractor, crawler, with bulldozer, torque converter, diesel 75 H.P.		10.97	365	1,100	3,300	307.75
4150	105 H.P.		14.68	500	1,500	4,500	417.45
4200	140 H.P.		17	615	1,850	5,550	506
4260	200 H.P.		29.87	1,000	3,000	9,000	838.95
4310	300 H.P.		38.83	1,375	4,100	12,300	1,131
4360	410 H.P.		47.48	1,675	5,000	15,000	1,380
4380	700 H.P.		95.58	3,325	10,000	30,000	2,765
4400	Loader, crawler, torque conv., diesel, 1-1/2 C.Y., 80 H.P.		12.51	415	1,250	3,750	350.10
4450	1-1/2 to 1-3/4 C.Y., 95 H.P.		14.96	465	1,400	4,200	399.70
4510	1-3/4 to 2-1/4 C.Y., 130 H.P.		18.80	600	1,800	5,400	510.40
4530	2-1/2 to 3-1/4 C.Y., 190 H.P.		30.64	1,025	3,100	9,300	865.10
4560	3-1/2 to 5 C.Y., 275 H.P.		42.81	1,325	4,000	12,000	1,142
4610	Tractor loader, wheel, torque conv., 4 x 4, 1 to 1-1/4 C.Y., 65 H.P.		9.33	283	850	2,550	244.65
4620	1-1/2 to 1-3/4 C.Y., 80 H.P.		11.89	375	1,125	3,375	320.10
4650	1-3/4 to 2 C.Y., 100 H.P.		12.47	405	1,215	3,650	342.75
4710	2-1/2 to 3-1/2 C.Y., 130 H.P.		18.81	500	1,500	4,500	450.50
4730	3 to 4-1/2 C.Y., 170 H.P.		20.86	685	2,050	6,150	576.90
4760	5-1/4 to 5-3/4 C.Y., 270 H.P.		38.31	1,025	3,100	9,300	926.50
4810	7 to 8 C.Y., 375 H.P.		56.88	1,375	4,125	12,400	1,280
4870	12-1/2 C.Y., 690 H.P.		115.60	2,400	7,200	21,600	2,365
4880	Wheeled, skid steer, 10 C.F., 30 H.P. gas		4.53	125	375	1,125	111.25
4890	1 C.Y., 78 H.P., diesel	↓	6.28	300	900	2,700	230.25
4891	Attachments for all skid steer loaders						
4892	Auger	Ea.	.12	83.50	250	750	50.95
4893	Backhoe		.15	110	330	990	67.20
4894	Broom		.16	107	320	960	65.30
4895	Forks		.08	38.50	115	345	23.65
4896	Grapple		.12	86.50	260	780	52.95
4897	Concrete hammer		.25	180	540	1,625	110
4898	Tree spade		.36	128	385	1,150	79.90
4899	Trencher	↓	.41	240	720	2,150	147.30

016 400 | Equipment Rental

				UNIT	HOURLY OPER. COST	RENT PER DAY	RENT PER WEEK	RENT PER MONTH	CREW EQUIPMENT COST/DAY
408	4900	Trencher, chain, boom type, gas, operator walking, 12 H.P.	R016 -410	Ea.	2.16	133	400	1,200	97.30
	4910	Operator riding, 40 H.P.			6.36	267	800	2,400	210.90
	5000	Wheel type, diesel, 4' deep, 12" wide	R022 -240		12.42	500	1,500	4,500	399.35
	5100	Diesel, 6' deep, 20" wide			15.65	735	2,200	6,600	565.20
	5150	Ladder type, diesel, 5' deep, 8" wide	R022 -250		9.41	375	1,125	3,375	300.30
	5200	Diesel, 8' deep, 16" wide			17.61	660	1,975	5,925	535.90
	5210	Tree spade, self-propelled			5.82	475	1,425	4,275	331.55
	5250	Truck, dump, tandem, 12 ton payload			17.41	375	1,125	3,375	364.30
	5300	Three axle dump, 16 ton payload			20.28	465	1,400	4,200	442.25
	5350	Dump trailer only, rear dump, 16-1/2 C.Y.			3.45	175	525	1,575	132.60
	5400	20 C.Y.			3.51	178	535	1,600	135.10
	5450	Flatbed, single axle, 1-1/2 ton rating			10.87	142	425	1,275	171.95
	5500	3 ton rating			11.43	147	440	1,325	179.45
	5550	Off highway rear dump, 25 ton capacity			20.50	890	2,675	8,025	699
	5600	35 ton capacity			32.27	1,300	3,900	11,700	1,038
	5610	50 ton capacity			41.30	1,775	5,350	16,100	1,400
	5620	65 ton capacity			55.53	1,925	5,770	17,300	1,598
	5630	100 ton capacity			73.47	2,750	8,250	24,800	2,238
	6000	Vibratory plow, 25 H.P., walking			1.34	137	410	1,225	92.70
420	0010	**GENERAL EQUIPMENT RENTAL** Without operators	R021 -620						
	0150	Aerial lift, scissor type, to 15' high, 1000 lb. cap., electric		Ea.	1.18	85	255	765	60.45
	0160	To 25' high, 2000 lb. capacity	R022 -220		1.80	128	385	1,150	91.40
	0170	Telescoping boom to 40' high, 750 lb. capacity, gas			6.42	385	1,150	3,450	281.35
	0180	1000 lb. capacity			7.98	435	1,300	3,900	323.85
	0190	To 60' high, 750 lb. capacity			8.47	525	1,575	4,725	382.75
	0200	Air compressor, portable, gas engine, 60 C.F.M.			5.41	55	165	495	76.30
	0300	160 C.F.M.			6.80	71.50	215	645	97.40
	0400	Diesel engine, rotary screw, 250 C.F.M.			6.49	122	365	1,100	124.90
	0500	365 C.F.M.			9.29	167	500	1,500	174.30
	0550	375 to 450 C.F.M. Compressor			9.80	233	700	2,100	218.40
	0600	600 C.F.M.			15.99	242	725	2,175	272.90
	0700	750 C.F.M.			17.57	247	740	2,225	288.55
	0800	For silenced models, small sizes, add			3%	5%	5%	5%	
	0900	Large sizes, add			5%	7%	7%	7%	
	0920	Air tools and accessories							
	0930	Breaker, pavement, 60 lb.		Ea.	.19	30	90	270	19.50
	0940	80 lb.			.21	35	105	315	22.70
	0950	Drills, hand (jackhammer) 65 lb.			.23	25	75	225	16.85
	0960	Track or wagon, swing boom, 4" drifter			10.57	380	1,135	3,400	311.55
	0970	5" drifter			11.38	650	1,955	5,875	482.05
	0975	Track mounted quarry drill, 6" diameter drill			22.81	1,100	3,295	9,875	841.50
	0980	Dust control per drill			2.11	11.65	35	105	23.90
	0990	Hammer, chipping, 12 lb.			.12	20	60	180	12.95
	1000	Hose, air with couplings, 50' long, 3/4" diameter			.15	5	15	45	4.20
	1100	1" diameter			.15	6.65	20	60	5.20
	1200	1-1/2" diameter			.15	11.65	35	105	8.20
	1300	2" diameter			.21	16.65	50	150	11.70
	1400	2-1/2" diameter			.22	20	60	180	13.75
	1410	3" diameter			.23	25	75	225	16.85
	1450	Drill, steel, 7/8" x 2'				3.33	10	30	2
	1460	7/8" x 6'				4	12	36	2.40
	1520	Moil points			.82	2	6	18	7.75
	1530	Sheeting driver for 60 lb. breaker			.15	10	30	90	7.20
	1540	For 90 lb. breaker			.15	16.65	50	150	11.20
	1550	Spade, 25 lb.			.08	8.35	25	75	5.65
	1560	Tamper, single, 35 lb.			.10	23.50	70	210	14.80
	1570	Triple, 140 lb.			1.80	41.50	125	375	39.40

016 400 | Equipment Rental

420	Description	UNIT	HOURLY OPER. COST	RENT PER DAY	RENT PER WEEK	RENT PER MONTH	CREW EQUIPMENT COST/DAY
1580	Wrenches, impact, air powered, up to 3/4" bolt (R021-620)	Ea.	.25	21.50	65	195	15
1590	Up to 1-1/4" bolt		.35	43.50	130	390	28.80
1600	Barricades, barrels, reflectorized, 1 to 50 barrels (R022-220)			2	6	18	1.20
1610	100 to 200 barrels			1.33	4	12	.80
1620	Barrels with flashers, 1 to 50 barrels			3	9	27	1.80
1630	100 to 200 barrels			2	6	18	1.20
1640	Barrels with steady burn type C lights			3.67	11	33	2.20
1650	Illuminated board, trailer mounted, with generator		.78	86.50	260	780	58.25
1670	Portable, stock, with flashers, 1 to 6 units			.77	2.30	6.90	.45
1680	25 to 50 units			.73	2.20	6.60	.45
1690	Butt fusion machine, electric		1.50	292	875	2,625	187
1695	Electro fusion machine		1.25	125	375	1,125	85
1700	Carts, brick, hand powered, 1000 lb. capacity		1.11	21.50	65	195	21.90
1800	Gas engine, 1500 lb., 7-1/2' lift		1.65	95	285	855	70.20
1830	Distributor, asphalt, trailer mtd, 2000 gal., 38 H.P. diesel		7.54	465	1,400	4,200	340.30
1840	3000 gal., 38 H.P. diesel		8.12	500	1,500	4,500	364.95
1850	Drill, rotary hammer, electric, 1-1/2" diameter		.15	23.50	70	210	15.20
1860	Carbide bit for above			5.65	17	51	3.40
1870	Emulsion sprayer, 65 gal., 5 H.P. gas engine		.63	55	165	495	38.05
1880	200 gal., 5 H.P. engine		.67	58.50	175	525	40.35
1900	Fencing, see division 015-304 & 028-300						
1920	Floodlight, mercury, vapor or quartz, on tripod						
1930	1000 watt	Ea.	.12	16.65	50	150	10.95
1940	2000 watt		.12	40	120	360	24.95
1960	Floodlights, trailer mounted with generator, 2-1000 watt lights		1.80	96.50	290	870	72.40
2020	Forklift, wheeled, for brick, 18', 3000 lb., 2 wheel drive, gas		9.28	167	500	1,500	174.25
2040	28', 4000 lb., 4 wheel drive, diesel		6.39	200	600	1,800	171.10
2060	For plant, 4 T. capacity, 80 H.P., 2 wheel drive, gas		10.04	192	575	1,725	195.30
2080	10 T. capacity, 120 H.P., 2 wheel drive, diesel		11.93	315	950	2,850	285.45
2100	Generator, electric, gas engine, 1.5 KW to 3 KW		1.13	38.50	115	345	32.05
2200	5 KW		1.49	56.50	170	510	45.90
2300	10 KW		2.39	137	410	1,225	101.10
2400	25 KW		6.81	152	455	1,375	145.50
2500	Diesel engine, 20 KW		4.35	110	330	990	100.80
2600	50 KW		6.86	123	370	1,100	128.90
2700	100 KW		11.34	188	565	1,700	203.70
2800	250 KW		28.99	325	980	2,950	427.90
2850	Hammer, hydraulic, for mounting on boom, to 500 ft.-lb.		.98	128	385	1,150	84.85
2860	500 to 1200 ft.-lb.		2.52	267	800	2,400	180.15
2900	Heaters, space, oil or electric, 50 MBH		.12	26.50	80	240	16.95
3000	100 MBH		.12	31.50	95	285	19.95
3100	300 MBH		.12	50	150	450	30.95
3150	500 MBH		.15	66.50	200	600	41.20
3200	Hose, water, suction with coupling, 20' long, 2" diameter		.06	10	30	90	6.50
3210	3" diameter		.06	15	45	135	9.50
3220	4" diameter		.06	20	60	180	12.50
3230	6" diameter		.06	35	105	315	21.50
3240	8" diameter		.07	45	135	405	27.55
3250	Discharge hose with coupling, 50' long, 2" diameter		.05	6.65	20	60	4.40
3260	3" diameter		.05	8.35	25	75	5.40
3270	4" diameter		.06	13.35	40	120	8.50
3280	6" diameter		.06	31.50	95	285	19.50
3290	8" diameter		.08	33.50	100	300	20.65
3300	Ladders, extension type, 16' to 36' long			15	45	135	9
3400	40' to 60' long			31.50	95	285	19
3405	Lance for cutting concrete		6.25	122	365	1,100	123
3410	Level, laser type, for pipe laying, self leveling			102	305	915	61
3430	Manual leveling			78.50	235	705	47

Important: See the Reference Section for critical supporting data - Reference Nos., Crews, & City Cost Indexe

016 400 | Equipment Rental

420

		UNIT	HOURLY OPER. COST	RENT PER DAY	RENT PER WEEK	RENT PER MONTH	CREW EQUIPMENT COST/DAY
3440	Rotary beacon with rod and sensor (R021-620)	Ea.		103	310	930	62
3460	Builders level with tripod and rod			30	90	270	18
3500	Light towers, towable, with diesel generator, 2000 watt (R022-220)		1.55	117	350	1,050	82.40
3600	4000 watt		1.97	137	410	1,225	97.75
3700	Mixer, powered, plaster and mortar, 6 C.F., 7 H.P.		.82	53.50	160	480	38.55
3800	10 C.F., 9 H.P.		1.22	76.50	230	690	55.75
3900	Paint sprayers complete, 8 CFM		.08	41.50	125	375	25.65
4000	17 CFM		.08	60	180	540	36.65
4020	Pavers, bituminous, rubber tires, 8' wide, 52 H.P., gas		14.94	575	1,725	5,175	464.50
4030	8' wide, 64 H.P., diesel		15.45	1,000	3,025	9,075	728.60
4050	Crawler, 10' wide, 78 H.P., gas		22.20	1,400	4,200	12,600	1,018
4060	10' wide, 87 H.P., diesel		22.80	1,075	3,200	9,600	822.40
4070	Concrete paver, 12' to 24' wide, 250 H.P.		24.27	1,525	4,550	13,700	1,104
4080	Placer-spreader-trimmer, 24' wide, 300 H.P.		32.02	1,775	5,325	16,000	1,321
4100	Pump, centrifugal gas pump, 1-1/2", 4 MGPH		.46	31.50	95	285	22.70
4200	2", 8 MGPH		.52	35	105	315	25.15
4300	3", 15 MGPH		1.24	46.50	140	420	37.90
4400	6", 90 MGPH		9.68	137	410	1,225	159.45
4500	Submersible electric pump, 1-1/4", 55 GPM		.36	36.50	110	330	24.90
4600	1-1/2", 83 GPM		.39	40	120	360	27.10
4700	2", 120 GPM		.41	45	135	405	30.30
4800	3", 300 GPM		.72	55	165	495	38.75
4900	4", 560 GPM		1.18	68.50	205	615	50.45
5000	6", 1590 GPM		5.36	207	620	1,850	166.90
5100	Diaphragm pump, gas, single, 1-1/2" diameter		.49	25	75	225	18.90
5200	2" diameter		.52	41.50	125	375	29.15
5300	3" diameter		.77	45	135	405	33.15
5400	Double, 4" diameter		1.55	86.50	260	780	64.40
5500	Trash pump, self-priming, gas, 2" diameter		1.24	40	120	360	33.90
5600	Diesel, 4" diameter		1.96	96.50	290	870	73.70
5650	Diesel, 6" diameter		5.03	162	485	1,450	137.25
5660	Rollers, see division 016-408						
5700	Salamanders, L.P. gas fired, 100,000 B.T.U.	Ea.	.70	21.50	65	195	18.60
5720	Sandblaster, portable, open top, 3 C.F. capacity		.19	50	150	450	31.50
5730	6 C.F. capacity		.33	60	180	540	38.65
5740	Accessories for above		.06	16.65	50	150	10.50
5800	Saw, chain, gas engine, 18" long		.54	38.50	115	345	27.30
5900	36" long		1.13	73.50	220	660	53.05
5950	60" long		1.15	83.50	250	750	59.20
6000	Masonry, table mounted, 14" diameter, 5 H.P.		1.82	50	150	450	44.55
6050	Saw, portable cut-off, 8 H.P.		.82	58.50	175	525	41.55
6100	Circular, hand held, electric, 7-1/4" diameter		.16	20	60	180	13.30
6200	12" diameter		.27	36.50	110	330	24.15
6275	Shot blaster, walk behind, 20" wide		1.49	200	600	1,800	131.90
6300	Steam cleaner, 100 gallons per hour		.43	51.50	155	465	34.45
6310	200 gallons per hour		.70	58.50	175	525	40.60
6350	Torch, cutting, acetylene-oxygen, 150' hose		7.53	25	75	225	75.25
6360	Hourly operating cost includes tips and gas		7.62				60.95
6410	Toilet, portable chemical			15	45	135	9
6420	Recycle flush type			18.35	55	165	11
6430	Toilet, fresh water flush, garden hose,			21.50	65	195	13
6440	Hoisted, non-flush, for high rise			18.35	55	165	11
6450	Toilet, trailers, minimum			33.50	100	300	20
6460	Maximum			100	300	900	60
6470	Trailer, office, see division 015-904						
6500	Trailers, platform, flush deck, 2 axle, 25 ton capacity	Ea.	1.39	123	370	1,100	85.10
6600	40 ton capacity		1.71	228	685	2,050	150.70
6700	3 axle, 50 ton capacity		2.85	243	730	2,200	168.80

420

GENERAL REQUIREMENTS 1

016 400 | Equipment Rental

		016 400 | Equipment Rental	Unit	Hourly Oper. Cost	Rent Per Day	Rent Per Week	Rent Per Month	Crew Equipment Cost/Day	
420	6800	75 ton capacity (R021-620)	Ea.	3.66	325	975	2,925	224.30	420
	6850	Trailer, storage, see division 015-904							
	6900	Water tank, engine driven discharge, 5000 gallons (R022-220)		8.14	233	700	2,100	205.10	
	7000	10,000 gallons		11.07	335	1,000	3,000	288.55	
	7020	Transit with tripod			33.50	100	300	20	
	7030	Trench box, 3000 lbs. 6'x8'		.80	100	300	900	66.40	
	7040	7200 lbs. 6'x20'		1.46	172	515	1,550	114.70	
	7050	8000 lbs., 8' x 16'		1.61	182	545	1,625	121.90	
	7060	9500 lbs., 8'x20'		1.82	188	565	1,700	127.55	
	7065	11,000 lbs., 8'x24'		1.87	242	725	2,175	159.95	
	7070	12,000 lbs., 10' x 20'		1.98	250	750	2,250	165.85	
	7100	Truck, pickup, 3/4 ton, 2 wheel drive		10.62	73.50	220	660	128.95	
	7200	4 wheel drive		12.06	81.50	245	735	145.50	
	7250	Crew carrier, 9 passenger		16.91	103	310	930	197.30	
	7290	Tool van, 24,000 G.V.W.		18.76	103	310	930	212.10	
	7300	Tractor, 4 x 2, 30 ton capacity, 195 H.P.		11.13	390	1,175	3,525	324.05	
	7410	250 H.P.		15	435	1,300	3,900	380	
	7500	6 x 2, 40 ton capacity, 240 H.P.		17.91	470	1,415	4,250	426.30	
	7600	6 x 4, 45 ton capacity, 240 H.P.		23.15	575	1,725	5,175	530.20	
	7620	Vacuum truck, hazardous material, 2500 gallon		13.89	300	900	2,700	291.10	
	7625	5,000 gallon		15.02	335	1,000	3,000	320.15	
	7650	Vacuum, H.E.P.A., 16 gal., wet/dry		.93	51.50	155	465	38.45	
	7660	Water tank, portable			20	60	180	12	
	7690	Large production vacuum loader, 3150 CFM		11.86	715	2,150	6,450	524.90	
	7700	Welder, electric, 200 amp		.81	31.50	95	285	25.50	
	7800	300 amp		1.09	68.50	205	615	49.70	
	7900	Gas engine, 200 amp		4.33	51.50	155	465	65.65	
	8000	300 amp		5.17	68.50	205	615	82.35	
	8100	Wheelbarrow, any size			8.35	25	75	5	
	8200	Wrecking ball, 4000 lb.	↓	.41	61.50	185	555	40.30	
440	0010	**HIGHWAY EQUIPMENT RENTAL**							440
	0050	Asphalt batch plant, portable drum mixer, 100 ton/hr.	Ea.	35.07	1,725	5,200	15,600	1,321	
	0060	200 ton/hr.		41.39	1,900	5,700	17,100	1,471	
	0070	300 ton/hr.		44.38	2,275	6,800	20,400	1,715	
	0100	Backhoe attachment, long stick, up to 185 HP, 10.5' long		.19	16.65	50	150	11.50	
	0140	Up to 250 HP, 12' long		.26	18.35	55	165	13.10	
	0180	Over 250 HP, 15' long		.31	23.50	70	210	16.50	
	0200	Special dipper arm, up to 100 HP, 32' long		.46	50	150	450	33.70	
	0240	Over 100 HP, 33' long		.77	63.50	190	570	44.15	
	0300	Concrete batch plant, portable, electric, 200 YPH	↓	66.83	715	2,150	6,450	964.65	
	0500	Grader attachment, ripper/scarifier, rear mounted							
	0520	Up to 135 HP	Ea.	.93	60	180	540	43.45	
	0540	Up to 180 HP		1.24	88.50	265	795	62.90	
	0580	Up to 250 HP		1.71	117	350	1,050	83.70	
	0700	Pvmt. removal bucket, for hyd. excavator, up to 90 HP		.29	38.50	115	345	25.30	
	0740	Up to 200 HP		.43	68.50	205	615	44.45	
	0780	Over 200 HP		.52	78.50	235	705	51.15	
	0900	Aggregate spreader, self-propelled, 187 HP		22.56	760	2,275	6,825	635.50	
	1000	Chemical spreader, 3 C.Y.		2.16	137	410	1,225	99.30	
	1900	Hammermill, traveling, 250 HP		31.42	1,550	4,625	13,900	1,176	
	2000	Horizontal borer, 3" diam, 13 HP gas driven		1.85	55	165	495	47.80	
	2200	Hydromulchers, gas power, 3000 gal., for truck mounting		12.72	300	900	2,700	281.75	
	2400	Joint & crack cleaner, walk behind, 25 HP		2.11	100	300	900	76.90	
	2500	Filler, trailer mounted, 400 gal., 20 HP		3.67	192	575	1,725	144.35	
	3000	Paint striper, self propelled, double line, 30 HP		4.22	283	850	2,550	203.75	
	3200	Post drivers, 6" I-Beam frame, for truck mounting		8.76	385	1,150	3,450	300.10	
	3400	Road sweeper, self propelled, 8' wide, 90 HP		9.79	193	580	1,750	194.30	
	4000	Road mixer, self-propelled, 130 HP	↓	14.92	565	1,700	5,100	459.35	

016 400 | Equipment Rental

			UNIT	HOURLY OPER. COST	RENT PER DAY	RENT PER WEEK	RENT PER MONTH	CREW EQUIPMENT COST/DAY	
440	4100	310 HP	Ea.	16.02	1,450	4,325	13,000	993.15	440
	4200	Cold mix paver, incl pug mill and bitumen tank,							
	4220	165 HP		32.45	1,950	5,875	17,600	1,435	
	4250	Paver, asphalt, wheel or crawler, 130 H.P., diesel		28.58	1,775	5,350	16,100	1,299	
	4300	Paver, road widener, gas 1' to 6', 67 HP		16.49	715	2,150	6,450	561.90	
	4400	Diesel, 2' to 14', 88 HP		18.75	865	2,600	7,800	670	
	4600	Slipform pavers, curb and gutter, 2 track, 75 HP		11.74	600	1,800	5,400	453.90	
	4700	4 track, 165 HP		19.47	1,250	3,750	11,300	905.75	
	4800	Median barrier, 215 HP		21.73	1,175	3,500	10,500	873.85	
	4901	Trailer, low bed, 75 ton capacity		19.47	380	1,140	3,425	383.75	
	5000	Road planer, walk behind, 10" cutting width, 10 HP		1.73	48.50	145	435	42.85	
	5100	Self propelled, 12" cutting width, 64 HP		14.03	615	1,850	5,550	482.25	
	5200	Pavement profiler, 4' to 6' wide, 450 HP		74.48	3,675	11,000	33,000	2,796	
	5300	8' to 10' wide, 750 HP		91.38	4,500	13,500	40,500	3,431	
	5400	Roadway plate, steel, 1"x8'x20'			6.65	20	60	4	
	5600	Stabilizer,self-propelled, 150 HP		17.28	805	2,420	7,250	622.25	
	5700	310 HP		17.13	935	2,800	8,400	697.05	
	5800	Striper, thermal, truck mounted 120 gal. paint, 150H.P.		8.59	292	875	2,625	243.70	
	6000	Tar kettle, 330 gal., trailer mounted		.62	30	90	270	22.95	
	7800	Windrow loader, elevating		18.73	1,275	3,850	11,600	919.85	
460	0010	**LIFTING AND HOISTING EQUIPMENT RENTAL** R015-010							460
	0100	without operators							
	0120	Aerial lift truck R022-250	Ea.	16.19	735	2,200	6,600	569.50	
	0140	Boom truck		15.35	220	660	1,975	254.80	
	0150	Crane, flatbed mntd, 3 ton cap.		15.35	267	800	2,400	282.80	
	0600	Crawler, cable, 1/2 C.Y., 15 tons at 12' radius		19	535	1,600	4,800	472	
	0700	3/4 C.Y., 20 tons at 12' radius		20.20	550	1,650	4,950	491.60	
	0800	1 C.Y., 25 tons at 12' radius		20.85	565	1,700	5,100	506.80	
	0900	Crawler, cable, 1-1/2 C.Y., 40 tons at 12' radius		30.54	765	2,300	6,900	704.30	
	1000	2 C.Y., 50 tons at 12' radius		35.59	935	2,800	8,400	844.70	
	1100	3 C.Y., 75 tons at 12' radius		43.66	965	2,900	8,700	929.30	
	1200	100 ton capacity, standard boom		41.43	1,325	4,000	12,000	1,131	
	1300	165 ton capacity, standard boom		64.77	2,175	6,500	19,500	1,818	
	1400	200 ton capacity, 150' boom		120.73	2,325	7,000	21,000	2,366	
	1500	450' boom		135.50	3,000	9,000	27,000	2,884	
	1600	Truck mounted, cable operated, 6 x 4, 20 tons at 10' radius		14.07	665	2,000	6,000	512.55	
	1700	25 tons at 10' radius		20.80	1,075	3,200	9,600	806.40	
	1800	8 x 4, 30 tons at 10' radius		28.61	600	1,800	5,400	588.90	
	1900	40 tons at 12' radius		29.34	765	2,300	6,900	694.70	
	2000	8 x 4, 60 tons at 15' radius		44.87	900	2,700	8,100	898.95	
	2050	82 tons at 15' radius		45.56	1,625	4,900	14,700	1,344	
	2100	90 tons at 15' radius		48.95	1,025	3,100	9,300	1,012	
	2200	115 tons at 15' radius		51.15	1,875	5,600	16,800	1,529	
	2300	150 tons at 18' radius		76.61	1,575	4,700	14,100	1,553	
	2350	165 tons at 18' radius		77.65	2,200	6,600	19,800	1,941	
	2400	Truck mounted, hydraulic, 12 ton capacity		22.93	435	1,300	3,900	443.45	
	2500	25 ton capacity		23.67	600	1,800	5,400	549.35	
	2550	33 ton capacity		24.37	835	2,500	7,500	694.95	
	2560	40 ton hydraulic truck crane		24.42	860	2,575	7,725	710.35	
	2600	55 ton capacity		34.14	865	2,600	7,800	793.10	
	2700	80 ton capacity		37.29	1,325	4,000	12,000	1,098	
	2800	Self-propelled, 4 x 4, with telescoping boom, 5 ton		10.18	315	950	2,850	271.45	
	2900	12-1/2 ton capacity		16.42	435	1,300	3,900	391.35	
	3000	15 ton capacity		18.20	485	1,450	4,350	435.60	
	3100	25 ton capacity		20.99	635	1,900	5,700	547.90	
	3200	Derricks, guy, 20 ton capacity, 60' boom, 75' mast		8.70	292	875	2,625	244.60	
	3300	100' boom, 115' mast		16.39	515	1,550	4,650	441.10	
	3400	Stiffleg, 20 ton capacity, 70' boom, 37' mast		11.56	385	1,150	3,450	322.50	

016 400 | Equipment Rental

				UNIT	HOURLY OPER. COST	RENT PER DAY	RENT PER WEEK	RENT PER MONTH	CREW EQUIPMENT COST/DAY
460	3500	100' boom, 47' mast	R015 -010	Ea.	18.06	625	1,875	5,625	519.50
	3550	Helicopter, small, lift to 1250 lbs. maximum			257.50	2,425	7,250	21,800	3,510
	3600	Hoists, chain type, overhead, manual, 3/4 ton	R022 -250		.06	6	18	54	4.10
	3900	10 ton			.25	25	75	225	17
	5200	Jacks, hydraulic, 20 ton			.15	2.67	8	24	2.80
	5500	100 ton			.19	20	60	180	13.50
	6000	Jacks, hydraulic, climbing with 50' jackrods							
	6010	and control consoles, minimum 3 mo. rental							
	6100	30 ton capacity		Ea.	.06	100	300	900	60.50
	6150	For each added 10' jackrod section, add				3	9	27	1.80
	6300	50 ton capacity				158	475	1,425	95
	6350	For each added 10' jackrod section, add				3.33	10	30	2
	6500	125 ton capacity				415	1,250	3,750	250
	6550	For each added 10' jackrod section, add				26.50	80	240	16
	6600	Cable jack, 10 ton capacity with 200' cable				83.50	250	750	50
	6650	For each added 50' of cable, add				8.35	25	75	5
465	0010	**MARINE EQUIPMENT RENTAL**							
	0200	Barge, 400 Ton, 30' wide x 90' long		Ea.	15.33	550	1,650	4,950	452.65
	0240	800 Ton, 45' wide x 90' long			25.04	735	2,200	6,600	640.30
	2000	Tugboat, diesel, 100 HP			11.23	192	575	1,725	204.85
	2040	250 HP			22.31	435	1,300	3,900	438.50
	2080	380 HP			51.24	965	2,900	8,700	989.90
490	0010	**WELLPOINT EQUIPMENT RENTAL** See also division 021-444	R021 -440						
	0020	Based on 2 months rental							
	0100	Combination jetting & wellpoint pump, 60 H.P. diesel		Ea.	3.48	242	725	2,175	172.85
	0200	High pressure gas jet pump, 200 H.P., 300 psi		"	9.69	207	620	1,850	201.50
	0300	Discharge pipe, 8" diameter		L.F.		.38	1.15	3.45	.25
	0350	12" diameter				.57	1.70	5.10	.35
	0400	Header pipe, flows up to 150 G.P.M., 4" diameter				.33	1	3	.20
	0500	400 G.P.M., 6" diameter				.33	1	3	.20
	0600	800 G.P.M., 8" diameter				.57	1.70	5.10	.35
	0700	1500 G.P.M., 10" diameter				.60	1.80	5.40	.35
	0800	2500 G.P.M., 12" diameter				1.17	3.50	10.50	.70
	0900	4500 G.P.M., 16" diameter				1.50	4.50	13.50	.90
	0950	For quick coupling aluminum and plastic pipe, add				1.53	4.60	13.80	.90
	1100	Wellpoint, 25' long, with fittings & riser pipe, 1-1/2" or 2" diameter		Ea.		3.08	9.25	28	1.85
	1200	Wellpoint pump, diesel powered, 4" diameter, 20 H.P.			3.49	140	420	1,250	111.90
	1300	6" diameter, 30 H.P.			5.39	177	530	1,600	149.10
	1400	8" suction, 40 H.P.			6.41	240	720	2,150	195.30
	1500	10" suction, 75 H.P.			6.91	280	840	2,525	223.30
	1600	12" suction, 100 H.P.			9.89	450	1,350	4,050	349.10
	1700	12" suction, 175 H.P.			10.83	500	1,500	4,500	386.65

017 100 | Final Cleaning

			CREW	DAILY OUTPUT	LABOR-HOURS	UNIT	1999 BARE COSTS MAT.	LABOR	EQUIP.	TOTAL	TOTAL INCL O&P	
104	0010	**CLEANING UP** After job completion, allow, minimum				Job					.30%	104
	0040	Maximum				"					1%	
	0200	Rubbish removal, see division 020-620										

Division Notes

	CREW	DAILY OUTPUT	LABOR-HOURS	UNIT	1999 BARE COSTS MAT.	LABOR	EQUIP.	TOTAL	TOTAL INCL O&P

Division 2
Site Work

Estimating Tips

020 Subsurface Investigation & Demolition

- In preparing estimates on structures involving earthwork or foundations, all information concerning soil characteristics should be obtained. Look particularly for hazardous waste, evidence of prior dumping of debris, and previous stream beds.
- The costs shown for selective demolition do not include rubbish handling or disposal. These items should be estimated separately using Means data or other sources.

021 Site Preparation & Excavation Support

- If possible visit the site and take an inventory of the type, quantity and size of the trees. Certain trees may have a landscape resale value or firewood value. Stump disposal can be very expensive, particularly if they cannot be buried at the site. Consider using a bulldozer in lieu of hand cutting trees.
- Estimators should visit the site to determine the need for haul road, access, storage of materials, and security considerations. When estimating for access roads on unstable soil, consider using a geotextile stabilization fabric. It can greatly reduce the quantity of crushed stone or gravel. Sites of limited size and access can cause cost overruns due to lost productivity. Theft and damage is another consideration if the location is isolated. A temporary fence or security guards may be required. Investigate the site thoroughly.

022 Earthwork

- Estimating the actual cost of performing earthwork requires careful consideration of the variables involved. This includes items such as type of soil, whether or not water will be encountered, dewatering, whether or not banks need bracing, disposal of excavated earth, length of haul to fill or spoil sites, etc. If the project has large quantities of cut or fill, consider raising or lowering the site to reduce costs while paying close attention to the effect on site drainage and utilities if doing this.
- If the project has large quantities of fill, creating a borrow pit on the site can significantly lower the costs. It is very important to consider what time of year the project is scheduled for completion. Bad weather can create large cost overruns from dewatering, site repair and lost productivity from cold weather.

025 Paving & Surfacing

- When estimating paving, keep in mind the project schedule. If an asphaltic paving project is in a colder climate and runs through to the spring, consider placing the base course in the autumn, then topping it in the spring just prior to completion. This could save considerable costs in spring repair. Keep in mind that prices for asphalt and concrete are generally higher in the cold seasons.

026 Piped Utilities
027 Sewerage & Drainage

- Never assume that the water, sewer and drainage lines will go in at the early stages of the project. Consider the site access needs before dividing the site in half with open trenches, loose pipe, and machinery obstructions. Always inspect the site to establish that the site drawings are complete. Check off all existing utilities on your drawing as you locate them. If you find any discrepancies, mark up the site plan for further research. Differing site conditions can be very costly if discovered later in the project.

029 Landscaping

- The timing of planting and guarantee specifications often dictate the costs for establishing tree and shrub growth and a stand of grass or ground cover. Establish the work performance schedule to coincide with the local planting season. Maintenance and growth guarantees can add from 20% to 100% to the total landscaping cost. The cost to replace trees and shrubs can be as high as 5% of the total cost depending on the planting zone, soil conditions and time of year.

Reference Numbers

Reference numbers are shown in bold squares at the beginning of some major classifications. These numbers refer to related items in the Reference Section. The reference information may be an estimating procedure, an alternate pricing method or technical information.

Note: Not all subdivisions listed here necessarily appear in this publication.

020 120 | Std Penetration Tests

		020 120 Std Penetration Tests	CREW	DAILY OUTPUT	LABOR-HOURS	UNIT	1999 BARE COSTS MAT.	LABOR	EQUIP.	TOTAL	TOTAL INCL O&P
123	0010	**BORINGS** Initial field stake out and determination of elevations	A-6	1	16	Day		410		410	625
	0100	Drawings showing boring details				Total		170		170	245
	0200	Report and recommendations from P.E.						375		375	540
	0300	Mobilization and demobilization, minimum	B-55	4	6			129	160	289	380
	0350	For over 100 miles, per added mile		450	.053	Mile		1.15	1.42	2.57	3.35
	0600	Auger holes in earth, no samples, 2-1/2″ diameter		78.60	.305	L.F.		6.60	8.15	14.75	19.20
	0650	4″ diameter		67.50	.356			7.65	9.45	17.10	22.50
	0800	Cased borings in earth, with samples, 2-1/2″ diameter		55.50	.432		12.50	9.30	11.50	33.30	41
	0850	4″ diameter		32.60	.736		20	15.85	19.60	55.45	68
	1000	Drilling in rock, "BX" core, no sampling	B-56	34.90	.458			11.15	17.25	28.40	36
	1050	With casing & sampling		31.70	.505		12.50	12.30	18.95	43.75	53.50
	1200	"NX" core, no sampling		25.92	.617			15	23	38	48.50
	1250	With casing and sampling		25	.640		16	15.55	24	55.55	68
	1400	Drill rig and crew with truck mounted auger	B-55	1	24	Day		515	640	1,155	1,500
	1450	With crawler type drill	B-56	1	16	"		390	600	990	1,250
	1500	For inner city borings add, minimum									10%
	1510	Maximum									20%
125	0010	**DRILLING, CORE** Reinforced concrete slab, up to 6″ thick slab									
	0020	Including bit, layout and set up									
	0100	1″ diameter core	B-89A	28	.571	Ea.	2.28	14.15	2.19	18.62	27.50
	0150	Each added inch thick, add		300	.053		.40	1.32	.20	1.92	2.75
	0300	3″ diameter core		23	.696		5	17.20	2.66	24.86	35.50
	0350	Each added inch thick, add		186	.086		.90	2.13	.33	3.36	4.70
	0500	4″ diameter core		19	.842		5	21	3.23	29.23	42
	0550	Each added inch thick, add		170	.094		1.15	2.33	.36	3.84	5.35
	0700	6″ diameter core		14	1.143		8.30	28.50	4.38	41.18	58.50
	0750	Each added inch thick, add		140	.114		1.40	2.83	.44	4.67	6.50
	0900	8″ diameter core		11	1.455		11.30	36	5.55	52.85	75
	0950	Each added inch thick, add		95	.168		1.90	4.17	.65	6.72	9.35
	1100	10″ diameter core		10	1.600		15.10	39.50	6.15	60.75	86
	1150	Each added inch thick, add		80	.200		2.50	4.95	.77	8.22	11.40
	1300	12″ diameter core		9	1.778		18.10	44	6.80	68.90	97
	1350	Each added inch thick, add		68	.235		3	5.80	.90	9.70	13.50
	1500	14″ diameter core		7	2.286		22	56.50	8.75	87.25	123
	1550	Each added inch thick, add		55	.291		3.80	7.20	1.11	12.11	16.75
	1700	18″ diameter core		4	4		28.50	99	15.30	142.80	204
	1750	Each added inch thick, add		28	.571		5	14.15	2.19	21.34	30.50
	1760	For horizontal holes, add to above								30%	30%
	1770	Prestressed hollow core plank, 6″ thick									
	1780	1″ diameter core	B-89A	52	.308	Ea.	1.50	7.60	1.18	10.28	14.95
	1790	Each added inch thick, add		350	.046		.25	1.13	.18	1.56	2.25
	1800	3″ diameter core		50	.320		3.30	7.90	1.23	12.43	17.50
	1810	Each added inch thick, add		240	.067		.55	1.65	.26	2.46	3.49
	1820	4″ diameter core		48	.333		4.40	8.25	1.28	13.93	19.25
	1830	Each added inch thick, add		216	.074		.75	1.83	.28	2.86	4.03
	1840	6″ diameter core		44	.364		5.45	9	1.39	15.84	21.50
	1850	Each added inch thick, add		175	.091		.90	2.26	.35	3.51	4.95
	1860	8″ diameter core		32	.500		7.30	12.40	1.92	21.62	29.50
	1870	Each added inch thick, add		118	.136		1.25	3.36	.52	5.13	7.25
	1880	10″ diameter core		28	.571		9.85	14.15	2.19	26.19	36
	1890	Each added inch thick, add		99	.162		1.35	4	.62	5.97	8.45
	1900	12″ diameter core		22	.727		12	18	2.79	32.79	45
	1910	Each added inch thick, add		85	.188		2	4.66	.72	7.38	10.35
	1950	Minimum charge for above, 3″ diameter core		7	2.286	Total		56.50	8.75	65.25	98.50
	2000	4″ diameter core		6.80	2.353			58	9	67	102

020 120 | Std Penetration Tests

			CREW	DAILY OUTPUT	LABOR-HOURS	UNIT	1999 BARE COSTS				TOTAL INCL O&P
							MAT.	LABOR	EQUIP.	TOTAL	
125	2050	6″ diameter core	B-89A	6	2.667	Total		66	10.20	76.20	115
	2100	8″ diameter core		5.50	2.909			72	11.15	83.15	125
	2150	10″ diameter core		4.75	3.368			83.50	12.90	96.40	145
	2200	12″ diameter core		3.90	4.103			102	15.70	117.70	177
	2250	14″ diameter core		3.38	4.734			117	18.15	135.15	205
	2300	18″ diameter core		3.15	5.079			126	19.45	145.45	220
	3010	Bits for core drill, diamond, premium, 1″ diameter				Ea.	112			112	123
	3020	3″ diameter					274			274	300
	3040	4″ diameter					305			305	335
	3050	6″ diameter					490			490	535
	3080	8″ diameter					715			715	785
	3120	12″ diameter					1,100			1,100	1,200
	3180	18″ diameter					1,800			1,800	1,975
	3240	24″ diameter					2,400			2,400	2,650
128	0010	**TEST PITS** Hand digging, light soil	1 Clab	4.50	1.778	C.Y.		38		38	60
	0100	Heavy soil	"	2.50	3.200			68.50		68.50	108
	0120	Loader-backhoe, light soil	B-11M	28	.571			14.25	10.35	24.60	33.50
	0130	Heavy soil	"	20	.800			19.95	14.50	34.45	47
	1000	Subsurface exploration, mobilization				Mile				5	5.75
	1010	Difficult access for rig, add				Hr.				100	115
	1020	Auger borings, drill rig, incl. samples				L.F.				11.50	13.25
	1030	Hand auger								17	19.55
	1050	Drill and sample every 5′, split spoon								15	17.25
	1060	Extra samples				Ea.				20	23

020 550 | Site Demolition

			CREW	DAILY OUTPUT	LABOR-HOURS	UNIT	MAT.	LABOR	EQUIP.	TOTAL	TOTAL INCL O&P
554	0010	**SITE DEMOLITION** No hauling, abandon catch basin or manhole	B-6	7	3.429	Ea.		80	31.50	111.50	159
	0020	Remove existing catch basin or manhole, masonry		4	6			140	54.50	194.50	278
	0030	Catch basin or manhole frames and covers, stored		13	1.846			43	16.85	59.85	85.50
	0040	Remove and reset		7	3.429			80	31.50	111.50	159
	0100	Roadside delineators, remove only	B-80	175	.183			4.29	3.08	7.37	10.05
	0110	Remove and reset	"	100	.320			7.50	5.40	12.90	17.60
	0600	Fencing, barbed wire, 3 strand	2 Clab	430	.037	L.F.		.80		.80	1.26
	0650	5 strand	"	280	.057			1.23		1.23	1.93
	0700	Chain link, posts & fabric, remove only, 8′ to 10′ high	B-6	445	.054			1.26	.49	1.75	2.50
	0750	Remove and reset		70	.343			8	3.13	11.13	15.90
	0755	Remove only, 6′ high		520	.046			1.08	.42	1.50	2.13
	0756	Remove and reset		84	.286			6.70	2.61	9.31	13.20
	0760	Wood fence to 6′, remove only, minimum	2 Clab	500	.032			.69		.69	1.08
	0770	Maximum		250	.064			1.37		1.37	2.16
	0775	Fencing, wood, all types, 4′to 6′ high		432	.037			.79		.79	1.25
	0780	Fence post, wood, 4″ x 4″, 6′ to 8′ high		96	.167	Ea.		3.58		3.58	5.65
	0785	Fence rail, 2″ x 4″, 8′ long		7,680	.002	L.F.		.04		.04	.07
	0790	Remove and store	B-80	235	.136			3.19	2.30	5.49	7.45
	0800	Guiderail, corrugated steel, remove only	B-80A	100	.240			5.15	1.80	6.95	10.05
	0850	Remove and reset	"	40	.600			12.85	4.49	17.34	25.50
	0860	Guide posts, remove only	B-80B	120	.267	Ea.		6.10	2.36	8.46	12.10
	0870	Remove and reset	B-55	50	.480			10.35	12.75	23.10	30
	0900	Hydrants, fire, remove only	B-21A	5	8			213	78	291	410
	0950	Remove and reset	"	2	20			530	196	726	1,025
	1000	Masonry walls, block or tile, solid, remove	B-5	1,800	.031	C.F.		.74	.58	1.32	1.79
	1100	Cavity wall		2,200	.025			.60	.48	1.08	1.46
	1200	Brick, solid		900	.062			1.48	1.16	2.64	3.58
	1300	With block back-up		1,130	.050			1.18	.93	2.11	2.85
	1400	Stone, with mortar		900	.062			1.48	1.16	2.64	3.58
	1500	Dry set		1,500	.037			.89	.70	1.59	2.15

020 550 | Site Demolition

			CREW	DAILY OUTPUT	LABOR-HOURS	UNIT	1999 BARE COSTS MAT.	LABOR	EQUIP.	TOTAL	TOTAL INCL O&P
554	1600	Median barrier, precast concrete, remove and store	B-3	430	.112	L.F.		2.59	4.07	6.66	8.50
	1610	Remove and reset	″	390	.123	″		2.85	4.49	7.34	9.35
	1710	Pavement removal, bituminous roads, 3″ thick	B-38	690	.058	S.Y.		1.41	1.48	2.89	3.82
	1750	4″ to 6″ thick		420	.095			2.32	2.43	4.75	6.25
	1800	Bituminous driveways		640	.063			1.52	1.59	3.11	4.11
	1900	Concrete to 6″ thick, hydraulic hammer, mesh reinforced		255	.157			3.83	4	7.83	10.35
	2000	Rod reinforced		200	.200			4.88	5.10	9.98	13.15
	2100	Concrete, 7″ to 24″ thick, plain		33	1.212	C.Y.		29.50	31	60.50	80
	2200	Reinforced		24	1.667	″		40.50	42.50	83	110
	2300	With hand held air equipment, bituminous, to 6″ thick	B-39	1,900	.025	S.F.		.57	.09	.66	1
	2320	Concrete to 6″ thick, no reinforcing		1,200	.040			.91	.15	1.06	1.59
	2340	Mesh reinforced		1,400	.034			.78	.13	.91	1.36
	2360	Rod reinforced		765	.063			1.43	.24	1.67	2.49
	2400	Curbs, concrete, plain	B-6	360	.067	L.F.		1.56	.61	2.17	3.09
	2500	Reinforced		275	.087			2.04	.80	2.84	4.05
	2600	Granite		360	.067			1.56	.61	2.17	3.09
	2700	Bituminous		528	.045			1.06	.41	1.47	2.11
	2800	Wood		570	.042			.98	.38	1.36	1.95
	2900	Pipe removal, sewer/water, no excavation, 12″ diameter		175	.137			3.21	1.25	4.46	6.35
	2930	15″ diameter		150	.160			3.74	1.46	5.20	7.40
	2960	24″ diameter		120	.200			4.67	1.82	6.49	9.25
	3000	36″ diameter		90	.267			6.25	2.43	8.68	12.30
	3200	Steel, welded connections, 4″ diameter		160	.150			3.51	1.37	4.88	6.95
	3300	10″ diameter		80	.300			7	2.74	9.74	13.90
	3500	Railroad track removal, ties and track	B-13	330	.170			3.94	1.66	5.60	8
	3600	Ballast	B-14	500	.096	C.Y.		2.18	.44	2.62	3.89
	3700	Remove and re-install, ties & track using new bolts & spikes		50	.960	L.F.		22	4.38	26.38	39
	3800	Turnouts using new bolts and spikes		1	48	Ea.		1,100	219	1,319	1,950
	4000	Sidewalk removal, bituminous, 2-1/2″ thick	B-6	325	.074	S.Y.		1.73	.67	2.40	3.42
	4050	Brick, set in mortar		185	.130			3.03	1.18	4.21	6
	4060	Dry set		270	.089			2.08	.81	2.89	4.11
	4100	Concrete, plain, 4″		160	.150			3.51	1.37	4.88	6.95
	4200	Mesh reinforced		150	.160			3.74	1.46	5.20	7.40
	5000	Slab on grade removal, plain	B-5	45	1.244	C.Y.		29.50	23	52.50	71.50
	5100	Mesh reinforced		33	1.697			40.50	31.50	72	97.50
	5200	Rod reinforced		25	2.240			53	42	95	129
	5500	For congested sites or small quantities, add up to								200%	200%
	5550	For disposal on site, add	B-11A	232	.069			1.72	3.62	5.34	6.65
	5600	To 5 miles, add	B-34D	76	.105			2.33	7.40	9.73	11.70
558	0010	**HYDRODEMOLITION**, concrete pavement, 4000 PSI, 2″ depth	B-5	500	.112	S.F.		2.66	2.09	4.75	6.45
	0120	4″ depth		450	.124			2.95	2.32	5.27	7.15
	0130	6″ depth		400	.140			3.32	2.61	5.93	8
	0410	6000 PSI, 2″ depth		410	.137			3.24	2.55	5.79	7.85
	0420	4″ depth		350	.160			3.80	2.99	6.79	9.20
	0430	6″ depth		300	.187			4.43	3.49	7.92	10.75
	0510	8000 PSI, 2″ depth		330	.170			4.03	3.17	7.20	9.75
	0520	4″ depth		280	.200			4.74	3.73	8.47	11.50
	0530	6″ depth		240	.233			5.55	4.36	9.91	13.40

020 600 | Building Demolition

			CREW	DAILY OUTPUT	LABOR-HOURS	UNIT	MAT.	LABOR	EQUIP.	TOTAL	TOTAL INCL O&P
604	0010	**BUILDING DEMOLITION** Large urban projects, incl. 20 Mi. haul									
	0012	No foundation or dump fees, C.F. is volume of building standing, steel	B-8	21,500	.003	C.F.		.07	.11	.18	.23
	0050	Concrete		15,300	.004			.10	.15	.25	.32
	0080	Masonry		20,100	.003			.08	.11	.19	.25

Important: See the Reference Section for critical supporting data - Reference Nos., Crews, & City Cost Indexes

020 600 | Building Demolition

			CREW	DAILY OUTPUT	LABOR-HOURS	UNIT	1999 BARE COSTS MAT.	LABOR	EQUIP.	TOTAL	TOTAL INCL O&P
604	0100	Mixture of types, average	B-8	20,100	.003	C.F.		.08	.11	.19	.25
	0500	Small bldgs, or single bldgs, no salvage included, steel	B-3	14,800	.003			.08	.12	.20	.25
	0600	Concrete		11,300	.004			.10	.15	.25	.32
	0650	Masonry		14,800	.003			.08	.12	.20	.25
	0700	Wood		14,800	.003			.08	.12	.20	.25
	1000	Single family, one story house, wood, minimum				Ea.				2,300	2,700
	1020	Maximum								4,000	4,800
	1200	Two family, two story house, wood, minimum								3,000	3,600
	1220	Maximum								5,800	7,000
	1300	Three family, three story house, wood, minimum								4,000	4,800
	1320	Maximum								7,000	8,400
	1400	Gutting building, see division 020-716									
	5000	For buildings with no interior walls, deduct				Ea.				50%	
608	0010	**DISPOSAL ONLY** Urban buildings with salvage value allowed									
	0020	Including loading and 5 mile haul to dump									
	0200	Steel frame	B-3	430	.112	C.Y.		2.59	4.07	6.66	8.50
	0300	Concrete frame		365	.132			3.05	4.79	7.84	9.95
	0400	Masonry construction		445	.108			2.50	3.93	6.43	8.20
	0500	Wood frame		247	.194			4.50	7.10	11.60	14.75
612	0010	**DUMP CHARGES** Typical urban city, tipping fees only									
	0100	Building construction materials				Ton					55
	0200	Trees, brush, lumber									45
	0300	Rubbish only									50
	0500	Reclamation station, usual charge									80
616	0010	**EXPLOSIVE/IMPLOSIVE DEMOLITION** Large projects, no disposal/fee									
	0020	based on building volume, steel building	B-5B	16,900	.003	C.F.		.07	.12	.19	.24
	0100	Concrete building		16,900	.003			.07	.12	.19	.24
	0200	Masonry building		16,900	.003			.07	.12	.19	.24
	0400	Disposal of material, minimum	B-3	445	.108	C.Y.		2.50	3.93	6.43	8.20
	0500	Maximum	″	365	.132	″		3.05	4.79	7.84	9.95
620	0010	**RUBBISH HANDLING** The following are to be added to the									
	0020	demolition prices									
	0400	Chute, circular, prefabricated steel, 18″ diameter	B-1	40	.600	L.F.	19.35	13.25		32.60	42.50
	0440	30″ diameter	″	30	.800	″	25	17.70		42.70	55.50
	0600	Dumpster, weekly rental, 1 dump/week, 6 C.Y. capacity (2 Tons)				Ea.					300
	0700	10 C.Y. capacity (4 Tons)									385
	0800	30 C.Y. capacity (10 Tons)									650
	0840	40 C.Y. capacity (13 Tons)									775
	0900	Alternate pricing for dumpsters									
	0910	Delivery, average for all sizes				Ea.					55
	0920	Haul, average for all sizes									130
	0930	Rent per day, average for all sizes									2.75
	0940	Rent per month, average for all sizes									25
	0950	Disposal fee per ton, average for all sizes				Ton					45
	1000	Dust partition, 6 mil polyethylene, 4′ x 8′ panels, 1″ x 3″ frame	2 Carp	2,000	.008	S.F.	.40	.22		.62	.78
	1080	2″ x 4″ frame	″	2,000	.008	″	.50	.22		.72	.89
	2000	Load, haul to chute & dumping into chute, 50′ haul	2 Clab	24	.667	C.Y.		14.30		14.30	22.50
	2040	100′ haul		16.50	.970			21		21	32.50
	2080	Over 100′ haul, add per 100 L.F.		35.50	.451			9.65		9.65	15.20
	2120	In elevators, per 10 floors, add		140	.114			2.45		2.45	3.86
	3000	Loading & trucking, including 2 mile haul, chute loaded	B-16	45	.711			15.70	9.85	25.55	35.50
	3040	Hand loading truck, 50′ haul	″	48	.667			14.75	9.20	23.95	33
	3080	Machine loading truck	B-17	120	.267			6.15	4.86	11.01	14.85
	5000	Haul, per mile, up to 8 C.Y. truck	B-34B	1,165	.007			.15	.38	.53	.65

		020 600 | Building Demolition	Crew	Daily Output	Labor-Hours	Unit	1999 Bare Costs Mat.	Labor	Equip.	Total	Total Incl O&P	
620	5100	Over 8 C.Y. truck	B-34B	1,550	.005	C.Y.		.11	.29	.40	.48	620

		020 700 | Selective Demolition	Crew	Daily Output	Labor-Hours	Unit	Mat.	Labor	Equip.	Total	Total Incl O&P	
702	0010	**CEILING DEMOLITION**										702
	0200	Drywall, furred and nailed	2 Clab	800	.020	S.F.		.43		.43	.68	
	0220	On metal frame		760	.021			.45		.45	.71	
	0240	On suspension system, including system		720	.022			.48		.48	.75	
	1000	Plaster, lime and horse hair, on wood lath, incl. lath		700	.023			.49		.49	.77	
	1020	On metal lath		570	.028			.60		.60	.95	
	1100	Gypsum, on gypsum lath		720	.022			.48		.48	.75	
	1120	On metal lath		500	.032			.69		.69	1.08	
	1200	Suspended ceiling, mineral fiber, 2′x2′ or 2′x4′		1,500	.011			.23		.23	.36	
	1250	On suspension system, incl. system		1,200	.013			.29		.29	.45	
	1500	Tile, wood fiber, 12″ x 12″, glued		900	.018			.38		.38	.60	
	1540	Stapled		1,500	.011			.23		.23	.36	
	1580	On suspension system, incl. system		760	.021			.45		.45	.71	
	2000	Wood, tongue and groove, 1″ x 4″		1,000	.016			.34		.34	.54	
	2040	1″ x 8″		1,100	.015			.31		.31	.49	
	2400	Plywood or wood fiberboard, 4′ x 8′ sheets		1,200	.013			.29		.29	.45	
704	0010	**CUTOUT DEMOLITION** Conc., elev. slab, light reinf., under 6 C.F.	B-9C	65	.615	C.F.		13.45	2.78	16.23	24	704
	0050	Light reinforcing, over 6 C.F.	″	75	.533	″		11.65	2.41	14.06	21	
	0200	Slab on grade to 6″ thick, not reinforced, under 8 S.F.	B-9	85	.471	S.F.		10.30	2.12	12.42	18.55	
	0250	Not reinforced, over 8 S.F.		175	.229	″		4.99	1.03	6.02	9	
	0600	Walls, not reinforced, under 6 C.F.		60	.667	C.F.		14.55	3.01	17.56	26.50	
	0650	Not reinforced, over 6 C.F.		65	.615			13.45	2.78	16.23	24	
	1000	Concrete, elevated slab, bar reinforced, under 6 C.F.	B-9C	45	.889			19.40	4.01	23.41	35	
	1050	Bar reinforced, over 6 C.F.	″	50	.800			17.50	3.61	21.11	31.50	
	1200	Slab on grade to 6″ thick, bar reinforced, under 8 S.F.	B-9	75	.533	S.F.		11.65	2.41	14.06	21	
	1250	Bar reinforced, over 8 S.F.	″	105	.381	″		8.30	1.72	10.02	15	
	1400	Walls, bar reinforced, under 6 C.F.	B-9C	50	.800	C.F.		17.50	3.61	21.11	31.50	
	1450	Bar reinforced, over 6 C.F.	″	55	.727	″		15.90	3.28	19.18	28.50	
	2000	Brick, to 4 S.F. opening, not including toothing										
	2040	4″ thick	B-9C	30	1.333	Ea.		29	6	35	52.50	
	2060	8″ thick		18	2.222			48.50	10	58.50	87.50	
	2080	12″ thick		10	4			87.50	18.05	105.55	158	
	2400	Concrete block, to 4 S.F. opening, 2″ thick		35	1.143			25	5.15	30.15	45	
	2420	4″ thick		30	1.333			29	6	35	52.50	
	2440	8″ thick		27	1.481			32.50	6.70	39.20	58.50	
	2460	12″ thick		24	1.667			36.50	7.50	44	66	
	2600	Gypsum block, to 4 S.F. opening, 2″ thick	B-9	80	.500			10.95	2.26	13.21	19.70	
	2620	4″ thick		70	.571			12.50	2.58	15.08	22.50	
	2640	8″ thick		55	.727			15.90	3.28	19.18	28.50	
	2800	Terra cotta, to 4 S.F. opening, 4″ thick		70	.571			12.50	2.58	15.08	22.50	
	2840	8″ thick		65	.615			13.45	2.78	16.23	24	
	2880	12″ thick		50	.800			17.50	3.61	21.11	31.50	
	3000	Toothing masonry cutouts, brick, soft old mortar	1 Brhe	40	.200	V.L.F.		4.32		4.32	6.70	
	3100	Hard mortar		30	.267			5.75		5.75	8.95	
	3200	Block, soft old mortar		70	.114			2.47		2.47	3.84	
	3400	Hard mortar		50	.160			3.46		3.46	5.40	
	6000	Walls, interior, not including re-framing,										
	6010	openings to 5 S.F.										
	6100	Drywall to 5/8″ thick	A-1	24	.333	Ea.		7.15	2.86	10.01	14.40	
	6200	Paneling to 3/4″ thick		20	.400			8.60	3.43	12.03	17.30	
	6300	Plaster, on gypsum lath		20	.400			8.60	3.43	12.03	17.30	
	6340	On wire lath		14	.571			12.25	4.90	17.15	24.50	

Important: See the Reference Section for critical supporting data - Reference Nos., Crews, & City Cost Indexes

020 700 | Selective Demolition

			CREW	DAILY OUTPUT	LABOR-HOURS	UNIT	1999 BARE COSTS MAT.	LABOR	EQUIP.	TOTAL	TOTAL INCL O&P	
704	7000	Wood frame, not including re-framing, openings to 5 S.F.										704
	7200	Floors, sheathing and flooring to 2″ thick	A-1	5	1.600	Ea.		34.50	13.75	48.25	69	
	7310	Roofs, sheathing to 1″ thick, not including roofing		6	1.333			28.50	11.45	39.95	57.50	
	7410	Walls, sheathing to 1″ thick, not including siding		7	1.143			24.50	9.80	34.30	49.50	
706	0010	**DOOR DEMOLITION**										706
	0200	Doors, exterior, 1-3/4″ thick, single, 3′ x 7′ high	1 Clab	16	.500	Ea.		10.75		10.75	16.90	
	0220	Double, 6′ x 7′ high		12	.667			14.30		14.30	22.50	
	0500	Interior, 1-3/8″ thick, single, 3′ x 7′ high		20	.400			8.60		8.60	13.50	
	0520	Double, 6′ x 7′ high		16	.500			10.75		10.75	16.90	
	0700	Bi-folding, 3′ x 6′-8″ high		20	.400			8.60		8.60	13.50	
	0720	6′ x 6′-8″ high		18	.444			9.55		9.55	15	
	0900	Bi-passing, 3′ x 6′-8″ high		16	.500			10.75		10.75	16.90	
	0940	6′ x 6′-8″ high		14	.571			12.25		12.25	19.30	
	1500	Remove and reset, minimum	1 Carp	8	1			27.50		27.50	43	
	1520	Maximum	″	6	1.333			36.50		36.50	57.50	
	2000	Frames, including trim, metal	A-1	8	1			21.50	8.60	30.10	43.50	
	2200	Wood	2 Carp	32	.500			13.65		13.65	21.50	
	3000	Special doors, counter doors		6	2.667			73		73	115	
	3100	Double acting		10	1.600			43.50		43.50	68.50	
	3200	Floor door (trap type)		8	2			54.50		54.50	86	
	3300	Glass, sliding, including frames		12	1.333			36.50		36.50	57.50	
	3400	Overhead, commercial, 12′ x 12′ high		4	4			109		109	172	
	3440	20′ x 16′ high		3	5.333			146		146	229	
	3500	Residential, 9′ x 7′ high		8	2			54.50		54.50	86	
	3540	16′ x 7′ high		7	2.286			62.50		62.50	98	
	3600	Remove and reset, minimum		4	4			109		109	172	
	3620	Maximum		2.50	6.400			175		175	275	
	3700	Roll-up grille		5	3.200			87.50		87.50	137	
	3800	Revolving door		2	8			218		218	345	
	3900	Storefront swing door		3	5.333			146		146	229	
708	0010	**ELECTRICAL DEMOLITION**										708
	0020	Conduit to 15′ high, including fittings & hangers										
	0100	Rigid galvanized steel, 1/2″ to 1″ diameter	1 Elec	242	.033	L.F.		1.05		1.05	1.58	
	0120	1-1/4″ to 2″		200	.040			1.28		1.28	1.91	
	0140	2″ to 4″		151	.053			1.69		1.69	2.53	
	0160	4″ to 6″		80	.100			3.19		3.19	4.77	
	0200	Electric metallic tubing (EMT) 1/2″ to 1″		394	.020			.65		.65	.97	
	0220	1-1/4″ to 1-1/2″		326	.025			.78		.78	1.17	
	0240	2″ to 3″		236	.034			1.08		1.08	1.62	
	0260	3-1/2″ to 4″		95	.084			2.69		2.69	4.02	
	1300	Transformer, dry type, 1 ph, incl. removal of										
	1320	supports, wire & pipe terminations										
	1340	1 kVA	1 Elec	7.70	1.039	Ea.		33		33	49.50	
	1420	75 kVA	″	1.25	6.400	″		204		204	305	
	1440	3 Phase to 600V, primary										
	1460	3 kVA	1 Elec	3.85	2.078	Ea.		66.50		66.50	99	
	1520	75 kVA	2 Elec	2.70	5.926			189		189	283	
	1550	300 kVA	R-3	1.80	11.111			350	75.50	425.50	610	
	1570	750 kVA		1.10	18.182			575	123	698	995	
	1575	150 KVA pad mtd, disconnect & load		3.20	6.250			197	42.50	239.50	345	
	1580	25 KVA pole mtd, disconnect & load		2	10			315	68	383	550	
712	0010	**FLOORING DEMOLITION**										712
	0200	Brick with mortar	2 Clab	475	.034	S.F.		.72		.72	1.14	
	0400	Carpet, bonded, including surface scraping		2,000	.008			.17		.17	.27	
	0440	Scrim applied		8,000	.002			.04		.04	.07	

020 700 | Selective Demolition

	020 700	Selective Demolition	CREW	DAILY OUTPUT	LABOR-HOURS	UNIT	1999 BARE COSTS MAT.	LABOR	EQUIP.	TOTAL	TOTAL INCL O&P	
712	0480	Tackless	2 Clab	9,000	.002	S.F.		.04		.04	.06	712
	0500	Carpet pad		3,500	.005	S.Y.		.10		.10	.15	
	0600	Composition, acrylic or epoxy		400	.040	S.F.		.86		.86	1.35	
	0700	Concrete, scarify skin	A-1	225	.036			.76	.31	1.07	1.54	
	0800	Resilient, sheet goods	2 Clab	1,400	.011			.25		.25	.39	
	0820	For gym floors		900	.018			.38		.38	.60	
	0900	Vinyl composition tile, 12″ x 12″		1,000	.016			.34		.34	.54	
	2000	Tile, ceramic, thin set		675	.024			.51		.51	.80	
	2020	Mud set		625	.026			.55		.55	.86	
	2200	Marble, slate, thin set		675	.024			.51		.51	.80	
	2220	Mud set		625	.026			.55		.55	.86	
	2600	Terrazzo, thin set		450	.036			.76		.76	1.20	
	2620	Mud set		425	.038			.81		.81	1.27	
	2640	Cast in place		300	.053			1.14		1.14	1.80	
	8000	Remove flooring, bead blast, minimum	A-1A	1,000	.008			.22	.13	.35	.50	
	8100	Maximum		400	.020			.56	.33	.89	1.24	
	8150	Mastic only		1,500	.005			.15	.09	.24	.34	
714	0010	**FRAMING DEMOLITION**										714
	1020	Concrete, average reinforcing, beams, 8″ x 10″	B-9	120	.333	L.F.		7.30	1.50	8.80	13.10	
	1040	10″ x 12″		110	.364			7.95	1.64	9.59	14.30	
	1060	12″ x 14″		90	.444			9.70	2	11.70	17.50	
	1200	Columns, 8″ x 8″		120	.333			7.30	1.50	8.80	13.10	
	1240	10″ x 10″		120	.333			7.30	1.50	8.80	13.10	
	1280	12″ x 12″		110	.364			7.95	1.64	9.59	14.30	
	1320	14″ x 14″		100	.400			8.75	1.80	10.55	15.75	
	1400	Girders, 14″ x 16″		55	.727			15.90	3.28	19.18	28.50	
	1440	16″ x 18″		40	1			22	4.51	26.51	39.50	
	1600	Slabs, elevated, 6″ thick		600	.067	S.F.		1.46	.30	1.76	2.62	
	1640	8″ thick		450	.089			1.94	.40	2.34	3.50	
	1680	10″ thick		360	.111			2.43	.50	2.93	4.37	
	1900	Add for heavy reinforcement									25%	
	2000	Steel framing, beams, 4″ x 6″	B-13	500	.112	L.F.		2.60	1.10	3.70	5.25	
	2020	4″ x 8″		400	.140			3.25	1.37	4.62	6.55	
	2080	8″ x 12″		250	.224			5.20	2.20	7.40	10.50	
	2200	Columns, 6″ x 6″		400	.140			3.25	1.37	4.62	6.55	
	2240	8″ x 8″		350	.160			3.72	1.57	5.29	7.55	
	2280	10″ x 10″		320	.175			4.07	1.72	5.79	8.25	
	2400	Girders, 10″ x 12″		225	.249			5.80	2.44	8.24	11.70	
	2440	10″ x 14″		200	.280			6.50	2.75	9.25	13.10	
	2480	10″ x 16″		165	.339			7.90	3.33	11.23	15.90	
	2520	10″ x 24″		125	.448			10.40	4.39	14.79	21	
	3000	Wood framing, beams, 6″ x 8″	B-2	275	.145			3.18		3.18	5	
	3040	6″ x 10″		220	.182			3.97		3.97	6.25	
	3080	6″ x 12″		185	.216			4.72		4.72	7.45	
	3120	8″ x 12″		140	.286			6.25		6.25	9.80	
	3160	10″ x 12″		110	.364			7.95		7.95	12.50	
	9500	See Div. 020-620 for rubbish handling										
716	0010	**GUTTING** Building interior, including disposal, dumpster fees not included										716
	0500	Residential building										
	0560	Minimum	B-16	400	.080	SF Flr.		1.77	1.11	2.88	3.98	
	0580	Maximum	″	360	.089	″		1.97	1.23	3.20	4.42	
	0900	Commercial building										
	1000	Minimum	B-16	350	.091	SF Flr.		2.02	1.26	3.28	4.55	
	1020	Maximum	″	250	.128	″		2.83	1.77	4.60	6.35	

	020 700 | Selective Demolition		CREW	DAILY OUTPUT	LABOR-HOURS	UNIT	1999 BARE COSTS MAT.	LABOR	EQUIP.	TOTAL	TOTAL INCL O&P	
717	0010	**HAZARDOUS WASTE CLEANUP/PICKUP/DISPOSAL**										717
	0100	For contractor equipment, i.e. dozer,										
	0110	front end loader, dump truck, etc., see div. 016-408										
	1000	Solid pickup										
	1100	55 gal. drums				Ea.					200	
	1120	Bulk material, minimum				Ton					150	
	1130	Maximum				″					500	
	1200	Transportation to disposal site										
	1220	Truckload = 80 drums or 25 C.Y. or 18 tons										
	1260	Minimum				Mile					2.30	
	1270	Maximum				″					4	
	3000	Liquid pickup, vacuum truck, stainless steel tank										
	3100	Minimum charge, 4 hours										
	3110	1 compartment, 2200 gallon				Hr.					100	
	3120	2 compartment, 5000 gallon				″					100	
	3400	Transportation in 6900 gallon bulk truck				Mile					4.30	
	3410	In teflon lined truck				″					5	
	5000	Heavy sludge or dry vacuumable material				Hr.					100	
	6000	Dumpsite disposal charge, minimum				Ton					100	
	6020	Maximum				″					400	
718	0010	**HVAC DEMOLITION**										718
	0100	Air conditioner, split unit, 3 ton	Q-5	2	8	Ea.		236		236	355	
	0150	Package unit, 3 ton	Q-6	3	8			245		245	370	
	0200	Air conditioner, rooftop, gas heat, remove & reset, 10 ton		.30	80			2,450		2,450	3,700	
	0240	Remove & reset, 20 ton	↓	.15	160			4,900		4,900	7,400	
	0300	Boiler, electric	Q-19	2	12			365		365	550	
	0340	Gas or oil, steel, under 150 MBH	Q-6	3	8			245		245	370	
	0380	Over 150 MBH	″	2	12	↓		365		365	555	
	1000	Ductwork, 4″ high, 8″ wide	1 Clab	200	.040	L.F.		.86		.86	1.35	
	1100	6″ high, 8″ wide		165	.048			1.04		1.04	1.64	
	1200	10″ high, 12″ wide		125	.064			1.37		1.37	2.16	
	1300	12″-14″ high, 16″-18″ wide		85	.094			2.02		2.02	3.18	
	1400	18″ high, 24″ wide		67	.119			2.56		2.56	4.03	
	1500	30″ high, 36″ wide		56	.143			3.06		3.06	4.82	
	1540	72″ wide	↓	50	.160	↓		3.43		3.43	5.40	
	3000	Mechanical equipment, light items. Unit is weight, not cooling.	Q-5	.90	17.778	Ton		525		525	795	
	3600	Heavy items	″	1.10	14.545	″		430		430	650	
	3700	Deduct for salvage (when applicable), minimum				Job					52	
	3750	Maximum				″					420	
720	0010	**MILLWORK AND TRIM DEMOLITION**										720
	1000	Cabinets, wood, base cabinets	2 Clab	80	.200	L.F.		4.29		4.29	6.75	
	1020	Wall cabinets		80	.200	″		4.29		4.29	6.75	
	1200	Casework, large area		320	.050	S.F.		1.07		1.07	1.69	
	1220	Selective		200	.080	″		1.72		1.72	2.70	
	1500	Counter top, minimum		200	.080	L.F.		1.72		1.72	2.70	
	1510	Maximum		120	.133	″		2.86		2.86	4.50	
	2000	Paneling, 4′ x 8′ sheets, 1/4″ thick		2,000	.008	S.F.		.17		.17	.27	
	2100	Boards, 1″ x 4″		700	.023			.49		.49	.77	
	2120	1″ x 6″		750	.021			.46		.46	.72	
	2140	1″ x 8″		800	.020	↓		.43		.43	.68	
	3000	Trim, baseboard, to 6″ wide		1,200	.013	L.F.		.29		.29	.45	
	3040	12″ wide	↓	1,000	.016	″		.34		.34	.54	
724	0010	**PLUMBING DEMOLITION**										724
	1020	Fixtures, including 10′ piping										

020 700 | Selective Demolition

		020 700 Selective Demolition	CREW	DAILY OUTPUT	LABOR-HOURS	UNIT	1999 BARE COSTS MAT.	LABOR	EQUIP.	TOTAL	TOTAL INCL O&P	
724	1100	Bath tubs, cast iron	1 Plum	4	2	Ea.		65		65	99	724
	1120	Fiberglass		6	1.333			43.50		43.50	66	
	1140	Steel		5	1.600			52		52	79	
	1200	Lavatory, wall hung		10	.800			26		26	39.50	
	1220	Counter top		8	1			32.50		32.50	49.50	
	1300	Sink, steel or cast iron, single		8	1			32.50		32.50	49.50	
	1320	Double		7	1.143			37.50		37.50	56.50	
	1400	Water closet, floor mounted		8	1			32.50		32.50	49.50	
	1420	Wall mounted		7	1.143			37.50		37.50	56.50	
	1500	Urinal, floor mounted		4	2			65		65	99	
	1520	Wall mounted		7	1.143			37.50		37.50	56.50	
	1600	Water fountains, free standing		8	1			32.50		32.50	49.50	
	1620	Recessed		6	1.333			43.50		43.50	66	
	2000	Piping, metal, to 2″ diameter		200	.040	L.F.		1.30		1.30	1.98	
	2050	2″ to 4″ diameter		150	.053			1.74		1.74	2.63	
	2100	4″ to 8″ diameter	2 Plum	100	.160			5.20		5.20	7.90	
	2150	8″ to 16″ diameter	″	60	.267			8.70		8.70	13.15	
	2160	Deduct for salvage, aluminum scrap				Ton					525	
	2170	Brass scrap									420	
	2180	Copper scrap									1,025	
	2200	Lead scrap									235	
	2220	Steel scrap									65	
	2250	Water heater, 40 gal.	1 Plum	6	1.333	Ea.		43.50		43.50	66	
726	0010	**ROOFING AND SIDING DEMOLITION**										726
	1000	Deck, roof, concrete plank	B-13	1,680	.033	S.F.		.77	.33	1.10	1.57	
	1100	Gypsum plank		3,900	.014			.33	.14	.47	.67	
	1150	Metal decking		3,500	.016			.37	.16	.53	.75	
	1200	Wood, boards, tongue and groove, 2″ x 6″	2 Clab	960	.017			.36		.36	.56	
	1220	2″ x 10″		1,040	.015			.33		.33	.52	
	1280	Standard planks, 1″ x 6″		1,080	.015			.32		.32	.50	
	1320	1″ x 8″		1,160	.014			.30		.30	.47	
	1340	1″ x 12″		1,200	.013			.29		.29	.45	
	1350	Plywood, to 1″ thick		2,000	.008			.17		.17	.27	
	1360	Flashing, aluminum	1 Clab	290	.028			.59		.59	.93	
	2000	Gutters, aluminum or wood, edge hung		240	.033	L.F.		.72		.72	1.13	
	2100	Built-in		100	.080	″		1.72		1.72	2.70	
	2145	Downspout storm water riser, remove		15	.533	Ea.		11.45		11.45	18	
	2500	Roof accessories, plumbing vent flashing		14	.571			12.25		12.25	19.30	
	2600	Adjustable metal chimney flashing		9	.889			19.05		19.05	30	
	2650	Coping, sheet metal, up to 12″ wide		240	.033	L.F.		.72		.72	1.13	
	2660	Concrete, up to 12″ wide	2 Clab	160	.100	″		2.15		2.15	3.38	
	2670	Roof hatch, 30″ x 36″, remove and reset	G-3	6.67	4.798	Ea.		128		128	199	
	3000	Roofing, built-up, 5 ply roof, no gravel	B-2	1,600	.025	S.F.		.55		.55	.86	
	3001	Including gravel		890	.045			.98		.98	1.55	
	3100	Gravel removal, minimum		5,000	.008			.17		.17	.28	
	3120	Maximum		2,000	.020			.44		.44	.69	
	3450	Roll roofing, cold adhesive	1 Clab	12	.667	Sq.		14.30		14.30	22.50	
	3500	Sheet metal roofing	B-2	2,150	.019	S.F.		.41		.41	.64	
	5000	Siding, metal, horizontal	1 Clab	444	.018			.39		.39	.61	
	5020	Vertical		400	.020			.43		.43	.68	
	5200	Wood, boards, vertical		400	.020			.43		.43	.68	
	5220	Clapboards, horizontal		380	.021			.45		.45	.71	
	5240	Shingles		350	.023			.49		.49	.77	
	5260	Textured plywood		725	.011			.24		.24	.37	

020 700 | Selective Demolition

		020 700 Selective Demolition	CREW	DAILY OUTPUT	LABOR-HOURS	UNIT	1999 BARE COSTS MAT.	LABOR	EQUIP.	TOTAL	TOTAL INCL O&P
728	0010	**SAW CUTTING**, Asphalt, up to 3" deep	B-89	1,050	.015	L.F.	.24	.37	.29	.90	1.15
	0020	Each additional inch of depth		1,800	.009		.06	.22	.17	.45	.58
	0400	Concrete slabs, mesh reinforcing, up to 3" deep		980	.016		.33	.40	.31	1.04	1.32
	0420	Each additional inch of depth		1,600	.010		.44	.24	.19	.87	1.06
	0800	Concrete walls, hydraulic saw, plain, per inch of depth	B-89B	250	.064		.30	1.57	2.11	3.98	5.05
	0820	Rod reinforcing, per inch of depth		150	.107		.42	2.61	3.52	6.55	8.30
	1200	Masonry walls, hydraulic saw, brick, per inch of depth		300	.053		.30	1.31	1.76	3.37	4.26
	1220	Block walls, solid, per inch of depth		250	.064		.31	1.57	2.11	3.99	5.05
	2000	Brick or masonry w/hand held saw, per inch of depth	A-1	125	.064		.24	1.37	.55	2.16	3.03
	3020	Blades for saw, diamond, 12" diameter				Ea.	620			620	680
	3040	18" diameter					910			910	1,000
	3080	24" diameter					1,400			1,400	1,550
	3120	30" diameter					1,850			1,850	2,025
	3160	36" diameter					2,225			2,225	2,450
	3200	42" diameter					3,800			3,800	4,175
	5000	Wood sheathing to 1" thick, on walls	1 Carp	200	.040	L.F.		1.09		1.09	1.72
	5020	On roof	"	250	.032	"		.87		.87	1.37
	9950	See also div. 020-125 core drilling									
730	0010	**TORCH CUTTING** Steel, 1" thick plate	1 Clab	32	.250	L.F.		5.35		5.35	8.45
	0040	1" diameter bar	"	210	.038	Ea.		.82		.82	1.29
	1000	Oxygen lance cutting, reinforced concrete walls									
	1040	12" to 16" thick walls	1 Clab	10	.800	L.F.		17.15		17.15	27
	1080	24" thick walls	"	6	1.333	"		28.50		28.50	45
732	0010	**WALLS AND PARTITIONS DEMOLITION**									
	0020	Concrete, reinforced	B-39	120	.400	C.F.		9.10	1.50	10.60	15.85
	0025	Plain	"	160	.300			6.80	1.13	7.93	11.90
	0100	Brick, 4" to 12" thick	B-9C	220	.182			3.97	.82	4.79	7.15
	0200	Concrete block, 4" thick		1,000	.040	S.F.		.87	.18	1.05	1.58
	0280	8" thick		810	.049			1.08	.22	1.30	1.94
	1000	Drywall, nailed	1 Clab	1,000	.008			.17		.17	.27
	1020	Glued and nailed	"	900	.009			.19		.19	.30
	2200	Metal or wood studs, finish 2 sides, fiberboard	B-1	520	.046			1.02		1.02	1.61
	2250	Lath and plaster		260	.092			2.04		2.04	3.21
	2300	Plasterboard (drywall)		520	.046			1.02		1.02	1.61
	2350	Plywood		450	.053			1.18		1.18	1.86
	3760	Tile, ceramic, on walls	1 Clab	300	.027			.57		.57	.90
	3800	Toilet partitions, slate or marble		5	1.600	Ea.		34.50		34.50	54
	3820	Hollow metal		8	1	"		21.50		21.50	34
	4000	Insulation, fiberglass batts or blankets, 15" wide		3,200	.002	S.F.		.05		.05	.08
	4010	Rigid insulation, 1" thick		2,000	.004			.09		.09	.13
	4060	Vapor barrier, polyethlene		7,400	.001			.02		.02	.04
	5000	Wallcovering, vinyl	1 Pape	700	.011			.29		.29	.44
	5040	Designer	"	480	.017			.42		.42	.64
734	0010	**WINDOW DEMOLITION**									
	0200	Aluminum, including trim, to 12 S.F.	1 Clab	16	.500	Ea.		10.75		10.75	16.90
	0240	To 25 S.F.		11	.727			15.60		15.60	24.50
	0280	To 50 S.F.		5	1.600			34.50		34.50	54
	0600	Glass, minimum		200	.040	S.F.		.86		.86	1.35
	0620	Maximum		150	.053	"		1.14		1.14	1.80
	1000	Steel, including trim, to 12 S.F.		13	.615	Ea.		13.20		13.20	21
	1020	To 25 S.F.		9	.889			19.05		19.05	30
	1040	To 50 S.F.		4	2			43		43	67.50
	2000	Wood, including trim, to 12 S.F.		22	.364			7.80		7.80	12.25
	2020	To 25 S.F.		18	.444			9.55		9.55	15
	2060	To 50 S.F.		13	.615			13.20		13.20	21

		020 700 | Selective Demolition	CREW	DAILY OUTPUT	LABOR-HOURS	UNIT	1999 BARE COSTS MAT.	LABOR	EQUIP.	TOTAL	TOTAL INCL O&P
734	5083	Remove unit skylight	1 Carp	1.54	5.202	Ea.		142		142	223
		020 750 | Concrete Removal									
754	0010	**FOOTINGS AND FOUNDATIONS DEMOLITION**									
	0200	Floors, concrete slab on grade,									
	0240	4" thick, plain concrete	B-9C	500	.080	S.F.		1.75	.36	2.11	3.15
	0280	Reinforced, wire mesh		470	.085			1.86	.38	2.24	3.35
	0300	Rods		400	.100			2.18	.45	2.63	3.94
	0400	6" thick, plain concrete		375	.107			2.33	.48	2.81	4.20
	0420	Reinforced, wire mesh		340	.118			2.57	.53	3.10	4.62
	0440	Rods		300	.133			2.91	.60	3.51	5.25
	1000	Footings, concrete, 1' thick, 2' wide	B-5	300	.187	L.F.		4.43	3.49	7.92	10.75
	1080	1'-6" thick, 2' wide		250	.224			5.30	4.18	9.48	12.85
	1120	3' wide		200	.280			6.65	5.25	11.90	16.10
	1140	2' thick, 3' wide		175	.320			7.60	5.95	13.55	18.35
	1200	Average reinforcing, add								10%	10%
	1220	Heavy reinforcing, add								20%	20%
	2000	Walls, block, 4" thick	1 Clab	180	.044	S.F.		.95		.95	1.50
	2040	6" thick		170	.047			1.01		1.01	1.59
	2080	8" thick		150	.053			1.14		1.14	1.80
	2100	12" thick		150	.053			1.14		1.14	1.80
	2200	For horizontal reinforcing, add								10%	10%
	2220	For vertical reinforcing, add								20%	20%
	2400	Concrete, plain concrete, 6" thick	B-9	160	.250			5.45	1.13	6.58	9.85
	2420	8" thick		140	.286			6.25	1.29	7.54	11.20
	2440	10" thick		120	.333			7.30	1.50	8.80	13.10
	2500	12" thick		100	.400			8.75	1.80	10.55	15.75
	2600	For average reinforcing, add								10%	10%
	2620	For heavy reinforcing, add								20%	20%
	4000	For congested sites or small quantities, add up to								200%	200%
	4200	Add for disposal, on site	B-11A	232	.069	C.Y.		1.72	3.62	5.34	6.65
	4250	To five miles	B-30	220	.109	"		2.64	7.25	9.89	12.05
758	0010	**MASONRY DEMOLITION**									
	1000	Chimney, 16" x 16", soft old mortar	A-1	24	.333	V.L.F.		7.15	2.86	10.01	14.40
	1020	Hard mortar		18	.444			9.55	3.81	13.36	19.20
	1080	20" x 20", soft old mortar		12	.667			14.30	5.70	20	29
	1100	Hard mortar		10	.800			17.15	6.85	24	34.50
	1140	20" x 32", soft old mortar		10	.800			17.15	6.85	24	34.50
	1160	Hard mortar		8	1			21.50	8.60	30.10	43.50
	1200	48" x 48", soft old mortar		5	1.600			34.50	13.75	48.25	69
	1220	Hard mortar		4	2			43	17.15	60.15	86.50
	4000	Fireplace, brick, 30" x 24" opening									
	4020	Soft old mortar	A-1	2	4	Ea.		86	34.50	120.50	173
	4040	Hard mortar		1.25	6.400			137	55	192	277
	4100	Stone, soft old mortar		1.50	5.333			114	46	160	231
	4120	Hard mortar		1	8			172	68.50	240.50	345
880	0010	**REMOVAL OF UNDERGROUND STORAGE TANKS** R020-880									
	0011	Petroleum storage tanks, non-leaking									
	0100	Excavate & load onto trailer									
	0110	3000 gal. to 5000 gal. tank	B-14	4	12	Ea.		273	54.50	327.50	485
	0120	6000 gal to 8000 gal tank	B-3A	3	13.333			305	238	543	735
	0130	9000 gal to 12000 gal tank	"	2	20			455	355	810	1,100
	0190	Known leaking tank add				%				100%	100%
	0200	Remove sludge, water and remaining product from bottom									

Important: See the Reference Section for critical supporting data - Reference Nos., Crews, & City Cost Indexes

020 | Subsurface Investigation & Demolition

020 750 | Concrete Removal

880		CREW	DAILY OUTPUT	LABOR-HOURS	UNIT	1999 BARE COSTS MAT.	LABOR	EQUIP.	TOTAL	TOTAL INCL O&P
0201	of tank with vacuum truck (R020-880)									
0300	3000 gal to 5000 gal tank	A-13	5	1.600	Ea.		43.50	105	148.50	181
0310	6000 gal to 8000 gal tank	↓	4	2	↓		54.50	131	185.50	227
0320	9000 gal to 12000 gal tank	↓	3	2.667	↓		72.50	175	247.50	300
0390	Dispose of sludge off-site, average				Gal.					4
0400	Insert solid carbon dioxide "dry ice" to produce inert gas									
0401	For cleaning/transporting tanks (1.5 lbs./100 gal. cap)	1 Clab	500	.016	Lb.	1.20	.34		1.54	1.86
1020	Haul tank to certified salvage dump, 100 miles round trip									
1023	3000 gal. to 5000 gal. tank				Ea.				550	630
1026	6000 gal. to 8000 gal. tank				↓				650	750
1029	9,000 gal. to 12,000 gal. tank				↓				875	1,000
1100	Disposal of contaminated soil to landfill									
1110	Minimum				C.Y.					100
1111	Maximum				"					300
1120	Disposal of contaminated soil to									
1121	bituminous concrete batch plant									
1130	Minimum				C.Y.					50
1131	Maximum				"					100
2010	Decontamination of soil on site incl poly tarp on top/bottom									
2011	Soil containment berm, and chemical treatment									
2020	Minimum	B-11C	100	.160	C.Y.	5.60	3.99	2.19	11.78	14.70
2021	Maximum	"	100	.160	↓	7.25	3.99	2.19	13.43	16.55
2050	Disposal of decontaminated soil, minimum				↓					60
2055	Maximum				↓					125

021 | Site Preparation & Excavation Support

021 100 | Site Clearing

104		CREW	DAILY OUTPUT	LABOR-HOURS	UNIT	1999 BARE COSTS MAT.	LABOR	EQUIP.	TOTAL	TOTAL INCL O&P
0010	**CLEAR AND GRUB** Cut & chip light, trees to 6" diam.	B-7	1	48	Acre		1,100	1,175	2,275	3,025
0150	Grub stumps and remove	B-30	2	12	↓		290	800	1,090	1,325
0160	Clear & grub brush including stumps	"	.58	41.379	↓		1,000	2,750	3,750	4,550
0200	Cut & chip medium, trees to 12" diam.	B-7	.70	68.571	↓		1,575	1,700	3,275	4,300
0250	Grub stumps and remove	B-30	1	24	↓		580	1,600	2,180	2,625
0260	Clear & grub dense brush including stumps	"	.47	51.064	↓		1,225	3,400	4,625	5,650
0300	Cut & chip heavy, trees to 24" diam.	B-7	.30	160	↓		3,675	3,950	7,625	10,100
0350	Grub stumps and remove	B-30	.50	48	↓		1,150	3,200	4,350	5,300
0400	If burning is allowed, reduce cut & chip				↓					40%
3000	Chipping stumps, to 18" deep, 12" diam.	B-86	20	.400	Ea.		11.35	8.10	19.45	26.50
3040	18" diameter	↓	16	.500	↓		14.20	10.15	24.35	32.50
3080	24" diameter	↓	14	.571	↓		16.25	11.60	27.85	37.50
3100	30" diameter	↓	12	.667	↓		18.95	13.55	32.50	44
3120	36" diameter	↓	10	.800	↓		22.50	16.25	38.75	52.50
3160	48" diameter	↓	8	1	↓		28.50	20.50	49	65.50
5000	Tree thinning, feller buncher, conifer									
5080	Up to 8" diameter	B-93	240	.033	Ea.		.95	1.62	2.57	3.23
5120	12" diameter	↓	160	.050	↓		1.42	2.44	3.86	4.84
5240	Hardwood, up to 4" diameter	↓	240	.033	↓		.95	1.62	2.57	3.23
5280	8" diameter	↓	180	.044	↓		1.26	2.17	3.43	4.30
5320	12" diameter	↓	120	.067	↓		1.89	3.25	5.14	6.45
7000	Tree removal, congested area, aerial lift truck									

		021 100 | Site Clearing	CREW	DAILY OUTPUT	LABOR-HOURS	UNIT	1999 BARE COSTS MAT.	LABOR	EQUIP.	TOTAL	TOTAL INCL O&P
104	7040	8" diameter	B-85	7	5.714	Ea.		131	115	246	330
	7080	12" diameter		6	6.667			153	135	288	385
	7120	18" diameter		5	8			184	162	346	465
	7160	24" diameter		4	10			230	202	432	575
	7240	36" diameter		3	13.333			305	269	574	770
	7280	48" diameter	↓	2	20	↓		460	405	865	1,150
108	0010	**CLEARING** Brush with brush saw	A-1	.25	32	Acre		685	275	960	1,375
	0100	By hand	"	.12	66.667			1,425	570	1,995	2,875
	0300	With dozer, ball and chain, light clearing	B-11A	2	8			199	420	619	770
	0400	Medium clearing		1.50	10.667			266	560	826	1,025
	0500	With dozer and brush rake, light		1	16			400	840	1,240	1,550
	0550	Medium brush to 4" diameter		.60	26.667			665	1,400	2,065	2,575
	0600	Heavy brush to 4" diameter	↓	.40	40	↓		995	2,100	3,095	3,850
	1000	Brush mowing, tractor w/rotary mower, no removal									
	1020	Light density	B-84	2	4	Acre		114	104	218	288
	1040	Medium density		1.50	5.333			151	139	290	385
	1080	Heavy density	↓	1	8	↓		227	208	435	575
116	0010	**FELLING TREES & PILING** With tractor, large tract, firm									
	0020	level terrain, no boulders, less than 12" diam. trees									
	0300	300 HP dozer, up to 400 trees/acre, 0 to 25% hardwoods	B-10M	.75	16	Acre		415	1,500	1,915	2,300
	0340	25% to 50% hardwoods		.60	20			520	1,875	2,395	2,875
	0370	75% to 100% hardwoods		.45	26.667			695	2,525	3,220	3,850
	0400	500 trees/acre, 0% to 25% hardwoods		.60	20			520	1,875	2,395	2,875
	0440	25% to 50% hardwoods		.48	25			650	2,350	3,000	3,600
	0470	75% to 100% hardwoods		.36	33.333			870	3,150	4,020	4,775
	0500	More than 600 trees/acre, 0 to 25% hardwoods		.52	23.077			600	2,175	2,775	3,325
	0540	25% to 50% hardwoods		.42	28.571			745	2,700	3,445	4,100
	0570	75% to 100% hardwoods	↓	.31	38.710	↓		1,000	3,650	4,650	5,575
	0900	Large tract clearing per tree									
	1500	300 HP dozer, to 12" diameter, softwood	B-10M	320	.038	Ea.		.98	3.53	4.51	5.40
	1550	Hardwood		100	.120			3.13	11.30	14.43	17.25
	1600	12" to 24" diameter, softwood		200	.060			1.56	5.65	7.21	8.60
	1650	Hardwood		80	.150			3.91	14.15	18.06	21.50
	1700	24" to 36" diameter, softwood		100	.120			3.13	11.30	14.43	17.25
	1750	Hardwood		50	.240			6.25	22.50	28.75	34.50
	1800	36" to 48" diameter, softwood		70	.171			4.47	16.15	20.62	24.50
	1850	Hardwood	↓	35	.343	↓		8.95	32.50	41.45	49.50
		021 140 | Stripping									
144	0010	**STRIPPING** Topsoil, and stockpiling, sandy loam									
	0020	200 H.P. dozer, ideal conditions	B-10B	2,300	.005	C.Y.		.14	.36	.50	.61
	0100	Adverse conditions	"	1,150	.010			.27	.73	1	1.22
	0200	300 HP dozer, ideal conditions	B-10M	3,000	.004			.10	.38	.48	.57
	0300	Adverse conditions	"	1,650	.007			.19	.69	.88	1.04
	0400	400 HP dozer, ideal conditions	B-10X	3,900	.003			.08	.35	.43	.51
	0500	Adverse conditions	"	2,000	.006			.16	.69	.85	1
	0600	Clay, dry and soft, 200 HP dozer, ideal conditions	B-10B	1,600	.008			.20	.52	.72	.88
	0601	Strip topsoil, clay, dry & soft, 200 HP dozer, ideal conditions		1,600	.008			.20	.52	.72	.88
	0700	Adverse conditions	↓	800	.015			.39	1.05	1.44	1.75
	1000	Medium hard, 300 HP dozer, ideal conditions	B-10M	2,000	.006			.16	.57	.73	.86
	1100	Adverse conditions	"	1,100	.011			.28	1.03	1.31	1.57
	1200	Very hard, 400 HP dozer, ideal conditions	B-10X	2,600	.005			.12	.53	.65	.76
	1300	Adverse conditions	"	1,340	.009	↓		.23	1.03	1.26	1.49

021 150 | Selective Clearing

		Description	Crew	Daily Output	Labor-Hours	Unit	1999 Bare Costs Mat.	Labor	Equip.	Total	Total Incl O&P	
154	0010	**SELECTIVE CLEARING**										154
	1000	Stump removal on site by hydraulic backhoe, 1-1/2 C.Y.										
	1040	4″ to 6″ diameter	B-17	60	.533	Ea.		12.30	9.70	22	29.50	
	1050	8″ to 12″ diameter	B-30	33	.727			17.60	48.50	66.10	80	
	1100	14″ to 24″ diameter		25	.960			23	64	87	106	
	1150	26″ to 36″ diameter	↓	16	1.500	↓		36.50	100	136.50	166	
	2000	Remove selective trees, on site using chain saws and chipper,										
	2050	not incl. stumps, up to 6″ diameter	B-7	18	2.667	Ea.		61	66	127	168	
	2100	8″ to 12″ diameter		12	4			92	99	191	252	
	2150	14″ to 24″ diameter		10	4.800			110	119	229	300	
	2200	26″ to 36″ diameter	↓	8	6			138	148	286	380	
	2300	Machine load, 2 mile haul to dump, 12″ diam. tree, add				↓				150	225	

021 200 | Structure Moving

		Description	Crew	Daily Output	Labor-Hours	Unit	1999 Bare Costs Mat.	Labor	Equip.	Total	Total Incl O&P	
204	0010	**MOVING BUILDINGS** One day move, up to 24′ wide										204
	0020	Reset on new foundation, patch & hook-up, average move				Total					8,700	
	0040	Wood or steel frame bldg., based on ground floor area	B-4	185	.259	S.F.		5.70	2.57	8.27	11.70	
	0060	Masonry bldg., based on ground floor area	″	137	.350			7.65	3.47	11.12	15.80	
	0200	For 24′ to 42′ wide, add				↓					15%	
	0220	For each additional day on road, add	B-4	1	48	Day		1,050	475	1,525	2,175	
	0240	Construct new basement, move building, 1 day										
	0300	move, patch & hook-up, based on ground floor area	B-3	155	.310	S.F.	5.75	7.15	11.30	24.20	30	

021 400 | Dewatering

		Description	Crew	Daily Output	Labor-Hours	Unit	1999 Bare Costs Mat.	Labor	Equip.	Total	Total Incl O&P	
404	0010	**DEWATERING** Excavate drainage trench, 2′ wide, 2′ deep	B-11C	90	.178	C.Y.		4.43	2.43	6.86	9.50	404
	0100	2′ wide, 3′ deep, with backhoe loader	″	135	.119			2.95	1.62	4.57	6.35	
	0200	Excavate sump pits by hand, light soil	1 Clab	7.10	1.127			24		24	38	
	0300	Heavy soil	″	3.50	2.286	↓		49		49	77	
	0500	Pumping 8 hr., attended 2 hrs. per day, including 20 L.F.										
	0550	of suction hose & 100 L.F. discharge hose										
	0600	2″ diaphragm pump used for 8 hours	B-10H	4	3	Day		78	11.10	89.10	132	
	0620	Add per additional pump							35	35	40	
	0650	4″ diaphragm pump used for 8 hours	B-10I	4	3			78	23.50	101.50	146	
	0670	Add per additional pump							75	75	85	
	0800	8 hrs. attended, 2″ diaphragm pump	B-10H	1	12			315	44.50	359.50	530	
	0820	Add per additional pump							35	35	40	
	0900	3″ centrifugal pump	B-10J	1	12			315	58	373	545	
	0920	Add per additional pump							49.50	49.50	54.50	
	1000	4″ diaphragm pump	B-10I	1	12			315	94	409	585	
	1020	Add per additional pump							85	85	98	
	1100	6″ centrifugal pump	B-10K	1	12			315	220	535	720	
	1120	Add per additional pump				↓			110	110	125	
	1300	CMP, incl. excavation 3′ deep, 12″ diameter	B-6	115	.209	L.F.	8.95	4.88	1.90	15.73	19.45	
	1400	18″ diameter		100	.240	″	9.95	5.60	2.19	17.74	22	
	1600	Sump hole construction, incl. excavation and gravel, pit		1,250	.019	C.F.	.56	.45	.18	1.19	1.51	
	1700	With 12″ gravel collar, 12″ pipe, corrugated, 16 ga.		70	.343	L.F.	11.90	8	3.13	23.03	29	
	1800	15″ pipe, corrugated, 16 ga.		55	.436		14.65	10.20	3.98	28.83	36.50	
	1900	18″ pipe, corrugated, 16 ga.		50	.480		13.30	11.20	4.38	28.88	37	
	2000	24″ pipe, corrugated, 14 ga.		40	.600	↓	21.50	14	5.45	40.95	52	
	2200	Wood lining, up to 4′ x 4′, add	↓	300	.080	SFCA	3	1.87	.73	5.60	7	
	9950	See div. 021-444 for wellpoints										
	9960	See div. 021-484 for deep well systems										
	9970	See div. 152-400 for pumps										

				DAILY	LABOR-		1999 BARE COSTS				TOTAL
		021 440 | Wellpoints	CREW	OUTPUT	HOURS	UNIT	MAT.	LABOR	EQUIP.	TOTAL	INCL O&P
444	0010	**WELLPOINTS** For wellpoint equipment rental, see div. 016-490 (R021-440)									
	0100	Installation and removal of single stage system									
	0110	Labor only, .75 labor-hours per L.F., minimum	1 Clab	10.70	.748	LF Hdr		16.05		16.05	25
	0200	2.0 labor-hours per L.F., maximum	"	4	2	"		43		43	67.50
	0400	Pump operation, 4 @ 6 hr. shifts									
	0410	Per 24 hour day	4 Eqlt	1.27	25.197	Day		685		685	1,050
	0500	Per 168 hour week, 160 hr. straight, 8 hr. double time	↓	.18	177	Week		4,825		4,825	7,350
	0550	Per 4.3 week month	↓	.04	800	Month		21,800		21,800	33,100
	0600	Complete installation, operation, equipment rental, fuel &									
	0610	removal of system with 2″ wellpoints 5′ O.C.									
	0700	100′ long header, 6″ diameter, first month	4 Eqlt	3.23	9.907	LF Hdr	100	269		369	520
	0800	Thereafter, per month	↓	4.13	7.748	↓	80	211		291	410
	1000	200′ long header, 8″ diameter, first month		6	5.333		100	145		245	330
	1100	Thereafter, per month		8.39	3.814		45	104		149	208
	1300	500′ long header, 8″ diameter, first month		10.63	3.010		35	82		117	163
	1400	Thereafter, per month		20.91	1.530		25	41.50		66.50	91
	1600	1,000′ long header, 10″ diameter, first month		11.62	2.754		30	75		105	147
	1700	Thereafter, per month	↓	41.81	.765	↓	15	21		36	48
	1900	Note: above figures include pumping 168 hrs. per week									
	1910	and include the pump operator and one stand-by pump.									
		021 480 | Relief Wells									
484	0010	**WELLS** For dewatering 10′ to 20′ deep, 2′ diameter									
	0020	with steel casing, minimum	B-6	165	.145	V.L.F.	2	3.40	1.33	6.73	8.95
	0050	Average	↓	98	.245	↓	4	5.70	2.23	11.93	15.75
	0100	Maximum	↓	49	.490	↓	10	11.45	4.47	25.92	33.50
	0300	For pumps for dewatering, see division 016-420-4100 to 4400									
	0500	For domestic water wells, see division 026-704									
		021 520 | Shores									
524	0010	**SHORING** Existing building, with timber, no salvage allowance	B-51	2.20	21.818	M.B.F.	645	475	78	1,198	1,550
	1000	With 35 ton screw jacks, per box and jack	"	3.60	13.333	Jack	40	291	47.50	378.50	550
		021 560 | Underpinning									
564	0010	**UNDERPINNING FOUNDATIONS** Including excavation,									
	0020	forming, reinforcing, concrete and equipment									
	0100	5′ to 16′ below grade, 100 to 500 C.Y.	B-52	2.30	24.348	C.Y.	168	615	188	971	1,350
	0200	Over 500 C.Y.	↓	2.50	22.400	↓	152	565	173	890	1,250
	0400	16′ to 25′ below grade, 100 to 500 C.Y.		2	28		185	705	216	1,106	1,550
	0500	Over 500 C.Y.		2.10	26.667		175	670	206	1,051	1,475
	0700	26′ to 40′ below grade, 100 to 500 C.Y.		1.60	35		202	880	270	1,352	1,900
	0800	Over 500 C.Y.	↓	1.80	31.111		185	785	240	1,210	1,700
	0900	For under 50 C.Y., add					10%	40%			
	1000	For 50 C.Y. to 100 C.Y., add				↓	5%	20%			
		021 610 | Sheet Piling									
614	0010	**SHEET PILING** Steel, not incl. wales, 22 psf, 15′ excav., left in place	B-40	10.81	5.920	Ton	795	163	178	1,136	1,350
	0100	Drive, extract & salvage	↓	6	10.667	↓	211	294	320	825	1,075
	0300	20′ deep excavation, 27 psf, left in place (R021-610)		12.95	4.942		795	136	149	1,080	1,275
	0400	Drive, extract & salvage		6.55	9.771		211	270	294	775	1,000
	0600	25′ deep excavation, 38 psf, left in place (R021-620)		19	3.368		795	93	101	989	1,150
	0700	Drive, extract & salvage		10.50	6.095		211	168	183	562	710
	0900	40′ deep excavation, 38 psf, left in place		21.20	3.019		795	83.50	90.50	969	1,100
	1000	Drive, extract & salvage		12.25	5.224	↓	211	144	157	512	645
	1200	15′ deep excavation, 22 psf, left in place		983	.065	S.F.	9.25	1.80	1.96	13.01	15.30
	1300	Drive, extract & salvage	↓	656	.098	↓	2.37	2.69	2.93	7.99	10.30

021 610 | Sheet Piling

			CREW	DAILY OUTPUT	LABOR-HOURS	UNIT	1999 BARE COSTS MAT.	LABOR	EQUIP.	TOTAL	TOTAL INCL O&P
614	1500	20′ deep excavation, 27 psf, left in place (R021-610)	B-40	960	.067	S.F.	11.60	1.84	2	15.44	18.05
	1600	Drive, extract & salvage		640	.100		3.08	2.76	3.01	8.85	11.25
	1800	25′ deep excavation, 38 psf, left in place (R021-620)		1,000	.064		17.10	1.77	1.92	20.79	24
	1900	Drive, extract & salvage		670	.096		4.22	2.64	2.87	9.73	12.15
	2100	Rent steel sheet piling and wales, first month				Ton	230			230	253
	2200	Per added month					23			23	25.50
	2300	Rental piling left in place, add to rental					450			450	495
	2500	Wales, connections & struts, 2/3 salvage					173			173	190
	2700	High strength piling, 50,000 psi, add					60			60	66
	2800	55,000 psi, add					65			65	71.50
	3000	Tie rod, not upset, 1-1/2″ to 4″ diameter with turnbuckle					1,200			1,200	1,325
	3100	No turnbuckle					1,000			1,000	1,100
	3300	Upset, 1-3/4″ to 4″ diameter with turnbuckle					1,500			1,500	1,650
	3400	No turnbuckle					1,300			1,300	1,425
	3600	Lightweight, 18″ to 28″ wide, 7 ga., 9.22 psf, and									
	3610	9 ga., 8.6 psf, minimum				Lb.	.50			.50	.55
	3700	Average					.55			.55	.61
	3750	Maximum					.62			.62	.68
	3900	Wood, solid sheeting, incl. wales, braces and spacers,									
	3910	drive, extract & salvage, 8′ deep excavation	B-31	330	.121	S.F.	1.52	2.79	.46	4.77	6.55
	4000	10′ deep, 50 S.F./hr. in & 150 S.F./hr. out		300	.133		1.56	3.07	.51	5.14	7.10
	4100	12′ deep, 45 S.F./hr. in & 135 S.F./hr. out		270	.148		1.61	3.41	.56	5.58	7.75
	4200	14′ deep, 42 S.F./hr. in & 126 S.F./hr. out		250	.160		1.66	3.68	.61	5.95	8.30
	4300	16′ deep, 40 S.F./hr. in & 120 S.F./hr. out		240	.167		1.71	3.84	.64	6.19	8.65
	4400	18′ deep, 38 S.F./hr. in & 114 S.F./hr. out		230	.174		1.76	4	.66	6.42	8.95
	4500	20′ deep, 35 S.F./hr. in & 105 S.F./hr. out		210	.190		1.82	4.38	.73	6.93	9.70
	4520	Left in place, 8′ deep, 55 S.F./hr.		440	.091		2.73	2.09	.35	5.17	6.70
	4540	10′ deep, 50 S.F./hr.		400	.100		2.88	2.30	.38	5.56	7.20
	4560	12′ deep, 45 S.F./hr.		360	.111		3.04	2.56	.42	6.02	7.85
	4565	14′ deep, 42 S.F./hr.		335	.119		3.22	2.75	.45	6.42	8.35
	4570	16′ deep, 40 S.F./hr.		320	.125		3.42	2.88	.48	6.78	8.80
	4580	18′ deep, 38 S.F./hr.		305	.131		3.65	3.02	.50	7.17	9.30
	4590	20′ deep, 35 S.F./hr.		280	.143		3.91	3.29	.54	7.74	10.05
	4700	Alternate pricing, left in place, 8′ deep		1.76	22.727	M.B.F.	615	525	86.50	1,226.50	1,600
	4800	Drive, extract and salvage, 8′ deep		1.32	30.303	″	545	700	115	1,360	1,825
	5000	For treated lumber add cost of treatment to lumber									
	5010	See division 063-102									

021 620 | Cribbing & Walers

			CREW	DAILY OUTPUT	LABOR-HOURS	UNIT	MAT.	LABOR	EQUIP.	TOTAL	TOTAL INCL O&P
624	0010	**SOLDIER BEAMS & LAGGING** H piles with 3″ wood sheeting									
	0020	horizontal between piles, including removal of wales & braces									
	0100	No hydrostatic head, 15′ deep, 1 line of braces, minimum	B-50	545	.206	S.F.	6.45	5.40	2.74	14.59	19.05
	0200	Maximum		495	.226		7.15	5.95	3.01	16.11	21
	0400	15′ to 22′ deep with 2 lines of braces, 10″ H, minimum		360	.311		7.60	8.20	4.14	19.94	26.50
	0500	Maximum		330	.339		8.60	8.95	4.52	22.07	29
	0700	23′ to 35′ deep with 3 lines of braces, 12″ H, minimum		325	.345		9.90	9.10	4.59	23.59	31
	0800	Maximum		295	.380		10.75	10	5.05	25.80	34
	1000	36′ to 45′ deep with 4 lines of braces, 14″ H, minimum		290	.386		11.10	10.15	5.15	26.40	35
	1100	Maximum		265	.423		11.75	11.15	5.65	28.55	37.50
	1300	No hydrostatic head, left in place, 15′ dp., 1 line of braces, min.		635	.176		8.60	4.65	2.35	15.60	19.75
	1400	Maximum		575	.195		9.20	5.15	2.59	16.94	21.50
	1600	15′ to 22′ deep with 2 lines of braces, minimum		455	.246		12.90	6.50	3.28	22.68	28.50
	1700	Maximum		415	.270		14.35	7.10	3.59	25.04	31.50
	1900	23′ to 35′ deep with 3 lines of braces, minimum		420	.267		15.35	7	3.55	25.90	32.50
	2000	Maximum		380	.295		16.95	7.75	3.92	28.62	36

2 SITE WORK

021 | Site Preparation & Excavation Support

	021 620 | Cribbing & Walers	CREW	DAILY OUTPUT	LABOR-HOURS	UNIT	1999 BARE COSTS MAT.	LABOR	EQUIP.	TOTAL	TOTAL INCL O&P
624	2200 36′ to 45′ deep with 4 lines of braces, minimum	B-50	385	.291	S.F.	18.45	7.65	3.87	29.97	37.50
	2300 Maximum	↓	350	.320		21.50	8.45	4.26	34.21	42
	2350 Lagging only, 3″ thick wood between piles 8′ O.C., minimum	B-46	400	.120		1.43	2.96	.13	4.52	6.65
	2370 Maximum		250	.192		2.15	4.73	.21	7.09	10.45
	2400 Open sheeting no bracing, for trenches to 10′ deep, min.		1,736	.028		.64	.68	.03	1.35	1.87
	2450 Maximum	↓	1,510	.032		.72	.78	.04	1.54	2.13
	2500 Tie-back method, add to open sheeting, add, minimum								20%	20%
	2550 Maximum				↓				60%	60%
	2700 Tie-backs only, based on tie-backs total length, minimum	B-46	86.80	.553	L.F.	8.15	13.65	.61	22.41	32
	2750 Maximum		38.50	1.247	"	14.30	30.50	1.38	46.18	68.50
	3500 Tie-backs only, typical average, 25′ long		2	24	Ea.	360	590	26.50	976.50	1,400
	3600 35′ long	↓	1.58	30.380	"	475	750	33.50	1,258.50	1,800
	4500 Trench box, 7′ deep, 16′ x 6′, see division 016-420-7050				Day				105	122
	4600 20′ x 10′, see division 016-420-7070				"				150	170
	5200 Wood sheeting, in trench, jacks at 4′ O.C., 8′ deep	B-1	800	.030	S.F.	.50	.66		1.16	1.59
	5250 12′ deep		700	.034		.59	.76		1.35	1.84
	5300 15′ deep	↓	600	.040	↓	.81	.88		1.69	2.28

	021 680 | Slurry Wall	CREW	DAILY OUTPUT	LABOR-HOURS	UNIT	MAT.	LABOR	EQUIP.	TOTAL	TOTAL INCL O&P
684	0010 **SLURRY TRENCH** Excavated slurry trench in wet soils									
	0020 backfilled with 3000 PSI concrete, no reinforcing steel									
	0050 Minimum	C-7	333	.216	C.F.	4.52	5.05	2.68	12.25	15.70
	0100 Maximum		200	.360	"	7.55	8.35	4.46	20.36	26.50
	0200 Alternate pricing method, minimum		150	.480	S.F.	9.05	11.15	5.95	26.15	34
	0300 Maximum	↓	120	.600		13.55	13.95	7.45	34.95	44.50
	0500 Reinforced slurry trench, minimum	B-48	177	.316		6.80	7.60	11.15	25.55	31.50
	0600 Maximum	"	69	.812	↓	22.50	19.55	28.50	70.55	86.50
	0800 Haul for disposal, 2 mile haul, excavated material, add	B-34B	99	.081	C.Y.		1.79	4.47	6.26	7.65
	0900 Haul bentonite castings for disposal, add	"	40	.200	"		4.42	11.05	15.47	18.90

	021 700 | Cofferdams	CREW	DAILY OUTPUT	LABOR-HOURS	UNIT	MAT.	LABOR	EQUIP.	TOTAL	TOTAL INCL O&P
704	0010 **COFFERDAMS** Incl. mobilization, temporary sheeting, shore driven	B-40	960	.067	S.F.	11.35	1.84	2	15.19	17.75
	0060 Barge driven	"	550	.116	"	11.35	3.21	3.50	18.06	21.50
	6000 See also div. 021-614									

022 | Earthwork

	022 100 | Grading	CREW	DAILY OUTPUT	LABOR-HOURS	UNIT	1999 BARE COSTS MAT.	LABOR	EQUIP.	TOTAL	TOTAL INCL O&P
104	0010 **GRADING** Site excav. & fill, see div 022-200									
	0020 Fine grading, see div 025-122									

	022 200 | Excav./Backfill/Compact.	CREW	DAILY OUTPUT	LABOR-HOURS	UNIT	MAT.	LABOR	EQUIP.	TOTAL	TOTAL INCL O&P
204	0010 **BACKFILL** By hand, no compaction, light soil (R022-220)	1 Clab	14	.571	C.Y.		12.25		12.25	19.30
	0100 Heavy soil		11	.727			15.60		15.60	24.50
	0300 Compaction in 6″ layers, hand tamp, add to above	↓	20.60	.388			8.35		8.35	13.10
	0400 Roller compaction operator walking, add	B-10A	100	.120			3.13	.93	4.06	5.85
	0500 Air tamp, add	B-9C	190	.211			4.60	.95	5.55	8.30
	0600 Vibrating plate, add	A-1	60	.133	↓		2.86	1.14	4	5.75

022 200 | Excav./Backfill/Compact.

			CREW	DAILY OUTPUT	LABOR-HOURS	UNIT	1999 BARE COSTS MAT.	LABOR	EQUIP.	TOTAL	TOTAL INCL O&P
204	0800	Compaction in 12" layers, hand tamp, add to above (R022-220)	1 Clab	34	.235	C.Y.		5.05		5.05	7.95
	0900	Roller compaction operator walking, add	B-10A	150	.080			2.09	.62	2.71	3.88
	1000	Air tamp, add	B-9	285	.140			3.07	.63	3.70	5.55
	1100	Vibrating plate, add	A-1	90	.089			1.91	.76	2.67	3.84
	3000	For flowable fill, see div. 033-126									
208	0010	**BACKFILL, STRUCTURAL** Dozer or F.E. loader									
	0020	From existing stockpile, no compaction									
	2000	75 H.P., 50' haul, sand & gravel	B-10L	1,100	.011	C.Y.		.28	.28	.56	.75
	2020	Common earth		975	.012			.32	.32	.64	.84
	2040	Clay		850	.014			.37	.36	.73	.97
	2200	150' haul, sand & gravel		550	.022			.57	.56	1.13	1.49
	2220	Common earth		490	.024			.64	.63	1.27	1.67
	2240	Clay		425	.028			.74	.72	1.46	1.93
	2400	300' haul, sand & gravel		370	.032			.85	.83	1.68	2.21
	2420	Common earth		330	.036			.95	.93	1.88	2.49
	2440	Clay		290	.041			1.08	1.06	2.14	2.83
	3000	105 H.P., 50' haul, sand & gravel	B-10W	1,350	.009			.23	.31	.54	.70
	3020	Common earth		1,225	.010			.26	.34	.60	.76
	3040	Clay		1,100	.011			.28	.38	.66	.86
	3200	150' haul, sand & gravel		670	.018			.47	.62	1.09	1.41
	3220	Common earth		610	.020			.51	.68	1.19	1.54
	3240	Clay		550	.022			.57	.76	1.33	1.70
	3300	300' haul, sand & gravel		465	.026			.67	.90	1.57	2.02
	3320	Common earth		415	.029			.75	1.01	1.76	2.27
	3340	Clay		370	.032			.85	1.13	1.98	2.54
	4000	200 H.P., 50' haul, sand & gravel	B-10B	2,500	.005			.13	.34	.47	.56
	4020	Common earth		2,200	.005			.14	.38	.52	.64
	4040	Clay		1,950	.006			.16	.43	.59	.72
	4200	150' haul, sand & gravel		1,225	.010			.26	.68	.94	1.14
	4220	Common earth		1,100	.011			.28	.76	1.04	1.28
	4240	Clay		975	.012			.32	.86	1.18	1.44
	4400	300' haul, sand & gravel		805	.015			.39	1.04	1.43	1.75
	4420	Common earth		735	.016			.43	1.14	1.57	1.91
	4440	Clay		660	.018			.47	1.27	1.74	2.13
	5000	300 H.P., 50' haul, sand & gravel	B-10M	3,170	.004			.10	.36	.46	.54
	5020	Common earth		2,900	.004			.11	.39	.50	.60
	5040	Clay		2,700	.004			.12	.42	.54	.64
	5200	150' haul, sand & gravel		2,200	.005			.14	.51	.65	.79
	5220	Common earth		1,950	.006			.16	.58	.74	.89
	5240	Clay		1,700	.007			.18	.67	.85	1.01
	5400	300' haul, sand & gravel		1,500	.008			.21	.75	.96	1.15
	5420	Common earth		1,350	.009			.23	.84	1.07	1.28
	5440	Clay		1,225	.010			.26	.92	1.18	1.41
	6000	For compaction, see div. 022-226									
	6010	For trench backfill, see div. 022-254 & 258									
216	0010	**BORROW, LOADING AND/OR SPREADING**									
	4000	Common earth, shovel, 1 C.Y. bucket	B-12N	840	.019	C.Y.	5.05	.51	.76	6.32	7.15
	4010	1-1/2 C.Y. bucket	B-12O	1,135	.014		5.05	.38	.76	6.19	6.95
	4020	3 C.Y. bucket	B-12T	1,800	.009		5.05	.24	.68	5.97	6.65
	4030	Front end loader, wheel mounted									
	4050	3/4 C.Y. bucket	B-10R	550	.022	C.Y.	5.05	.57	.44	6.06	6.90
	4060	1-1/2 C.Y. bucket	B-10S	970	.012		5.05	.32	.33	5.70	6.40
	4070	3 C.Y. bucket	B-10T	1,575	.008		5.05	.20	.29	5.54	6.15
	4080	5 C.Y. bucket	B-10U	2,600	.005		5.05	.12	.36	5.53	6.10
	5000	Select granular fill, shovel, 1 C.Y. bucket	B-12N	925	.017		7.50	.46	.69	8.65	9.70

		022 200 | Excav./Backfill/Compact.	CREW	DAILY OUTPUT	LABOR-HOURS	UNIT	1999 BARE COSTS MAT.	LABOR	EQUIP.	TOTAL	TOTAL INCL O&P	
216	5010	1-1/2 C.Y. bucket	B-12O	1,250	.013	C.Y.	7.50	.34	.69	8.53	9.55	216
	5020	3 C.Y. bucket	B-12T	1,980	.008	↓	7.50	.22	.62	8.34	9.25	
	5030	Front end loader, wheel mounted										
	5050	3/4 C.Y. bucket	B-10R	800	.015	C.Y.	7.50	.39	.31	8.20	9.20	
	5060	1-1/2 C.Y. bucket	B-10S	1,065	.011		7.50	.29	.30	8.09	9.05	
	5070	3 C.Y. bucket	B-10T	1,735	.007		7.50	.18	.26	7.94	8.80	
	5080	5 C.Y. bucket	B-10U	2,850	.004		7.50	.11	.33	7.94	8.80	
	6000	Clay, till, or blasted rock, shovel, 1 C.Y. bucket	B-12N	715	.022		3.75	.60	.89	5.24	6	
	6010	1-1/2 C.Y. bucket	B-12O	965	.017		3.75	.44	.90	5.09	5.80	
	6020	3 C.Y. bucket	B-12T	1,530	.010	↓	3.75	.28	.80	4.83	5.45	
	6030	Front end loader, wheel mounted										
	6035	3/4 C.Y. bucket	B-10R	465	.026	C.Y.	3.75	.67	.53	4.95	5.75	
	6040	1-1/2 C.Y. bucket	B-10S	825	.015		3.75	.38	.39	4.52	5.15	
	6045	3 C.Y. bucket	B-10T	1,340	.009		3.75	.23	.34	4.32	4.86	
	6050	5 C.Y. bucket	B-10U	2,200	.005	↓	3.75	.14	.42	4.31	4.81	
	6060	Front end loader, track mounted										
	6065	1-1/2 C.Y. bucket	B-10N	715	.017	C.Y.	3.75	.44	.49	4.68	5.35	
	6070	3 C.Y. bucket	B-10P	1,190	.010		3.75	.26	.73	4.74	5.35	
	6075	5 C.Y. bucket	B-10Q	1,835	.007		3.75	.17	.62	4.54	5.05	
	7000	Topsoil or loam from stockpile, shovel, 1 C.Y. bucket	B-12N	840	.019		14.15	.51	.76	15.42	17.20	
	7010	1-1/2 C.Y. bucket	B-12O	1,135	.014		14.15	.38	.76	15.29	17	
	7020	3 C.Y. bucket	B-12T	1,800	.009	↓	14.15	.24	.68	15.07	16.70	
	7030	Front end loader, wheel mounted										
	7050	3/4 C.Y. bucket	B-10R	550	.022	C.Y.	14.15	.57	.44	15.16	16.95	
	7060	1-1/2 C.Y. bucket	B-10S	970	.012		14.15	.32	.33	14.80	16.45	
	7070	3 C.Y. bucket	B-10T	1,575	.008		14.15	.20	.29	14.64	16.20	
	7080	5 C.Y. bucket	B-10U	2,600	.005	↓	14.15	.12	.36	14.63	16.15	
	8900	For larger hauling units, deduct from above								30%		
	9200	For flowable fill, see section 033-126										
226	0010	**COMPACTION** R022-220										226
	5000	Riding, vibrating roller, 6" lifts, 2 passes	B-10Y	3,000	.004	C.Y.		.10	.12	.22	.29	
	5020	3 passes		2,300	.005			.14	.15	.29	.38	
	5040	4 passes		1,900	.006			.16	.19	.35	.45	
	5060	12" lifts, 2 passes		5,200	.002			.06	.07	.13	.16	
	5080	3 passes		3,500	.003			.09	.10	.19	.25	
	5100	4 passes	↓	2,600	.005			.12	.14	.26	.33	
	5600	Sheepsfoot or wobbly wheel roller, 6" lifts, 2 passes	B-10G	2,400	.005			.13	.23	.36	.46	
	5620	3 passes		1,735	.007			.18	.32	.50	.63	
	5640	4 passes		1,300	.009			.24	.43	.67	.84	
	5680	12" lifts, 2 passes		5,200	.002			.06	.11	.17	.21	
	5700	3 passes		3,500	.003			.09	.16	.25	.32	
	5720	4 passes	↓	2,600	.005			.12	.21	.33	.42	
	6000	Towed sheepsfoot or wobbly wheel roller, 6" lifts, 2 passes	B-10D	10,000	.001			.03	.10	.13	.16	
	6020	3 passes		2,000	.006			.16	.48	.64	.77	
	6030	4 passes		1,500	.008			.21	.64	.85	1.03	
	6050	12" lifts, 2 passes		6,000	.002			.05	.16	.21	.26	
	6060	3 passes		4,000	.003			.08	.24	.32	.39	
	6070	4 passes	↓	3,000	.004			.10	.32	.42	.51	
	6200	Vibrating roller, 6" lifts, 2 passes	B-10C	2,600	.005			.12	.37	.49	.58	
	6210	3 passes		1,735	.007			.18	.55	.73	.88	
	6220	4 passes		1,300	.009			.24	.73	.97	1.18	
	6250	12" lifts, 2 passes		5,200	.002			.06	.18	.24	.29	
	6260	3 passes		3,465	.003			.09	.27	.36	.44	
	6270	4 passes	↓	2,600	.005			.12	.37	.49	.58	
	7000	Walk behind, vibrating plate 18" wide, 6" lifts, 2 passes	B-18	200	.120	↓		2.65	.26	2.91	4.47	

022 200 | Excav./Backfill/Compact.

			CREW	DAILY OUTPUT	LABOR-HOURS	UNIT	1999 BARE COSTS MAT.	LABOR	EQUIP.	TOTAL	TOTAL INCL O&P
226	7020	3 passes R022-220	A-1	185	.043	C.Y.		.93	.37	1.30	1.87
	7040	4 passes		140	.057			1.23	.49	1.72	2.47
	7200	12" lifts, 2 passes		560	.014			.31	.12	.43	.61
	7220	3 passes		375	.021			.46	.18	.64	.92
	7240	4 passes		280	.029			.61	.25	.86	1.23
	7500	Vibrating roller 24" wide, 6" lifts, 2 passes	B-10A	420	.029			.75	.22	.97	1.38
	7520	3 passes		280	.043			1.12	.33	1.45	2.09
	7540	4 passes		210	.057			1.49	.44	1.93	2.78
	7600	12" lifts, 2 passes		840	.014			.37	.11	.48	.69
	7620	3 passes		560	.021			.56	.17	.73	1.04
	7640	4 passes		420	.029			.75	.22	.97	1.38
	8000	Rammer tamper, 6" to 11", 4" lifts, 2 passes	A-1	130	.062			1.32	.53	1.85	2.66
	8050	3 passes		97	.082			1.77	.71	2.48	3.56
	8100	4 passes		65	.123			2.64	1.06	3.70	5.30
	8200	8" lifts, 2 passes		260	.031			.66	.26	.92	1.33
	8250	3 passes		195	.041			.88	.35	1.23	1.77
	8300	4 passes		130	.062			1.32	.53	1.85	2.66
	8400	13" to 18", 4" lifts, 2 passes		390	.021			.44	.18	.62	.88
	8450	3 passes		290	.028			.59	.24	.83	1.19
	8500	4 passes		195	.041			.88	.35	1.23	1.77
	8600	8" lifts, 2 passes		780	.010			.22	.09	.31	.45
	8650	3 passes		585	.014			.29	.12	.41	.59
	8700	4 passes		390	.021			.44	.18	.62	.88
	9000	Water, 3000 gal. truck, 3 mile haul	B-45	1,888	.008		.20	.21	.39	.80	.98
	9010	6 mile haul		1,444	.011		.20	.28	.52	1	1.22
	9020	12 mile haul		1,000	.016		.20	.40	.74	1.34	1.66
	9030	6000 gal. wagon, 3 mile haul	B-59	2,000	.004		.20	.09	.26	.55	.65
	9040	6 mile haul	"	1,600	.005		.20	.11	.33	.64	.75
230	0010	**DRILLING ONLY** 2" hole for rock bolts, average	B-47	316	.076	L.F.		1.83	1.96	3.79	4.99
	0800	2-1/2" hole for pre-splitting, average		600	.040			.96	1.03	1.99	2.62
	1600	Quarry operations, 2-1/2" to 3-1/2" diameter		715	.034			.81	.86	1.67	2.20
	1610	6" diameter drill holes	B-47A	1,350	.018			.46	.36	.82	1.09
234	0010	**DRILLING AND BLASTING** Only, rock, open face, under 1500 C.Y.	B-47	225	.107	C.Y.	1.60	2.56	2.75	6.91	8.75
	0100	Over 1500 C.Y.	"	300	.080		1.60	1.92	2.06	5.58	7
	0200	Areas where blasting mats are required, under 1500 C.Y.	B-47	175	.137		1.60	3.30	3.53	8.43	10.75
	0250	Over 1500 C.Y.	"	250	.096		1.60	2.31	2.47	6.38	8.05
	0300	Bulk drilling and blasting, can vary greatly, average									5
	0500	Pits, average									20
	1300	Deep hole method, up to 1500 C.Y.	B-47	50	.480		1.60	11.55	12.35	25.50	33.50
	1400	Over 1500 C.Y.		66	.364		1.60	8.75	9.35	19.70	25.50
	1900	Restricted areas, up to 1500 C.Y.		13	1.846		1.60	44.50	47.50	93.60	123
	2000	Over 1500 C.Y.		20	1.200		1.60	29	31	61.60	81
	2200	Trenches, up to 1500 C.Y.		22	1.091		4.64	26	28	58.64	76.50
	2300	Over 1500 C.Y.		26	.923		4.64	22	24	50.64	65.50
	2500	Pier holes, up to 1500 C.Y.		22	1.091		1.60	26	28	55.60	73.50
	2600	Over 1500 C.Y.		31	.774		1.60	18.60	19.95	40.15	53
	2800	Boulders under 1/2 C.Y., loaded on truck, no hauling	B-100	80	.150			3.91	6.40	10.31	13
	2900	Boulders, drilled, blasted	B-47	100	.240		1.60	5.75	6.20	13.55	17.50
	3100	Jackhammer operators with foreman compressor, air tools	B-9	1	40	Day		875	180	1,055	1,575
	3300	Track drill, compressor, operator and foreman	B-47	1	24	"		575	620	1,195	1,575
	3500	Blasting caps				Ea.	3			3	3.30
	3900	Blasting mats, rent, for first day					90			90	99
	4000	Per added day					30			30	33
	4200	Preblast survey for 6 room house, individual lot, minimum	A-6	2.40	6.667			171		171	260
	4300	Maximum	"	1.35	11.852			305		305	460
	4500	City block within zone of influence, minimum	A-8	25,200	.001	S.F.		.03		.03	.05

	022 200 | Excav./Backfill/Compact.		CREW	DAILY OUTPUT	LABOR-HOURS	UNIT	1999 BARE COSTS MAT.	LABOR	EQUIP.	TOTAL	TOTAL INCL O&P
234	4600	Maximum	A-8	15,100	.002	S.F.		.05		.05	.08
	5000	Excavate and load boulders, less than 0.5 C.Y.	B-10T	80	.150	C.Y.		3.91	5.65	9.56	12.20
	5020	0.5 C.Y. to 1 C.Y.	B-10U	100	.120			3.13	9.25	12.38	15
	5200	Excavate and load blasted rock, 3 C.Y. power shovel	B-12T	1,530	.010			.28	.80	1.08	1.30
	5400	Haul boulders, 25 Ton off-highway dump, 1 mile round trip	B-34E	330	.024			.54	2.12	2.66	3.15
	5420	2 mile round trip		275	.029			.64	2.54	3.18	3.78
	5440	3 mile round trip		225	.036			.79	3.11	3.90	4.62
	5460	4 mile round trip		200	.040			.88	3.50	4.38	5.20
	5600	Bury boulders on site, less than 0.5 C.Y., 300 H.P. dozer									
	5620	150′ haul	B-10M	310	.039	C.Y.		1.01	3.65	4.66	5.55
	5640	300′ haul		210	.057			1.49	5.40	6.89	8.20
	5800	0.5 to 1 C.Y., 300 H.P. dozer, 150′ haul		300	.040			1.04	3.77	4.81	5.75
	5820	300′ haul		200	.060			1.56	5.65	7.21	8.60
238	0010	**EXCAVATING, BULK BANK MEASURE** Common earth piled (R022-240)									
	0020	For loading onto trucks, add								15%	15%
	0050	For mobilization and demobilization, see division 022-274 (R022-250)									
	0100	For hauling, see division 022-266									
	0200	Backhoe, hydraulic, crawler mtd., 1 C.Y. cap. = 75 C.Y./hr.	B-12A	600	.027	C.Y.		.71	.91	1.62	2.08
	0250	1-1/2 C.Y. cap. = 100 C.Y./hr.	B-12B	800	.020			.53	.89	1.42	1.79
	0260	2 C.Y. cap. = 130 C.Y./hr.	B-12C	1,040	.015			.41	.98	1.39	1.71
	0300	3 C.Y. cap. = 160 C.Y./hr.	B-12D	1,280	.013			.33	1.68	2.01	2.36
	0310	Wheel mounted, 1/2 C.Y. cap. = 30 C.Y./hr.	B-12E	240	.067			1.78	1.40	3.18	4.25
	0360	3/4 C.Y. cap. = 45 C.Y./hr.	B-12F	360	.044			1.19	1.25	2.44	3.18
	0500	Clamshell, 1/2 C.Y. cap. = 20 C.Y./hr.	B-12G	160	.100			2.67	3.26	5.93	7.65
	0550	1 C.Y. cap. = 35 C.Y./hr.	B-12H	280	.057			1.53	2.05	3.58	4.57
	0950	Dragline, 1/2 C.Y. cap. = 30 C.Y./hr.	B-12I	240	.067			1.78	2.19	3.97	5.10
	1000	3/4 C.Y. cap. = 35 C.Y./hr.	"	280	.057			1.53	1.88	3.41	4.38
	1050	1-1/2 C.Y. cap. = 65 C.Y./hr.	B-12P	520	.031			.82	1.45	2.27	2.84
	1100	3 C.Y. cap. = 112 C.Y./hr.	B-12V	900	.018			.47	1.12	1.59	1.96
	1200	Front end loader, track mtd., 1-1/2 C.Y. cap. = 70 C.Y./hr.	B-10N	560	.021			.56	.63	1.19	1.55
	1250	2-1/2 C.Y. cap. = 95 C.Y./hr.	B-10O	760	.016			.41	.67	1.08	1.37
	1300	3 C.Y. cap. = 130 C.Y./hr.	B-10P	1,040	.012			.30	.83	1.13	1.38
	1350	5 C.Y. cap. = 160 C.Y./hr.	B-10Q	1,280	.009			.24	.89	1.13	1.36
	1500	Wheel mounted, 3/4 C.Y. cap. = 45 C.Y./hr.	B-10R	360	.033			.87	.68	1.55	2.09
	1550	1-1/2 C.Y. cap. = 80 C.Y./hr.	B-10S	640	.019			.49	.50	.99	1.30
	1600	2-1/4 C.Y. cap. = 100 C.Y./hr.	B-10T	800	.015			.39	.56	.95	1.22
	1601	3 C.Y. cap. = 140 C.Y./hr.	"	1,120	.011			.28	.40	.68	.87
	1650	5 C.Y. cap. = 185 C.Y./hr.	B-10U	1,480	.008			.21	.63	.84	1.01
	1800	Hydraulic excavator, truck mtd, 1/2 C.Y. = 30 C.Y./hr.	B-12J	240	.067			1.78	2.65	4.43	5.65
	1850	48 inch bucket, 1 C.Y. = 45 C.Y./hr.	B-12K	360	.044			1.19	2.24	3.43	4.27
	3700	Shovel, 1/2 C.Y. capacity = 55 C.Y./hr.	B-12L	440	.036			.97	1.20	2.17	2.80
	3750	3/4 C.Y. capacity = 85 C.Y./hr.	B-12M	680	.024			.63	.87	1.50	1.92
	3800	1 C.Y. capacity = 120 C.Y./hr.	B-12N	960	.017			.45	.66	1.11	1.41
	3850	1-1/2 C.Y. capacity = 160 C.Y./hr.	B-12O	1,280	.013			.33	.68	1.01	1.25
	3900	3 C.Y. cap. = 250 C.Y./hr.	B-12T	2,000	.008			.21	.61	.82	1.01
	4000	For soft soil or sand, deduct								15%	15%
	4100	For heavy soil or stiff clay, add								60%	60%
	4200	For wet excavation with clamshell or dragline, add								100%	100%
	4250	All other equipment, add								50%	50%
	4400	Clamshell in sheeting or cofferdam, minimum	B-12H	160	.100			2.67	3.59	6.26	8
	4450	Maximum	"	60	.267			7.10	9.55	16.65	21.50
	8000	For hauling excavated material, see div. 022-266									
242	0010	**EXCAVATING, BULK, DOZER** Open site									
	2000	75 H.P., 50′ haul, sand & gravel	B-10L	460	.026	C.Y.		.68	.67	1.35	1.78

022 200 | Excav./Backfill/Compact.

			CREW	DAILY OUTPUT	LABOR-HOURS	UNIT	1999 BARE COSTS				TOTAL INCL O&P
							MAT.	LABOR	EQUIP.	TOTAL	
242	2020	Common earth	B-10L	400	.030	C.Y.		.78	.77	1.55	2.05
	2040	Clay		250	.048			1.25	1.23	2.48	3.27
	2200	150' haul, sand & gravel		230	.052			1.36	1.34	2.70	3.56
	2220	Common earth		200	.060			1.56	1.54	3.10	4.09
	2240	Clay		125	.096			2.50	2.46	4.96	6.55
	2400	300' haul, sand & gravel		120	.100			2.61	2.57	5.18	6.85
	2420	Common earth		100	.120			3.13	3.08	6.21	8.20
	2440	Clay	↓	65	.185			4.81	4.74	9.55	12.60
	3000	105 H.P., 50' haul, sand & gravel	B-10W	700	.017			.45	.60	1.05	1.35
	3020	Common earth		610	.020			.51	.68	1.19	1.54
	3040	Clay		385	.031			.81	1.08	1.89	2.44
	3200	150' haul, sand & gravel		310	.039			1.01	1.35	2.36	3.03
	3220	Common earth		270	.044			1.16	1.55	2.71	3.48
	3240	Clay		170	.071			1.84	2.46	4.30	5.55
	3300	300' haul, sand & gravel		140	.086			2.24	2.98	5.22	6.70
	3320	Common earth		120	.100			2.61	3.48	6.09	7.85
	3340	Clay	↓	100	.120			3.13	4.17	7.30	9.40
	4000	200 H.P., 50' haul, sand & gravel	B-10B	1,400	.009			.22	.60	.82	1
	4020	Common earth		1,230	.010			.25	.68	.93	1.14
	4040	Clay		770	.016			.41	1.09	1.50	1.82
	4200	150' haul, sand & gravel		595	.020			.53	1.41	1.94	2.36
	4220	Common earth		516	.023			.61	1.63	2.24	2.72
	4240	Clay		325	.037			.96	2.58	3.54	4.32
	4400	300' haul, sand & gravel		310	.039			1.01	2.71	3.72	4.53
	4420	Common earth		270	.044			1.16	3.11	4.27	5.20
	4440	Clay	↓	170	.071			1.84	4.93	6.77	8.30
	5000	300 H.P., 50' haul, sand & gravel	B-10M	1,900	.006			.16	.60	.76	.90
	5020	Common earth		1,650	.007			.19	.69	.88	1.04
	5040	Clay		1,025	.012			.31	1.10	1.41	1.68
	5200	150' haul, sand & gravel		920	.013			.34	1.23	1.57	1.87
	5220	Common earth		800	.015			.39	1.41	1.80	2.16
	5240	Clay		500	.024			.63	2.26	2.89	3.45
	5400	300' haul, sand & gravel		470	.026			.67	2.41	3.08	3.67
	5420	Common earth		410	.029			.76	2.76	3.52	4.20
	5440	Clay	↓	250	.048			1.25	4.52	5.77	6.90
	5500	460 H.P., 50' haul, sand & gravel	B-10X	1,930	.006			.16	.72	.88	1.04
	5510	Common earth		1,680	.007			.19	.82	1.01	1.19
	5520	Clay		1,050	.011			.30	1.31	1.61	1.91
	5530	150' haul, sand & gravel		1,290	.009			.24	1.07	1.31	1.55
	5540	Common earth		1,120	.011			.28	1.23	1.51	1.79
	5550	Clay		700	.017			.45	1.97	2.42	2.86
	5560	300' haul, sand & gravel		660	.018			.47	2.09	2.56	3.03
	5570	Common earth		575	.021			.54	2.40	2.94	3.48
	5580	Clay	↓	350	.034			.89	3.94	4.83	5.70
	6000	700 H.P., 50' haul, sand & gravel	B-10V	3,500	.003			.09	.79	.88	1.01
	6010	Common earth		3,035	.004			.10	.91	1.01	1.16
	6020	Clay		1,925	.006			.16	1.44	1.60	1.83
	6030	150' haul, sand & gravel		2,025	.006			.15	1.37	1.52	1.74
	6040	Common earth		1,750	.007			.18	1.58	1.76	2.01
	6050	Clay		1,100	.011			.28	2.51	2.79	3.21
	6060	300' haul, sand & gravel		1,030	.012			.30	2.68	2.98	3.42
	6070	Common earth		900	.013			.35	3.07	3.42	3.91
	6080	Clay	↓	550	.022	↓		.57	5.05	5.62	6.40
	6090	For dozer with ripper, see div. 022-278									
244	0010	**EXCAVATION, BULK, DRAG LINE**									
	0011	Excavate and load on truck, bank measure									

022 200 | Excav./Backfill/Compact.

			CREW	DAILY OUTPUT	LABOR-HOURS	UNIT	1999 BARE COSTS MAT.	LABOR	EQUIP.	TOTAL	TOTAL INCL O&P
244	0012	Bucket drag line, 3/4 C.Y., sand/gravel	B-12I	440	.036	C.Y.		.97	1.19	2.16	2.79
	0100	Light clay		310	.052			1.38	1.69	3.07	3.96
	0110	Heavy clay		250	.064			1.71	2.10	3.81	4.91
	0120	Unclassified soil	↓	280	.057			1.53	1.88	3.41	4.38
	0200	1-1/2 C.Y. bucket, sand/gravel	B-12P	575	.028			.74	1.31	2.05	2.57
	0210	Light clay		440	.036			.97	1.71	2.68	3.36
	0220	Heavy clay		352	.045			1.21	2.14	3.35	4.20
	0230	Unclassified soil	↓	300	.053			1.42	2.51	3.93	4.93
	0300	3 C.Y., sand/gravel	B-12V	720	.022			.59	1.41	2	2.45
	0310	Light clay		700	.023			.61	1.45	2.06	2.52
	0320	Heavy clay		600	.027			.71	1.69	2.40	2.94
	0330	Unclassified soil	↓	550	.029	↓		.78	1.84	2.62	3.20
246	0010	**EXCAVATION, BULK, SCRAPERS** R022-240									
	0100	Elevating scraper 11 C.Y., sand & gravel 1500′ haul	B-33F	690	.020	C.Y.		.54	1.41	1.95	2.37
	0150	3000′ haul R022-250		610	.023			.61	1.59	2.20	2.68
	0200	5000′ haul		505	.028			.73	1.92	2.65	3.24
	0300	Common earth, 1500′ haul		600	.023			.62	1.62	2.24	2.73
	0350	3000′ haul		530	.026			.70	1.83	2.53	3.09
	0400	5000′ haul		440	.032			.84	2.21	3.05	3.72
	0500	Clay, 1500′ haul		375	.037			.99	2.59	3.58	4.36
	0550	3000′ haul		330	.042			1.12	2.94	4.06	4.96
	0600	5000′ haul	↓	275	.051	↓		1.34	3.53	4.87	5.95
	1000	Self propelled scraper, 14 C.Y. 1/4 push dozer, sand									
	1050	and gravel, 1500′ haul	B-33D	920	.015	C.Y.		.40	2.07	2.47	2.89
	1100	3000′ haul		805	.017			.46	2.36	2.82	3.30
	1200	5000′ haul		645	.022			.57	2.95	3.52	4.12
	1300	Common earth, 1500′ haul		800	.017			.46	2.37	2.83	3.32
	1350	3000′ haul		700	.020			.53	2.71	3.24	3.80
	1400	5000′ haul		560	.025			.66	3.39	4.05	4.74
	1500	Clay, 1500′ haul		500	.028			.74	3.80	4.54	5.30
	1550	3000′ haul		440	.032			.84	4.32	5.16	6.05
	1600	5000′ haul	↓	350	.040			1.06	5.45	6.51	7.55
	2000	21 C.Y., 1/4 push dozer, sand & gravel, 1500′ haul	B-33E	1,180	.012			.31	1.86	2.17	2.53
	2100	3000′ haul		910	.015			.41	2.41	2.82	3.27
	2200	5000′ haul		750	.019			.49	2.93	3.42	3.98
	2300	Common earth, 1500′ haul		1,030	.014			.36	2.13	2.49	2.89
	2350	3000′ haul		790	.018			.47	2.78	3.25	3.78
	2400	5000′ haul		650	.022			.57	3.38	3.95	4.58
	2500	Clay, 1500′ haul		645	.022			.57	3.40	3.97	4.62
	2550	3000′ haul		495	.028			.75	4.43	5.18	6.05
	2600	5000′ haul	↓	405	.035			.91	5.40	6.31	7.35
	2700	Towed, 10 C.Y., 1/4 push dozer, sand & gravel, 1500′ haul	B-33B	560	.025			.66	2.89	3.55	4.18
	2720	3000′ haul		450	.031			.82	3.59	4.41	5.20
	2730	5000′ haul		365	.038			1.01	4.43	5.44	6.40
	2750	Common earth, 1500′ haul		420	.033			.88	3.85	4.73	5.60
	2770	3000′ haul		400	.035			.92	4.04	4.96	5.85
	2780	5000′ haul		310	.045			1.19	5.20	6.39	7.60
	2800	Clay, 1500′ haul		315	.044			1.17	5.15	6.32	7.45
	2820	3000′ haul		300	.047			1.23	5.40	6.63	7.85
	2840	5000′ haul	↓	225	.062			1.64	7.20	8.84	10.40
	2900	15 C.Y., 1/4 push dozer, sand & gravel, 1500′ haul	B-33C	800	.017			.46	2.02	2.48	2.93
	2920	3000′ haul		640	.022			.58	2.53	3.11	3.67
	2940	5000′ haul		520	.027			.71	3.11	3.82	4.51
	2960	Common earth, 1500′ haul		600	.023			.62	2.69	3.31	3.91
	2980	3000′ haul		560	.025			.66	2.89	3.55	4.18
	3000	5000′ haul	↓	440	.032	↓		.84	3.67	4.51	5.35

022 200 | Excav./Backfill/Compact.

		022 200 Excav./Backfill/Compact.	CREW	DAILY OUTPUT	LABOR-HOURS	UNIT	1999 BARE COSTS MAT.	LABOR	EQUIP.	TOTAL	TOTAL INCL O&P
246	3020	Clay, 1500' haul (R022-240)	B-33C	450	.031	C.Y.		.82	3.59	4.41	5.20
	3040	3000' haul		420	.033			.88	3.85	4.73	5.60
	3060	5000' haul (R022-250)		320	.044			1.16	5.05	6.21	7.30
250	0010	**EXCAVATING, STRUCTURAL** Hand, pits to 6' deep, sandy soil (R022-240)	1 Clab	8	1	C.Y.		21.50		21.50	34
	0020	Normal soil	B-2	16	2.500			54.50		54.50	86
	0030	Medium clay		12	3.333			73		73	115
	0040	Heavy clay		8	5			109		109	172
	0050	Loose rock		6	6.667			146		146	229
	0100	Heavy soil or clay	1 Clab	4	2			43		43	67.50
	0200	Pits to 2' deep, normal soil	B-2	24	1.667			36.50		36.50	57.50
	0210	Sand and gravel		24	1.667			36.50		36.50	57.50
	0220	Medium clay		18	2.222			48.50		48.50	76.50
	0230	Heavy clay		12	3.333			73		73	115
	0300	Pits 6' to 12' deep, sandy soil	1 Clab	5	1.600			34.50		34.50	54
	0500	Heavy soil or clay		3	2.667			57		57	90
	0700	Pits 12' to 18' deep, sandy soil		4	2			43		43	67.50
	0900	Heavy soil or clay		2	4			86		86	135
	1000	Hand trimming, bottom of excavation	B-2	2,400	.017	S.F.		.36		.36	.57
	1010	Slopes and sides		2,400	.017	"		.36		.36	.57
	1030	Around obstructions		8	5	C.Y.		109		109	172
	1100	Hand loading trucks from stock pile, sandy soil	1 Clab	12	.667			14.30		14.30	22.50
	1300	Heavy soil or clay	"	8	1			21.50		21.50	34
	1500	For wet or muck hand excavation, add to above				%				50%	50%
	1550	Excavation rock by hand/air tool	B-9	3.40	11.765	C.Y.		257	53	310	465
	2000	Machine excavation, for spread and mat footings, elevator pits,									
	2001	and small building foundations									
	2035	Common earth, hydraulic backhoe, 3/4 C.Y. bucket	B-12F	90	.178	C.Y.		4.75	4.99	9.74	12.70
	2040	1 C.Y. bucket	B-12A	108	.148			3.96	5.05	9.01	11.60
	2050	1-1/2 C.Y. bucket	B-12B	144	.111			2.97	4.95	7.92	9.95
	2060	2 C.Y. bucket	B-12C	200	.080			2.14	5.10	7.24	8.90
	2070	Sand and gravel, 3/4 C.Y. bucket	B-12F	100	.160			4.27	4.49	8.76	11.45
	2080	1 C.Y. bucket	B-12A	120	.133			3.56	4.56	8.12	10.40
	2090	1-1/2 C.Y. bucket	B-12B	160	.100			2.67	4.45	7.12	8.95
	3000	2 C.Y. bucket	B-12C	220	.073			1.94	4.65	6.59	8.05
	3010	Clay, till, or blasted rock, 3/4 C.Y. bucket	B-12F	80	.200			5.35	5.60	10.95	14.35
	3020	1 C.Y. bucket	B-12A	95	.168			4.50	5.75	10.25	13.20
	3030	1-1/2 C.Y. bucket	B-12B	130	.123			3.29	5.50	8.79	11.05
	3040	2 C.Y. bucket	B-12C	175	.091			2.44	5.85	8.29	10.15
	9010	For mobilization or demobilization, see div. 022-274									
	9020	For dewatering, see div. 021-404									
	9022	For larger structures, see Bulk Excavation, div. 022-238									
	9024	For loading onto trucks, add								15%	
	9026	For hauling, see div. 022-266									
	9030	For sheeting or soldier beams & lagging, see div. 021-614 & 624									
	9040	For trench excavation of strip footings, see div. 022-254									
254	0010	**EXCAVATING, TRENCH** or continuous footing, common earth (R022-240)									
	0020	No sheeting or dewatering included									
	0050	1' to 4' deep, 3/8 C.Y. tractor loader/backhoe	B-11C	150	.107	C.Y.		2.66	1.46	4.12	5.70
	0060	1/2 C.Y. tractor loader/backhoe	B-11M	200	.080			1.99	1.45	3.44	4.68
	0062	3/4 C.Y. hydraulic backhoe	B-12F	270	.059			1.58	1.66	3.24	4.24
	0090	4' to 6' deep, 1/2 C.Y. tractor loader/backhoe	B-11M	200	.080			1.99	1.45	3.44	4.68
	0100	5/8 C.Y. hydraulic backhoe	B-12Q	250	.064			1.71	1.58	3.29	4.33
	0110	3/4 C.Y. hydraulic backhoe	B-12F	300	.053			1.42	1.50	2.92	3.82
	0120	1 C.Y. hydraulic backhoe	B-12A	400	.040			1.07	1.37	2.44	3.14
	0130	1-1/2 C.Y. hydraulic backhoe	B-12B	540	.030			.79	1.32	2.11	2.65

022 200 | Excav./Backfill/Compact.

		Description	CREW	DAILY OUTPUT	LABOR-HOURS	UNIT	1999 BARE COSTS MAT.	LABOR	EQUIP.	TOTAL	TOTAL INCL O&P
254	0300	1/2 C.Y. hydraulic excavator, truck mounted (R022-240)	B-12J	200	.080	C.Y.		2.14	3.18	5.32	6.75
	0500	6' to 10' deep, 3/4 C.Y. hydraulic backhoe, 6' to 10' deep	B-12F	225	.071			1.90	2	3.90	5.10
	0510	1 C.Y. hydraulic backhoe	B-12A	400	.040			1.07	1.37	2.44	3.14
	0600	1 C.Y. hydraulic excavator, truck mounted	B-12K	400	.040			1.07	2.01	3.08	3.85
	0610	1-1/2 C.Y. hydraulic backhoe	B-12B	600	.027			.71	1.19	1.90	2.39
	0620	2-1/2 C.Y. hydraulic backhoe	B-12S	1,000	.016			.43	1.70	2.13	2.51
	0900	10' to 14' deep, 3/4 C.Y. hydraulic backhoe	B-12F	200	.080			2.14	2.25	4.39	5.70
	0910	1 C.Y. hydraulic backhoe	B-12A	360	.044			1.19	1.52	2.71	3.48
	1000	1-1/2 C.Y. hydraulic backhoe	B-12B	540	.030			.79	1.32	2.11	2.65
	1020	2-1/2 C.Y. hydraulic backhoe	B-12S	1,000	.016			.43	1.70	2.13	2.51
	1030	3 C.Y. hydraulic backhoe	B-12D	1,400	.011			.31	1.54	1.85	2.15
	1300	14' to 20' deep, 1 C.Y. hydraulic backhoe	B-12A	320	.050			1.34	1.71	3.05	3.91
	1310	1-1/2 C.Y. hydraulic backhoe	B-12B	480	.033			.89	1.48	2.37	2.98
	1320	2-1/2 C.Y. hydraulic backhoe	B-12S	850	.019			.50	1.99	2.49	2.95
	1330	3 C.Y. hydraulic backhoe	B-12D	1,000	.016			.43	2.16	2.59	3.02
	1400	By hand with pick and shovel 2' to 6' deep, light soil	1 Clab	8	1			21.50		21.50	34
	1500	Heavy soil	"	4	2			43		43	67.50
	1700	For tamping backfilled trenches, air tamp, add	A-1	100	.080			1.72	.69	2.41	3.46
	1900	Vibrating plate, add	B-18	230	.104			2.31	.23	2.54	3.88
	2100	Trim sides and bottom for concrete pours, common earth		1,500	.016	S.F.		.35	.03	.38	.60
	2300	Hardpan		600	.040	"		.88	.09	.97	1.49
	2400	Pier and spread footing excavation, add to above				C.Y.				30%	30%
	3000	Backfill trench, F.E. loader, wheel mtd., 1 C.Y. bucket									
	3020	Minimal haul	B-10R	400	.030	C.Y.		.78	.61	1.39	1.87
	3040	100' haul		200	.060			1.56	1.22	2.78	3.75
	3060	200' haul		100	.120			3.13	2.45	5.58	7.50
	3080	2-1/4 C.Y. bucket, minimum haul	B-10T	600	.020			.52	.75	1.27	1.63
	3090	100' haul		300	.040			1.04	1.50	2.54	3.25
	3100	200' haul		150	.080			2.09	3	5.09	6.50
	4000	For backfill with dozer, see div. 022-208									
	4010	For compaction of backfill, see div. 022-226									
258	0010	**EXCAVATING, UTILITY TRENCH** Common earth									
	0050	Trenching with chain trencher, 12 H.P., operator walking									
	0100	4" wide trench, 12" deep	B-53	800	.010	L.F.		.27	.12	.39	.54
	0150	18" deep		750	.011			.29	.13	.42	.58
	0200	24" deep		700	.011			.31	.14	.45	.62
	0300	6" wide trench, 12" deep		650	.012			.33	.15	.48	.67
	0350	18" deep		600	.013			.36	.16	.52	.73
	0400	24" deep		550	.015			.40	.18	.58	.79
	0450	36" deep		450	.018			.48	.22	.70	.98
	0600	8" wide trench, 12" deep		475	.017			.46	.20	.66	.93
	0650	18" deep		400	.020			.54	.24	.78	1.10
	0700	24" deep		350	.023			.62	.28	.90	1.26
	0750	36" deep		300	.027			.73	.32	1.05	1.46
	0830	Fly wheel trencher, 18" wide trench, 6' deep, light soil	B-54A	1,992	.005	C.Y.		.13	.20	.33	.42
	0840	Medium soil		1,594	.006			.16	.25	.41	.53
	0850	Heavy soil		1,295	.007			.20	.31	.51	.65
	0860	24" wide trench, 9' deep, light soil	B-54B	4,981	.002			.06	.11	.17	.20
	0870	Medium soil		4,000	.002			.07	.14	.21	.26
	0880	Heavy soil		3,237	.003			.08	.17	.25	.32
	1000	Backfill by hand including compaction, add									
	1050	4" wide trench, 12" deep	A-1	800	.010	L.F.		.21	.09	.30	.43
	1100	18" deep		530	.015			.32	.13	.45	.65
	1150	24" deep		400	.020			.43	.17	.60	.87
	1300	6" wide trench, 12" deep		540	.015			.32	.13	.45	.64

022 200 | Excav./Backfill/Compact.

		022 200 Excav./Backfill/Compact.	CREW	DAILY OUTPUT	LABOR-HOURS	UNIT	1999 BARE COSTS MAT.	LABOR	EQUIP.	TOTAL	TOTAL INCL O&P
258	1350	18″ deep	A-1	405	.020	L.F.		.42	.17	.59	.86
	1400	24″ deep		270	.030			.64	.25	.89	1.28
	1450	36″ deep		180	.044			.95	.38	1.33	1.92
	1600	8″ wide trench, 12″ deep		400	.020			.43	.17	.60	.87
	1650	18″ deep		265	.030			.65	.26	.91	1.30
	1700	24″ deep		200	.040			.86	.34	1.20	1.73
	1750	36″ deep		135	.059			1.27	.51	1.78	2.56
	2000	Chain trencher, 40 H.P. operator riding									
	2050	6″ wide trench and backfill, 12″ deep	B-54	1,200	.007	L.F.		.18	.18	.36	.47
	2100	18″ deep		1,000	.008			.22	.21	.43	.56
	2150	24″ deep		975	.008			.22	.22	.44	.58
	2200	36″ deep		900	.009			.24	.23	.47	.63
	2250	48″ deep		750	.011			.29	.28	.57	.75
	2300	60″ deep		650	.012			.33	.32	.65	.87
	2400	8″ wide trench and backfill, 12″ deep		1,000	.008			.22	.21	.43	.56
	2450	18″ deep		950	.008			.23	.22	.45	.59
	2500	24″ deep		900	.009			.24	.23	.47	.63
	2550	36″ deep		800	.010			.27	.26	.53	.70
	2600	48″ deep		650	.012			.33	.32	.65	.87
	2700	12″ wide trench and backfill, 12″ deep		975	.008			.22	.22	.44	.58
	2750	18″ deep		860	.009			.25	.25	.50	.65
	2800	24″ deep		800	.010			.27	.26	.53	.70
	2850	36″ deep		725	.011			.30	.29	.59	.78
	3000	16″ wide trench and backfill, 12″ deep		835	.010			.26	.25	.51	.68
	3050	18″ deep		750	.011			.29	.28	.57	.75
	3100	24″ deep		700	.011			.31	.30	.61	.80
	3200	Compaction with vibratory plate, add								50%	50%
	9100	For clay or till, add up to								150%	150%
262	0010	**FILL** Spread dumped material, by dozer, no compaction	B-10B	1,000	.012	C.Y.		.31	.84	1.15	1.40
	0100	By hand	1 Clab	12	.667	″		14.30		14.30	22.50
	0150	Spread fill, from stockpile with 2-1/2 C.Y. F.E. loader									
	0170	130 H.P., 300′ haul	B-10P	600	.020	C.Y.		.52	1.44	1.96	2.39
	0190	With dozer 300 H.P., 300′ haul	B-10M	600	.020	″		.52	1.89	2.41	2.87
	0400	For compaction of embankment, see div. 022-226									
	0500	Gravel fill, compacted, under floor slabs, 4″ deep	B-37	10,000	.005	S.F.	.15	.11	.02	.28	.36
	0600	6″ deep		8,600	.006		.23	.13	.02	.38	.47
	0700	9″ deep		7,200	.007		.38	.15	.02	.55	.67
	0800	12″ deep		6,000	.008		.52	.18	.03	.73	.89
	1000	Alternate pricing method, 4″ deep		120	.400	C.Y.	11.25	9.10	1.27	21.62	28
	1100	6″ deep		160	.300		11.25	6.80	.95	19	24
	1200	9″ deep		200	.240		11.25	5.45	.76	17.46	22
	1300	12″ deep		220	.218		11.25	4.96	.69	16.90	21
	1500	For fill under exterior paving, see division 022-308									
266	0011	**HAULING** Excavated or borrow material, loose cubic yards (R022-240)									
	0015	no loading included, highway haulers									
	0020	6 C.Y. dump truck, 1/4 mile round trip, 5.0 loads/hr.	B-34A	195	.041	C.Y.		.91	1.87	2.78	3.44
	0030	1/2 mile round trip, 4.1 loads/hr.		160	.050			1.11	2.28	3.39	4.19
	0040	1 mile round trip, 3.3 loads/hr.		130	.062			1.36	2.80	4.16	5.15
	0100	2 mile round trip, 2.6 loads/hr.		100	.080			1.77	3.64	5.41	6.70
	0150	3 mile round trip, 2.1 loads/hr.		80	.100			2.21	4.55	6.76	8.40
	0200	4 mile round trip, 1.8 loads/hr.		70	.114			2.53	5.20	7.73	9.55
	0310	12 C.Y. dump truck, 1/4 mile round trip 3.7 loads/hr.	B-34B	288	.028			.61	1.54	2.15	2.63
	0320	1/2 mile round trip, 3.2 loads/hr.		250	.032			.71	1.77	2.48	3.03
	0330	1 mile round trip 2.7, loads/hr.		210	.038			.84	2.11	2.95	3.61
	0400	2 mile round trip, 2.2 loads/hr.		180	.044			.98	2.46	3.44	4.20

	022 200 | Excav./Backfill/Compact.		CREW	DAILY OUTPUT	LABOR-HOURS	UNIT	1999 BARE COSTS MAT.	LABOR	EQUIP.	TOTAL	TOTAL INCL O&P
266	0450	3 mile round trip, 1.9 loads/hr. R022-240	B-34B	170	.047	C.Y.		1.04	2.60	3.64	4.45
	0500	4 mile round trip, 1.6 loads/hr.		125	.064			1.41	3.54	4.95	6.05
	0540	5 mile round trip, 1 load/hr.		78	.103			2.27	5.65	7.92	9.70
	0550	10 mile round trip, 0.60 load/hr.		58	.138			3.05	7.60	10.65	13.05
	0560	20 mile round trip, 0.4 load/hr.		39	.205			4.53	11.35	15.88	19.40
	0600	16.5 C.Y. dump trailer, 1 mile round trip, 2.6 loads/hr.	B-34C	280	.029			.63	2	2.63	3.17
	0700	2 mile round trip, 2.1 loads/hr.		225	.036			.79	2.48	3.27	3.93
	1000	3 mile round trip, 1.8 loads/hr.		193	.041			.92	2.90	3.82	4.59
	1100	4 mile round trip, 1.6 loads/hr.		172	.047			1.03	3.25	4.28	5.15
	1110	5 mile round trip, 1 load/hr.		108	.074			1.64	5.15	6.79	8.20
	1120	10 mile round trip, .60 load/hr.		80	.100			2.21	7	9.21	11.10
	1130	20 mile round trip, .4 load/hr.		54	.148			3.27	10.35	13.62	16.40
	1150	20 C.Y. dump trailer, 1 mile round trip, 2.5 loads/hr.	B-34D	325	.025			.54	1.73	2.27	2.73
	1200	2 mile round trip, 2 loads/hr.		260	.031			.68	2.16	2.84	3.42
	1220	3 mile round trip, 1.7 loads/hr.		221	.036			.80	2.54	3.34	4.02
	1240	4 mile round trip, 1.5 loads/hr.		195	.041			.91	2.88	3.79	4.56
	1245	5 mile round trip, 1.1 load/hr.		143	.056			1.24	3.93	5.17	6.20
	1250	10 mile round trip, .75 load/hr.		110	.073			1.61	5.10	6.71	8.05
	1255	20 mile round trip, .5 load/hr.		78	.103			2.27	7.20	9.47	11.35
	1300	Hauling in medium traffic, add								20%	20%
	1400	Heavy traffic, add								30%	30%
	1600	Grading at dump, or embankment if required, by dozer	B-10B	1,000	.012			.31	.84	1.15	1.40
	1800	Spotter at fill or cut, if required	1 Clab	8	1	Hr.		21.50		21.50	34
	2000	Off highway haulers									
	2010	22 C.Y. rear/bottom dump, 1000′ rnd trip, 4.5 loads/hr.	B-34F	645	.012	C.Y.		.27	1.61	1.88	2.19
	2020	1/2 mile round trip, 4.2 loads/hr.		600	.013			.29	1.73	2.02	2.35
	2030	1 mile round trip, 3.9 loads/hr.		555	.014			.32	1.87	2.19	2.55
	2040	2 mile round trip, 3.3 loads/hr.		470	.017			.38	2.21	2.59	3.01
	2050	34 C.Y. rear or bottom dump, 1000′ round trip, 4 loads/hr.	B-34G	885	.009			.20	1.58	1.78	2.05
	2060	1/2 mile round trip, 3.8 loads/hr.		840	.010			.21	1.67	1.88	2.15
	2070	1 mile round trip, 3.5 loads/hr.		775	.010			.23	1.81	2.04	2.34
	2080	2 mile round trip, 3.0 loads/hr.		665	.012			.27	2.11	2.38	2.73
	2090	42 C.Y. rear or bottom bump, 1000′ round trip, 3.8 loads/hr.	B-34H	1,040	.008			.17	1.54	1.71	1.95
	2100	1/2 mile round trip, 3.6 loads/hr.		980	.008			.18	1.63	1.81	2.07
	2110	1 mile round trip, 3.3 loads/hr.		900	.009			.20	1.78	1.98	2.25
	2120	2 mile round trip, 2.8 loads/hr.		765	.010			.23	2.09	2.32	2.65
	2130	60 C.Y. rear or bottom dump, 1000′ round trip, 3.6 loads/hr.	B-34J	1,400	.006			.13	1.60	1.73	1.95
	2140	1/2 mile round trip, 3.4 loads/hr.		1,325	.006			.13	1.69	1.82	2.06
	2150	1 mile round trip, 3.1 loads/hr.		1,200	.007			.15	1.86	2.01	2.28
	2160	2 mile round trip, 2.6 loads/hr.		1,015	.008			.17	2.20	2.37	2.70
	3000	Rough terrain or steep grades, add to above								100%	
	4500	Dust control, light	B-59	1	8	Day		177	530	707	850
	4501	Heavy	"	.50	16			355	1,050	1,405	1,725
	4600	Haul road maintenance	B-86A	1	8			227	575	802	980
	4700	Highway hauling beyond 20 miles, per loaded mile, minimum				Mile					1
	4750	Maximum				"					2
270	0010	**HORIZONTAL BORING** Casing only, 100′ minimum,									
	0020	not incl. jacking pits or dewatering									
	0100	Roadwork, 1/2″ thick wall, 24″ diameter casing	B-42	14	4.571	L.F.	45	110	80	235	315
	0200	36″ diameter		12	5.333		75	129	93.50	297.50	390
	0300	48″ diameter		12	5.333		105	129	93.50	327.50	425
	0500	Railroad work, 24″ diameter		10	6.400		45	155	112	312	420
	0600	36″ diameter		9.50	6.737		75	163	118	356	475
	0700	48″ diameter		9	7.111		105	172	125	402	530
	0900	For ledge, add								145	175
	1000	Small diameter boring, 3″, sandy soil	B-82	900	.018		13	.43	.05	13.48	15.05

Important: See the Reference Section for critical supporting data - Reference Nos., Crews, & City Cost Indexes

022 200 | Excav./Backfill/Compact.

		022 200 Excav./Backfill/Compact.	CREW	DAILY OUTPUT	LABOR-HOURS	UNIT	1999 BARE COSTS MAT.	LABOR	EQUIP.	TOTAL	TOTAL INCL O&P
270	1040	Rocky soil	B-82	500	.032	L.F.	13	.78	.10	13.88	15.60
	1100	Prepare jacking pits, incl. mobilization & demobilization, minimum				Ea.				2,650	3,150
	1101	Maximum				″				15,000	18,000
274	0010	**MOBILIZATION OR DEMOBILIZATION** Up to 50 miles									
	0020	Dozer, loader, backhoe or excavator, 70 H.P.- 250 H.P.	B-34K	6	1.333	Ea.		29.50	152	181.50	213
	0100	Above 250 H.P		4	2			44	228	272	320
	0300	Scraper, towed type (incl. tractor), 6 C.Y. capacity		3.75	2.133			47	244	291	340
	0400	10 C.Y.		3.50	2.286			50.50	261	311.50	365
	0600	Self-propelled scraper, 15 C.Y.		3.30	2.424			53.50	277	330.50	385
	0700	24 C.Y.		3	2.667			59	305	364	425
	0900	Shovel or dragline, 3/4 C.Y.		3.60	2.222			49	254	303	355
	1000	1-1/2 C.Y.		3	2.667			59	305	364	425
	1100	Delivery charge for small equipment on flatbed trailer, minimum									40
	1150	Maximum									100
	3000	For large pieces of equipment, allow for knockdown, assembly									
	3001	and lead and tail vehicles for highway transport									
278	0010	**RIPPING** Trap rock, soft, 200 HP dozer, ideal conditions	B-10B	710	.017	C.Y.		.44	1.18	1.62	1.98
	1500	Adverse conditions	″	660	.018			.47	1.27	1.74	2.13
	1600	Medium hard, 300 HP dozer, ideal conditons	B-10M	600	.020			.52	1.89	2.41	2.87
	1700	Adverse conditions	″	540	.022			.58	2.09	2.67	3.19
	2000	Very hard, 460 HP dozer, ideal conditions	B-10X	300	.040			1.04	4.60	5.64	6.65
	2100	Adverse conditions	″	270	.044			1.16	5.10	6.26	7.40
	2200	Shale, soft, 200 HP dozer, ideal conditons	B-10B	850	.014			.37	.99	1.36	1.66
	2300	Adverse conditions	″	790	.015			.40	1.06	1.46	1.78
	2310	Grader rear ripper, 180 H.P. ideal conditions	B-11L	740	.022			.54	.78	1.32	1.69
	2320	Adverse conditions	″	630	.025			.63	.91	1.54	1.98
	2400	Medium hard, 300 HP dozer, ideal conditons	B-10M	720	.017			.43	1.57	2	2.40
	2500	Adverse conditions	″	650	.018			.48	1.74	2.22	2.65
	2510	Grader rear ripper, 180 H.P. ideal conditions	B-11L	625	.026			.64	.92	1.56	2
	2520	Adverse conditions	″	530	.030			.75	1.09	1.84	2.35
	2600	Very hard, 460 HP dozer, ideal conditons	B-10X	360	.033			.87	3.83	4.70	5.55
	2700	Adverse conditions	″	320	.038			.98	4.31	5.29	6.25
	2800	Till, boulder clay/hardpan, soft, 200 H.P. dozer, ideal conditions	B-10B	1,400	.009			.22	.60	.82	1
	2810	Adverse conditions	″	1,315	.009			.24	.64	.88	1.07
	2815	Grader rear ripper, 180 H.P. ideal conditions	B-11L	1,500	.011			.27	.38	.65	.83
	2816	Adverse conditions	″	1,275	.013			.31	.45	.76	.98
	2820	Medium hard, 300 H.P. dozer, ideal conditions	B-10M	1,200	.010			.26	.94	1.20	1.44
	2830	Adverse conditions	″	1,080	.011			.29	1.05	1.34	1.59
	2835	Grader rear ripper, 180 H.P. ideal conditions	B-11L	1,300	.012			.31	.44	.75	.96
	2836	Adverse conditions	″	1,100	.015			.36	.52	.88	1.14
	2840	Very hard, 460 H.P. dozer, ideal conditions	B-10X	600	.020			.52	2.30	2.82	3.33
	2850	Adverse conditions	″	530	.023			.59	2.60	3.19	3.77
	3000	Dozing ripped material, 200 HP, 100′ haul	B-10B	700	.017			.45	1.20	1.65	2.01
	3050	300′ haul	″	250	.048			1.25	3.36	4.61	5.60
	3200	300 HP, 100′ haul	B-10M	1,150	.010			.27	.98	1.25	1.50
	3250	300′ haul	″	400	.030			.78	2.83	3.61	4.31
	3400	460 HP, 100′ haul	B-10X	1,680	.007			.19	.82	1.01	1.19
	3450	300′ haul	″	600	.020			.52	2.30	2.82	3.33
86	0010	**LOAM OR TOPSOIL** Remove and stockpile on site									
	0020	6″ deep, 200′ haul	B-10B	865	.014	C.Y.		.36	.97	1.33	1.63
	0100	300′ haul		520	.023			.60	1.61	2.21	2.69
	0150	500′ haul		225	.053			1.39	3.73	5.12	[illegible]
	0200	Alternate method: 6″ deep, 200′ haul		5,090	.002	S.Y.		.06	.16	.22	[illegible]
	0250	500′ haul		1,325	.009	″		.24	.63	[illegible]	[illegible]

022 | Earthwork

		022 200 | Excav./Backfill/Compact.	Crew	Daily Output	Labor-Hours	Unit	1999 Bare Costs Mat.	Labor	Equip.	Total	Total Incl O&P	
286	0400	Spread from pile to rough finish grade, F.E. loader, 1.5 C.Y.	B-10S	200	.060	C.Y.		1.56	1.60	3.16	4.16	286
	0500	Up to 200′ radius, by hand	1 Clab	14	.571			12.25		12.25	19.30	
	0600	Top dress by hand, 1 C.Y. for 600 S.F.	"	11.50	.696	↓	18.25	14.90		33.15	43.50	
	0700	Furnish and place, truck dumped, screened, 4″ deep	B-10S	1,300	.009	S.Y.	2.28	.24	.25	2.77	3.15	
	0800	6″ deep	"	820	.015	"	2.92	.38	.39	3.69	4.23	
	0900	Fine grading and seeding, incl. lime, fertilizer & seed,										
	1000	With equipment	B-14	1,000	.048	S.Y.	.30	1.09	.22	1.61	2.28	

		022 300 | Pavement Base	Crew	Daily Output	Labor-Hours	Unit	Mat.	Labor	Equip.	Total	Total Incl O&P	
304	0010	**BASE** Prepare and roll sub-base, small areas to 2500 S.Y.	B-32A	1,500	.016	S.Y.		.42	.66	1.08	1.37	304
	0100	Large areas over 2500 S.Y.	B-32	3,700	.009	"		.23	.45	.68	.84	
308	0010	**BASE COURSE** For roadways and large paved areas										308
	0050	Crushed 3/4″ stone base, compacted, 3″ deep	B-36B	5,200	.012	S.Y.	3.07	.31	.63	4.01	4.55	
	0100	6″ deep		5,000	.013		6.15	.32	.65	7.12	8	
	0200	9″ deep		4,600	.014		9.25	.35	.71	10.31	11.45	
	0300	12″ deep		4,200	.015		12.55	.38	.78	13.71	15.30	
	0301	Crushed 1-1/2″ stone base, compacted to 4″ deep		6,000	.011		3.78	.27	.54	4.59	5.15	
	0302	6″ deep		5,400	.012		5.80	.30	.60	6.70	7.45	
	0303	8″ deep		4,500	.014		7.55	.36	.73	8.64	9.70	
	0304	12″ deep	↓	3,800	.017	↓	11.60	.43	.86	12.89	14.35	
	0350	Bank run gravel, spread and compacted										
	0370	6″ deep	B-32	6,000	.005	S.Y.	2.34	.14	.28	2.76	3.10	
	0390	9″ deep		4,900	.007		3.44	.17	.34	3.95	4.43	
	0400	12″ deep	↓	4,200	.008		4.68	.20	.40	5.28	5.90	
	0500	Bituminous concrete, 4″ thick	B-25	4,545	.019		6	.46	.39	6.85	7.75	
	0550	6″ thick		3,700	.024		8.80	.56	.48	9.84	11.10	
	0560	8″ thick		3,000	.029		11.75	.69	.60	13.04	14.65	
	0570	10″ thick	↓	2,545	.035	↓	14.55	.81	.70	16.06	18.05	
	0600	Cold laid asphalt pavement, see div. 025-116										
	0700	Liquid application to gravel base, asphalt emulsion	B-45	6,000	.003	Gal.	1.50	.07	.12	1.69	1.89	
	0800	Prime and seal, cut back asphalt		6,000	.003	"	1.77	.07	.12	1.96	2.19	
	1000	Macadam penetration crushed stone, 2 gal. per S.Y., 4″ thick		6,000	.003	S.Y.	3	.07	.12	3.19	3.54	
	1100	6″ thick, 3 gal. per S.Y.		4,000	.004		4.50	.10	.19	4.79	5.30	
	1200	8″ thick, 4 gal. per S.Y.	↓	3,000	.005	↓	6	.13	.25	6.38	7.10	
	1500	Alternate method to figure base course										
	1510	Crushed stone, 3/4″ maximum size, 3″ deep	B-36B	800	.080	C.Y.	37.50	2.02	4.08	43.60	48.50	
	1511	6″ deep		1,500	.043		37.50	1.08	2.18	40.76	45	
	1512	9″ deep		1,650	.039		37.50	.98	1.98	40.46	44.50	
	1513	12″ deep		1,800	.036		37.50	.90	1.81	40.21	44.50	
	1520	Crushed stone, 1-1/2″ maximum size, 4″ deep		670	.096		34.50	2.41	4.87	41.78	47	
	1521	6″ deep		750	.085		34.50	2.16	4.35	41.01	46	
	1522	8″ deep		765	.084		34.50	2.11	4.27	40.88	46	
	1523	12″ deep	↓	785	.082		34.50	2.06	4.16	40.72	45.50	
	1530	Gravel, bank run, 6″ deep	B-32A	650	.037		13.80	.96	1.53	16.29	18.30	
	1531	9″ deep		720	.033		13.80	.87	1.39	16.06	18	
	1532	[illegible]ep	↓	785	.031	↓	13.80	.80	1.27	15.87	17.75	
	2000	[illegible] to figure base course										
	[illegible]	[illegible]crete, 4″ thick	B-25	1,000	.088	Ton	26.50	2.07	1.79	30.36	34.50	
	[illegible]	[illegible]		1,220	.072		26.50	1.70	1.47	29.67	34	
	[illegible]	[illegible]		1,320	.067		26.50	1.57	1.35	29.42	33.50	
	[illegible]	[illegible]	↓	1,400	.063		26.50	1.48	1.28	29.26	33	
	[illegible]	[illegible]aximum size, 3″ deep	B-36	540	.074		17.05	1.82	2.14	21.01	24	
	[illegible]	[illegible]		1,625	.025		17.05	.61	.71	18.37	20.50	
	[illegible]	[illegible]		1,785	.022		17.05	.55	.65	18.25	20.50	
	[illegible]	[illegible]	↓	1,950	.021	↓	17.05	.51	.59	18.15	20	

022 300 | Pavement Base

				DAILY	LABOR-		1999 BARE COSTS				TOTAL
			CREW	OUTPUT	HOURS	UNIT	MAT.	LABOR	EQUIP.	TOTAL	INCL O&P
308	2020	Crushed stone, 1-1/2" maximum size, 4" deep	B-36	720	.056	Ton	15.70	1.37	1.60	18.67	21
	2021	6" deep		815	.049		15.70	1.21	1.42	18.33	20.50
	2022	8" deep		835	.048		15.70	1.18	1.38	18.26	20.50
	2023	12" deep		975	.041		15.70	1.01	1.18	17.89	20
	2030	Bank run gravel, 6" deep	B-32A	875	.027		13.80	.72	1.14	15.66	17.50
	2031	9" deep		970	.025		13.80	.65	1.03	15.48	17.25
	2032	12" deep		1,060	.023		13.80	.59	.94	15.33	17.10
	6000	Stabilization fabric, polypropylene, 6 oz./S.Y.	B-6	10,000	.002	S.Y.	1.20	.06	.02	1.28	1.43
	8900	For small and irregular areas, add						50%	50%		

022 400 | Soil Stabilization

			CREW	DAILY OUTPUT	LABOR-HOURS	UNIT	MAT.	LABOR	EQUIP.	TOTAL	TOTAL INCL O&P
408	0010	**GROUTING, PRESSURE** Cement and sand, 1:1 mix, minimum	B-61	124	.323	Bag	7.90	7.40	2.88	18.18	23.50
	0100	Maximum		51	.784	"	7.90	18.05	7	32.95	44.50
	0200	Cement and sand, 1:1 mix, minimum		250	.160	C.F.	15.85	3.68	1.43	20.96	24.50
	0300	Maximum		100	.400		24	9.20	3.57	36.77	44.50
	0400	Epoxy cement grout, minimum		137	.292		100	6.70	2.60	109.30	123
	0500	Maximum		57	.702		100	16.15	6.25	122.40	142
	0600	Structural epoxy grout				Gal.	45			45	49.50
	0700	Alternate pricing method: (Add for materials)									
	0710	5 person crew and equipment	B-61	1	40	Day		920	355	1,275	1,825
412	0010	**SOIL STABILIZATION** Including scarifying and compaction									
	0020	Asphalt, 1-1/2" deep, 1/2 gal/S.Y.	B-75	4,000	.014	S.Y.	.73	.36	.70	1.79	2.13
	0040	3/4 gal/S.Y.		4,000	.014		1.10	.36	.70	2.16	2.53
	0100	3" deep, 1 gal/S.Y.		3,500	.016		1.46	.41	.81	2.68	3.13
	0140	1-1/2 gal/S.Y.		3,500	.016		2.19	.41	.81	3.41	3.93
	0200	6" deep, 2 gal/S.Y.		3,000	.019		2.92	.48	.94	4.34	4.98
	0240	3 gal/S.Y.		3,000	.019		4.38	.48	.94	5.80	6.60
	0300	8" deep, 2-2/3 gal/S.Y.		2,800	.020		3.90	.52	1.01	5.43	6.20
	0340	4 gal/S.Y.		2,800	.020		5.85	.52	1.01	7.38	8.30
	0500	12" deep, 4 gal/S.Y.		5,000	.011		5.85	.29	.56	6.70	7.45
	0540	6 gal/S.Y.		2,600	.022		8.75	.56	1.08	10.39	11.70
	1020	Cement, 4% mix, by volume, 6" deep	B-74	1,100	.058		.93	1.47	2.97	5.37	6.55
	1030	8" deep		1,050	.061		1.21	1.54	3.11	5.86	7.15
	1060	12" deep		960	.067		1.82	1.69	3.40	6.91	8.35
	1100	6% mix, 6" deep		1,100	.058		1.34	1.47	2.97	5.78	7
	1120	8" deep		1,050	.061		1.74	1.54	3.11	6.39	7.70
	1160	12" deep		960	.067		2.63	1.69	3.40	7.72	9.25
	1200	9% mix, 6" deep		1,100	.058		2.02	1.47	2.97	6.46	7.75
	1220	8" deep		1,050	.061		2.63	1.54	3.11	7.28	8.70
	1260	12" deep		960	.067		3.97	1.69	3.40	9.06	10.70
	1300	12% mix, 6" deep		1,100	.058		2.63	1.47	2.97	7.07	8.40
	1320	8" deep		1,050	.061		3.52	1.54	3.11	8.17	9.65
	1360	12" deep		960	.067		5.25	1.69	3.40	10.34	12.15
	1500	Geotextile fabric, woven, 200 lb. tensile strength	2 Clab	2,500	.006		1.46	.14		1.60	1.83
	1510	Heavy Duty, 600 lb. tensile strength		2,400	.007		1.55	.14		1.69	1.92
	1550	Non-woven, 120 lb. tensile strength		2,500	.006		.65	.14		.79	.93
	2020	Hydrated lime, for base, 2% mix by weight, 6" deep	B-74	1,800	.036		2.25	.90	1.81	4.96	5.85
	2030	8" deep		1,700	.038		3	.95	1.92	5.87	6.85
	2060	12" deep		1,550	.041		2.10	1.05	2.10	5.25	6.25
	2100	4% mix, 6" deep		1,800	.036		2.13	.90	1.81	4.84	5.70
	2120	8" deep		1,700	.038		2.88	.95	1.92	5.75	6.75
	2160	12" deep		1,550	.041		4.31	1.05	2.10	7.46	8.65
	2200	6% mix, 6" deep		1,800	.036		3.23	.90	1.81	5.94	6.95
	2220	8" deep		1,700	.038		4.31	.95	1.92	7.18	8.30

		022 400 | Soil Stabilization	CREW	DAILY OUTPUT	LABOR-HOURS	UNIT	1999 BARE COSTS MAT.	LABOR	EQUIP.	TOTAL	TOTAL INCL O&P	
412	2260	12" deep	B-74	1,550	.041	S.Y.	6.40	1.05	2.10	9.55	11	412
	5000	Soil poisoning (sterilization)	1 Clab	4,496	.002	S.F.	.03	.04		.07	.09	
	5100	Herbicide application from truck	B-59	19,000	.001	S.Y.		.01	.03	.04	.04	
		022 500 | Vibroflotation										
504	0010	**VIBROFLOTATION**										504
	0900	Vibroflotation compacted sand cylinder, minimum	B-60	750	.075	V.L.F.		1.86	1.34	3.20	4.33	
	0950	Maximum		325	.172			4.29	3.08	7.37	10	
	1100	Vibro replacement compacted stone cylinder, minimum		500	.112			2.79	2	4.79	6.50	
	1150	Maximum		250	.224	↓		5.55	4.01	9.56	13	
	1300	Mobilization and demobilization, minimum		.47	119	Total		2,975	2,125	5,100	6,925	
	1400	Maximum	↓	.14	400	"		9,950	7,150	17,100	23,200	
		022 700 | Slope/Erosion Control										
702	0010	**CUT DRAINAGE DITCH** Common earth, 30'w x 1'deep	B-11L	6,000	.003	L.F.		.07	.10	.17	.21	702
	0200	Clay and till		4,200	.004			.09	.14	.23	.30	
	0250	Clean wet drainage ditch, 30' wide	↓	10,000	.002	↓		.04	.06	.10	.12	
704	0010	**EROSION CONTROL** Jute mesh, 100 S.Y. per roll, 4' wide, stapled	B-80A	2,400	.010	S.Y.	.72	.21	.07	1	1.21	704
	0060	Nylon, 3 dimensional geomatrix, 9 mil thick		700	.034		3.85	.74	.26	4.85	5.70	
	0062	12 mil thick		515	.047		4	1	.35	5.35	6.35	
	0064	18 mil thick	↓	460	.052		5	1.12	.39	6.51	7.70	
	0070	Paper biodegradable mesh	B-1	2,500	.010		.07	.21		.28	.41	
	0080	Paper mulch	B-64	20,000	.001		.05	.02	.01	.08	.11	
	0100	Plastic netting, stapled, 2" x 1" mesh, 20 mil	B-1	2,500	.010		.40	.21		.61	.77	
	0120	Revegetation mat, webbed	2 Clab	1,000	.016	↓	2.88	.34		3.22	3.70	
	0160	Underdrain fabric, 18" x 100' roll	"	32	.500	Roll	16.50	10.75		27.25	35	
	0200	Polypropylene mesh, stapled, 6.5 oz./S.Y.	B-1	2,500	.010	S.Y.	1	.21		1.21	1.43	
	0300	Tobacco netting, or jute mesh #2, stapled	"	2,500	.010		.07	.21		.28	.41	
	0400	Soil sealant, liquid sprayed from truck	B-81	5,000	.005	↓	.40	.12	.12	.64	.75	
	1000	Silt fence, polypropylene, 3' high, ideal conditions	2 Clab	1,600	.010	L.F.	.35	.21		.56	.73	
	1100	Adverse conditions	"	950	.017	"	.28	.36		.64	.88	
	1130	Cellular confinement, poly, 3-dimen, 8' x 20' panels, 4" deep cell	B-6	1,600	.015	S.F.	.89	.35	.14	1.38	1.67	
	1140	8" deep cells	"	1,200	.020	"	1.60	.47	.18	2.25	2.69	
	1200	Place and remove hay bales	A-2	3	8	Ton	50	172	57.50	279.50	385	
	1250	Hay bales, staked	"	1,800	.013	L.F.	6	.29	.10	6.39	7.15	
	1305	For less than 3 To 1 slope, add						15%				
	1310	For greater than 3 To 1 slope, add						25%				
706	0010	**MEMBRANE LINING SYSTEMS**, HDPE, 100,000 S.F. or more										706
	0100	30 mil thick	3 Skwk	1,850	.013	S.F.	.30	.36		.66	.90	
	0200	60 mil thick		1,600	.015		.60	.42		1.02	1.32	
	0300	120 mil thick	↓	1,440	.017	↓	1.20	.47		1.67	2.06	
708	0010	**RETAINING WALLS** Aluminized steel bin, excavation										708
	0020	and backfill not included, 10' wide										
	0100	4' high, 5.5' deep	B-13	650	.086	S.F.	14.50	2	.85	17.35	20	
	0200	8' high, 5.5' deep		615	.091		16.70	2.12	.89	19.71	22.50	
	0300	10' high, 7.7' deep		580	.097		17.55	2.24	.95	20.74	24	
	0400	12' high, 7.7' deep		530	.106		18.95	2.46	1.04	22.45	26	
	0500	16' high, 7.7' deep		515	.109		20	2.53	1.07	23.60	27	
	0600	16' high, 9.9' deep		500	.112		21	2.60	1.10	24.70	29	
	0700	20' high, 9.9' deep		470	.119		24	2.77	1.17	27.94	31.50	
	0800	20' high, 12.1' deep		460	.122		25.50	2.83	1.19	29.52	33.50	
	0900	24' high, 12.1' deep		455	.123		27	2.86	1.21	31.07	36	
	1000	24' high, 14.3' deep	↓	450	.124	↓	28	2.89	1.22	32.11	37	

022 700 | Slope/Erosion Control

708		CREW	DAILY OUTPUT	LABOR-HOURS	UNIT	1999 BARE COSTS MAT.	LABOR	EQUIP.	TOTAL	TOTAL INCL O&P
1100	28′ high, 14.3′ deep	B-13	440	.127	S.F.	29.50	2.96	1.25	33.71	38
1300	For plain galvanized bin type walls, deduct					10%				
1800	Concrete gravity wall with vertical face including excavation & backfill									
1850	No reinforcing									
1900	6′ high, level embankment	C-17C	36	2.306	L.F.	43	65.50	11.45	119.95	163
2000	33° slope embankment		32	2.594		40.50	74	12.85	127.35	175
2200	8′ high, no surcharge		27	3.074		55.50	87.50	15.25	158.25	216
2300	33° slope embankment		24	3.458		67	98.50	17.15	182.65	248
2500	10′ high, level embankment		19	4.368		79.50	124	21.50	225	305
2600	33° slope embankment		18	4.611		110	131	23	264	355
2800	Reinforced concrete cantilever, incl. excavation, backfill & reinf.									
2900	6′ high, 33° slope embankment	C-17C	35	2.371	L.F.	40.50	67.50	11.75	119.75	163
3000	8′ high, 33° slope embankment		29	2.862		47	81.50	14.20	142.70	195
3100	10′ high, 33° slope embankment		20	4.150		61	118	20.50	199.50	276
3200	20′ high, 500 lb. per L.F. surcharge		7.50	11.067		183	315	55	553	755
3500	Concrete cribbing, incl. excavation and backfill									
3700	12′ high, open face	B-13	210	.267	S.F.	20.50	6.20	2.62	29.32	35
3900	Closed face	″	210	.267	″	19.05	6.20	2.62	27.87	33.50
4100	Concrete filled slurry trench, see division 021-684									
4300	Stone filled gabions, not incl. excavation,									
4310	Stone, delivered, 3′ wide									
4340	Galvanized, 6′ long, 1′ high	B-13	200	.280	Ea.	50	6.50	2.75	59.25	68
4350	Galvanized, 6′ high, 33° slope embankment		49	1.143	L.F.	20	26.50	11.20	57.70	76
4400	1′-6″ high		130	.431	Ea.	66	10	4.23	80.23	93
4490	3′-0″ high		55	1.018	″	120	23.50	10	153.50	180
4500	Highway surcharge		27	2.074	L.F.	40	48	20.50	108.50	142
4590	9′ long, 1′ high		130	.431	Ea.	70	10	4.23	84.23	97.50
4600	9′ high, up to 33° slope embankment		24	2.333	L.F.	45	54	23	122	159
4650	1′-6″ high		87	.644	Ea.	96.50	14.95	6.30	117.75	136
4690	3′-0″ high		37	1.514	″	165	35	14.85	214.85	253
4700	Highway surcharge		16	3.500	L.F.	70	81.50	34.50	186	242
4890	12′ long, 1′ high		100	.560	Ea.	95.50	13	5.50	114	132
4900	12′ high, up to 33° slope embankment		14	4	L.F.	70	93	39	202	265
4950	1′-6″ high		55	1.018	Ea.	125	23.50	10	158.50	186
4990	3′-0″ high		25	2.240	″	225	52	22	299	355
5000	Highway surcharge		11	5.091	L.F.	100	118	50	268	350
5200	PVC coated, 6′ long, 1′ high		200	.280	Ea.	55	6.50	2.75	64.25	73.50
5250	1′-6″ high		130	.431		72.50	10	4.23	86.73	100
5300	3′ high		55	1.018		129	23.50	10	162.50	190
5500	9′ long, 1′ high		130	.431		83	10	4.23	97.23	112
5550	1′-6″ high		87	.644		107	14.95	6.30	128.25	148
5600	3′ high		37	1.514		191	35	14.85	240.85	281
5800	12′ long, 1′ high		100	.560		108	13	5.50	126.50	146
5850	1′-6″ high		55	1.018		141	23.50	10	174.50	203
5900	3′ high		25	2.240		248	52	22	322	375
5950	For PVC coating, add				L.F.	20%				
7100	Segmental wall system, incl backfill, compaction, to 8′									
7120	interlocking pins, no scaffolding, base, or tiebacks									
7140	8″ x 18″ x 21.5″, 95 lbs	D-12	250	.128	S.F.	6.40	3.21		9.61	12.05
7160	8″ x 18″ x 12.5″, 85 lbs		330	.097		6.20	2.43		8.63	10.60
7180	6″ x 17.25″ x 12″, 72 lbs		375	.085		5.90	2.14		8.04	9.85
7200	Beveled, 6″ x 6″ x 12″, 68 lbs		350	.091		5.65	2.29		7.94	9.75
7220	Step, 6″ x 16″ x 12″, 85 lbs		300	.107		5.10	2.68		7.78	9.80
7240	Caps, to 40 lbs		600	.053	L.F.	6.40	1.34		7.74	9.15
7260	For reinforcing, add				S.F.				2	2.50
8000	For higher walls, add components as necessary									

022 | Earthwork

022 700 | Slope/Erosion Control

		022 700 Slope/Erosion Control	CREW	DAILY OUTPUT	LABOR-HOURS	UNIT	1999 BARE COSTS MAT.	LABOR	EQUIP.	TOTAL	TOTAL INCL O&P
712	0010	**RIP-RAP** Random, broken stone									
	0100	Machine placed for slope protection	B-12G	62	.258	C.Y.	17.75	6.90	8.40	33.05	39.50
	0110	3/8 to 1/4 C.Y. pieces, grouted	B-13	80	.700	S.Y.	29	16.25	6.85	52.10	65
	0200	18" minimum thickness, not grouted	"	53	1.057	"	13.15	24.50	10.35	48	64
	0300	Dumped, 50 lb. average	B-11A	800	.020	Ton	10	.50	1.05	11.55	12.90
	0350	100 lb. average		700	.023		14.30	.57	1.20	16.07	17.95
	0370	300 lb. average	↓	600	.027	↓	16.65	.66	1.40	18.71	21
	0400	Gabions, galvanized steel mesh mats or boxes, stone filled, 6" deep	B-13	200	.280	S.Y.	11.85	6.50	2.75	21.10	26
	0500	9" deep		163	.344		13.50	8	3.37	24.87	31
	0600	12" deep		153	.366		15.65	8.50	3.59	27.74	34.50
	0700	18" deep		102	.549		25.50	12.75	5.40	43.65	54
	0800	36" deep	↓	60	.933	↓	42	21.50	9.15	72.65	90.50
716	0010	**STONE WALL** Including excavation, concrete footing and									
	0020	stone 3' below grade. Price is exposed face area.									
	0200	Decorative random stone, to 6' high, 1'-6" thick, dry set	D-1	35	.457	S.F.	7.60	11.25		18.85	26
	0300	Mortar set		40	.400		9.25	9.85		19.10	25.50
	0600	Mortar set		40	.400		13.50	9.85		23.35	30
	0800	Retaining wall, random stone, 6' to 10' high, 2' thick, drv set		45	.356		9.50	8.75		18.25	24
	0900	Mortar set		50	.320		11.50	7.85		19.35	25
	1100	Cut stone, 6' to 10' high, 2' thick, dry set		45	.356		14.75	8.75		23.50	30
	1200	Mortar set		50	.320	↓	15.75	7.85		23.60	29.50
	5100	Setting stone, dry		100	.160	C.F.		3.94		3.94	6.10
	5600	With mortar	↓	120	.133	"		3.28		3.28	5.10

022 800 | Soil Treatment

		022 800 Soil Treatment	CREW	DAILY OUTPUT	LABOR-HOURS	UNIT	MAT.	LABOR	EQUIP.	TOTAL	TOTAL INCL O&P
804	0010	**TERMITE PRETREATMENT**									
	0020	Slab and walls, residential	1 Skwk	1,200	.007	SF Flr.	.19	.19		.38	.49
	0100	Commercial, minimum		2,496	.003		.16	.09		.25	.31
	0200	Maximum		1,645	.005	↓	.13	.14		.27	.36
	0400	Insecticides for termite control, minimum		14.20	.563	Gal.	10	15.80		25.80	36
	0500	Maximum	↓	11	.727	"	17.10	20.50		37.60	51

023 | Tunneling, Piles & Caissons

023 100 | Tunnel Construction

		023 100 Tunnel Construction	CREW	DAILY OUTPUT	LABOR-HOURS	UNIT	1999 BARE COSTS MAT.	LABOR	EQUIP.	TOTAL	TOTAL INCL O&P
150	0010	**MICROTUNNELING** Not including excavation, backfill, shoring,									
	0020	or dewatering, average 50'/day, slurry method									
	0100	24" to 48" outside diameter, minimum				L.F.					600
	0110	Adverse conditions, add				%					50%
	1000	Rent microtunneling machine, average monthly lease				Month					80,000
	1010	Operating technician				Day					600
	1100	Mobilization and demobilization, minimum				Job					40,000
	1110	Maximum				"					400,000

023 400 | Tunnel Support Systems

		023 400 Tunnel Support Systems	CREW	DAILY OUTPUT	LABOR-HOURS	UNIT	MAT.	LABOR	EQUIP.	TOTAL	TOTAL INCL O&P
404	0011	**ROCK BOLTS**									
	2020	Hollow core, prestressable anchor, 1" diameter, 5' long	2 Skwk	32	.500	Ea.	54	14		68	81.50
	2025	10' long		24	.667		101	18.70		119.70	142
	2060	2" diameter, 5' long	↓	32	.500	↓	203	14		217	245

023 400 | Tunnel Support Systems

			CREW	DAILY OUTPUT	LABOR-HOURS	UNIT	1999 BARE COSTS MAT.	LABOR	EQUIP.	TOTAL	TOTAL INCL O&P	
404	2065	10′ long	2 Skwk	24	.667	Ea.	390	18.70		408.70	460	404
	2100	Super high-tensile, 3/4″ diameter, 5′ long		32	.500		12.80	14		26.80	36	
	2105	10′ long		24	.667		23.50	18.70		42.20	55.50	
	2160	2″ diameter, 5′ long		32	.500		113	14		127	147	
	2165	10′ long		24	.667		204	18.70		222.70	254	
	4400	Drill hole for rock bolt, 1-3/4″ diam., 5′ long (for 3/4″ bolt)	B-56	17	.941			23	35.50	58.50	74.50	
	4405	10′ long		9	1.778			43.50	67	110.50	141	
	4420	2″ diameter, 5′ long (for 1″ bolt)		13	1.231			30	46.50	76.50	97	
	4425	10′ long		7	2.286			55.50	86	141.50	181	
	4460	3-1/2″ diameter, 5′ long (for 2″ bolt)		10	1.600			39	60	99	126	
	4465	10′ long		5	3.200			78	120	198	252	

023 550 | Pile Driving

			CREW	DAILY OUTPUT	LABOR-HOURS	UNIT	MAT.	LABOR	EQUIP.	TOTAL	TOTAL INCL O&P	
554	0010	**MOBILIZATION** Set up & remove, air compressor, 600 C.F.M.	A-5	3.30	5.455	Ea.		117	13.05	130.05	198	554
	0100	1200 C.F.M.	"	2.20	8.182			176	19.55	195.55	298	
	0200	Crane, with pile leads and pile hammer, 75 ton	B-19	.60	106			2,950	2,400	5,350	7,475	
	0300	150 ton	"	.36	177			4,900	4,000	8,900	12,500	
	0500	Drill rig, for caissons, to 36″, minimum	B-43	2	24			565	890	1,455	1,850	
	0520	Maximum		.50	96			2,250	3,550	5,800	7,400	
	0600	Up to 84″		1	48			1,125	1,775	2,900	3,700	
	0800	Auxiliary boiler, for steam small	A-5	1.66	10.843			233	26	259	395	
	0900	Large	"	.83	21.687			465	52	517	785	
	1100	Rule of thumb: complete pile driving set up, small	B-19	.45	142			3,925	3,200	7,125	10,000	
	1200	Large	"	.27	237			6,550	5,325	11,875	16,700	
	1300	Mobilization by water for barge driving rig										
	1310	Minimum				Ea.				5,850	6,450	
	1320	Maximum				"				38,000	42,000	
558	0011	**PILING SPECIAL COSTS** pile caps, see Division 033-130										558
	0500	Cutoffs, concrete piles, plain	1 Pile	5.50	1.455	Ea.		39.50		39.50	68.50	
	0600	With steel thin shell, add		38	.211			5.75		5.75	9.90	
	0700	Steel pile or "H" piles		19	.421			11.45		11.45	19.80	
	0800	Wood piles		38	.211			5.75		5.75	9.90	
	0900	Pre-augering up to 30′ deep, average soil, 24″ diameter	B-43	180	.267	L.F.		6.25	9.85	16.10	20.50	
	0920	36″ diameter		115	.417			9.80	15.45	25.25	32.50	
	0960	48″ diameter		70	.686			16.15	25.50	41.65	53	
	0980	60″ diameter		50	.960			22.50	35.50	58	74	
	1000	Testing, any type piles, test load is twice the design load										
	1050	50 ton design load, 100 ton test				Ea.				14,000	15,000	
	1100	100 ton design load, 200 ton test								18,000	19,000	
	1150	150 ton design load, 300 ton test								22,500	24,000	
	1200	200 ton design load, 400 ton test								24,500	27,000	
	1250	400 ton design load, 800 ton test								28,350	31,500	
	1500	Wet conditions, soft damp ground										
	1600	Requiring mats for crane, add								40%	40%	
	1700	Barge mounted driving rig, add								30%	30%	

023 600 | Driven Piles

			CREW	DAILY OUTPUT	LABOR-HOURS	UNIT	MAT.	LABOR	EQUIP.	TOTAL	TOTAL INCL O&P	
604	0010	**PILES, CONCRETE** 200 piles, 60′ long										604
	0020	unless specified otherwise, not incl. pile caps or mobilization										
	0050	Cast in place augered piles, no casing or reinforcing										
	0060	8″ diameter	B-43	540	.089	V.L.F.	2.03	2.09	3.29	7.41	9.10	
	0065	10″ diameter		480	.100		3.23	2.35	3.70	9.28	11.30	
	0070	12″ diameter		420	.114		4.55	2.69	4.23	11.47	13.85	
	0075	14″ diameter		360	.133		6.15	3.14	4.93	14.22	17.10	
	0080	16″ diameter		300	.160		8.25	3.76	5.90	17.91	21.50	

023 600 | Driven Piles

		023 600 Driven Piles	CREW	DAILY OUTPUT	LABOR-HOURS	UNIT	1999 BARE COSTS MAT.	LABOR	EQUIP.	TOTAL	TOTAL INCL O&P
604	0085	18″ diameter	B-43	240	.200	V.L.F.	10.25	4.71	7.40	22.36	26.50
	0100	Cast in place, thin wall shell pile, straight sided,									
	0110	not incl. reinforcing, 8″ diam., 16 ga., 5.8 lb./L.F.	B-19	700	.091	V.L.F.	3.81	2.52	2.05	8.38	10.60
	0200	10″ diameter, 16 ga. corrugated, 7.3 lb./L.F.		650	.098		4.99	2.72	2.21	9.92	12.40
	0300	12″ diameter, 16 ga. corrugated, 8.7 lb./L.F.		600	.107		6.50	2.94	2.40	11.84	14.60
	0400	14″ diameter, 16 ga. corrugated, 10.0 lb./L.F.		550	.116		7.60	3.21	2.61	13.42	16.60
	0500	16″ diameter, 16 ga. corrugated, 11.6 lb./L.F.		500	.128		9.35	3.53	2.87	15.75	19.25
	0800	Cast in place friction pile, 50′ long, fluted,									
	0810	tapered steel, 4000 psi concrete, no reinforcing									
	0900	12″ diameter, 7 ga.	B-19	600	.107	V.L.F.	11.65	2.94	2.40	16.99	20.50
	1000	14″ diameter, 7 ga.		560	.114		12.70	3.15	2.57	18.42	22
	1100	16″ diameter, 7 ga.		520	.123		14.95	3.40	2.76	21.11	25
	1200	18″ diameter, 7 ga.		480	.133		17.50	3.68	2.99	24.17	28.50
	1300	End bearing, fluted, constant diameter,									
	1320	4000 psi concrete, no reinforcing									
	1340	12″ diameter, 7 ga.	B-19	600	.107	V.L.F.	12.20	2.94	2.40	17.54	21
	1360	14″ diameter, 7 ga.		560	.114		15.20	3.15	2.57	20.92	25
	1380	16″ diameter, 7 ga.		520	.123		17.65	3.40	2.76	23.81	28
	1400	18″ diameter, 7 ga.		480	.133		19.50	3.68	2.99	26.17	31
	1500	For reinforcing steel, add				Lb.	.52			.52	.57
	1700	For ball or pedestal end, add	B-19	11	5.818	C.Y.	76	161	131	368	495
	1900	For lengths above 60′, concrete, add	″	11	5.818	″	79	161	131	371	495
	2000	For steel thin shell, pipe only				Lb.	.50			.50	.55
	2200	Precast, prestressed, 50′ long, 12″ diam., 2-3/8″ wall	B-19	720	.089	V.L.F.	16	2.45	2	20.45	24
	2300	14″ diameter, 2-1/2″ wall		680	.094		21	2.60	2.11	25.71	29.50
	2500	16″ diameter, 3″ wall		640	.100		29	2.76	2.25	34.01	38.50
	2600	18″ diameter, 3″ wall	B-19A	600	.107		36.50	2.94	2.35	41.79	47.50
	2800	20″ diameter, 3-1/2″ wall		560	.114		43.50	3.15	2.52	49.17	56
	2900	24″ diameter, 3-1/2″ wall		520	.123		51.50	3.40	2.71	57.61	65
	2920	36″ diameter, 4-1/2″ wall	B-19	400	.160		35	4.41	3.59	43	50
	2940	54″ diameter, 5″ wall		340	.188		50	5.20	4.23	59.43	68
	2960	66″ diameter, 6″ wall		220	.291		80	8.05	6.55	94.60	108
	3100	Precast, prestressed, 40′ long, 10″ thick, square		700	.091		7.70	2.52	2.05	12.27	14.90
	3200	12″ thick, square		680	.094		8.55	2.60	2.11	13.26	16
	3400	14″ thick, square		600	.107		9.80	2.94	2.40	15.14	18.25
	3500	Octagonal		640	.100		14	2.76	2.25	19.01	22.50
	3700	16″ thick, square		560	.114		15.60	3.15	2.57	21.32	25
	3800	Octagonal		600	.107		16.80	2.94	2.40	22.14	26
	4000	18″ thick, square	B-19A	520	.123		17.80	3.40	2.71	23.91	28
	4100	Octagonal	B-19	560	.114		19.90	3.15	2.57	25.62	30
	4300	20″ thick, square	B-19A	480	.133		28	3.68	2.94	34.62	40.50
	4400	Octagonal	B-19	520	.123		22	3.40	2.76	28.16	32.50
	4600	24″ thick, square	B-19A	440	.145		34.50	4.01	3.21	41.72	48
	4700	Octagonal	B-19	480	.133		31.50	3.68	2.99	38.17	44.50
	4750	Mobilization for 10,000 L.F. pile job, add		3,300	.019			.54	.44	.98	1.36
	4800	25,000 L.F. pile job, add		8,500	.008			.21	.17	.38	.53
	5000	Pressure grouted pin pile, 5″ diam., cased, up to 50 ton,									
	5040	End bearing, less than 20′	B-48	90	.622	V.L.F.	15.50	14.95	22	52.45	64
	5080	More than 40′		135	.415		14.60	10	14.65	39.25	47.50
	5120	Friction, loose sand and gravel		107	.523		15.50	12.60	18.45	46.55	57
	5160	Dense sand and gravel		135	.415		14.60	10	14.65	39.25	47.50
	5200	Uncased, up to 10 ton capacity, 20′		135	.415		4.75	10	14.65	29.40	37
608	0010	**PILES, STEEL** Not including mobilization or demobilization									
	0100	Step tapered, round, concrete filled									
	0110	8″ tip, 60 ton capacity, 30′ depth	B-19	760	.084	V.L.F.	5	2.32	1.89	9.21	11.40
	0120	60′ depth		740	.086		5.65	2.39	1.94	9.98	12.30

023 600 | Driven Piles

608		CREW	DAILY OUTPUT	LABOR-HOURS	UNIT	1999 BARE COSTS MAT.	LABOR	EQUIP.	TOTAL	TOTAL INCL O&P
0130	80′ depth	B-19	700	.091	V.L.F.	5.85	2.52	2.05	10.42	12.85
0150	10″ tip, 90 ton capacity, 30′ depth		700	.091		6.15	2.52	2.05	10.72	13.15
0160	60′ depth		690	.093		6.35	2.56	2.08	10.99	13.45
0170	80′ depth		670	.096		6.85	2.64	2.15	11.64	14.20
0190	12″ tip, 120 ton capacity, 30′ depth		660	.097		8.75	2.68	2.18	13.61	16.40
0200	60′ depth, 12″ diameter		630	.102		7.75	2.80	2.28	12.83	15.70
0210	80′ depth		590	.108		7.75	2.99	2.44	13.18	16.10
0250	"H" Sections, 50′ long, HP8 x 36		640	.100		10.60	2.76	2.25	15.61	18.65
0400	HP10 X 42		610	.105		10.95	2.89	2.36	16.20	19.40
0500	HP10 X 57		610	.105		14.85	2.89	2.36	20.10	23.50
0700	HP12 X 53		590	.108		13.90	2.99	2.44	19.33	23
0800	HP12 X 74	B-19A	590	.108		19.25	2.99	2.39	24.63	28.50
1000	HP14 X 73		540	.119		19	3.27	2.61	24.88	29.50
1100	HP14 X 89		540	.119		23.50	3.27	2.61	29.38	34
1300	HP14 X 102		510	.125		26.50	3.46	2.77	32.73	37.50
1400	HP14 X 117		510	.125		30.50	3.46	2.77	36.73	42
1600	Splice on standard points, not in leads, 8″ or 10″	1 Sswl	5	1.600	Ea.	55	49		104	151
1700	12″ or 14″		4	2		80	61		141	200
1900	Heavy duty points, not in leads, 10″ wide		4	2		85	61		146	206
2100	14″ wide		3.50	2.286		110	70		180	249
2600	Pipe piles, 50′ lg. 8″ diam., 29 lb. per L.F., no concrete	B-19	500	.128	V.L.F.	9.45	3.53	2.87	15.85	19.40
2700	Concrete filled		460	.139		10.05	3.84	3.12	17.01	21
2900	10″ diameter, 34 lb. per L.F., no concrete		500	.128		11.90	3.53	2.87	18.30	22
3000	Concrete filled		450	.142		13.15	3.92	3.19	20.26	24.50
3200	12″ diameter, 44 lb. per L.F., no concrete		475	.135		15.15	3.72	3.03	21.90	26
3300	Concrete filled		415	.154		17.15	4.25	3.46	24.86	29.50
3500	14″ diameter, 46 lb. per L.F., no concrete		430	.149		16.15	4.11	3.34	23.60	28
3600	Concrete filled		355	.180		19.05	4.97	4.05	28.07	33.50
3800	16″ diameter, 52 lb. per L.F., no concrete		385	.166		17.95	4.59	3.73	26.27	31.50
3900	Concrete filled		335	.191		22	5.25	4.29	31.54	38
4100	18″ diameter, 59 lb. per L.F., no concrete		355	.180		21.50	4.97	4.05	30.52	36
4200	Concrete filled		310	.206		28	5.70	4.64	38.34	45.50
4400	Splices for pipe piles, not in leads, 8″ diameter	1 Sswl	4.67	1.713	Ea.	31.50	52.50		84	131
4500	14″ diameter		3.79	2.111		41.50	64.50		106	164
4600	16″ diameter		3.03	2.640		51	81		132	204
4650	18″ diameter		4.50	1.778		67.50	54.50		122	174
4800	Points, standard, 8″ diameter		4.61	1.735		24.50	53		77.50	125
4900	14″ diameter		4.05	1.975		34	60.50		94.50	148
5000	16″ diameter		3.37	2.374		41	72.50		113.50	179
5050	18″ diameter		5	1.600		64.50	49		113.50	161
5200	Points, heavy duty, 10″ diameter		2.89	2.768		33	84.50		117.50	191
5300	14″ or 16″ diameter		2.02	3.960		52.50	121		173.50	280
5500	For reinforcing steel, add		1,150	.007	Lb.	.40	.21		.61	.83
5700	For thick wall sections, add				"	.45			.45	.50
6020	Steel pipe pile end plates, 8″ diameter	1 Sswl	14	.571	Ea.	35	17.50		52.50	70.50
6050	10″ diameter		14	.571		37	17.50		54.50	72.50
6100	12″ diameter		12	.667		39	20.50		59.50	80.50
6150	14″ diameter		10	.800		40.50	24.50		65	89.50
6200	16″ diameter		9	.889		43	27		70	97.50
6250	18″ diameter		8	1		48.50	30.50		79	110
6300	Steel pipe pile shoes, 8″ diameter		12	.667		36	20.50		56.50	77
6350	10″ diameter		12	.667		37.50	20.50		58	78.50
6400	12″ diameter		10	.800		40	24.50		64.50	89
6450	14″ diameter		9	.889		46	27		73	101
6500	16″ diameter		8	1		51	30.50		81.50	112
6550	18″ diameter		6	1.333		57	41		98	138

023 600 | Driven Piles

612

		Description	Crew	Daily Output	Labor-Hours	Unit	1999 Bare Costs Mat.	Labor	Equip.	Total	Total Incl O&P
612	0010	**PILES, WOOD** Friction or end bearing, not including (R023-620)									
	0050	mobilization or demobilization									
	0100	Untreated piles, up to 30′ long, 12″ butts, 8″ points	B-19	625	.102	V.L.F.	5.25	2.83	2.30	10.38	13
	0200	30′ to 39′ long, 12″ butts, 8″ points		700	.091		5.25	2.52	2.05	9.82	12.20
	0300	40′ to 49′ long, 12″ butts, 7″ points		720	.089		5.50	2.45	2	9.95	12.30
	0400	50′ to 59′ long, 13″butts, 7″ points		800	.080		5	2.21	1.80	9.01	11.10
	0500	60′ to 69′ long, 13″ butts, 7″ points		840	.076		5.65	2.10	1.71	9.46	11.55
	0600	70′ to 80′ long, 13″ butts, 6″ points		840	.076		6.25	2.10	1.71	10.06	12.25
	0800	Treated piles, 12 lb. per C.F.,									
	0810	friction or end bearing, ASTM class B									
	1000	Up to 30′ long, 12″ butts, 8″ points	B-19	625	.102	V.L.F.	9.45	2.83	2.30	14.58	17.55
	1100	30′ to 39′ long, 12″ butts, 8″ points		700	.091		9.35	2.52	2.05	13.92	16.70
	1200	40′ to 49′ long, 12″ butts, 7″ points		720	.089		8.95	2.45	2	13.40	16.10
	1300	50′ to 59′ long, 13″ butts, 7″ points		800	.080		10.90	2.21	1.80	14.91	17.60
	1400	60′ to 69′ long, 13″ butts, 6″ points	B-19A	840	.076		13.50	2.10	1.68	17.28	20
	1500	70′ to 80′ long, 13″ butts, 6″ points	"	840	.076		16.10	2.10	1.68	19.88	23
	1600	Treated piles, C.C.A., 2.5# per C.F.									
	1610	8″ butts, 10′ long	B-19	400	.160	V.L.F.	6	4.41	3.59	14	17.85
	1620	11′ to 16′ long		500	.128		6	3.53	2.87	12.40	15.60
	1630	17′ to 20′ long		575	.111		6	3.07	2.50	11.57	14.40
	1640	10″ butts, 10′ to 16′ long		500	.128		6.50	3.53	2.87	12.90	16.15
	1650	17′ to 20′ long		575	.111		6.50	3.07	2.50	12.07	14.95
	1660	21′ to 40′ long		700	.091		6.50	2.52	2.05	11.07	13.55
	1670	12″ butts, 10′ to 20′ long		575	.111		7	3.07	2.50	12.57	15.50
	1680	21′ to 35′ long		650	.098		7	2.72	2.21	11.93	14.60
	1690	36′ to 40′ long		700	.091		7	2.52	2.05	11.57	14.10
	1695	14″ butts. to 40′ long		700	.091		10	2.52	2.05	14.57	17.40
	1700	Boot for pile tip, minimum	1 Pile	27	.296	Ea.	14	8.05		22.05	29.50
	1800	Maximum		21	.381		42	10.35		52.35	64
	2000	Point for pile tip, minimum		20	.400		16	10.90		26.90	36.50
	2100	Maximum		15	.533		57.50	14.50		72	88.50
	2300	Splice for piles over 50′ long, minimum	B-46	35	1.371		40	34	1.52	75.52	102
	2400	Maximum		20	2.400		48	59	2.66	109.66	154
	2600	Concrete encasement with wire mesh and tube		331	.145	V.L.F.	7.50	3.58	.16	11.24	14.40
	2700	Mobilization for 10,000 L.F. pile job, add	B-19	3,300	.019			.54	.44	.98	1.36
	2800	25,000 L.F. pile job, add	"	8,500	.008			.21	.17	.38	.53

023 700 | Bored/Augered Piles

704

		Description	Crew	Daily Output	Labor-Hours	Unit	1999 Bare Costs Mat.	Labor	Equip.	Total	Total Incl O&P
704	0010	**PRESSURE INJECTED FOOTINGS** or Displacement Caissons									
	0100	incl. mobilization and demobilization, up to 50 miles									
	0200	Uncased shafts, 30 to 80 tons cap., 17″ diam., 10′ depth (R023-710)	B-44	88	.727	V.L.F.	12.20	19.85	8.90	40.95	56
	0300	25′ depth		165	.388		8.70	10.60	4.75	24.05	32.50
	0400	80-150 ton capacity, 22″ diameter, 10′ depth		80	.800		15.25	22	9.80	47.05	63.50
	0500	20′ depth		130	.492		12.20	13.45	6.05	31.70	42.50
	0700	Cased shafts, 10 to 30 ton capacity, 10-5/8″ diam., 20′ depth		175	.366		8.70	9.95	4.48	23.13	31
	0800	30′ depth		240	.267		8.15	7.25	3.26	18.66	24.50
	0850	30 to 60 ton capacity, 12″ diameter, 20′ depth		160	.400		12.20	10.90	4.90	28	37
	0900	40′ depth		230	.278		9.40	7.60	3.41	20.41	26.50
	1000	80 to 100 ton capacity, 16″ diameter, 20′ depth		160	.400		17.45	10.90	4.90	33.25	42.50
	1100	40′ depth		230	.278		16.25	7.60	3.41	27.26	34
	1200	110 to 140 ton capacity, 17-5/8″ diameter, 20′ depth		160	.400		18.75	10.90	4.90	34.55	44
	1300	40′ depth		230	.278		17.45	7.60	3.41	28.46	35.50
	1400	140 to 175 ton capacity, 19″ diameter, 20′ depth		130	.492		20.50	13.45	6.05	40	51.50
	1500	40′ depth		210	.305		18.75	8.30	3.73	30.78	38.50
	1700	Over 30′ long, L.F. cost tends to be lower									
	1900	Maximum depth is about 90′									

023 800 | Caissons

	023 800 Caissons	CREW	DAILY OUTPUT	LABOR-HOURS	UNIT	1999 BARE COSTS MAT.	LABOR	EQUIP.	TOTAL	TOTAL INCL O&P
804 0010	**CAISSONS** Incl. excav., concrete, 50 lbs. reinf. per C.Y., not (R023-810)									
0020	incl. mobilization, boulder removal, disposal									
0100	Open style, machine drilled, to 50′ deep, in stable ground, no									
0110	casings or ground water, 18″ diam., 0.065 C.Y./L.F.	B-43	200	.240	V.L.F.	4.82	5.65	8.90	19.37	24
0200	24″ diameter, 0.116 C.Y./L.F.		190	.253		8.65	5.95	9.35	23.95	29
0300	30″ diameter, 0.182 C.Y./L.F.		150	.320		13.50	7.55	11.85	32.90	39.50
0400	36″ diameter, 0.262 C.Y./L.F.		125	.384		19.45	9.05	14.20	42.70	51
0500	48″ diameter, 0.465 C.Y./L.F.		100	.480		34.50	11.30	17.75	63.55	75
0600	60″ diameter, 0.727 C.Y./L.F.		90	.533		54	12.55	19.75	86.30	101
0700	72″ diameter, 1.05 C.Y./L.F.		80	.600		78	14.10	22	114.10	133
0800	84″ diameter, 1.43 C.Y./L.F.		75	.640		106	15.05	23.50	144.55	167
1000	For bell excavation and concrete, add									
1020	4′ bell diameter, 24″ shaft, 0.444 C.Y.	B-43	20	2.400	Ea.	27	56.50	89	172.50	216
1040	6′ bell diameter, 30″ shaft, 1.57 C.Y.		5.70	8.421		96	198	310	604	760
1060	8′ bell diameter, 36″ shaft, 3.72 C.Y.		2.40	20		227	470	740	1,437	1,800
1080	9′ bell diameter, 48″ shaft, 4.48 C.Y.		2	24		273	565	890	1,728	2,150
1100	10′ bell diameter, 60″ shaft, 5.24 C.Y.		1.70	28.235		320	665	1,050	2,035	2,525
1120	12′ bell diameter, 72″ shaft, 8.74 C.Y.		1	48		535	1,125	1,775	3,435	4,275
1140	14′ bell diameter, 84″ shaft, 13.6 C.Y.		.70	68.571		830	1,625	2,525	4,980	6,225
1200	Open style, machine drilled, to 50′ deep, in wet ground, pulled									
1300	casing and pumping, 18″ diameter, 0.065 C.Y./L.F.	B-48	160	.350	V.L.F.	4.82	8.40	12.35	25.57	32
1400	24″ diameter, 0.116 C.Y./L.F.		125	.448		8.65	10.80	15.80	35.25	43.50
1500	30″ diameter, 0.182 C.Y./L.F.		85	.659		13.50	15.85	23.50	52.85	65
1600	36″ diameter, 0.262 C.Y./L.F.		60	.933		19.45	22.50	33	74.95	92.50
1700	48″ diameter, 0.465 C.Y./L.F.	B-49	55	1.600		34.50	40	46	120.50	152
1800	60″ diameter, 0.727 C.Y./L.F.		35	2.514		54	63	72	189	239
1900	72″ diameter, 1.05 C.Y./L.F.		30	2.933		78	73.50	84	235.50	295
2000	84″ diameter, 1.43 C.Y./L.F.		25	3.520		106	88.50	101	295.50	370
2100	For bell excavation and concrete, add									
2120	4′ bell diameter, 24″ shaft, 0.444 C.Y.	B-48	19.80	2.828	Ea.	27	68	100	195	245
2140	6′ bell diameter, 30″ shaft, 1.57 C.Y.		5.70	9.825		96	236	345	677	850
2160	8′ bell diameter, 36″ shaft, 3.72 C.Y.		2.40	23.333		227	560	825	1,612	2,025
2180	9′ bell diameter, 48″ shaft, 4.48 C.Y.	B-49	3.30	26.667		273	670	765	1,708	2,200
2200	10′ bell diameter, 60″ shaft, 5.24 C.Y.		2.80	31.429		320	790	900	2,010	2,600
2220	12′ bell diameter, 72″ shaft, 8.74 C.Y.		1.60	55		535	1,375	1,575	3,485	4,475
2240	14′ bell diameter, 84″ shaft, 13.6 C.Y.		1	88		830	2,200	2,525	5,555	7,200
2300	Open style, machine drilled, to 50′ deep, in soft rocks and									
2400	medium hard shales, 18″ diameter, 0.065 C.Y./L.F.	B-49	50	1.760	V.L.F.	4.82	44	50.50	99.32	131
2500	24″ diameter, 0.116 C.Y./L.F.		30	2.933		8.65	73.50	84	166.15	218
2600	30″ diameter, 0.182 C.Y./L.F.		20	4.400		13.50	110	126	249.50	330
2700	36″ diameter, 0.262 C.Y./L.F.		15	5.867		19.45	147	168	334.45	440
2800	48″ diameter, 0.465 C.Y./L.F.		10	8.800		34.50	221	253	508.50	665
2900	60″ diameter, 0.727 C.Y./L.F.		7	12.571		54	315	360	729	955
3000	72″ diameter, 1.05 C.Y./L.F.		6	14.667		78	370	420	868	1,125
3100	84″ diameter, 1.43 C.Y./L.F.		5	17.600		106	440	505	1,051	1,375
3200	For bell excavation and concrete, add									
3220	4′ bell diameter, 24″ shaft, 0.444 C.Y.	B-49	10.90	8.073	Ea.	27	203	232	462	605
3240	6′ bell diameter, 30″ shaft, 1.57 C.Y.		3.10	28.387		96	715	815	1,626	2,125
3260	8′ bell diameter, 36″ shaft, 3.72 C.Y.		1.30	67.692		227	1,700	1,950	3,877	5,050
3280	9′ bell diameter, 48″ shaft, 4.48 C.Y.		1.10	80		273	2,000	2,300	4,573	6,000
3300	10′ bell diameter, 60″ shaft, 5.24 C.Y.		.90	97.778		320	2,450	2,800	5,570	7,300
3320	12′ bell diameter, 72″ shaft, 8.74 C.Y.		.60	146		535	3,675	4,200	8,410	11,000
3340	14′ bell diameter, 84″ shaft, 13.6 C.Y.		.40	220		830	5,525	6,325	12,680	16,600
3600	For rock excavation, sockets, add, minimum		120	.733	C.F.		18.40	21	39.40	52
3650	Average		95	.926			23.50	26.50	50	65.50
3700	Maximum		48	1.833			46	52.50	98.50	131

023 | Tunneling, Piles & Caissons

023 800 | Caissons

			CREW	DAILY OUTPUT	LABOR-HOURS	UNIT	1999 BARE COSTS MAT.	LABOR	EQUIP.	TOTAL	TOTAL INCL O&P
804	3900	For 50′ to 100′ deep, add R023-810				V.L.F.				7%	7%
	4000	For 100′ to 150′ deep, add								25%	25%
	4100	For 150′ to 200′ deep, add				↓				30%	30%
	4200	For casings left in place, add				Lb.	.50			.50	.55
	4300	For other than 50 lb. reinf. per C.Y., add or deduct				"	.50			.50	.55
	4400	For steel "I" beam cores, add	B-49	8.30	10.602	Ton	960	266	305	1,531	1,800
	4500	Load and haul excess excavation, 2 miles	B-34B	178	.045	C.Y.		.99	2.48	3.47	4.25
	4600	For mobilization, 50 mile radius, rig to 36″	B-43	2	24	Ea.		565	890	1,455	1,850
	4650	Rig to 84″	B-48	1.75	32			770	1,125	1,895	2,450
	4700	For low headroom, add								50%	
	5000	Bottom inspection	1 Skwk	1.20	6.667	↓		187		187	295

024 | Railroad & Marine Work

024 520 | Railroad Trackwork

			CREW	DAILY OUTPUT	LABOR-HOURS	UNIT	1999 BARE COSTS MAT.	LABOR	EQUIP.	TOTAL	TOTAL INCL O&P
524	0010	**RAILROAD** Car bumpers, standard R024-520	B-14	2	24	Ea.	2,000	545	109	2,654	3,175
	0100	Heavy duty		2	24		3,775	545	109	4,429	5,150
	0200	Derails hand throw (sliding) R024-530		10	4.800		700	109	22	831	965
	0300	Hand throw with standard timbers, open stand & target		8	6	↓	755	136	27.50	918.50	1,075
	0400	Resurface and realign existing track		200	.240	L.F.		5.45	1.09	6.54	9.75
	0600	For crushed stone ballast, add	↓	500	.096	"	10.90	2.18	.44	13.52	15.90
	0800	Siding, yard spur, level grade									
	0808	Wood ties and ballast, 80 lb. rail	B-14	57	.842	L.F.	50.50	19.15	3.84	73.49	89.50
	0809	80 lb. relay rail		57	.842		38	19.15	3.84	60.99	75.50
	0812	90 lb. rail		57	.842		42	19.15	3.84	64.99	80.50
	0813	90 lb. relay rail		57	.842		42	19.15	3.84	64.99	80
	0820	100 lb. rail		57	.842		60	19.15	3.84	82.99	100
	0822	100 lb. relay rail		57	.842		43	19.15	3.84	65.99	81
	0830	110 lb. rail		57	.842		46	19.15	3.84	68.99	84.50
	0832	110 lb. relay rail	↓	57	.842	↓	45.50	19.15	3.84	68.49	84
	1002	Steel ties in concrete, incl. fasteners & plates									
	1003	80 lb. rails	B-14	22	2.182	L.F.	95	49.50	9.95	154.45	193
	1005	80 lb. relay rail		22	2.182		75	49.50	9.95	134.45	171
	1012	90 lb. rail		22	2.182		88	49.50	9.95	147.45	185
	1015	90 lb. relay rail		22	2.182		74	49.50	9.95	133.45	170
	1020	100 lb. rail		22	2.182		95	49.50	9.95	154.45	193
	1025	100 lb. relay rail		22	2.182		74	49.50	9.95	133.45	170
	1030	110 lb. rail		22	2.182		98	49.50	9.95	157.45	196
	1035	110 lb. relay rail		22	2.182	↓	82	49.50	9.95	141.45	178
	1200	Switch timber, for a #8 switch, pressure treated		3.70	12.973	M.B.F.	580	295	59	934	1,175
	1300	Complete set of timbers, 3.7 M.B.F. for #8 switch		1	48	Total	2,500	1,100	219	3,819	4,700
	1400	Ties, concrete, 8′-6″ long, 30″ O.C.		80	.600	Ea.	89.50	13.65	2.74	105.89	123
	1600	Wood, pressure treated, 6″ x 8″ x 8′-6″, C.L. lots		90	.533		27	12.15	2.43	41.58	51
	1700	L.C.L. lots		90	.533		28	12.15	2.43	42.58	52.50
	1900	Heavy duty, 7″ x 9″ x 8′-6″, C.L. lots		70	.686		29.50	15.60	3.13	48.23	60.50
	2000	L.C.L. lots	↓	70	.686	↓	29	15.60	3.13	47.73	60
	2200	Turnouts, #8, incl. 100 lb. rails, plates, bars, frog, switch pt.									
	2300	Timbers and ballast 6″ below bottom of tie	B-14	.50	96	Ea.	17,400	2,175	440	20,015	23,000
	2400	Wheel stops, fixed	↓	18	2.667	Pr.	430	60.50	12.15	502.65	585

024 520 | Railroad Trackwork

	Line	Description	Crew	Daily Output	Labor-Hours	Unit	1999 Bare Costs Mat.	Labor	Equip.	Total	Total Incl O&P	
524	2450	Hinged (R024-520)	B-14	14	3.429	Pr.	505	78	15.65	598.65	695	524

024 820 | Dredging

	Line	Description	Crew	Daily Output	Labor-Hours	Unit	1999 Bare Costs Mat.	Labor	Equip.	Total	Total Incl O&P	
824	0010	**DREDGING** Mobilization and demobilization., add to below, minimum	B-8	.53	120	Total		2,900	4,350	7,250	9,225	824
	0100	Maximum	″	.10	640	″		15,300	23,000	38,300	48,900	
	0300	Barge mounted clamshell excavation into scows,										
	0310	Dumped 20 miles at sea, minimum	B-57	310	.155	C.Y.		3.79	3.69	7.48	9.90	
	0400	Maximum	″	213	.225	″		5.50	5.35	10.85	14.45	
	0500	Barge mounted dragline or clamshell, hopper dumped,										
	0510	pumped 1000′ to shore dump, minimum	B-57	340	.141	C.Y.		3.46	3.37	6.83	9.05	
	0525	All pumping uses 2000 gallons of water per cubic yard										
	0600	Maximum	B-57	243	.198	C.Y.		4.84	4.71	9.55	12.65	
	1000	Hydraulic method, pumped 1000′ to shore dump, minimum		460	.104			2.56	2.49	5.05	6.70	
	1100	Maximum		310	.155			3.79	3.69	7.48	9.90	
	1400	Into scows dumped 20 miles, minimum		425	.113			2.77	2.69	5.46	7.25	
	1500	Maximum		243	.198			4.84	4.71	9.55	12.65	
	1600	For inland rivers and canals in South, deduct								30%	30%	

024 840 | Seawall & Bulkheads

	Line	Description	Crew	Daily Output	Labor-Hours	Unit	1999 Bare Costs Mat.	Labor	Equip.	Total	Total Incl O&P	
844	0010	**BULKHEADS** Reinforced concrete, include footing and tie-backs										844
	0020	Up to 6′ high, minimum	C-17C	28	2.964	L.F.	30.50	84.50	14.70	129.70	183	
	0060	Maximum		24.25	3.423		49	97.50	17	163.50	226	
	0100	12′ high, minimum		20	4.150		79.50	118	20.50	218	296	
	0160	Maximum		18.50	4.486		91.50	128	22.50	242	325	
	0180	Precast bulkhead, complete, including										
	0190	vertical and battered piles, face panels, and cap										
	0195	Using 16′ vertical piles				L.F.				225	260	
	0196	Using 20′ vertical piles				″				240	275	
	0200	Steel sheeting, with 4′ x 4′ x 8″ concrete deadmen, @ 10′ O.C.										
	0210	12′ high, shore driven	B-40	27	2.370	L.F.	53	65.50	71	189.50	245	
	0260	Barge driven	B-76	15	4.800	″	79.50	132	135	346.50	455	
	6000	Crushed stone placed behind bulkhead by clam bucket	B-12H	120	.133	C.Y.	16.35	3.56	4.78	24.69	28.50	
848	0010	**BULKHEADS, RESIDENTIAL CANAL**										848
	0020	Aluminum panel sheeting, incl. concrete cap and anchor										
	0030	Coarse compact sand, 4′-0″ high, 2′-0″ embedment	B-40	200	.320	L.F.	50	8.85	9.60	68.45	80	
	0040	3′-6″ embedment		140	.457		59	12.60	13.75	85.35	101	
	0060	6′-0″ embedment		90	.711		75	19.60	21.50	116.10	139	
	0120	6′-0″ high, 2′-6″ embedment		170	.376		55	10.40	11.30	76.70	90	
	0140	4′-0″ embedment		125	.512		65	14.15	15.40	94.55	112	
	0160	5′-6″ embedment		95	.674		85.50	18.60	20	124.10	147	
	0220	8′-0″ high, 3′-6″ embedment		140	.457		85	12.60	13.75	111.35	130	
	0240	5′-0″ embedment		100	.640		85	17.65	19.25	121.90	144	
	0420	Medium compact sand, 3′-0″ high, 2′-0″ embedment		235	.272		42	7.50	8.20	57.70	67.50	
	0440	4′-0″ embedment		150	.427		51.50	11.75	12.80	76.05	90.50	
	0460	5′-6″ embedment		115	.557		62.50	15.35	16.70	94.55	113	
	0520	5′-0″ high, 3′-6″ embedment		165	.388		60	10.70	11.65	82.35	96.50	
	0540	5′-0″ embedment		120	.533		69.50	14.70	16.05	100.25	118	
	0560	6′-6″ embedment		105	.610		90	16.80	18.30	125.10	147	
	0620	7′-0″ high, 4′-6″ embedment		135	.474		90	13.10	14.25	117.35	136	
	0640	6′-0″ embedment		110	.582		99.50	16.05	17.50	133.05	155	
	0720	Loose silty sand, 3′-0″ high, 3′-0″ embedment		205	.312		50	8.60	9.40	68	79.50	
	0740	4′-6″ embedment		155	.413		60.50	11.40	12.40	84.30	99	
	0760	6′-0″ embedment		125	.512		70.50	14.15	15.40	100.05	118	
	0820	4′-6″ high, 4′-6″ embedment		155	.413		65	11.40	12.40	88.80	104	

024 840 | Seawall & Bulkheads

		024 840 Seawall & Bulkheads	Crew	Daily Output	Labor-Hours	Unit	1999 Bare Costs Mat.	Labor	Equip.	Total	Total Incl O&P	
848	0840	6′-0″ embedment	B-40	125	.512	L.F.	75.50	14.15	15.40	105.05	123	848
	0860	7′-0″ embedment		115	.557		91	15.35	16.70	123.05	144	
	0920	6′-0″ high, 5′-6″ embedment		130	.492		90	13.60	14.80	118.40	138	
	0940	7′-0″ embedment		115	.557		99.50	15.35	16.70	131.55	153	

024 880 | Docks & Facilities

		024 880 Docks & Facilities	Crew	Daily Output	Labor-Hours	Unit	Mat.	Labor	Equip.	Total	Total Incl O&P	
884	0010	DOCKS, FIXED										884
	0020	Pile supported, treated wood,										
	0030	5′ x 20′ platform, minimum	F-3	115	.348	S.F.	7.25	9.65	3.86	20.76	27	
	0060	Maximum		75	.533		15.30	14.80	5.90	36	46.50	
	0100	6′ x 20′ platform, minimum		130	.308		6.40	8.55	3.41	18.36	24	
	0160	Maximum		85	.471		15.05	13.05	5.20	33.30	43	
	0200	8′ x 20′ platform, minimum		150	.267		6	7.40	2.96	16.36	21.50	
	0260	Maximum		100	.400		13.85	11.10	4.44	29.39	37.50	
	0420	5′ x 30′ platform, minimum		155	.258		6.75	7.15	2.86	16.76	22	
	0460	Maximum		100	.400		15.65	11.10	4.44	31.19	39.50	
	0500	For Greenhart lumber, add								40%		
	0550	Diagonal planking, add								25%		
	1000	Pipe supported dock, 1″ aluminum pipe										
	1020	Aluminum planks, galv. stl. framing, 2′ 8″ wide, minimum	F-3	320	.125	S.F.	15	3.46	1.39	19.85	23.50	
	1060	Maximum		120	.333		18.85	9.25	3.70	31.80	39	
	1100	5′ wide, minimum		250	.160		10	4.43	1.77	16.20	19.90	
	1160	Maximum		160	.250		11	6.95	2.77	20.72	26	
	1200	Wood deck and galv. steel framing										
	1220	3′ wide, minimum	F-3	320	.125	S.F.	20	3.46	1.39	24.85	29	
	1260	Maximum		120	.333		24	9.25	3.70	36.95	45	
	1300	6′ wide, minimum		260	.154		15	4.26	1.71	20.97	25	
	1360	Maximum		100	.400		18.45	11.10	4.44	33.99	42.50	
	1400	10′ wide, minimum		200	.200		15	5.55	2.22	22.77	27.50	
	1460	Maximum		80	.500		18.45	13.85	5.55	37.85	48	
	1800	1-1/2″ galv. steel pipe, treated wood dock and framing										
	1820	4′ wide, minimum	F-3	275	.145	S.F.	10	4.03	1.61	15.64	19.05	
	1860	Maximum		170	.235		16.35	6.50	2.61	25.46	31	
	1900	5′ wide, minimum		250	.160		8	4.43	1.77	14.20	17.70	
	1960	Maximum		160	.250		13.10	6.95	2.77	22.82	28.50	
	2000	6′ wide, minimum		225	.178		8	4.93	1.97	14.90	18.65	
	2060	Maximum		150	.267		11.35	7.40	2.96	21.71	27.50	
886	0010	DOCK ACCESSORIES										886
	0100	Cleats, aluminum, "S" type, 6″ long	1 Clab	8	1	Ea.	9.50	21.50		31	44.50	
	0140	10″ long		6.70	1.194		23	25.50		48.50	66	
	0180	15″ long		6	1.333		30.50	28.50		59	78.50	
	0400	Dock wheel for corners and piles, vinyl, 12″ diameter		4	2		90	43		133	167	
	1000	Electrical receptacle with circuit breaker,										
	1020	Pile mounted, double 30 amp, 125 volt	1 Elec	2	4	Unit	375	128		503	605	
	1060	Double 50 amp, 125/240 volt		1.60	5		505	160		665	795	
	1120	Free standing, add		4	2		100	64		164	206	
	1140	Double free standing, add		2.70	2.963		500	94.50		594.50	690	
	1160	Light, 2 louvered, with photo electric switch, add		8	1		80	32		112	136	
	1180	Telephone jack on stanchion		8	1		55	32		87	108	
	1300	Fender, Vinyl, 4″ high	1 Clab	160	.050	L.F.	2.50	1.07		3.57	4.44	
	1380	Corner piece		80	.100	Ea.	12	2.15		14.15	16.60	
	1400	Hose holder, cast aluminum		16	.500	″	25	10.75		35.75	44.50	
	1500	Ladder, aluminum, heavy duty										
	1520	Crown top, 5 to 7 step, minimum	1 Clab	5.30	1.509	Ea.	140	32.50		172.50	205	
	1560	Maximum		2	4		250	86		336	410	

024 880 | Docks & Facilities

			CREW	DAILY OUTPUT	LABOR-HOURS	UNIT	1999 BARE COSTS MAT.	LABOR	EQUIP.	TOTAL	TOTAL INCL O&P
886	1580	Bracket for portable clamp mounting	1 Clab	8	1	Ea.	90	21.50		111.50	133
	1800	Line holder, treated wood, small		16	.500		8	10.75		18.75	25.50
	1840	Large		13.30	.602		13.60	12.90		26.50	35.50
	2000	Mooring whip, fiberglass bolted to dock,									
	2020	1200 lb. boat	1 Clab	8.80	.909	Pr.	225	19.50		244.50	279
	2040	10,000 lb. boat		6.70	1.194		455	25.50		480.50	540
	2080	60,000 lb. boat		4	2		535	43		578	655
	2400	Shock absorbing tubing, vertical bumpers									
	2420	3" diam., vinyl, white	1 Clab	80	.100	L.F.	15	2.15		17.15	19.90
	2440	Polybutyl, clear		80	.100	"	16.65	2.15		18.80	21.50
	2480	Mounts, polybutyl		20	.400	Ea.	17.55	8.60		26.15	33
	2490	Deluxe		20	.400	"	39	8.60		47.60	56.50
888	0010	**DOCKS, FLOATING** including anchors									
	1000	Polystyrene flotation, minimum	F-3	200	.200	S.F.	10	5.55	2.22	17.77	22
	1040	Maximum		135	.296	"	13	8.20	3.29	24.49	31
	1100	Alternate method of figuring, minimum		1.13	35.398	Slip	2,500	980	395	3,875	4,700
	1140	Maximum		.70	57.143	"	3,575	1,575	635	5,785	7,125
	1200	Galv. steel frame and wood deck, 3' wide, minimum		320	.125	S.F.	8.50	3.46	1.39	13.35	16.25
	1240	Maximum		200	.200		13.70	5.55	2.22	21.47	26
	1300	4' wide, minimum		320	.125		8	3.46	1.39	12.85	15.70
	1340	Maximum		200	.200		12.80	5.55	2.22	20.57	25
	1500	8' wide, minimum		250	.160		6.75	4.43	1.77	12.95	16.35
	1540	Maximum		160	.250		10.25	6.95	2.77	19.97	25
	1700	Treated wood frames and deck, 3' wide, minimum		250	.160		8.80	4.43	1.77	15	18.60
	1740	Maximum		125	.320		30	8.85	3.55	42.40	51
	2000	Polyethylene drums, treated wood frame and deck									
	2100	6' wide, minimum	F-3	250	.160	S.F.	10	4.43	1.77	16.20	19.90
	2140	Maximum		125	.320		13.30	8.85	3.55	25.70	32.50
	2200	8' wide, minimum		233	.172		9	4.76	1.90	15.66	19.45
	2240	Maximum		120	.333		12.35	9.25	3.70	25.30	32
	2300	10' wide, minimum		200	.200		7.80	5.55	2.22	15.57	19.70
	2340	Maximum		110	.364		10.80	10.10	4.03	24.93	32
890	0010	**FLOATING DOCK ACCESSORIES**									
	0200	Dock connectors, stressed cables with rubber spacers									
	0220	25" long, 3' wide dock	1 Clab	2	4	Joint	400	86		486	575
	0240	5' wide dock		2	4		465	86		551	650
	0400	38" long, 4' wide dock		1.75	4.571		440	98		538	635
	0440	6' wide dock		1.45	5.517		500	118		618	735
	1000	Gangway, aluminum, one end rolling, no hand rails									
	1020	3' wide, minimum	1 Clab	67	.119	L.F.	75	2.56		77.56	86.50
	1040	Maximum		32	.250		86.50	5.35		91.85	103
	1100	4' wide, minimum		40	.200		79.50	4.29		83.79	94.50
	1140	Maximum		24	.333		91.50	7.15		98.65	112
	1180	For handrails, add					15			15	16.50
	2000	Pile guides, beads on stainless cable	1 Clab	4	2	Ea.	250	43		293	345
	2020	Rod type, 8" diameter piles, minimum		4	2		25	43		68	95
	2040	Maximum		2	4		36.50	86		122.50	175
	2100	10" to 14" diameter piles minimum		3.20	2.500		60	53.50		113.50	151
	2140	Maximum		1.75	4.571		105	98		203	270
	2200	Roller type, 4 rollers, minimum		4	2		225	43		268	315
	2240	Maximum		1.75	4.571		320	98		418	510
892	0010	**PIERS**, Municipal with 3" x 12" framing and 3" decking, wood piles									
	0020	and cross bracing, alternate bents battered	B-76	60	1.200	S.F.	61.50	33	34	128.50	160
	0200	Treated piles, not including mobilization									
	0210	50' long, 20 lb. creosote, shore driven	B-19	540	.119	V.L.F.	8.95	3.27	2.66	14.88	18.20

024 | Railroad & Marine Work

024 880 | Docks & Facilities

			CREW	DAILY OUTPUT	LABOR-HOURS	UNIT	1999 BARE COSTS MAT.	LABOR	EQUIP.	TOTAL	TOTAL INCL O&P
892	0220	Barge driven	B-76	320	.225	V.L.F.	8.95	6.20	6.35	21.50	27
	0230	2.5 lb. CCA, shore driven	B-19	540	.119		10	3.27	2.66	15.93	19.35
	0240	Barge driven	B-76	320	.225		10	6.20	6.35	22.55	28.50
	0250	30′ long, 20 lb. creosote, shore driven	B-19	540	.119		9.35	3.27	2.66	15.28	18.65
	0260	Barge driven	B-76	320	.225		9.35	6.20	6.35	21.90	27.50
	0270	2.5 lb. CCA, shore driven	B-19	540	.119		6	3.27	2.66	11.93	14.95
	0280	Barge driven	B-76	320	.225	↓	6	6.20	6.35	18.55	24
	0300	Mobilization, barge, by tug boat	B-83	25	.640	Mile		15.95	17.55	33.50	44
	0350	Standby time for shore pile driving crew				Hr.				400	500
	0360	Standby time for barge driving rig				″				550	650

025 | Paving & Surfacing

025 100 | Walk/Rd/Parkng Paving

			CREW	DAILY OUTPUT	LABOR-HOURS	UNIT	1999 BARE COSTS MAT.	LABOR	EQUIP.	TOTAL	TOTAL INCL O&P
104	0010	**ASPHALTIC CONCRETE PAVEMENT** for highways (R025-110)									
	0020	and large paved areas									
	0080	Binder course, 1-1/2″ thick	B-25	7,725	.011	S.Y.	2.03	.27	.23	2.53	2.91
	0120	2″ thick		6,345	.014		2.71	.33	.28	3.32	3.80
	0160	3″ thick		4,905	.018		4.02	.42	.36	4.80	5.50
	0200	4″ thick	↓	4,140	.021		5.40	.50	.43	6.33	7.20
	0300	Wearing course, 1″ thick	B-25B	10,575	.009		1.56	.22	.19	1.97	2.27
	0340	1-1/2″ thick		7,725	.012		2.38	.30	.26	2.94	3.37
	0380	2″ thick		6,345	.015		3.20	.36	.32	3.88	4.43
	0420	2-1/2″ thick		5,480	.018		3.95	.42	.37	4.74	5.40
	0460	3″ thick	↓	4,900	.020	↓	4.70	.47	.41	5.58	6.35
	0800	Alternate method of figuring paving costs									
	0810	Binder course, 1-1/2″ thick	B-25	630	.140	Ton	26.50	3.29	2.84	32.63	37.50
	0811	2″ thick		690	.128		26.50	3	2.59	32.09	37
	0812	3″ thick		800	.110		26.50	2.59	2.24	31.33	36
	0813	4″ thick	↓	850	.104		26.50	2.44	2.10	31.04	35.50
	0850	Wearing course, 1″ thick	B-25B	575	.167		29	4	3.54	36.54	42
	0851	1-1/2″ thick		630	.152		29	3.65	3.23	35.88	41
	0852	2″ thick		690	.139		29	3.33	2.95	35.28	40.50
	0853	2-1/2″ thick		745	.129		29	3.08	2.73	34.81	40
	0854	3″ thick	↓	800	.120	↓	29	2.87	2.54	34.41	39.50
	1000	Pavement replacement over trench, 2″ thick	B-37	90	.533	S.Y.	3.25	12.15	1.70	17.10	24.50
	1050	4″ thick		70	.686		6.45	15.60	2.18	24.23	34
	1080	6″ thick	↓	55	.873	↓	10.25	19.85	2.78	32.88	45.50
108	0010	**ASPHALTIC CONCRETE** plant mix (145 lb. per C.F.) (R025-110)				Ton	26			26	28.50
	0200	All weather patching mix, hot					31			31	34
	0250	Cold patch					36			36	39.50
	0300	Berm mix					31			31	34
	0400	Base mix					23.50			23.50	25.50
	0500	Binder mix					26			26	28.50
	0600	Sand or sheet mix					32			32	35
	2000	Reclaimed pavement in stockpile					12			12	13.20
	2100	Recycled pavement, at plant, ratio old: new, 70:30					23			23	25.50
	2120	Ratio old: new, 30:70				↓	27.50			27.50	30.50
112	0010	**CALCIUM CHLORIDE** Delivered, 100 lb. bags, truckload lots				Ton	400			400	440
	0200	Solution, 4 lb. flake per gallon, tank truck delivery				Gal.	.85			.85	.94

025 100 | Walk/Rd/Parkng Paving

			CREW	DAILY OUTPUT	LABOR-HOURS	UNIT	1999 BARE COSTS MAT.	LABOR	EQUIP.	TOTAL	TOTAL INCL O&P
116	0010	**COLD LAID ASPHALT PAVEMENT** 0.5 gal. asphalt/S.Y. per in. depth									
	0020	Well graded granular aggregate									
	0100	Blade mixed in windrows, spread & compacted 4″ course	B-90A	1,600	.035	S.Y.	3.80	.90	1.03	5.73	6.70
	0200	Traveling plant mixed in windrows, compacted 4″ course	B-90B	3,000	.016		3.80	.40	.49	4.69	5.35
	0300	Rotary plant mixed in place, compacted 4″ course	"	3,500	.014		3.80	.35	.42	4.57	5.20
	0400	Central stationary plant, mixed, compacted 4″ course	B-36	7,200	.006		7.60	.14	.16	7.90	8.75
120	0010	**CONCRETE PAVEMENT** Including joints, finishing, and curing									
	0020	Fixed form, 12′ pass, unreinforced, 6″ thick	B-26	3,000	.029	S.Y.	14.55	.71	.63	15.89	17.80
	0030	7″ thick		2,850	.031		18.35	.75	.67	19.77	22
	0100	8″ thick		2,750	.032		20.50	.77	.69	21.96	24.50
	0200	9″ thick		2,500	.035		23.50	.85	.76	25.11	28
	0300	10″ thick		2,100	.042		25.50	1.01	.90	27.41	30.50
	0400	12″ thick		1,800	.049		27	1.18	1.05	29.23	33
	0500	15″ thick		1,500	.059		30.50	1.42	1.26	33.18	37
	0510	For small irregular areas, add						100%		100%	
	0600	For continuous welded steel reinforcement over 10′ wide, add				S.Y.				3.59	
	0610	Under 10′ wide pass, add								6.80	
	0700	Finishing, broom finish small areas	2 Cefi	120	.133			3.49		3.49	5.20
	0730	Transverse expansion joints, incl. premolded bit. jt. filler	C-1	150	.213	L.F.	1.20	5.50		6.70	9.95
	0740	Transverse construction joint using bulkhead	"	73	.438	"	1.74	11.35		13.09	19.70
	0750	Longitudinal joint tie bars, grouted	B-23	70	.571	Ea.	2.70	12.50	28	43.20	53
	1000	Curing, with sprayed membrane by hand	2 Clab	1,500	.011	S.Y.	.22	.23		.45	.60
	1650	For integral coloring, see div. 033-126									
122	0010	**FINE GRADE** Area to be paved with grader, small area	B-11L	400	.040	S.Y.		1	1.44	2.44	3.12
	0100	Large area		2,000	.008			.20	.29	.49	.63
	0200	Grade subgrade for base course, roadways		3,500	.005			.11	.16	.27	.36
	0300	Fine grade, base course for paving, see div. 022-308									
	1020	For large parking lots	B-32C	5,000	.010	S.Y.		.24	.33	.57	.74
	1050	For small irregular areas	"	2,000	.024			.61	.83	1.44	1.85
	1100	Fine grade for slab on grade, machine	B-11L	1,040	.015			.38	.55	.93	1.20
	1150	Hand grading	B-18	700	.034			.76	.07	.83	1.27
	1200	Fine grade granular base for sidewalks and bikeways	B-62	1,200	.020			.47	.09	.56	.83
	2550	Hand grade select gravel	2 Clab	60	.267	C.S.F.		5.70		5.70	9
	3000	Hand grade select gravel, including compaction, 4″ deep	B-18	555	.043	S.Y.		.96	.09	1.05	1.60
	3100	6″ deep		400	.060			1.33	.13	1.46	2.23
	3120	8″ deep		300	.080			1.77	.17	1.94	2.97
	3300	Finishing grading slopes, gentle	B-11L	8,900	.002			.04	.06	.10	.14
	3310	Steep slopes	"	7,100	.002			.06	.08	.14	.18
124	0011	**PAVING** Asphaltic concrete, parking lots & driveways									
	0020	6″ stone base, 2″ binder course, 1″ topping	B-25C	9,000	.005	S.F.	1.13	.13	.17	1.43	1.64
	0300	Binder course, 1-1/2″ thick		35,000	.001		.23	.03	.04	.30	.36
	0400	2″ thick		25,000	.002		.30	.05	.06	.41	.47
	0500	3″ thick		15,000	.003		.47	.08	.10	.65	.75
	0600	4″ thick		10,800	.004		.62	.11	.14	.87	1.01
	0800	Sand finish course, 3/4″ thick		41,000	.001		.15	.03	.04	.22	.25
	0900	1″ thick		34,000	.001		.19	.03	.05	.27	.31
	1000	Fill pot holes, hot mix, 2″ thick	B-16	4,200	.008		.38	.17	.11	.66	.80
	1100	4″ thick		3,500	.009		.56	.20	.13	.89	1.08
	1120	6″ thick		3,100	.010		.75	.23	.14	1.12	1.35
	1140	Cold patch, 2″ thick	B-51	3,000	.016		.45	.35	.06	.86	1.11
	1160	4″ thick		2,700	.018		.86	.39	.06	1.31	1.62
	1180	6″ thick		1,900	.025		1.33	.55	.09	1.97	2.43
128	0010	**SIDEWALKS, DRIVEWAYS, & PATIOS** No base									
	0020	Asphaltic concrete, 2″ thick	B-37	720	.067	S.Y.	2.74	1.52	.21	4.47	5.60

025 100 | Walk/Rd/Parkng Paving

		025 100 Walk/Rd/Parkng Paving	CREW	DAILY OUTPUT	LABOR-HOURS	UNIT	1999 BARE COSTS MAT.	LABOR	EQUIP.	TOTAL	TOTAL INCL O&P
128	0100	2-1/2" thick	B-37	660	.073	S.Y.	3.47	1.65	.23	5.35	6.65
	0110	Bedding for brick or stone, mortar, 1" thick	D-1	300	.053	S.F.	.32	1.31		1.63	2.39
	0120	2" thick	"	200	.080		.79	1.97		2.76	3.93
	0130	Sand, 2" thick	B-18	8,000	.003		.16	.07	.01	.24	.29
	0140	4" thick	"	4,000	.006		.32	.13	.01	.46	.57
	0300	Concrete, 3000 psi, CIP, 6 x 6 - W1.4 x W1.4 mesh,									
	0310	broomed finish, no base, 4" thick	B-24	600	.040	S.F.	1.02	1		2.02	2.66
	0350	5" thick		545	.044		1.36	1.10		2.46	3.19
	0400	6" thick		510	.047		1.58	1.18		2.76	3.56
	0440	For other finishes, see Div. 033-450									
	0450	For bank run gravel base, 4" thick, add	B-18	2,500	.010	S.F.	.14	.21	.02	.37	.51
	0520	8" thick, add	"	1,600	.015		.28	.33	.03	.64	.87
	0550	Exposed aggregate finish, add to above, minimum	B-24	1,875	.013		.11	.32		.43	.61
	0600	Maximum		455	.053		.35	1.32		1.67	2.42
	0700	Patterned surface, add to above min.		1,200	.020			.50		.50	.77
	0710	Maximum		500	.048			1.20		1.20	1.85
	0800	For integral colors, see Div. 033-126									
	0850	Splash block, precast concrete	1 Clab	150	.053	Ea.	5.15	1.14		6.29	7.45
	0950	Concrete tree grate, 5' square	B-6	25	.960		265	22.50	8.75	296.25	335
	0960	Cast iron tree grate with frame, 2 piece, round, 5' diameter		25	.960		685	22.50	8.75	716.25	800
	0980	Square, 5' side		25	.960		700	22.50	8.75	731.25	815
	1000	Crushed stone, 1" thick, white marble	2 Clab	1,700	.009	S.F.	.18	.20		.38	.51
	1050	Bluestone		1,700	.009		.18	.20		.38	.52
	1070	Granite chips		1,700	.009		.16	.20		.36	.50
	1200	For 2" asphaltic conc base and tack coat, add to above	B-37	7,200	.007		.43	.15	.02	.60	.74
	1660	Limestone pavers, 3" thick	D-1	72	.222		5.75	5.45		11.20	14.85
	1670	4" thick		70	.229		7.60	5.60		13.20	17.15
	1680	5" thick		68	.235		9.55	5.80		15.35	19.50
	1700	Redwood, prefabricated, 4' x 4' sections	2 Carp	316	.051		4.53	1.38		5.91	7.15
	1750	Redwood planks, 1" thick, on sleepers	"	240	.067		3.17	1.82		4.99	6.35
	1830	1-1/2" thick	B-28	167	.144		3.49	3.64		7.13	9.60
	1840	2" thick		167	.144		4.50	3.64		8.14	10.70
	1850	3" thick		150	.160		6.65	4.06		10.71	13.70
	1860	4" thick		150	.160		8.70	4.06		12.76	16
	1870	5" thick		150	.160		11.05	4.06		15.11	18.60
	2100	River or beach stone, stock	B-1	18	1.333	Ton	23	29.50		52.50	72
	2150	Quarried	"	18	1.333	"	36.50	29.50		66	86.50
	2160	Load, dump, and spread stone with skid steer, 100' haul	B-62	24	1	C.Y.		23.50	4.64	28.14	41.50
	2165	200' haul		18	1.333			31	6.20	37.20	55.50
	2168	300' haul		12	2			46.50	9.30	55.80	82.50
	2170	Shale paver, 2-1/4" thick	D-1	200	.080	S.F.	2.50	1.97		4.47	5.80
	2200	Coarse washed sand bed, 1"	B-62	1,350	.018	S.Y.	.80	.42	.08	1.30	1.61
	2300	Tile thinset pavers, 3/8" thick	D-1	300	.053	S.F.	2.75	1.31		4.06	5.05
	2350	3/4" thick	"	280	.057	"	4.39	1.41		5.80	7
	2400	Wood rounds, cypress	B-1	175	.137	Ea.	8	3.03		11.03	13.55

025 150 | Unit Pavers

		025 150 Unit Pavers	CREW	DAILY OUTPUT	LABOR-HOURS	UNIT	MAT.	LABOR	EQUIP.	TOTAL	TOTAL INCL O&P
154	0010	**ASPHALT BLOCKS**, 6"x12"x1-1/4", w/bed & neopr. adhesive	D-1	135	.119	S.F.	3	2.92		5.92	7.85
	0100	3" thick		130	.123		4.20	3.03		7.23	9.35
	0300	Hexagonal tile, 8" wide, 1-1/4" thick		135	.119		3.20	2.92		6.12	8.05
	0400	2" thick		130	.123		4.48	3.03		7.51	9.65
	0500	Square, 8" x 8", 1-1/4" thick		135	.119		3	2.92		5.92	7.85
	0600	2" thick		130	.123		4.20	3.03		7.23	9.35
	0900	For exposed aggregate (ground finish) add					.50			.50	.55
	0910	For colors, add					.35			.35	.39

025 150 | Unit Pavers

		025 150 Unit Pavers	CREW	DAILY OUTPUT	LABOR-HOURS	UNIT	1999 BARE COSTS MAT.	LABOR	EQUIP.	TOTAL	TOTAL INCL O&P
158	0010	**BRICK PAVING** 4" x 8" x 1-1/2", without joints (4.5 brick/S.F.)	D-1	110	.145	S.F.	2.04	3.58		5.62	7.80
	0100	Grouted, 3/8" joint (3.9 brick/S.F.)		90	.178		2.43	4.37		6.80	9.45
	0200	4" x 8" x 2-1/4", without joints (4.5 bricks/S.F.)		110	.145		2.69	3.58		6.27	8.50
	0300	Grouted, 3/8" joint (3.9 brick/S.F.)		90	.178		2.48	4.37		6.85	9.55
	0400	4" x 8" x 2-1/4", dry, on edge (8/S.F.)		140	.114		4.70	2.81		7.51	9.50
	0450	Grouted, 3/8" joint (6.5/S.F.)	↓	85	.188		4.13	4.63		8.76	11.75
	0500	Bedding, asphalt, 3/4" thick	B-25	5,130	.017		.26	.40	.35	1.01	1.29
	0540	Course washed sand bed, 1" thick	B-18	5,000	.005		.18	.11	.01	.30	.38
	0580	Mortar, 1" thick	D-1	300	.053		.36	1.31		1.67	2.44
	0620	2" thick		200	.080		.36	1.97		2.33	3.46
	2000	Brick pavers, laid on edge, 7.2 per S.F.		70	.229		2.06	5.60		7.66	11
	2500	For 4" thick concrete bed and joints, add	↓	595	.027		.75	.66		1.41	1.86
	2800	For steam cleaning, add	A-1	950	.008	↓	.05	.18	.07	.30	.42
162	0010	**PRECAST CONCRETE PAVING SLABS**									
	0710	Precast concrete patio blocks, 2-3/8" thick, colors, 8" x 16"	D-1	265	.060	S.F.	.95	1.49		2.44	3.36
	0715	12" x 12"		300	.053		1.24	1.31		2.55	3.40
	0720	16" x 16"		335	.048		1.36	1.17		2.53	3.32
	0730	24" x 24"		510	.031		1.69	.77		2.46	3.06
	0740	Green, 8" x 16"	↓	265	.060		1.23	1.49		2.72	3.66
	0750	Exposed local aggregate, natural	2 Bric	250	.064		4.63	1.77		6.40	7.85
	0800	Colors		250	.064		4.75	1.77		6.52	7.95
	0850	Exposed granite or limestone aggregate		250	.064		5.50	1.77		7.27	8.80
	0900	Exposed white tumblestone aggregate	↓	250	.064	↓	2.92	1.77		4.69	5.95
166	0010	**STONE PAVERS**									
	1100	Flagging, bluestone, irregular, 1" thick,	D-1	81	.198	S.F.	1.85	4.86		6.71	9.60
	1110	1-1/2" thick		90	.178		2.18	4.37		6.55	9.20
	1120	Pavers, 1/2" thick		110	.145		3.08	3.58		6.66	8.95
	1130	3/4" thick		95	.168		3.92	4.14		8.06	10.75
	1140	1" thick		81	.198		4.20	4.86		9.06	12.15
	1150	Snapped random rectangular, 1" thick		92	.174		2.80	4.28		7.08	9.75
	1200	1-1/2" thick		85	.188		3.37	4.63		8	10.90
	1250	2" thick		83	.193		3.92	4.74		8.66	11.65
	1300	Slate, natural cleft, irregular, 3/4" thick		92	.174		1.80	4.28		6.08	8.65
	1310	1" thick		85	.188		2.10	4.63		6.73	9.50
	1350	Random rectangular, gauged, 1/2" thick		105	.152		3.90	3.75		7.65	10.15
	1400	Random rectangular, butt joint, gauged, 1/4" thick	↓	150	.107		4.19	2.62		6.81	8.70
	1450	For sand rubbed finish, add				↓	2.50			2.50	2.75
	1500	For interior setting, add								25%	25%
	1550	Granite blocks, 3-1/2" x 3-1/2" x 3-1/2"	D-1	92	.174	S.F.	5.25	4.28		9.53	12.45
	1560	4" x 4" x 4"		95	.168		5.55	4.14		9.69	12.55
	1600	4" to 12" long, 3" to 5" wide, 3" to 5" thick		98	.163		4.38	4.02		8.40	11.05
	1650	6" to 15" long, 3" to 6" wide, 3" to 5" thick	↓	105	.152		2.34	3.75		6.09	8.40
	2000	Granite paver, sawn with thermal finish, white, 4" x 4" x 2"	D-2	80	.550		17.55	13.95		31.50	41
	2005	4" x 4" x 3"		80	.550		22	13.95		35.95	45.50
	2010	4" x 4" x 4"		80	.550		29.50	13.95		43.45	54
	2015	4" x 8" x 2"		170	.259		16.85	6.55		23.40	29
	2020	4" x 8" x 3"		170	.259		21	6.55		27.55	33.50
	2025	4" x 8" x 4"		170	.259		26	6.55		32.55	39.50
	2030	8" x 8" x 2"		300	.147		14.25	3.72		17.97	21.50
	2035	8" x 8" x 3"		300	.147		17.70	3.72		21.42	25.50
	2040	8" x 8" x 4"		300	.147		21.50	3.72		25.22	29.50
	2045	12" x 12" x 2"		725	.061		12.55	1.54		14.09	16.20
	2050	12" x 12" x 3"		725	.061		15.30	1.54		16.84	19.20
	2055	12" x 12" x 4"	↓	725	.061	↓	18.20	1.54		19.74	22.50

025 250 | Curbs

			CREW	DAILY OUTPUT	LABOR-HOURS	UNIT	1999 BARE COSTS MAT.	LABOR	EQUIP.	TOTAL	TOTAL INCL O&P
254	0010	**CURBS** Asphaltic, machine formed, 8″ wide, 6″ high, 40 L.F./ton	B-27	1,000	.032	L.F.	.58	.70	.07	1.35	1.83
	0100	8″ wide, 8″ high, 30 L.F. per ton		900	.036		.67	.78	.08	1.53	2.05
	0150	Asphaltic berm, 12″ W, 3″-6″ H, 35 L.F./ton, before pavement		700	.046		.80	1	.10	1.90	2.58
	0200	12″ W, 1-1/2″ to 4″ H, 60 L.F. per ton, laid with pavement	B-2	1,050	.038		.49	.83		1.32	1.85
	0300	Concrete, wood forms, 6″ x 18″, straight	C-2A	500	.096		2.11	2.54		4.65	6.30
	0400	6″ x 18″, radius		200	.240		2.22	6.35		8.57	12.35
	0410	Steel forms, 6″ x 18″, straight		700	.069		2.57	1.82		4.39	5.65
	0411	6″ x 18″, radius		400	.120		3.46	3.18		6.64	8.75
	0415	Machine formed, 6″ x 18″, straight	B-69A	2,000	.024		3.04	.57	.23	3.84	4.47
	0416	6″ x 18″, radius	″	900	.053		3.17	1.27	.50	4.94	6
	0421	Curb and gutter, straight									
	0422	with 6″ high curb and 6″ thick gutter, wood forms									
	0430	24″ wide, .055 C.Y. per L.F.	C-2A	375	.128	L.F.	10.25	3.39		13.64	16.60
	0435	30″ wide, .066 C.Y. per L.F.		340	.141		11.15	3.74		14.89	18.10
	0440	Steel forms, 24″ wide, straight		700	.069		4.95	1.82		6.77	8.30
	0441	Radius		300	.160		4.95	4.24		9.19	12.05
	0442	30″ wide, straight		700	.069		5.95	1.82		7.77	9.40
	0443	Radius		300	.160		5.95	4.24		10.19	13.15
	0445	Machine formed, 24″ wide, straight	B-69A	2,000	.024		4.95	.57	.23	5.75	6.60
	0446	Radius		900	.053		4.95	1.27	.50	6.72	7.95
	0447	30″ wide, straight		2,000	.024		5.95	.57	.23	6.75	7.70
	0448	Radius		900	.053		5.95	1.27	.50	7.72	9.05
	0451	Median mall, 2′ x 9″ high, straight		2,200	.022		5.20	.52	.21	5.93	6.75
	0452	Radius	B-69B	900	.053		5.20	1.27	1.01	7.48	8.75
	0453	4′ x 9″ high, straight		2,000	.024		10.40	.57	.45	11.42	12.85
	0454	Radius		800	.060		10.40	1.42	1.13	12.95	14.90
	0550	Precast, 6″ x 18″, straight	B-29	700	.080		6.25	1.86	.91	9.02	10.80
	0600	6″ x 18″, radius	″	325	.172		7.75	4	1.96	13.71	16.90
	1000	Granite, split face, straight, 5″ x 16″	D-13	500	.096		10	2.51	.78	13.29	15.75
	1100	6″ x 18″	″	450	.107		13.15	2.79	.87	16.81	19.75
	1300	Radius curbing, 6″ x 18″, over 10′ radius	B-29	260	.215		16.10	5	2.45	23.55	28
	1400	Corners, 2′ radius		80	.700	Ea.	54	16.25	7.95	78.20	94
	1600	Edging, 4-1/2″ x 12″, straight		300	.187	L.F.	5	4.34	2.12	11.46	14.60
	1800	Curb inlets, (guttermouth) straight		41	1.366	Ea.	120	31.50	15.55	167.05	199
	2000	Indian granite (belgian block)									
	2100	Jumbo, 10-1/2″x7-1/2″x4″, grey	D-1	150	.107	L.F.	1.75	2.62		4.37	6
	2150	Pink		150	.107		2.15	2.62		4.77	6.45
	2200	Regular, 9″x4-1/2″x4-1/2″, grey		160	.100		1.70	2.46		4.16	5.70
	2250	Pink		160	.100		2	2.46		4.46	6.05
	2300	Cubes, 4″x4″x4″, grey		175	.091		1.65	2.25		3.90	5.30
	2350	Pink		175	.091		1.75	2.25		4	5.45
	2400	6″x6″x6″, pink		155	.103		3.60	2.54		6.14	7.90
	2500	Alternate pricing method for indian granite									
	2550	Jumbo, 10-1/2″x7-1/2″x4″ (30lb), grey				Ton	100			100	110
	2600	Pink					125			125	138
	2650	Regular, 9″x4-1/2″x4-1/2″ (20lb), grey					120			120	132
	2700	Pink					140			140	154
	2750	Cubes, 4″x4″x4″ (5lb), grey					200			200	220
	2800	Pink					225			225	248
	2850	6″x6″x6″ (25lb), pink					140			140	154
	2900	For pallets, add					15			15	16.50
258	0010	**EDGING**									
	0050	Aluminum alloy, including stakes, 1/8″ x 4″, mill finish	B-1	390	.062	L.F.	1.75	1.36		3.11	4.07
	0051	Black paint		390	.062		2.03	1.36		3.39	4.37
	0052	Black anodized		390	.062		2.34	1.36		3.70	4.72

025 250 | Curbs

			CREW	DAILY OUTPUT	LABOR-HOURS	UNIT	1999 BARE COSTS MAT.	LABOR	EQUIP.	TOTAL	TOTAL INCL O&P
258	0060	3/16" x 4", mill finish	B-1	380	.063	L.F.	2.52	1.40		3.92	4.97
	0061	Black paint		380	.063		2.91	1.40		4.31	5.40
	0062	Black anodized		380	.063		3.40	1.40		4.80	5.95
	0070	1/8" x 5-1/2" mill finish		370	.065		2.54	1.43		3.97	5.05
	0071	Black paint		370	.065		3.01	1.43		4.44	5.55
	0072	Black anodized		370	.065		3.45	1.43		4.88	6.05
	0080	3/16" x 5-1/2" mill finish		360	.067		3.40	1.47		4.87	6.05
	0081	Black paint		360	.067		3.83	1.47		5.30	6.55
	0082	Black anodized	↓	360	.067		4.43	1.47		5.90	7.20
	0100	Brick, set horizontally, 1-1/2 bricks per L.F.	D-1	370	.043		.90	1.06		1.96	2.64
	0150	Set vertically, 3 bricks per L.F.	"	135	.119		1.90	2.92		4.82	6.60
	0200	Corrugated aluminum, roll, 4" wide	1 Carp	650	.012		.28	.34		.62	.84
	0250	6" wide	"	550	.015	↓	.35	.40		.75	1.01
	0300	Concrete, cast in place, see 025-254									
	0350	Granite, 5" x 16", straight	B-29	300	.187	L.F.	10	4.34	2.12	16.46	20
	0400	Polyethylene grass barrier, 5" x 1/8"	D-1	400	.040		.70	.98		1.68	2.30
	0410	5" x 1/4"		400	.040		.80	.98		1.78	2.42
	0420	5" x 5/32"		400	.040		.93	.98		1.91	2.55
	0430	6" x 3/32"		400	.040		.75	.98		1.73	2.35
	0500	Precast scallops, green, 2" x 8" x 16"		400	.040		1.10	.98		2.08	2.74
	0550	2" x 8" x 16" other than green	↓	400	.040		.88	.98		1.86	2.50
	0600	Railroad ties, 6" x 8"	2 Carp	170	.094		2.30	2.57		4.87	6.55
	0650	7" x 9"		136	.118		2.55	3.21		5.76	7.85
	0750	2" x 4"	↓	330	.048		2.67	1.32		3.99	5
	0800	Steel edge strips, incl. stakes, 1/4" x 5"	B-1	390	.062		2.75	1.36		4.11	5.15
	0850	3/16" x 4"	"	390	.062		2.17	1.36		3.53	4.53
	0900	Hardwood, pressure treated, 4" x 6"	2 Carp	250	.064		1.80	1.75		3.55	4.73
	0940	6" x 6"		200	.080		2.57	2.18		4.75	6.25
	0980	6" x 8"		170	.094		2.93	2.57		5.50	7.25
	1000	Pine, pressure treated, 1" x 4"		500	.032		.35	.87		1.22	1.76
	1040	2" x 4"		330	.048		.59	1.32		1.91	2.73
	1080	4" x 6"		250	.064		1.96	1.75		3.71	4.91
	1100	6" x 6"		200	.080		2.94	2.18		5.12	6.65
	1140	6" x 8"	↓	170	.094	↓	3.85	2.57		6.42	8.25

025 300 | Athletic Pave/Surfacing

			CREW	DAILY OUTPUT	LABOR-HOURS	UNIT	MAT.	LABOR	EQUIP.	TOTAL	TOTAL INCL O&P
304	0010	**RUNNING TRACK** Asphalt, incl base, 3" thick	B-37	300	.160	S.Y.	10.35	3.64	.51	14.50	17.60
	0100	Surface, latex rubber system, 3/8" thick, black	B-20	125	.192		4.65	4.67		9.32	12.45
	0150	Colors		125	.192		8.25	4.67		12.92	16.45
	0300	Urethane rubber system, 3/8" thick, black		120	.200		12.40	4.86		17.26	21.50
	0400	Color coating	↓	115	.209	↓	13.50	5.10		18.60	23
308	0010	**TENNIS COURT** Asphalt, incl. base, 2-1/2" thick, one court	B-37	450	.107	S.Y.	9.25	2.43	.34	12.02	14.30
	0200	Two courts		675	.071		7	1.62	.23	8.85	10.50
	0300	Clay courts		360	.133		26.50	3.03	.42	29.95	34
	0400	Pulverized natural greenstone with 4" base, fast dry		250	.192		26	4.37	.61	30.98	36
	0800	Rubber-acrylic base resilient pavement	↓	600	.080		26	1.82	.25	28.07	32
	1000	Colored sealer, acrylic emulsion, 3 coats	2 Clab	800	.020		3.73	.43		4.16	4.78
	1100	3 coat, 2 colors	"	900	.018		4.60	.38		4.98	5.65
	1200	For preparing old courts, add	1 Clab	825	.010	↓		.21		.21	.33
	1400	Posts for nets, 3-1/2" diameter with eye bolts	B-1	3.40	7.059	Pr.	138	156		294	395
	1500	With pulley & reel		3.40	7.059	"	189	156		345	455
	1700	Net, 42' long, nylon thread with binder		50	.480	Ea.	170	10.60		180.60	204
	1800	All metal	↓	6.50	3.692	"	287	81.50		368.50	445
	2000	Paint markings on asphalt, 2 coats	1 Pord	1.78	4.494	Court	58.50	112		170.50	235
	2200	Complete court with fence, etc., asphaltic conc., minimum	B-37	.20	240	↓	11,700	5,450	765	17,915	22,200

025 300 | Athletic Pave/Surfacing

			CREW	DAILY OUTPUT	LABOR-HOURS	UNIT	1999 BARE COSTS MAT.	LABOR	EQUIP.	TOTAL	TOTAL INCL O&P
308	2300	Maximum	B-37	.16	300	Court	14,200	6,825	955	21,980	27,400
	2800	Clay courts, minimum		.20	240		13,600	5,450	765	19,815	24,400
	2900	Maximum		.16	300		16,700	6,825	955	24,480	30,200

025 400 | Synthetic Surfacing

			CREW	DAILY OUTPUT	LABOR-HOURS	UNIT	MAT.	LABOR	EQUIP.	TOTAL	TOTAL INCL O&P
404	0010	**TURF, ARTIFICIAL** Not including asphalt base or drainage, but									
	0020	including cushion pad, over 50,000 S.F.									
	0200	1/2" pile and 5/16" cushion pad, standard	C-17	3,200	.025	S.F.	5.50	.71		6.21	7.15
	0300	Deluxe		2,560	.031		6.50	.89		7.39	8.55
	0500	1/2" pile and 5/8" cushion pad, standard		2,844	.028		8	.80		8.80	10.05
	0600	Deluxe		2,327	.034		8.75	.98		9.73	11.20
	0700	Acrylic emulsion texture coat, sand filled	B-1	5,200	.005	S.Y.	.80	.10		.90	1.04
	0710	Rubber filled	"	5,200	.005	"	.90	.10		1	1.15
	0720	Acrylic emulsion color coat									
	0730	Brown, red, or tan	B-1	6,000	.004	S.Y.	.40	.09		.49	.58
	0740	Green		6,000	.004		.40	.09		.49	.58
	0750	Blue		6,000	.004		.35	.09		.44	.53
	0760	Rubber acrylic cushion base coat		2,600	.009		1.50	.20		1.70	1.97
	0800	For asphaltic concrete base, 2-1/2" thick,									
	0900	with 6" crushed stone sub-base, add	B-25	12,000	.007	S.F.	1.07	.17	.15	1.39	1.61

025 450 | Surfacing

			CREW	DAILY OUTPUT	LABOR-HOURS	UNIT	MAT.	LABOR	EQUIP.	TOTAL	TOTAL INCL O&P
454	0010	**SURFACE TREATMENT**									
	3000	Pavement overlay, polypropylene									
	3040	6 oz. per S.Y., ideal conditions	B-63	10,000	.004	S.Y.	1	.09	.01	1.10	1.25
	3080	Adverse conditions		1,000	.040		1.34	.90	.11	2.35	3
	3120	4 oz. per S.Y., ideal conditions		10,000	.004		.70	.09	.01	.80	.92
	3160	Adverse conditions		1,000	.040		.90	.90	.11	1.91	2.52
	3200	Tack coat, emulsion, .05 gal per S.Y., 1000 S.Y	B-45	2,500	.006		.15	.16	.30	.61	.75
	3240	10,000 S.Y.		10,000	.002		.12	.04	.07	.23	.27
	3280	.15 gal per S.Y., 1000 S.Y.		2,500	.006		.41	.16	.30	.87	1.03
	3320	10,000 S.Y.		10,000	.002		.33	.04	.07	.44	.50
	5000	Reclamation, pulverizing and blending with existing base									
	5040	Aggregate base, 4" thick pavement, over 15,000 S.Y.	B-73	2,400	.027	S.Y.		.69	1.42	2.11	2.64
	5080	5,000 S.Y. to 15,000 S.Y.		2,200	.029			.76	1.55	2.31	2.88
	5120	8" thick pavement, over 15,000 S.Y.		2,200	.029			.76	1.55	2.31	2.88
	5160	5,000 S.Y. to 15,000 S.Y.		2,000	.032			.83	1.71	2.54	3.16
	5180	Add for mobilization and demobilization				Ea.				1,600	
	5200	Cold planing & cleaning, 1" to 3" asphalt pavmt., over 25,000 S.Y.	B-71	6,000	.009	S.Y.		.23	.66	.89	1.08
	5280	5,000 S.Y. to 10,000 S.Y.	"	4,000	.014	"		.35	.99	1.34	1.62
	5291	See also concrete pavement, div. 025-120									
	5300	Asphalt pavement removal from conc. base, no haul									
	5320	Rip, load & sweep 1" to 3"	B-70	8,000	.007	S.Y.		.17	.14	.31	.43
	5330	3" to 6" deep	"	5,000	.011			.28	.23	.51	.68
	5340	Profile grooving, asphalt pavement load & sweep, 1" deep	B-71	12,500	.004			.11	.32	.43	.52
	5350	3" deep		9,000	.006			.15	.44	.59	.72
	5360	6" deep		5,000	.011			.28	.79	1.07	1.30
	5400	Mixing material in windrow, 180 H.P. grader	B-11L	9,400	.002	C.Y.		.04	.06	.10	.14
	5450	See also 025-116									
	5500	Recycle asphalt pavement at site									
	5520	Remove, rejuvenate and spread 4" deep	B-72	2,500	.026	S.Y.	2	.64	2.88	5.52	6.35
	5521	6" deep	"	2,000	.032	"	3	.81	3.60	7.41	8.50
	6010	For subbase treatment, see Div. 022-412									

	025 450 | Surfacing	CREW	DAILY OUTPUT	LABOR-HOURS	UNIT	1999 BARE COSTS MAT.	LABOR	EQUIP.	TOTAL	TOTAL INCL O&P	
458	0010 **SEALCOATING** 2 coat coal tar pitch emulsion over 10,000 S.Y.	B-45	5,000	.003	S.Y.	.43	.08	.15	.66	.76	458
	0030 1000 to 10,000 S.Y.	"	3,000	.005		.43	.13	.25	.81	.96	
	0100 Under 1000 S.Y.	B-1	1,050	.023		.43	.51		.94	1.28	
	0300 Petroleum resistant, over 10,000 S.Y.	B-45	5,000	.003		.50	.08	.15	.73	.83	
	0320 1000 to 10,000 S.Y.	"	3,000	.005		.50	.13	.25	.88	1.03	
	0400 Under 1000 S.Y.	B-1	1,050	.023		.50	.51		1.01	1.35	
	0600 Non-skid pavement renewal, over 10,000 S.Y.	B-45	5,000	.003		.60	.08	.15	.83	.94	
	0620 1000 to 10,000 S.Y.	"	3,000	.005		.60	.13	.25	.98	1.14	
	0700 Under 1000 S.Y.	B-1	1,050	.023		.60	.51		1.11	1.46	
	0800 Prepare and clean surface for above	A-2	8,545	.003	↓		.06	.02	.08	.11	
	1000 Hand seal asphalt curbing	B-1	4,420	.005	L.F.	.30	.12		.42	.52	
	1900 Asphalt surface treatment, single course, small area										
	1901 0.30 gal/S.Y. asphalt material, 20#/S.Y. aggregate	B-91	5,000	.013	S.Y.	.70	.32	.30	1.32	1.60	
	1910 Roadway or large area		10,000	.006		.64	.16	.15	.95	1.12	
	1950 Asphalt surface treatment, dbl. course for small area		3,000	.021		1.30	.54	.50	2.34	2.81	
	1960 Roadway or large area		6,000	.011		1.17	.27	.25	1.69	1.97	
	1980 Asphalt surface treatment,single course, for shoulders		7,500	.009		.75	.22	.20	1.17	1.38	
	2080 Sand sealing, sharp sand, asphalt emulsion, small area		10,000	.006		.50	.16	.15	.81	.96	
	2120 Roadway or large area	↓	18,000	.004	↓	.43	.09	.08	.60	.70	
	3000 Sealing random cracks, min 1/2" wide, to 1-1/2", 1,000 L.F.	B-77	2,800	.014	L.F.	.40	.31	.14	.85	1.09	
	3040 10,000 L.F.		4,000	.010	"	.31	.22	.10	.63	.79	
	3080 Alternate method, 1,000 L.F.		200	.200	Gal.	9	4.38	2	15.38	18.95	
	3120 10,000 L.F.	↓	325	.123	"	7.25	2.70	1.23	11.18	13.55	
	3200 Multi-cracks (flooding), 1 coat, small area	B-92	460	.070	S.Y.	1.92	1.53	.77	4.22	5.35	
	3240 Large area		2,850	.011		1.70	.25	.12	2.07	2.40	
	3280 2 coat, small area		230	.139		6	3.05	1.55	10.60	13.10	
	3320 Large area		1,425	.022	↓	5.45	.49	.25	6.19	7.05	
	3360 Alternate method, small area		115	.278	Gal.	8.50	6.10	3.09	17.69	22.50	
	3400 Large area	↓	715	.045	"	8	.98	.50	9.48	10.90	
	3600 Waterproofing, membrane, tar and fabric, small area	B-63	233	.172	S.Y.	4.80	3.88	.48	9.16	11.90	
	3640 Large area		1,435	.028		4.42	.63	.08	5.13	5.95	
	3680 Preformed rubberized asphalt, small area		100	.400		6.50	9.05	1.11	16.66	22.50	
	3720 Large area	↓	367	.109	↓	5.90	2.46	.30	8.66	10.65	
459	0010 **SLURRY SEAL**										459
	0100 Slurry seal, type I, 8 lbs agg./S.Y., 1 coat, small or irregular area	B-90	2,800	.023	S.Y.	.70	.53	.48	1.71	2.11	
	0150 Roadway or large area		10,000	.006		.70	.15	.13	.98	1.15	
	0200 Type II, 12 lbs aggregate/S.Y., 2 coats,small or irregular area		2,000	.032		1.40	.75	.67	2.82	3.42	
	0250 Roadway or large area		8,000	.008		1.40	.19	.17	1.76	2.01	
	0300 Type III, 20 lbs aggregate/S.Y., 2 coats, small or irregular area		1,800	.036		1.65	.83	.74	3.22	3.91	
	0350 Roadway or large area		6,000	.011		1.65	.25	.22	2.12	2.44	
	0400 Slurry seal, thermoplastic coal-tar, type I, small or irregular area		2,400	.027		1.45	.62	.56	2.63	3.17	
	0450 Roadway or large area		8,000	.008		1.45	.19	.17	1.81	2.07	
	0500 Type II, small or irregular area		2,400	.027		1.85	.62	.56	3.03	3.61	
	0550 Roadway or large area	↓	7,800	.008	↓	1.85	.19	.17	2.21	2.53	
	0600 Average mobilization cost				Ea.				3,500	3,500	
	025 500 | Bridges										
504	0010 **BRIDGES** Pedestrian, spans over streams, roadways, etc.										504
	0020 including erection, not including foundations										
	0050 Precast concrete, complete in place, 8' wide, 60' span	E-2	215	.260	S.F.	25	7.75	4.71	37.46	46.50	
	0100 100' span		185	.303		27.50	9	5.45	41.95	52	
	0150 120' span		160	.350		30	10.40	6.30	46.70	57.50	
	0200 150' span		145	.386		31	11.50	7	49.50	61.50	
	0300 Steel, trussed or arch spans, compl. in place, 8' wide, 40' span		320	.175		35	5.20	3.16	43.36	51	
	0400 50' span	↓	395	.142	↓	31.50	4.22	2.56	38.28	44.50	

025 500 | Bridges

			CREW	DAILY OUTPUT	LABOR-HOURS	UNIT	1999 BARE COSTS MAT.	LABOR	EQUIP.	TOTAL	TOTAL INCL O&P
504	0500	60′ span	E-2	465	.120	S.F.	31.50	3.59	2.18	37.27	43
	0600	80′ span		570	.098		37.50	2.92	1.78	42.20	48
	0700	100′ span		465	.120		52.50	3.59	2.18	58.27	66.50
	0800	120′ span		365	.153		66.50	4.57	2.77	73.84	84
	0900	150′ span		310	.181		70.50	5.40	3.26	79.16	90.50
	1000	160′ span		255	.220		70.50	6.55	3.97	81.02	93.50
	1100	10′ wide, 80′ span		640	.087		36	2.60	1.58	40.18	46
	1200	120′ span		415	.135		46.50	4.02	2.44	52.96	61
	1300	150′ span		445	.126		52.50	3.75	2.27	58.52	66.50
	1400	200′ span	↓	205	.273		56	8.15	4.94	69.09	81
	1600	Wood, laminated type, complete in place, 80′ span	C-12	203	.236		30	6.40	2.18	38.58	45.50
	1700	130′ span	″	153	.314	↓	31	8.45	2.90	42.35	51
508	0010	**BRIDGES, HIGHWAY**									
	0020	Structural steel, rolled beams	E-5	8.50	9.412	Ton	1,200	284	129	1,613	1,975
	0500	Built up, plate girders	E-6	10.50	12.190	″	1,500	370	119	1,989	2,425
	1000	Concrete in place, no reinforcing, abutment footings	C-17B	30	2.733	C.Y.	85	78	9.55	172.55	227
	1100	Walls, stems and wing walls		20	4.100		114	117	14.35	245.35	325
	1200	Decks		17	4.824		189	137	16.90	342.90	445
	1300	Sidewalks and parapets	↓	11	7.455	↓	300	212	26	538	695
	2000	Reinforcing, in place	4 Rodm	3	10.667	Ton	675	325		1,000	1,300
	2050	Galvanized coated		3	10.667		845	325		1,170	1,500
	2100	Epoxy coated	↓	3	10.667	↓	1,250	325		1,575	1,925
	3000	Expansion dams, steel, double upset 4″ x 8″ angles welded to									
	3010	double 8″ x 8″ angles, 1-3/4″ compression seal	C-22	30	1.400	L.F.	230	43	2.28	275.28	330
	3040	Double 8″ x 8″ angles only, 1-3/4″ compression seal		35	1.200		173	36.50	1.96	211.46	255
	3050	Galvanized		35	1.200		213	36.50	1.96	251.46	299
	3060	Double 8″ x 6″ angles only, 1-3/4″ compression seal		35	1.200		132	36.50	1.96	170.46	210
	3100	Double 10″ channels, 1-3/4″ compression seal	↓	35	1.200		121	36.50	1.96	159.46	198
	3420	For 3″ compression seal, add					30			30	33
	3440	For double slotted extrusions with seal strip, add				↓	82			82	90
	3490	For galvanizing, add				Lb.	.50			.50	.55
	4000	Approach railings, steel, galv. pipe, 2 line	C-22	140	.300	L.F.	65	9.20	.49	74.69	88
	4200	Bridge railings, steel, galv. pipe, 3 line w/screen		85	.494		145	15.10	.81	160.91	187
	4220	4 line w/screen		75	.560		175	17.15	.91	193.06	224
	4300	Aluminum, pipe, 3 line w/screen	↓	95	.442	↓	65	13.55	.72	79.27	96
	8000	For structural excavation, see div. 022-250									
	8010	For dewatering, see div. 021-414									

025 800 | Pavement Marking

			CREW	DAILY OUTPUT	LABOR-HOURS	UNIT	MAT.	LABOR	EQUIP.	TOTAL	TOTAL INCL O&P
804	0010	**LINES ON PAV'T** Acrylic waterborne, white or yellow, 4″ wide	B-78	20,000	.002	L.F.	.09	.05	.03	.17	.20
	0200	6″ wide		11,000	.004		.09	.10	.05	.24	.30
	0500	8″ wide		10,000	.005		.11	.10	.05	.26	.34
	0600	12″ wide		4,000	.012	↓	.20	.26	.13	.59	.77
	0620	Arrows or gore lines		2,300	.021	S.F.	.50	.46	.22	1.18	1.51
	0640	Temporary paint, white or yellow	↓	15,000	.003	L.F.	.15	.07	.03	.25	.32
	0660	Removal	1 Clab	300	.027			.57		.57	.90
	0680	Temporary tape	2 Clab	1,500	.011		1.12	.23		1.35	1.59
	0710	Thermoplastic, white or yellow, 4″ wide	B-79	15,000	.003		.56	.06	.05	.67	.76
	0730	6″ wide		14,000	.003		.81	.06	.05	.92	1.04
	0740	8″ wide		12,000	.003		1.09	.07	.06	1.22	1.37
	0750	12″ wide		6,000	.007	↓	1.62	.15	.11	1.88	2.14
	0760	Arrows		660	.061	S.F.	1.50	1.33	1.03	3.86	4.87
	0770	Gore lines		2,500	.016		1	.35	.27	1.62	1.95
	0780	Letters	↓	660	.061	↓	1.25	1.33	1.03	3.61	4.60
	0782	Thermoplastic material				Ton	800			800	880

025 | Paving & Surfacing

025 800 | Pavement Marking

		025 800 | Pavement Marking	CREW	DAILY OUTPUT	LABOR-HOURS	UNIT	1999 BARE COSTS MAT.	LABOR	EQUIP.	TOTAL	TOTAL INCL O&P	
804	0784	Glass beads, add				M.L.F.	7.50			7.50	8.25	804
	0790	Layout of pavement marking	A-2	25,000	.001	L.F.		.02	.01	.03	.04	
	0800	Parking stall, paint, white	B-78	440	.109	Stall	1.84	2.38	1.16	5.38	7.05	
	1000	Street letters and numbers	"	1,600	.030	S.F.	.50	.65	.32	1.47	1.93	
	1100	Pavement marking letter, 6"	2 Pord	400	.040	Ea.	1.02	1		2.02	2.64	
	1110	12" letter		272	.059		1.43	1.46		2.89	3.81	
	1120	24" letter		160	.100		2.04	2.49		4.53	6.05	
	1130	36" letter		84	.190		3.37	4.74		8.11	10.95	
	1140	42" letter		84	.190		10.20	4.74		14.94	18.45	
	1150	72" letter		40	.400		29	9.95		38.95	47	
	1200	Handicap symbol		40	.400		18.35	9.95		28.30	35.50	
810	0010	**TRAFFIC CONTROL DEVICES**										810
	0100	Traffic channelizing pavement markers, layout only	A-7	2,000	.012	Ea.		.29		.29	.44	
	0110	13" x 7-1/2" x 2-1/2" high, non-plowable install	2 Clab	96	.167		16.50	3.58		20.08	24	
	0200	8" x 8"x 3-1/4" high, non-plowable, install		96	.167		18	3.58		21.58	25.50	
	0230	4" x 4" x 3/4" high, non-plowable, install		120	.133		2	2.86		4.86	6.70	
	0240	9-1/4" x 5-7/8" x 1/4" high, plowable, concrete pav't	A-2A	70	.343		12	7.40	4.04	23.44	29	
	0250	9-1/4" x 5-7/8" x 1/4" high, plowable, asphalt pav't	"	120	.200		2.30	4.31	2.36	8.97	11.80	
	0300	Barrier and curb delineators,reflectorized, 2" x 4"	2 Clab	150	.107		1.30	2.29		3.59	5.05	
	0310	3" x 5"	"	150	.107		2.65	2.29		4.94	6.50	
	0500	Rumble strip, polycarbonate										
	0510	24" x 3-1/2" x 1/2" high	2 Clab	50	.320	Ea.	5	6.85		11.85	16.30	

026 | Piped Utilities

026 010 | Piped Utilities

		026 010 | Piped Utilities	CREW	DAILY OUTPUT	LABOR-HOURS	UNIT	1999 BARE COSTS MAT.	LABOR	EQUIP.	TOTAL	TOTAL INCL O&P	
012	0010	**BEDDING** For pipe and conduit, not incl. compaction										012
	0050	Crushed or screened bank run gravel	B-6	150	.160	C.Y.	6.90	3.74	1.46	12.10	15	
	0100	Crushed stone 3/4" to 1/2"		150	.160		17.75	3.74	1.46	22.95	27	
	0200	Sand, dead or bank		150	.160		3.75	3.74	1.46	8.95	11.55	
	0500	Compacting bedding in trench	A-1	90	.089			1.91	.76	2.67	3.84	
014	0010	**EXCAVATION AND BACKFILL** See division 022-204 & 254										014
	0100	Hand excavate and trim for pipe bells after trench excavation										
	0200	8" pipe	1 Clab	155	.052	L.F.		1.11		1.11	1.74	
	0300	18" pipe		130	.062	"		1.32		1.32	2.08	
	0400	Underground tape, detectable aluminum, 2"		150	.053	C.L.F.	3.70	1.14		4.84	5.85	
	0500	6"		140	.057	"	9.25	1.23		10.48	12.15	

026 050 | Manholes & Cleanouts

		026 050 | Manholes & Cleanouts	CREW	DAILY OUTPUT	LABOR-HOURS	UNIT	MAT.	LABOR	EQUIP.	TOTAL	TOTAL INCL O&P	
054	0010	**UTILITY VAULTS** Precast concrete, 6" thick										054
	0040	4' x 6' x 6' high, I.D.	B-13	2	28	Ea.	1,025	650	275	1,950	2,425	
	0050	5' x 10' x 6' high, I.D.		2	28		1,250	650	275	2,175	2,700	
	0100	6' x 10' x 6' high, I.D.		2	28		1,300	650	275	2,225	2,725	
	0150	5' x 12' x 6' high, I.D.		2	28		1,375	650	275	2,300	2,825	
	0200	6' x 12' x 6' high, I.D.		1.80	31.111		1,550	725	305	2,580	3,150	
	0250	6' x 13' x 6' high, I.D.		1.50	37.333		2,025	870	365	3,260	4,000	
	0300	8' x 14' x 7' high, I.D.		1	56		2,200	1,300	550	4,050	5,050	

026 050 | Manholes & Cleanouts

			CREW	DAILY OUTPUT	LABOR-HOURS	UNIT	1999 BARE COSTS MAT.	LABOR	EQUIP.	TOTAL	TOTAL INCL O&P
054	0350	Hand hole, precast concrete, 1-1/2" thick									
	0400	1'-0" x 2'-0" x 1'-9", I.D., light duty	B-1	4	6	Ea.	330	133		463	575
	0450	4'-6" x 3'-2" x 2'-0", O.D., heavy duty	B-6	3	8		870	187	73	1,130	1,325
	0460	Meter pit, 4' x 4', 4' deep		2	12		545	280	109	934	1,150
	0470	6' deep		1.60	15		775	350	137	1,262	1,550
	0480	8' deep		1.40	17.143		1,025	400	156	1,581	1,925
	0490	10' deep		1.20	20		1,300	465	182	1,947	2,350
	0500	15' deep		1	24		1,900	560	219	2,679	3,200
	0510	6' x 6', 4' deep		1.40	17.143		860	400	156	1,416	1,725
	0520	6' deep		1.20	20		1,300	465	182	1,947	2,350
	0530	8' deep		1	24		1,725	560	219	2,504	3,000
	0540	10' deep		.80	30		2,150	700	274	3,124	3,775
	0550	15' deep		.60	40		3,275	935	365	4,575	5,450

026 100 | Pipe & Fittings

			CREW	DAILY OUTPUT	LABOR-HOURS	UNIT	1999 BARE COSTS MAT.	LABOR	EQUIP.	TOTAL	TOTAL INCL O&P
104	0010	**PIPE INSULATION**									
	0100	Calcium silicate, 1" thick, 4" diameter	Q-14	150	.107	L.F.	1.33	2.92		4.25	6.20
	0120	5" diameter		145	.110		1.46	3.02		4.48	6.50
	0140	6" diameter		140	.114		1.53	3.13		4.66	6.75
	0160	1-1/2" thick, 8" diameter		130	.123		2.70	3.37		6.07	8.40
	0180	10" diameter		125	.128		3.38	3.51		6.89	9.35
	0200	12" diameter		120	.133		4.05	3.65		7.70	10.35
	0240	16" diameter		100	.160		4.72	4.38		9.10	12.30
	0400	2" thick, 4" diameter		125	.128		2.66	3.51		6.17	8.60
	0440	6" diameter		115	.139		3.32	3.81		7.13	9.80
	0460	8" diameter		100	.160		3.99	4.38		8.37	11.50
	0480	10" diameter		95	.168		4.65	4.61		9.26	12.55
	0500	12" diameter		90	.178		5.30	4.87		10.17	13.70
	0540	16" diameter		80	.200		6.65	5.50		12.15	16.15
	0600	24" diameter		55	.291		9.30	7.95		17.25	23
	0700	3" thick, 4" diameter		95	.168		4.59	4.61		9.20	12.50
	0740	6" diameter		90	.178		5.60	4.87		10.47	14
	0760	8" diameter		85	.188		6.65	5.15		11.80	15.65
	0780	10" diameter		75	.213		8.10	5.85		13.95	18.35
	0800	12" diameter		70	.229		8.95	6.25		15.20	19.95
	0840	16" diameter		60	.267		10.80	7.30		18.10	23.50
	0900	24" diameter		40	.400		15.90	10.95		26.85	35
	1020	Fiberglass, 1" thick, 4" diameter		150	.107		1.14	2.92		4.06	5.95
	1060	6" diameter		120	.133		1.45	3.65		5.10	7.50
	1080	8" diameter		100	.160		2.01	4.38		6.39	9.30
	1100	10" diameter		90	.178		2.34	4.87		7.21	10.45
	1120	12" diameter		80	.200		2.79	5.50		8.29	11.90
	1160	16" diameter		70	.229		3.24	6.25		9.49	13.65
	1220	24" diameter		60	.267		5.15	7.30		12.45	17.45
	1300	1-1/2" thick, 4" diameter		130	.123		1.63	3.37		5	7.25
	1340	6" diameter		100	.160		1.79	4.38		6.17	9.05
	1360	8" diameter		80	.200		2.19	5.50		7.69	11.25
	1380	10" diameter		70	.229		2.54	6.25		8.79	12.90
	1400	12" diameter		65	.246		2.98	6.75		9.73	14.20
	1440	16" diameter		55	.291		3.88	7.95		11.83	17.10
	1500	24" diameter		40	.400		5.10	10.95		16.05	23.50
	1600	2" thick, 4" diameter		110	.145		2.49	3.99		6.48	9.20
	1640	6" diameter		80	.200		2.94	5.50		8.44	12.10
	1660	8" diameter		60	.267		3.73	7.30		11.03	15.90
	1680	10" diameter		55	.291		4.53	7.95		12.48	17.85

Important: See the Reference Section for critical supporting data - Reference Nos., Crews, & City Cost Indexe

026 100 | Pipe & Fittings

		026 100 Pipe & Fittings	CREW	DAILY OUTPUT	LABOR-HOURS	UNIT	1999 BARE COSTS MAT.	LABOR	EQUIP.	TOTAL	TOTAL INCL O&P	
104	1700	12″ diameter	Q-14	50	.320	L.F.	4.87	8.75		13.62	19.50	104
	1740	16″ diameter		40	.400		5.90	10.95		16.85	24	
	1760	18″ diameter		35	.457		6.70	12.55		19.25	27.50	
	1800	24″ diameter	↓	25	.640	↓	8.15	17.55		25.70	37.50	
108	0010	**PIPE INTERNAL CLEANING & INSPECTION**										108
	0100	Cleaning, pressure pipe systems										
	0120	Pig method, lengths 1000′ to 10,000′										
	0140	4″ diameter thru 24″ diameter, minimum				L.F.				2	2.30	
	0160	Maximum				″				5	6	
	6000	Sewage/sanitary systems										
	6100	Power rodder with header & cutters										
	6110	Mobilization charge, minimum				Total				300	350	
	6120	Mobilization charge, maximum				″				700	800	
	6140	4″ diameter				L.F.				1	1.15	
	6150	6″ diameter								1.25	1.45	
	6160	8″ diameter								1.50	1.75	
	6170	10″ diameter								1.75	2	
	6180	12″ diameter								1.90	2.20	
	6190	14″ diameter								2	2.30	
	6200	16″ diameter								2.25	2.60	
	6210	18″ diameter								2.50	2.85	
	6220	20″ diameter								2.60	3	
	6230	24″ diameter								2.90	3.35	
	6240	30″ diameter								3.50	4	
	6250	36″ diameter								4	4.60	
	6260	48″ diameter								4.30	4.95	
	6270	60″ diameter								5.60	6.45	
	6280	72″ diameter				↓				6.20	7.15	
	9000	Inspection, television camera with film										
	9060	500 linear feet				Total				1,000	1,150	

026 400 | Valves & Cocks

		026 400 Valves & Cocks	CREW	DAILY OUTPUT	LABOR-HOURS	UNIT	MAT.	LABOR	EQUIP.	TOTAL	TOTAL INCL O&P	
404	0010	**PIPING VALVES** Water distribution, see also div. 151-900										404
	3000	Butterfly valves with boxes, cast iron, mech. jt.										
	3100	4″ diameter	B-20	6	4	Ea.	330	97.50		427.50	520	
	3140	6″ diameter	″	5	4.800		380	117		497	600	
	3180	8″ diameter	B-21	4	7		460	175	34	669	815	
	3300	10″ diameter		3.50	8		675	200	39	914	1,100	
	3340	12″ diameter		3	9.333		885	234	45.50	1,164.50	1,375	
	3400	14″ diameter		2	14		1,525	350	68	1,943	2,300	
	3440	16″ diameter		2	14		2,025	350	68	2,443	2,850	
	3460	18″ diameter		1.50	18.667		2,550	465	90.50	3,105.50	3,625	
	3480	20″ diameter		1	28		3,200	700	136	4,036	4,750	
	3500	24″ diameter	↓	.50	56	↓	4,450	1,400	272	6,122	7,400	
	3600	With lever operator										
	3610	4″ diameter	B-20	6	4	Ea.	225	97.50		322.50	400	
	3614	6″ diameter	″	5	4.800		315	117		432	530	
	3616	8″ diameter	B-21	4	7		455	175	34	664	810	
	3618	10″ diameter		3.50	8		730	200	39	969	1,175	
	3620	12″ diameter		3	9.333		1,350	234	45.50	1,629.50	1,925	
	3622	14″ diameter		2	14		2,175	350	68	2,593	3,000	
	3624	16″ diameter		2	14		2,950	350	68	3,368	3,850	
	3626	18″ diameter		1.50	18.667		3,375	465	90.50	3,930.50	4,550	
	3628	20″ diameter		1	28		4,500	700	136	5,336	6,200	
	3630	24″ diameter	↓	.50	56	↓	4,850	1,400	272	6,522	7,825	
	3700	Check valves, flanged										

026 400 | Valves & Cocks

			CREW	DAILY OUTPUT	LABOR-HOURS	UNIT	1999 BARE COSTS MAT.	LABOR	EQUIP.	TOTAL	TOTAL INCL O&P	
404	3710	4″ diameter	B-20	6	4	Ea.	250	97.50		347.50	430	404
	3712	5″ diameter		5	4.800		425	117		542	655	
	3714	6″ diameter		5	4.800		425	117		542	650	
	3716	8″ diameter	B-21	4	7		865	175	34	1,074	1,250	
	3718	10″ diameter		3.50	8		1,375	200	39	1,614	1,875	
	3720	12″ diameter		3	9.333		2,125	234	45.50	2,404.50	2,750	
	3722	14″ diameter		2	14		3,375	350	68	3,793	4,350	
	3724	16″ diameter		2	14		4,650	350	68	5,068	5,725	
	3726	18″ diameter		1.50	18.667		6,875	465	90.50	7,430.50	8,400	
	3728	20″ diameter		1	28		8,000	700	136	8,836	10,000	
	3730	24″ diameter		.50	56		11,500	1,400	272	13,172	15,200	
	3800	Gate valves, C.I., 250 PSI, mechanical joint, w/boxes										
	3810	4″ diameter	B-21A	8	5	Ea.	147	133	49	329	420	
	3814	6″ diameter		6.80	5.882		179	156	57.50	392.50	500	
	3816	8″ diameter		5.60	7.143		258	190	70	518	655	
	3818	10″ diameter		4.80	8.333		420	222	81.50	723.50	890	
	3820	12″ diameter		4.08	9.804		500	261	96	857	1,050	
	3822	14″ diameter	B-21	2	14		675	350	68	1,093	1,375	
	3824	16″ diameter		1	28		795	700	136	1,631	2,125	
	3826	18″ diameter		.80	35		980	875	170	2,025	2,625	
	3828	20″ diameter		.80	35		1,250	875	170	2,295	2,925	
	3830	24″ diameter		.50	56		1,525	1,400	272	3,197	4,175	
	3831	30″ diameter		.35	80		1,850	2,000	390	4,240	5,575	
	3832	36″ diameter		.30	93.333		2,200	2,325	455	4,980	6,575	
	3880	Sleeve, for tapping mains, 8″ x 4″, add					430			430	475	
	3900	Globe valves, flanged, iron body, class 125										
	3910	4″ diameter	B-20	10	2.400	Ea.	340	58.50		398.50	465	
	3912	5″ diameter		10	2.400		560	58.50		618.50	705	
	3914	6″ diameter		9	2.667		590	65		655	745	
	3916	8″ diameter	B-21	6	4.667		1,100	117	22.50	1,239.50	1,400	
	3918	10″ diameter		5	5.600		1,800	140	27	1,967	2,225	
	3920	12″ diameter		4	7		2,525	175	34	2,734	3,075	
	3922	14″ diameter		3	9.333		2,700	234	45.50	2,979.50	3,400	
	3924	16″ diameter		2	14		4,200	350	68	4,618	5,250	
	3926	18″ diameter		.80	35		4,625	875	170	5,670	6,625	
	3928	20″ diameter		.60	46.667		7,200	1,175	226	8,601	10,000	
	3930	24″ diameter		.50	56		8,525	1,400	272	10,197	11,900	
	5010	Piping, site utility, fittings, corporations, brass, 3/4″ diameter	1 Plum	19	.421		15.50	13.75		29.25	38	
	5040	1″ diameter		16	.500		25.50	16.30		41.80	52.50	
	5060	1-1/2″ diameter		13	.615		69.50	20		89.50	107	
	5080	2″ diameter		11	.727		119	23.50		142.50	167	
	5200	Curb stops, brass, 3/4″ diameter		19	.421		33.50	13.75		47.25	57.50	
	5220	1″ diameter		16	.500		51.50	16.30		67.80	81.50	
	5240	1-1/2″ diameter		13	.615		130	20		150	174	
	5260	2″ diameter		11	.727		209	23.50		232.50	265	
	5400	Curb box, cast iron, 3/4″ diameter		12	.667		46.50	21.50		68	84	
	5420	2″ diameter		8	1		85.50	32.50		118	144	
	5500	Saddles, cast iron, 3/4″, add					30			30	33	
	5550	2″, add					40			40	44	

026 450 | Hydrants

			CREW	DAILY OUTPUT	LABOR-HOURS	UNIT	MAT.	LABOR	EQUIP.	TOTAL	TOTAL INCL O&P	
454	0010	**PIPING, WATER DISTRIBUTION** Mech. joints unless noted										454
	1000	Fire hydrants, two way; excavation and backfill not incl.										
	1100	4-1/2″ valve size, depth 2′-0″	B-21	10	2.800	Ea.	730	70	13.60	813.60	925	
	1120	2′-6″		10	2.800		770	70	13.60	853.60	970	

026 450 | Hydrants

454	Line	Description	Crew	Daily Output	Labor-Hours	Unit	1999 Bare Costs Mat.	1999 Bare Costs Labor	1999 Bare Costs Equip.	1999 Bare Costs Total	Total Incl O&P
454	1140	3'-0"	B-21	10	2.800	Ea.	815	70	13.60	898.60	1,025
	1160	3'-6"		9	3.111		805	78	15.10	898.10	1,025
	1200	4'-6"		9	3.111		830	78	15.10	923.10	1,050
	1220	5'-0"		8	3.500		855	87.50	17	959.50	1,100
	1240	5'-6"		8	3.500		905	87.50	17	1,009.50	1,150
	1260	6'-0"		7	4		1,000	100	19.40	1,119.40	1,275
	1280	6'-6"		7	4		1,050	100	19.40	1,169.40	1,325
	1300	7'-0"		6	4.667		985	117	22.50	1,124.50	1,275
	1340	8'-0"		6	4.667		1,175	117	22.50	1,314.50	1,500
	1420	10'-0"		5	5.600		1,250	140	27	1,417	1,625
	2000	5-1/4" valve size, depth 2'-0"		10	2.800		825	70	13.60	908.60	1,025
	2080	4'-0"		9	3.111		930	78	15.10	1,023.10	1,175
	2160	6'-0"		7	4		1,050	100	19.40	1,169.40	1,325
	2240	8'-0"		6	4.667		1,150	117	22.50	1,289.50	1,475
	2320	10'-0"		5	5.600		1,250	140	27	1,417	1,650
	2350	For threeway valves, add					7%				
	2400	Lower barrel extensions with stems, 1'-0"	B-20	14	1.714		242	41.50		283.50	330
	2440	2'-0"		13	1.846		296	45		341	395
	2480	3'-0"		12	2		735	48.50		783.50	880
	2520	4'-0"		10	2.400		400	58.50		458.50	530
	5000	Indicator post									
	5020	Adjustable, valve size 4" to 14", 4' bury	B-21	10	2.800	Ea.	530	70	13.60	613.60	710
	5060	8' bury		7	4		600	100	19.40	719.40	840
	5080	10' bury		6	4.667		675	117	22.50	814.50	955
	5100	12' bury		5	5.600		730	140	27	897	1,050
	5120	14' bury		4	7		800	175	34	1,009	1,200
	5500	Non-adjustable, valve size 4" to 14", 3' bury		10	2.800		475	70	13.60	558.60	650
	5520	3'-6" bury		10	2.800		500	70	13.60	583.60	675
	5540	4' bury		9	3.111		560	78	15.10	653.10	755

026 650 | Water Systems

652	Line	Description	Crew	Daily Output	Labor-Hours	Unit	1999 Bare Costs Mat.	1999 Bare Costs Labor	1999 Bare Costs Equip.	1999 Bare Costs Total	Total Incl O&P
652	0010	**CORROSION RESISTANCE** Wrap & coat, add to pipe, 4" dia.				L.F.	1.31			1.31	1.44
	0020	5" diameter					1.38			1.38	1.52
	0040	6" diameter					1.38			1.38	1.52
	0060	8" diameter					2.13			2.13	2.34
	0080	10" diameter					3.17			3.17	3.49
	0100	12" diameter					3.17			3.17	3.49
	0120	14" diameter					4.65			4.65	5.10
	0140	16" diameter					4.65			4.65	5.10
	0160	18" diameter					4.70			4.70	5.15
	0180	20" diameter					4.92			4.92	5.40
	0200	24" diameter					6			6	6.60
	0220	Small diameter pipe, 1" diameter, add					.99			.99	1.09
	0240	2" diameter					1.09			1.09	1.20
	0260	2-1/2" diameter					1.19			1.19	1.31
	0280	3" diameter					1.19			1.19	1.31
	0300	Fittings, field covered, add				S.F.	6.30			6.30	6.90
	0500	Coating, bituminous, per diameter inch, 1 coat, add				L.F.	.23			.23	.25
	0540	3 coat					.37			.37	.41
	0560	Coal tar epoxy, per diameter inch, 1 coat, add					.15			.15	.17
	0600	3 coat					.29			.29	.32
	1000	Polyethylene H.D. extruded, .025" thk., 1/2" diameter add					.21			.21	.23
	1020	3/4" diameter					.22			.22	.24
	1040	1" diameter					.25			.25	.28
	1060	1-1/4" diameter					.30			.30	.33

026 650 | Water Systems

		026 650 Water Systems	CREW	DAILY OUTPUT	LABOR-HOURS	UNIT	1999 BARE COSTS MAT.	LABOR	EQUIP.	TOTAL	TOTAL INCL O&P
652	1080	1-1/2" diameter				L.F.	.34			.34	.37
	1100	.030" thk., 2" diameter					.37			.37	.41
	1120	2-1/2" diameter					.44			.44	.48
	1140	.035" thk., 3" diameter					.53			.53	.58
	1160	3-1/2" diameter					.61			.61	.67
	1180	4" diameter					.68			.68	.75
	1200	.040" thk, 5" diameter					.94			.94	1.03
	1220	6" diameter					.98			.98	1.08
	1240	8" diameter					1.28			1.28	1.41
	1260	10" diameter					1.56			1.56	1.72
	1280	12" diameter					1.86			1.86	2.05
	1300	.060" thk., 14" diameter					2.57			2.57	2.83
	1320	16" diameter					3.02			3.02	3.32
	1340	18" diameter					3.56			3.56	3.92
	1360	20" diameter					4.01			4.01	4.41
	1380	Fittings, field wrapped, add				S.F.	4.13			4.13	4.54
658	0010	**PIPING, WATER DISTRIBUTION, CONCRETE PIPE**									
	0020	Not including excavation or backfill, without gaskets									
	3000	Conc. cylinder pipe (CCP), 150 PSI, 40' L, 15" diam.	B-13	190	.295	L.F.	48.50	6.85	2.89	58.24	67.50
	3010	24" diameter	"	130	.431		60	10	4.23	74.23	86.50
	3040	36" diameter	B-13B	75	.747		70.50	17.35	10.55	98.40	116
	3050	48" diameter		68	.824		101	19.15	11.65	131.80	154
	3060	Prestressed (PCCP), 150 PSI, 24' L, 60" diameter		52	1.077		147	25	15.25	187.25	218
	3070	72" diameter		46	1.217		194	28.50	17.25	239.75	277
	3080	84" diameter		40	1.400		250	32.50	19.80	302.30	350
	3090	96" diameter		32	1.750		365	40.50	25	430.50	490
	3100	108" diameter	B-13C	32	1.750		510	40.50	35.50	586	670
	3102	120" diameter		16	3.500		745	81.50	70.50	897	1,025
	3104	144" diameter		16	3.500		960	81.50	70.50	1,112	1,250
	3110	Conc. cylinder pipe (CCP), 150 PSI, elbow, 90°, 12" diameter	B-13	32	1.750	Ea.	143	40.50	17.15	200.65	240
	3140	24" diameter	"	6	9.333		294	217	91.50	602.50	760
	3150	36" diameter	B-13B	15	3.733		615	87	53	755	870
	3160	48" diameter		12	4.667		725	108	66	899	1,025
	3170	Prestressed (PCCP), 150 PSI, elbow, 90°, 60" diameter		10	5.600		1,150	130	79.50	1,359.50	1,550
	3180	72" diameter		4	14		1,700	325	198	2,223	2,600
	3190	84" diameter		3.20	17.500		1,350	405	248	2,003	2,400
	3200	96" diameter		2	28		16,500	650	395	17,545	19,600
	3210	108" diameter	B-13C	1.20	46.667		19,800	1,075	945	21,820	24,500
	3220	120" diameter		.40	140		23,300	3,250	2,825	29,375	33,800
	3225	144" diameter		.30	184		22,000	4,275	3,725	30,000	35,000
	3230	Concrete cylinder pipe (CCP), 150 PSI, elbow, 45°, 12" diameter	B-13	24	2.333		135	54	23	212	259
	3250	24" diameter	"	6	9.333		292	217	91.50	600.50	755
	3260	36" diameter	B-13B	4	14		515	325	198	1,038	1,300
	3270	48" diameter		3	18.667		600	435	264	1,299	1,625
	3280	Prestressed, (PCCP), 150 PSI, elbow, 45°, 60" diameter		2	28		875	650	395	1,920	2,400
	3290	72" diameter		1.60	35		1,325	815	495	2,635	3,275
	3300	84" diameter		1.30	42.945		1,625	1,000	610	3,235	4,025
	3310	96" diameter		1	56		10,500	1,300	795	12,595	14,400
	3320	108" diameter	B-13C	.66	84.337		11,900	1,950	1,700	15,550	18,000
	3330	120" diameter		.40	140		13,600	3,250	2,825	19,675	23,100
	3340	144" diameter		.30	184		15,100	4,275	3,725	23,100	27,400
660	0010	**PIPING, WATER DISTRIBUTION, BLACK STEEL**									
	0011	Not including excavation or backfill									
	1000	Pipe, black steel, plain end, welded, 1/4" wall thk, 8" diam.	B-35A	108	.519	L.F.	7.25	13.35	9.35	29.95	39
	1010	10" diameter		104	.538		9	13.90	9.75	32.65	42

026 650 | Water Systems

			CREW	DAILY OUTPUT	LABOR-HOURS	UNIT	1999 BARE COSTS MAT.	LABOR	EQUIP.	TOTAL	TOTAL INCL O&P
660	1020	12″ diameter	B-35A	195	.287	L.F.	10.80	7.40	5.20	23.40	29
	1030	18″ diameter		175	.320		16.55	8.25	5.80	30.60	37.50
	1040	5/16″ wall thickness, 12″ diameter		195	.287		13.95	7.40	5.20	26.55	32.50
	1050	18″ diameter		59.20	.946		24.50	24.50	17.10	66.10	83.50
	1060	36″ diameter		28.96	1.934		48.50	50	35	133.50	169
	1070	3/8″ wall thickness, 18″ diameter		43.20	1.296		24.50	33.50	23.50	81.50	105
	1080	24″ diameter		36	1.556		32.50	40	28	100.50	129
	1090	30″ diameter		30.40	1.842		43	47.50	33.50	124	158
	1100	1/2″ wall thickness, 36″ diameter		26.08	2.147		69.50	55.50	39	164	205
	1110	48″ diameter		21.68	2.583		103	66.50	46.50	216	269
	1120	60″ diameter		16	3.500		135	90	63	288	360
	1130	72″ diameter		10.16	5.512		154	142	99.50	395.50	500
	1135	7/16″ wall thickness, 48″ diameter		20.80	2.692		90.50	69.50	48.50	208.50	260
	1140	5/8″ wall thickness, 48″ diameter		21.68	2.583		129	66.50	46.50	242	297
	1150	60″ diameter		16	3.500		161	90	63	314	385
	1170	84″ diameter		10	5.600		227	144	101	472	585
	1180	96″ diameter		9.84	5.691		259	147	103	509	625
	1190	3/4″ wall thickness, 60″ diameter		16	3.500		193	90	63	346	420
	1200	72″ diameter		10.16	5.512		232	142	99.50	473.50	585
	1210	84″ diameter		10	5.600		271	144	101	516	630
	1220	96″ diameter		8.64	6.481		310	167	117	594	730
	1230	108″ diameter		8.48	6.604		350	170	119	639	780
	1240	120″ diameter		8	7		390	180	126	696	850
	1250	7/8″ wall thickness, 72″ diameter		10	5.600		271	144	101	516	630
	1260	84″ diameter		9.84	5.691		310	147	103	560	680
	1270	96″ diameter		8	7		360	180	126	666	820
	1290	120″ diameter		7.20	7.778		455	200	141	796	965
	1300	1″ wall thickness, 84″ diameter		10	5.600		365	144	101	610	735
	1310	96″ diameter		8	7		415	180	126	721	880
	1320	132″ diameter		7.20	7.778		590	200	141	931	1,100
	1330	144″ diameter		6.80	8.235		645	212	149	1,006	1,200
	1340	1-1/8″ wall thickness, 108″		6	9.333		540	241	169	950	1,150
	1350	120″ diameter		5.76	9.722		600	251	176	1,027	1,250
	1360	1-1/4″ wall thickness, 132″ diameter		5.60	10		775	258	181	1,214	1,450
	1370	144″ diameter	↓	5.20	10.769	↓	775	278	195	1,248	1,500
662	0010	**PIPING, WATER DISTRIBUTION, COPPER**									
	0020	Not including excavation or backfill									
	2000	Tubing, type K, 20′ joints, 3/4″ diameter	Q-1	150	.107	L.F.	1.82	3.13		4.95	6.75
	2200	1″ diameter	1 Plum	136	.059		2.32	1.92		4.24	5.45
	3000	1-1/2″ diameter	Q-1	120	.133		3.73	3.91		7.64	10.05
	3020	2″ diameter		105	.152		5.95	4.47		10.42	13.35
	3040	2-1/2″ diameter		146	.110		8.60	3.22		11.82	14.35
	3060	3″ diameter		134	.119		11.90	3.50		15.40	18.40
	4012	4″ diameter		105	.152		18.55	4.47		23.02	27.50
	4014	5″ diameter	↓	19	.842		47.50	24.50		72	90
	4016	6″ diameter	Q-2	24	1	↓	55	30.50		85.50	107
	5000	Tubing, type L									
	5108	2″ diameter	1 Plum	106	.075	L.F.	6.40	2.46		8.86	10.80
	6010	3″ diameter	Q-1	140	.114		12.40	3.35		15.75	18.75
	6012	4″ diameter	″	105	.152		19.60	4.47		24.07	28.50
	6016	6″ diameter	Q-2	24	1	↓	48	30.50		78.50	98.50
	7020	Fittings, brass, corporation stops, 3/4″ diameter	1 Plum	19	.421	Ea.	16	13.75		29.75	38.50
	7040	1″ diameter		16	.500		24	16.30		40.30	51
	7060	1-1/2″ diameter		13	.615		65	20		85	102
	7080	2″ diameter	↓	11	.727	↓	115	23.50		138.50	163

	026 650 | Water Systems	CREW	DAILY OUTPUT	LABOR-HOURS	UNIT	1999 BARE COSTS MAT.	LABOR	EQUIP.	TOTAL	TOTAL INCL O&P
662 7100	Curb stops, 3/4" diameter	1 Plum	19	.421	Ea.	35	13.75		48.75	59.50
7120	1" diameter		16	.500		46	16.30		62.30	75
7140	1-1/2" diameter		13	.615		102	20		122	144
7160	2" diameter		11	.727		168	23.50		191.50	220
7180	Curb box, cast iron, 1/2" to 1" curb stops		12	.667		48	21.50		69.50	86
7200	1-1/4" to 2" curb stops		8	1		88	32.50		120.50	147
7220	Saddles, 3/4" diameter, add					30			30	33
7240	2" diameter, add					40			40	44
7250	For copper fittings, see Div. 151-430									
666 0010	**PIPING, WATER DISTRIBUTION, DUCTILE IRON** cement lined									
0020	Not including excavation or backfill									
2000	Pipe, class 50 water piping, 18' lengths									
2020	Mechanical joint, 4" diameter	B-20A	192	.167	L.F.	6.95	4.32		11.27	14.30
2040	6" diameter		168	.190		7.90	4.93		12.83	16.30
2060	8" diameter		144	.222		10.35	5.75		16.10	20
2080	10" diameter		90	.356		14	9.20		23.20	29.50
2100	12" diameter		96	.333		17.05	8.65		25.70	32
2120	14" diameter	B-21A	54	.741		22	19.70	7.25	48.95	62.50
2140	16" diameter		46	.870		24	23	8.50	55.50	71.50
2160	18" diameter		42	.952		30.50	25.50	9.30	65.30	83
2170	20" diameter		40	1		35.50	26.50	9.80	71.80	91
2180	24" diameter		35	1.143		47	30.50	11.20	88.70	111
3000	Tyton, Push-on joint, 4" diameter	B-20	158	.152		6.35	3.69		10.04	12.75
3020	6" diameter		138	.174		7.25	4.23		11.48	14.65
3040	8" diameter		118	.203		9.95	4.95		14.90	18.75
3060	10" diameter		100	.240		15.60	5.85		21.45	26.50
3080	12" diameter		80	.300		16.45	7.30		23.75	29.50
3100	14" diameter	B-21	60	.467		18.10	11.70	2.26	32.06	40.50
3120	16" diameter		54	.519		25.50	13	2.51	41.01	51.50
3140	18" diameter		44	.636		28.50	15.95	3.09	47.54	59.50
3160	20" diameter		42	.667		31	16.70	3.23	50.93	63.50
3180	24" diameter		40	.700		40	17.55	3.39	60.94	75
4000	Drill and tap pressurized main (labor only)									
4100	6" main, 1" to 2" service	Q-1	3	5.333	Ea.		157		157	237
4150	8" main, 1" to 2" service	"	2.75	5.818	"		171		171	259
4500	Tap and insert gate valve									
4600	8" main, 4" branch	B-21	3.20	8.750	Ea.		219	42.50	261.50	390
4650	6" branch		2.70	10.370			260	50.50	310.50	460
4651	Piping, drill, tap & insert gate valve, 8" main, 6" branch		2.70	10.370			260	50.50	310.50	460
4700	10" main, 4" branch		2.70	10.370			260	50.50	310.50	460
4750	6" branch		2.35	11.915			298	58	356	530
4800	12" main, 6" branch		2.35	11.915			298	58	356	530
4850	8" branch		2.35	11.915			298	58	356	530
8000	Fittings, mechanical joint									
8006	90° bend or elbow, 4" diameter	B-20A	50.32	.636	Ea.	97.50	16.45		113.95	133
8020	6" diameter		34	.941		159	24.50		183.50	213
8040	8" diameter		28.56	1.120		230	29		259	298
8060	10" diameter		21	1.524		258	39.50		297.50	345
8080	12" diameter	B-21A	24.48	1.634		390	43.50	16	449.50	510
8100	14" diameter		16	2.500		550	66.50	24.50	641	735
8120	16" diameter		14	2.857		625	76	28	729	835
8140	18" diameter		13.60	2.941		1,050	78	29	1,157	1,300
8160	20" diameter		8	5		1,950	133	49	2,132	2,400
8180	24" diameter		8.16	4.902		2,150	130	48	2,328	2,625
8200	Wye or tee, 4" diameter	B-20	25	.960		157	23.50		180.50	210

026 650 | Water Systems

			CREW	DAILY OUTPUT	LABOR-HOURS	UNIT	1999 BARE COSTS MAT.	LABOR	EQUIP.	TOTAL	TOTAL INCL O&P
666	8220	6" diameter	B-20	17	1.412	Ea.	169	34.50		203.50	240
	8240	8" diameter		14	1.714		243	41.50		284.50	335
	8260	10" diameter	B-21	14	2		510	50	9.70	569.70	650
	8280	12" diameter		12	2.333		700	58.50	11.30	769.80	870
	8300	14" diameter		10	2.800		670	70	13.60	753.60	860
	8320	16" diameter		8	3.500		935	87.50	17	1,039.50	1,175
	8340	18" diameter		6	4.667		1,825	117	22.50	1,964.50	2,200
	8360	20" diameter		4	7		1,600	175	34	1,809	2,050
	8380	24" diameter		3	9.333		2,800	234	45.50	3,079.50	3,525
	8400	45° bend or elbow, 4" diameter	B-20A	32.64	.980		129	25.50		154.50	181
	8410	12" diameter	"	21.76	1.471		315	38		353	410
	8420	16" diameter	B-21A	12	3.333		360	88.50	32.50	481	565
	8430	20" diameter		6	6.667		855	177	65	1,097	1,300
	8440	24" diameter		5.44	7.353		2,100	196	72	2,368	2,675
	8450	Decreaser, 6" x 4" diameter	B-20A	16.32	1.961		99.50	51		150.50	187
	8460	8" x 6" diameter		13.60	2.353		149	61		210	258
	8470	10" x 6 " diameter		12.24	2.614		179	67.50		246.50	300
	8480	12" x 6" diameter		10.88	2.941		225	76		301	365
	8490	6" x 16" diameter	B-21A	6	6.667		355	177	65	597	740
	8500	6" x 20" diameter	"	5	8		565	213	78	856	1,025
	9600	Steel sleeve and tap, 4" diameter	B-20	3	8		340	195		535	680
	9620	6" diameter		2	12		400	292		692	900
	9630	8" diameter		2	12		540	292		832	1,050
674	0010	**PIPING, WATER DISTRIBUTION, POLYETHYLENE, C901**									
	0020	Not including excavation or backfill									
	1000	Piping, 160 P.S.I., 3/4" diameter	B-20	525	.046	L.F.	.23	1.11		1.34	2
	1120	1" diameter		485	.049		.37	1.20		1.57	2.31
	1140	1-1/2" diameter		450	.053		.74	1.30		2.04	2.85
	1160	2" diameter		365	.066		1.26	1.60		2.86	3.91
	2000	Fittings, insert type, nylon, 160 & 250 psi, cold water									
	2220	Clamp ring, stainless steel, 3/4" diameter	B-20	345	.070	Ea.	.71	1.69		2.40	3.44
	2240	1" diameter		321	.075		.72	1.82		2.54	3.65
	2260	1-1/2" diameter		285	.084		.73	2.05		2.78	4.03
	2280	2" diameter		255	.094		.73	2.29		3.02	4.40
	2300	Coupling, 3/4" diameter		66	.364		.57	8.85		9.42	14.60
	2320	1" diameter		57	.421		.76	10.25		11.01	17
	2340	1-1/2" diameter		51	.471		1.77	11.45		13.22	19.95
	2360	2" diameter		48	.500		2.13	12.15		14.28	21.50
	2400	Elbow, 90°, 3/4" diameter		66	.364		1.02	8.85		9.87	15.05
	2420	1" diameter		57	.421		1.09	10.25		11.34	17.35
	2440	1-1/2" diameter		51	.471		2.71	11.45		14.16	21
	2460	2" diameter		48	.500		4.52	12.15		16.67	24
	2500	Tee, 3/4" diameter		42	.571		1.10	13.90		15	23
	2520	1" diameter		39	.615		1.78	14.95		16.73	25.50
	2540	1-1/2" diameter		33	.727		4.03	17.70		21.73	32.50
	2560	2" diameter		30	.800		5.45	19.45		24.90	36.50
678	0010	**PIPING, WATER DISTRIBUTION, POLYVINYL CHLORIDE**									
	0020	Not including excavation or backfill, unless specified									
	1000	AWWA C900, Class 150, SDR 18									
	1040	10" diameter	B-20	125	.192	L.F.	8.75	4.67		13.42	17
	1050	12" diameter		125	.192		10.65	4.67		15.32	19.05
	2100	Class 160, S.D.R. 26, 1-1/2" diameter		350	.069		1	1.67		2.67	3.73
	2120	2" diameter		300	.080		1.35	1.95		3.30	4.55
	2140	2-1/2" diameter		300	.080		1.65	1.95		3.60	4.88

026 650 | Water Systems

		CREW	DAILY OUTPUT	LABOR-HOURS	UNIT	1999 BARE COSTS MAT.	LABOR	EQUIP.	TOTAL	TOTAL INCL O&P
678 2160	3" diameter	B-20	275	.087	L.F.	2.15	2.12		4.27	5.70
2180	4" diameter		200	.120		2.95	2.92		5.87	7.85
2200	6" diameter		180	.133		5.50	3.24		8.74	11.15
2210	8" diameter	↓	180	.133	↓	7.75	3.24		10.99	13.65
3010	AWWA C905, PR 100, DR 41									
3030	14" diameter	B-20	160	.150	L.F.	15.35	3.65		19	22.50
3040	16" diameter		150	.160		19.75	3.89		23.64	27.50
3050	18" diameter		120	.200		25	4.86		29.86	35
3060	20" diameter		100	.240		33	5.85		38.85	45.50
3070	24" diameter		80	.300		45	7.30		52.30	61
3080	30" diameter		60	.400		60	9.75		69.75	81.50
3090	36" diameter	↓	40	.600		74	14.60		88.60	105
3960	Pressure pipe, class 200, SDR 21, 3/4"	Q-1	125	.128		.09	3.76		3.85	5.80
3980	1"		150	.107		.11	3.13		3.24	4.86
4000	1-1/2"		600	.027		.25	.78		1.03	1.47
4010	2"		550	.029		.38	.85		1.23	1.70
4020	2-1/2"		500	.032		.55	.94		1.49	2.03
4030	3"		425	.038		.80	1.10		1.90	2.55
4040	4"		400	.040		1.35	1.17		2.52	3.27
4050	6"		375	.043		2.76	1.25		4.01	4.94
4060	8"	↓	350	.046		4.77	1.34		6.11	7.30
4090	Including trenching to 3' deep, 3/4"	Q-1C	300	.080		.09	2.32	1.79	4.20	5.60
4100	1"		280	.086		.11	2.49	1.91	4.51	6
4110	1-1/2"		260	.092		.25	2.68	2.06	4.99	6.60
4120	2"		220	.109		.38	3.17	2.44	5.99	7.90
4130	2-1/2"		200	.120		.55	3.48	2.68	6.71	8.85
4140	3"		175	.137		.80	3.98	3.06	7.84	10.30
4150	4"		150	.160		1.35	4.64	3.57	9.56	12.45
4160	6"	↓	125	.192	↓	2.76	5.55	4.29	12.60	16.20
4165	Fittings									
4170	Elbow, 90°, 3/4"	Q-1	90	.178	Ea.	.11	5.20		5.31	8.05
4180	1"		90	.178		.19	5.20		5.39	8.10
4190	1-1/2"		74.96	.213		.36	6.25		6.61	9.90
4200	2"		57	.281		.59	8.25		8.84	13.15
4210	3"		12	1.333		.91	39		39.91	60.50
4220	4"		10	1.600		5.15	47		52.15	76.50
4230	6"		8.85	1.808		3.96	53		56.96	85
4240	8"		8.85	1.808		5.60	53		58.60	86.50
4250	Elbow, 45°, 3/4"		90	.178		.31	5.20		5.51	8.25
4260	1"		90	.178		.37	5.20		5.57	8.30
4270	1-1/2"		74.96	.213		3.18	6.25		9.43	13
4280	2"		57	.281		3.45	8.25		11.70	16.30
4290	2-1/2"		45	.356		4.56	10.45		15.01	21
4300	3"		12	1.333		6.10	39		45.10	66
4310	4"		10	1.600		10.55	47		57.55	82.50
4320	6"		8.85	1.808		21.50	53		74.50	104
4330	8"		8.85	1.808		45	53		98	130
4340	Tee, 3/4"		43	.372		.17	10.90		11.07	16.75
4350	1"		43	.372		.32	10.90		11.22	16.90
4360	1-1/2"		28	.571		5.10	16.75		21.85	31
4370	2"		23	.696		6.15	20.50		26.65	38
4380	2-1/2"		17.50	.914		7.80	27		34.80	49
4390	3"		10	1.600		8.60	47		55.60	80.50
4400	4"		7.50	2.133		13.75	62.50		76.25	110
4410	6"		5.85	2.735		23	80.50		103.50	148
4420	8"	↓	5.85	2.735	↓	67	80.50		147.50	196

026 650 | Water Systems

			CREW	DAILY OUTPUT	LABOR-HOURS	UNIT	1999 BARE COSTS MAT.	LABOR	EQUIP.	TOTAL	TOTAL INCL O&P
678	4430	Coupling, 3/4"	Q-1	90	.178	Ea.	.11	5.20		5.31	8
	4440	1"		90	.178		.19	5.20		5.39	8.10
	4450	1-1/2"		74.96	.213		2.03	6.25		8.28	11.75
	4460	2"		57	.281		2.15	8.25		10.40	14.85
	4470	2-1/2"		45	.356		2.58	10.45		13.03	18.65
	4480	3"		12	1.333		3.41	39		42.41	63.50
	4490	4"		10	1.600		5.90	47		52.90	77.50
	4500	6"		8.85	1.808		10.60	53		63.60	92
	4510	8"		8.85	1.808		22	53		75	105
	8000	Fittings with rubber gasket									
	8003	Class 150, D.R. 18									
	8006	90° Bend , 4" diameter	B-20	100	.240	Ea.	22	5.85		27.85	33
	8020	6" diameter		90	.267		45	6.50		51.50	59.50
	8040	8" diameter		80	.300		85	7.30		92.30	105
	8060	10" diameter		50	.480		170	11.65		181.65	205
	8080	12" diameter		30	.800		230	19.45		249.45	284
	8100	Tee, 4" diameter		90	.267		26	6.50		32.50	38.50
	8120	6" diameter		80	.300		45	7.30		52.30	61
	8140	8" diameter		70	.343		95	8.35		103.35	118
	8160	10" diameter		40	.600		200	14.60		214.60	243
	8180	12" diameter		20	1.200		290	29		319	365
	8200	45° Bend, 4" diameter		100	.240		26	5.85		31.85	37.50
	8220	6" diameter		90	.267		45	6.50		51.50	59.50
	8240	8" diameter		50	.480		89	11.65		100.65	116
	8260	10" diameter		50	.480		200	11.65		211.65	238
	8280	12" diameter		30	.800		295	19.45		314.45	355
	8300	Reducing tee 6"x4"		100	.240		80	5.85		85.85	97
	8320	8" x 6"		90	.267		130	6.50		136.50	153
	8330	10" x 6"		90	.267		180	6.50		186.50	208
	8340	10"x8"		90	.267		260	6.50		266.50	296
	8350	12" x 6"		90	.267		300	6.50		306.50	340
	8360	12" x 8"		90	.267		385	6.50		391.50	435
	8400	Tapped service tee (threaded type) 6" x 6" x 3/4		100	.240		58	5.85		63.85	73
	8420	6" x 6" x 3/4"		90	.267		58	6.50		64.50	74
	8430	6" x 6" x 1"		90	.267		58	6.50		64.50	74
	8440	6" x 6" x 1 1/2"		90	.267		58	6.50		64.50	74
	8450	6" x 6" x 2"		90	.267		58	6.50		64.50	74
	8460	8" x 8" x 3/4"		90	.267		58	6.50		64.50	74
	8470	8" x 8" x 1"		90	.267		80	6.50		86.50	98
	8480	8" x 8" x 1 1/2"		90	.267		85	6.50		91.50	104
	8490	8" x 8" x 2"		90	.267		85	6.50		91.50	104
	8500	Repair coupling 4"		100	.240		15	5.85		20.85	25.50
	8520	6" diameter		90	.267		30	6.50		36.50	43
	8540	8" diameter		50	.480		55	11.65		66.65	79
	8560	10" diameter		50	.480		150	11.65		161.65	183
	8580	12" diameter		50	.480		180	11.65		191.65	216
	8600	Plug end 4"		100	.240		15	5.85		20.85	25.50
	8620	6" diameter		90	.267		20	6.50		26.50	32
	8640	8" diameter		50	.480		45	11.65		56.65	68
	8660	10" diameter		50	.480		58	11.65		69.65	82.50
	8680	12" diameter		50	.480		90	11.65		101.65	117
680	0010	**PIPING, HDPE BUTT FUSION JOINTS**, SDR 21, 40' lengths									
	0100	4" diameter	B-22A	400	.095	L.F.	1.62	2.33	.63	4.58	6.10
	0200	6" diameter		380	.100		2.68	2.45	.66	5.79	7.50
	0300	8" diameter		320	.119		4.55	2.91	.78	8.24	10.40

026 650 | Water Systems			CREW	DAILY OUTPUT	LABOR-HOURS	UNIT	1999 BARE COSTS MAT.	LABOR	EQUIP.	TOTAL	TOTAL INCL O&P	
680	0400	10" diameter	B-22A	300	.127	L.F.	7.10	3.10	.83	11.03	13.60	680
	0500	12" diameter		260	.146		9.95	3.58	.96	14.49	17.55	
	0600	14" diameter		220	.173		12	4.23	1.14	17.37	21	
	0700	16" diameter		180	.211		15.65	5.15	1.39	22.19	27	
	0800	18" diameter		140	.271		19.80	6.65	1.79	28.24	34.50	
	0900	24" diameter	↓	100	.380	↓	35.50	9.30	2.50	47.30	56.50	
	1000	Fittings										
	1100	Elbows, 90 degrees										
	1200	4" diameter	B-22B	32	.500	Ea.	25	12.40	.02	37.42	47	
	1300	6" diameter		28	.571		66.50	14.15	.02	80.67	95.50	
	1400	8" diameter		24	.667		139	16.50	.02	155.52	179	
	1500	10" diameter		18	.889		232	22	.03	254.03	290	
	1600	12" diameter		12	1.333		395	33	.04	428.04	480	
	1700	14" diameter		9	1.778		405	44	.05	449.05	515	
	1800	16" diameter		6	2.667		485	66	.08	551.08	640	
	1900	18" diameter		4	4		660	99	.12	759.12	880	
	2000	24" diameter	↓	3	5.333	↓	1,325	132	.16	1,457.16	1,675	
	2100	Tees										
	2200	4" diameter	B-22B	30	.533	Ea.	28.50	13.20	.02	41.72	52	
	2300	6" diameter		26	.615		74.50	15.25	.02	89.77	106	
	2400	8" diameter		22	.727		189	18	.02	207.02	237	
	2500	10" diameter		15	1.067		249	26.50	.03	275.53	315	
	2600	12" diameter		10	1.600		340	39.50	.05	379.55	440	
	2700	14" diameter		8	2		400	49.50	.06	449.56	520	
	2800	16" diameter		6	2.667		475	66	.08	541.08	625	
	2900	18" diameter		4	4		660	99	.12	759.12	880	
	3000	24" diameter	↓	2	8	↓	1,375	198	.24	1,573.24	1,800	
690	0010	**DISTRIBUTION CONNECTION**										690
	7000	Tapping crosses, sleeves, valves; with rubber gaskets										
	7020	Crosses, 4" x 4"	B-21	37	.757	Ea.	400	18.95	3.67	422.62	475	
	7040	6" x 6"		25	1.120		480	28	5.45	513.45	580	
	7060	8" x 6"		21	1.333		595	33.50	6.45	634.95	715	
	7080	8" x 8"		21	1.333		640	33.50	6.45	679.95	765	
	7100	10" x 6"		21	1.333		1,175	33.50	6.45	1,214.95	1,350	
	7120	10" x 10"		21	1.333		1,275	33.50	6.45	1,314.95	1,450	
	7140	12" x 6"		18	1.556		1,200	39	7.55	1,246.55	1,400	
	7160	12" x 12"		18	1.556		1,500	39	7.55	1,546.55	1,725	
	7180	14" x 6"		16	1.750		2,950	44	8.50	3,002.50	3,325	
	7200	14" x 14"		16	1.750		3,075	44	8.50	3,127.50	3,475	
	7220	16" x 6"		14	2		3,200	50	9.70	3,259.70	3,625	
	7240	16" x 10"		14	2		3,250	50	9.70	3,309.70	3,675	
	7260	16" x 16"		14	2		3,325	50	9.70	3,384.70	3,775	
	7280	18" x 6"		10	2.800		4,650	70	13.60	4,733.60	5,250	
	7300	18" x 12"		10	2.800		4,700	70	13.60	4,783.60	5,300	
	7320	18" x 18"		10	2.800		4,850	70	13.60	4,933.60	5,450	
	7340	20" x 6"		8	3.500		3,775	87.50	17	3,879.50	4,300	
	7360	20" x 12"		8	3.500		4,000	87.50	17	4,104.50	4,550	
	7380	20" x 20"		8	3.500		5,925	87.50	17	6,029.50	6,650	
	7400	24" x 6"		6	4.667		5,000	117	22.50	5,139.50	5,700	
	7420	24" x 12"		6	4.667		5,100	117	22.50	5,239.50	5,800	
	7440	24" x 18"		6	4.667		7,450	117	22.50	7,589.50	8,400	
	7460	24" x 24"		6	4.667		7,700	117	22.50	7,839.50	8,675	
	7600	Cut-in sleeves with rubber gaskets, 4"		18	1.556		140	39	7.55	186.55	223	
	7620	6"		12	2.333		180	58.50	11.30	249.80	300	
	7640	8"	↓	10	2.800	↓	240	70	13.60	323.60	390	

Important: See the Reference Section for critical supporting data - Reference Nos., Crews, & City Cost Indexes

026 650 | Water Systems

		CREW	DAILY OUTPUT	LABOR-HOURS	UNIT	1999 BARE COSTS MAT.	LABOR	EQUIP.	TOTAL	TOTAL INCL O&P
690 7660	10″	B-21	10	2.800	Ea.	330	70	13.60	413.60	490
7680	12″		9	3.111		390	78	15.10	483.10	570
7800	Cut-in valves with rubber gaskets, 4″		18	1.556		400	39	7.55	446.55	510
7820	6″		12	2.333		530	58.50	11.30	599.80	690
7840	8″		10	2.800		800	70	13.60	883.60	1,000
7860	10″		10	2.800		1,200	70	13.60	1,283.60	1,450
7880	12″		9	3.111		1,500	78	15.10	1,593.10	1,800
8000	Sleeves with rubber gaskets, 4″ x 4″		37	.757		350	18.95	3.67	372.62	420
8020	6″ x 6″		25	1.120		400	28	5.45	433.45	490
8040	8″ x 6″		21	1.333		500	33.50	6.45	539.95	610
8060	8″ x 8″		21	1.333		550	33.50	6.45	589.95	665
8080	10″ x 6″		21	1.333		1,000	33.50	6.45	1,039.95	1,150
8100	10″ x 10″		21	1.333		1,100	33.50	6.45	1,139.95	1,250
8120	12″ x 6″		18	1.556		1,000	39	7.55	1,046.55	1,175
8140	12″ x 12″		18	1.556		1,275	39	7.55	1,321.55	1,475
8160	14″ x 6″		16	1.750		2,500	44	8.50	2,552.50	2,825
8180	14″ x 14″		16	1.750		2,650	44	8.50	2,702.50	3,000
8200	16″ x 6″		14	2		2,650	50	9.70	2,709.70	3,025
8220	16″ x 10″		14	2		2,725	50	9.70	2,784.70	3,100
8240	16″ x 16″		14	2		2,900	50	9.70	2,959.70	3,300
8260	18″ x 6″		10	2.800		3,975	70	13.60	4,058.60	4,500
8280	18″ x 12″		10	2.800		4,000	70	13.60	4,083.60	4,525
8300	18″ x 18″		10	2.800		4,150	70	13.60	4,233.60	4,700
8320	20″ x 6″		8	3.500		3,225	87.50	17	3,329.50	3,700
8340	20″ x 12″		8	3.500		3,500	87.50	17	3,604.50	4,000
8360	20″ x 20″		8	3.500		5,000	87.50	17	5,104.50	5,650
8380	24″ x 6″		6	4.667		4,150	117	22.50	4,289.50	4,775
8400	24″ x 12″		6	4.667		4,500	117	22.50	4,639.50	5,150
8420	24″ x 18″		6	4.667		6,300	117	22.50	6,439.50	7,125
8440	24″ x 24″	↓	6	4.667		6,500	117	22.50	6,639.50	7,350
8800	Curb box, 6′ long	B-20	20	1.200		85	29		114	140
8820	8′ long	″	18	1.333		95	32.50		127.50	156
9000	Valves, gate valve, N.R.S. post type, 4″ diameter	B-21	32	.875		295	22	4.24	321.24	365
9020	6″ diameter		20	1.400		450	35	6.80	491.80	555
9040	8″ diameter		16	1.750		750	44	8.50	802.50	905
9060	10″ diameter		16	1.750		1,200	44	8.50	1,252.50	1,400
9080	12″ diameter		13	2.154		2,000	54	10.45	2,064.45	2,300
9100	14″ diameter		11	2.545		2,500	63.50	12.35	2,575.85	2,875
9120	O.S.&Y., 4″ diameter		32	.875		380	22	4.24	406.24	460
9140	6″ diameter		20	1.400		500	35	6.80	541.80	610
9160	8″ diameter		16	1.750		825	44	8.50	877.50	990
9180	10″ diameter		16	1.750		1,300	44	8.50	1,352.50	1,500
9200	12″ diameter		13	2.154		2,000	54	10.45	2,064.45	2,300
9220	14″ diameter	↓	11	2.545		2,675	63.50	12.35	2,750.85	3,075
9400	Check valves, rubber disc, 2-1/2″ diameter	B-20	44	.545		300	13.25		313.25	350
9420	3″ diameter	″	38	.632		340	15.35		355.35	400
9440	4″ diameter	B-21	32	.875		425	22	4.24	451.24	510
9480	6″ diameter		20	1.400		600	35	6.80	641.80	720
9500	8″ diameter		16	1.750		890	44	8.50	942.50	1,050
9520	10″ diameter		16	1.750		1,550	44	8.50	1,602.50	1,775
9540	12″ diameter		13	2.154		2,475	54	10.45	2,539.45	2,825
9542	14″ diameter		11	2.545		3,950	63.50	12.35	4,025.85	4,475
9700	Detector check valves, red, 4″ diameter		32	.875		700	22	4.24	726.24	810
9720	6″ diameter		20	1.400		1,075	35	6.80	1,116.80	1,225
9740	8″ diameter		16	1.750		1,800	44	8.50	1,852.50	2,050
9760	10″ diameter	↓	16	1.750	↓	3,475	44	8.50	3,527.50	3,900

		026 650 | Water Systems	CREW	DAILY OUTPUT	LABOR-HOURS	UNIT	1999 BARE COSTS MAT.	LABOR	EQUIP.	TOTAL	TOTAL INCL O&P	
690	9800	Galvanized, 4″ diameter	B-21	32	.875	Ea.	850	22	4.24	876.24	975	690
	9820	6″ diameter		20	1.400		1,200	35	6.80	1,241.80	1,375	
	9840	8″ diameter		16	1.750		1,950	44	8.50	2,002.50	2,225	
	9860	10″ diameter		16	1.750		3,800	44	8.50	3,852.50	4,250	

		026 700 | Water Wells	CREW	DAILY OUTPUT	LABOR-HOURS	UNIT	MAT.	LABOR	EQUIP.	TOTAL	TOTAL INCL O&P	
704	0010	**WELLS** Domestic water										704
	0100	Drilled, 4″ to 6″ diameter	B-23	120	.333	L.F.		7.30	16.30	23.60	29.50	
	0200	8″ diameter	″	95.20	.420	″		9.20	20.50	29.70	37	
	0400	Gravel pack well, 40′ deep, incl. gravel & casing, complete										
	0500	24″ diameter casing x 18″ diameter screen	B-23	.13	307	Total	20,000	6,725	15,000	41,725	49,100	
	0600	36″ diameter casing x 18″ diameter screen		.12	333		21,500	7,275	16,300	45,075	53,000	
	0601	Gravel pack well, 36″ diam. casing x 18″ diam. screen		.12	333		21,500	7,275	16,300	45,075	53,000	
	0800	Observation wells, 1-1/4″ riser pipe		163	.245	V.L.F.	11	5.35	12	28.35	34	
	0900	For flush Buffalo roadway box, add	1 Skwk	16.60	.482	Ea.	30	13.50		43.50	54.50	
	1200	Test well, 2-1/2″ diameter, up to 50′ deep (15 to 50 GPM)	B-23	1.51	26.490	″	450	580	1,300	2,330	2,825	
	1300	Over 50′ deep, add	″	121.80	.328	L.F.	12	7.20	16.05	35.25	42	
	1400	Remove & reset pump, minimum	B-21	4	7	Ea.		175	34	209	310	
	1420	Maximum	″	2	14	″		350	68	418	625	
	1500	Pumps, installed in wells to 100′ deep, 4″ submersible										
	1510	1/2 H.P.	Q-1	3.22	4.969	Ea.	425	146		571	690	
	1520	3/4 H.P.		2.66	6.015		475	177		652	795	
	1600	1 H.P.		2.29	6.987		525	205		730	890	
	1700	1-1/2 H.P.	Q-22	1.60	10		580	294	277	1,151	1,400	
	1800	2 H.P.		1.33	12.030		620	355	335	1,310	1,575	
	1900	3 H.P.		1.14	14.035		775	410	390	1,575	1,900	
	2000	5 H.P.		1.14	14.035		1,250	410	390	2,050	2,425	
	2050	Remove and install motor only, 4 H.P.		1.14	14.035		585	410	390	1,385	1,700	
	3000	Pump, 6″ submersible, 25′ to 150′ deep, 25 H.P., 249 to 297 GPM		.89	17.978		3,775	530	500	4,805	5,500	
	3100	25′ to 500′ deep, 30 H.P., 100 to 300 GPM		.73	21.918		3,875	645	610	5,130	5,900	
	5000	Wells to 180 ft. deep, 4″ submersible, 1 HP	B-21	1.10	25.455		400	635	123	1,158	1,575	
	5500	2 HP		1.10	25.455		675	635	123	1,433	1,875	
	6000	3 HP		1	28		700	700	136	1,536	2,025	
	7000	5 HP		.90	31.111		725	780	151	1,656	2,200	
	8000	Steel well casing	B-23A	3,020	.008	Lb.	.40	.19	.63	1.22	1.43	
	8110	Well screen assembly, stainless steel, 2″ diameter		273	.088	L.F.	34	2.15	7	43.15	48.50	
	8120	3″ diameter		253	.095		47.50	2.32	7.55	57.37	64.50	
	8130	4″ diameter		200	.120		54	2.93	9.50	66.43	74.50	
	8140	5″ diameter		168	.143		63.50	3.49	11.35	78.34	88	
	8150	6″ diameter		126	.190		76	4.65	15.10	95.75	107	
	8160	8″ diameter		98.50	.244		99.50	5.95	19.35	124.80	140	
	8170	10″ diameter		73	.329		125	8.05	26	159.05	178	
	8180	12″ diameter		62.50	.384		147	9.40	30.50	186.90	209	
	8190	14″ diameter		54.30	.442		166	10.80	35	211.80	238	
	8200	16″ diameter		48.30	.497		184	12.15	39.50	235.65	264	
	8210	18″ diameter		39.20	.612		228	14.95	48.50	291.45	330	
	8220	20″ diameter		31.20	.769		261	18.80	61	340.80	385	
	8230	24″ diameter		23.80	1.008		325	24.50	80	429.50	480	
	8240	26″ diameter		21	1.143		355	28	90.50	473.50	535	
	8300	Slotted PVC, 1-1/4″ diameter		521	.046		1.72	1.13	3.66	6.51	7.65	
	8310	1-1/2″ diameter		488	.049		2.29	1.20	3.90	7.39	8.70	
	8320	2″ diameter		273	.088		2.39	2.15	7	11.54	13.65	
	8330	3″ diameter		253	.095		3.45	2.32	7.55	13.32	15.70	
	8340	4″ diameter		200	.120		1.85	2.93	9.50	14.28	17.10	
	8350	5″ diameter		168	.143		2.92	3.49	11.35	17.76	21	
	8360	6″ diameter		126	.190		3.03	4.65	15.10	22.78	27	

Important: See the Reference Section for critical supporting data - Reference Nos., Crews, & City Cost Indexe

026 700 | Water Wells

		Description	Crew	Daily Output	Labor-Hours	Unit	1999 Bare Costs Mat.	1999 Bare Costs Labor	1999 Bare Costs Equip.	1999 Bare Costs Total	Total Incl O&P	
704	8370	8" diameter	B-23A	98.50	.244	L.F.	5.30	5.95	19.35	30.60	36.50	704
	8400	Artificial gravel pack, 2" screen, 6" casing	B-23B	174	.138		1.89	3.37	11.85	17.11	20.50	
	8405	8" casing		111	.216		2.58	5.30	18.60	26.48	31.50	
	8410	10" casing		74.50	.322		3.24	7.85	27.50	38.59	46.50	
	8415	12" casing		60	.400		4.38	9.75	34.50	48.63	58	
	8420	14" casing		50.20	.478		5.65	11.70	41	58.35	69.50	
	8425	16" casing		40.70	.590		7.80	14.40	50.50	72.70	87	
	8430	18" casing		36	.667		9.10	16.30	57.50	82.90	98.50	
	8435	20" casing		29.50	.814		10.45	19.90	70	100.35	120	
	8440	24" casing		25.70	.934		11.60	23	80.50	115.10	137	
	8445	26" casing		24.60	.976		13	24	84	121	144	
	8450	30" casing		20	1.200		14.90	29.50	103	147.40	176	
	8455	36" casing		16.40	1.463	↓	16	36	126	178	211	
	8500	Develop well		8	3	Hr.	165	73.50	258	496.50	580	
	8550	Pump test well	↓	8	3		44.50	73.50	258	376	445	
	8560	Standby well	B-23A	8	3	↓	43.50	73.50	238	355	425	
	8570	Standby, drill rig		8	3	Ea.		73.50	238	311.50	375	
	8580	Surface seal well, concrete filled	↓	1	24		410	585	1,900	2,895	3,450	
	8590	Well test pump, install & remove	B-23	1	40			875	1,950	2,825	3,525	
	8600	Well sterilization, chlorine	2 Clab	1	16	↓	345	345		690	920	
	9950	See div. 021-444 for wellpoints										
	9960	See div. 021-484 for drainage wells										

026 800 | Fuel Distribution

		Description	Crew	Daily Output	Labor-Hours	Unit	1999 Bare Costs Mat.	1999 Bare Costs Labor	1999 Bare Costs Equip.	1999 Bare Costs Total	Total Incl O&P	
804	0010	**GAS STATION PRODUCT LINE**										804
	0020	Primary containment pipe, fiberglass-reinforced										
	0030	Plastic pipe 15' & 30' lengths										
	0040	2" diameter	Q-6	425	.056	L.F.	2.97	1.73		4.70	5.90	
	0050	3" diameter		400	.060		4.40	1.83		6.23	7.60	
	0060	4" diameter	↓	375	.064	↓	5.75	1.96		7.71	9.30	
	0100	Fittings										
	0110	Elbows, 90° & 45°, bell-ends, 2"	Q-6	24	1	Ea.	31	30.50		61.50	80.50	
	0120	3" diameter		22	1.091		35	33.50		68.50	89	
	0130	4" diameter		20	1.200		42	36.50		78.50	102	
	0200	Tees, bell ends, 2"		21	1.143		36.50	35		71.50	93.50	
	0210	3" diameter		18	1.333		40	41		81	106	
	0220	4" diameter		15	1.600		52	49		101	131	
	0230	Flanges bell ends, 2"		24	1		20	30.50		50.50	68.50	
	0240	3" diameter		22	1.091		22	33.50		55.50	74.50	
	0250	4" diameter		20	1.200		25	36.50		61.50	83	
	0260	Sleeve couplings, 2"		21	1.143		7.75	35		42.75	61.50	
	0270	3" diameter		18	1.333		10.75	41		51.75	73.50	
	0280	4" diameter		15	1.600		15.50	49		64.50	91	
	0290	Threaded adapters 2"		21	1.143		10.95	35		45.95	65	
	0300	3" diameter		18	1.333		16.50	41		57.50	79.50	
	0310	4" diameter		15	1.600		23	49		72	99.50	
	0320	Reducers, 2"		27	.889		14.15	27		41.15	56.50	
	0330	3" diameter		22	1.091		16.50	33.50		50	68.50	
	0340	4" diameter	↓	20	1.200	↓	25	36.50		61.50	83	
	1010	Gas station product line for secondary containment (double wall)										
	1100	Fiberglass reinforced plastic pipe 25' lengths										
	1120	Pipe, plain end, 3"	Q-6	375	.064	L.F.	3.98	1.96		5.94	7.35	
	1130	4" diameter		350	.069		4.95	2.10		7.05	8.60	
	1140	5" diameter		325	.074		5.50	2.26		7.76	9.45	
	1150	6" diameter	↓	300	.080	↓	9.45	2.45		11.90	14.10	
	1200	Fittings										

		026 800 | Fuel Distribution	CREW	DAILY OUTPUT	LABOR-HOURS	UNIT	1999 BARE COSTS MAT.	LABOR	EQUIP.	TOTAL	TOTAL INCL O&P	
804	1230	Elbows, 90° & 45°, 3"	Q-6	18	1.333	Ea.	37	41		78	102	804
	1240	4" diameter		16	1.500		70	46		116	147	
	1250	5" diameter		14	1.714		145	52.50		197.50	240	
	1260	6" diameter		12	2		150	61		211	258	
	1270	Tees, 3"		15	1.600		53	49		102	133	
	1280	4" diameter		12	2		85	61		146	186	
	1290	5" diameter		9	2.667		158	81.50		239.50	297	
	1300	6" diameter		6	4		165	122		287	365	
	1310	Couplings, 3"		18	1.333		24	41		65	88	
	1320	4" diameter		16	1.500		65	46		111	141	
	1330	5" diameter		14	1.714		135	52.50		187.50	229	
	1340	6" diameter		12	2		140	61		201	247	
	1350	Cross-over nipples, 3"		18	1.333		5.75	41		46.75	68	
	1360	4" diameter		16	1.500		6.75	46		52.75	77	
	1370	5" diameter		14	1.714		10	52.50		62.50	90.50	
	1380	6" diameter		12	2		10.50	61		71.50	104	
	1400	Telescoping, reducers, concentric 4" x 3"		18	1.333		19.15	41		60.15	82.50	
	1410	5" x 4"		17	1.412		50	43		93	121	
	1420	6" x 5"	↓	16	1.500	↓	120	46		166	202	
		026 850 | Gas Distribution System										
854	0010	**PIPING, GAS SERVICE & DISTRIBUTION, POLYETHYLENE**										854
	0020	not including excavation or backfill										
	1000	60 psi coils, comp cplg @ 100', 1/2" diameter, SDR 9.3	B-20A	608	.053	L.F.	.35	1.36		1.71	2.49	
	1040	1-1/4" diameter, SDR 11		544	.059		.57	1.52		2.09	2.98	
	1100	2" diameter, SDR 11		488	.066		.72	1.70		2.42	3.41	
	1160	3" diameter, SDR 11	↓	408	.078		1.51	2.03		3.54	4.79	
	1500	60 PSI 40' joints with coupling, 3" diameter, SDR 11	B-21A	408	.098		1.51	2.61	.96	5.08	6.70	
	1540	4" diameter, SDR 11		352	.114		3.33	3.02	1.11	7.46	9.55	
	1600	6" diameter, SDR 11		328	.122		10.90	3.24	1.19	15.33	18.30	
	1640	8" diameter, SDR 11	↓	272	.147	↓	14.55	3.91	1.44	19.90	23.50	
856	0010	**PIPING, GAS SERVICE & DISTRIBUTION, STEEL**										856
	0020	not including excavation or backfill, tar coated and wrapped										
	4000	Schedule 40, plain end										
	4040	1" diameter	Q-4	300	.107	L.F.	2.32	3.32	.17	5.81	7.80	
	4080	2" diameter		280	.114		3.64	3.55	.18	7.37	9.60	
	4120	3" diameter	↓	260	.123		6	3.83	.19	10.02	12.65	
	4160	4" diameter	B-35	255	.188		7.85	4.99	1.96	14.80	18.45	
	4200	5" diameter		220	.218		11.35	5.80	2.27	19.42	24	
	4240	6" diameter		180	.267		13.90	7.05	2.77	23.72	29.50	
	4280	8" diameter		140	.343		22	9.10	3.56	34.66	42	
	4320	10" diameter		100	.480		35.50	12.70	4.99	53.19	64	
	4360	12" diameter		80	.600		49.50	15.90	6.25	71.65	86	
	4400	14" diameter		75	.640		49.50	16.95	6.65	73.10	88	
	4440	16" diameter		70	.686		58	18.15	7.10	83.25	100	
	4480	18" diameter		65	.738		74.50	19.55	7.65	101.70	120	
	4520	20" diameter		60	.800		116	21	8.30	145.30	170	
	4560	24" diameter	↓	50	.960	↓	133	25.50	9.95	168.45	196	
	5000	Threaded and coupled										
	5002	4" diameter	B-20	144	.167	L.F.	8	4.05		12.05	15.20	
	5004	5" diameter	"	140	.171		10.65	4.17		14.82	18.25	
	5006	6" diameter	B-21	126	.222		21.50	5.55	1.08	28.13	33.50	
	5008	8" diameter		108	.259		32	6.50	1.26	39.76	46.50	
	5010	10" diameter		90	.311		45	7.80	1.51	54.31	63.50	
	5012	12" diameter	↓	72	.389	↓	63.50	9.75	1.89	75.14	87.50	

026 850 | Gas Distribution System		CREW	DAILY OUTPUT	LABOR-HOURS	UNIT	1999 BARE COSTS MAT.	LABOR	EQUIP.	TOTAL	TOTAL INCL O&P
856 6000	Schedule 80, plain end									
6002	4" diameter	B-35	144	.333	L.F.	11.65	8.85	3.46	23.96	30.50
6004	5" diameter		140	.343		12.65	9.10	3.56	25.31	32
6006	6" diameter		126	.381		15.90	10.10	3.96	29.96	37.50
6008	8" diameter		108	.444		21	11.75	4.62	37.37	47
6010	10" diameter		90	.533		32	14.15	5.55	51.70	63
6012	12" diameter		72	.667		42.50	17.65	6.95	67.10	81.50
6014	14" diameter 1/2" wall		54	.889		98	23.50	9.25	130.75	155
6016	16" diameter, 1/2" wall		48	1		105	26.50	10.40	141.90	168
6018	18" diameter 1/2" wall		40	1.200		158	32	12.45	202.45	236
6020	20" diameter, 1/2" wall		38	1.263		184	33.50	13.10	230.60	268
6022	24" diameter, 1/2" wall		36	1.333		210	35.50	13.85	259.35	300
7000	Threaded and coupled									
7002	4" diameter	B-20	144	.167	L.F.	11.65	4.05		15.70	19.20
7004	5" diameter	"	140	.171		12.65	4.17		16.82	20.50
7006	6" diameter	B-21	126	.222		15.90	5.55	1.08	22.53	27.50
7008	8" diameter		108	.259		50.50	6.50	1.26	58.26	67.50
7010	10" diameter		90	.311		78.50	7.80	1.51	87.81	100
7012	12" diameter		72	.389		86.50	9.75	1.89	98.14	112
8008	Elbow, weld joint, standard weight									
8020	4" diameter	Q-16	6.80	3.529	Ea.	34	107	7.30	148.30	209
8022	5" diameter		5.40	4.444		71	135	9.20	215.20	293
8024	6" diameter		4.50	5.333		80.50	162	11.05	253.55	345
8026	8" diameter		3.40	7.059		152	215	14.60	381.60	510
8028	10" diameter		2.70	8.889		208	270	18.40	496.40	660
8030	12" diameter		2.30	10.435		299	320	21.50	640.50	835
8032	14" diameter		1.80	13.333		415	405	27.50	847.50	1,100
8034	16" diameter		1.50	16		565	485	33	1,083	1,400
8036	18" diameter		1.40	17.143		725	520	35.50	1,280.50	1,625
8038	20" diameter		1.20	20		1,075	610	41.50	1,726.50	2,150
8040	24" diameter		1.02	23.529		1,425	715	48.50	2,188.50	2,675
8100	Extra heavy									
8102	4" diameter	Q-16	5.30	4.528	Ea.	68	138	9.35	215.35	294
8104	5" diameter		4.20	5.714		142	174	11.85	327.85	435
8106	6" diameter		3.50	6.857		161	209	14.20	384.20	510
8108	8" diameter		2.60	9.231		305	281	19.10	605.10	780
8110	10" diameter		2.10	11.429		277	350	23.50	650.50	855
8112	12" diameter		1.80	13.333		400	405	27.50	832.50	1,075
8114	14" diameter		1.40	17.143		550	520	35.50	1,105.50	1,425
8116	16" diameter		1.20	20		755	610	41.50	1,406.50	1,800
8118	18" diameter		1.10	21.818		970	665	45	1,680	2,125
8120	20" diameter		.94	25.532		1,425	775	53	2,253	2,800
8122	24" diameter		.80	30		1,900	915	62	2,877	3,525
8200	Malleable, standard weight									
8202	4" diameter	B-20	12	2	Ea.	44.50	48.50		93	126
8204	5" diameter		9	2.667		107	65		172	220
8206	6" diameter		8	3		126	73		199	254
8208	8" diameter		6	4		264	97.50		361.50	445
8210	10" diameter		5	4.800		480	117		597	715
8212	12" diameter		4	6		580	146		726	865
8300	Extra heavy									
8302	4" diameter	B-20	12	2	Ea.	89	48.50		137.50	175
8304	5" diameter	B-21	9	3.111		161	78	15.10	254.10	315
8306	6" diameter		8	3.500		252	87.50	17	356.50	435
8308	8" diameter		6	4.667		395	117	22.50	534.50	645
8310	10" diameter		5	5.600		480	140	27	647	780

		026 850 | Gas Distribution System	CREW	DAILY OUTPUT	LABOR-HOURS	UNIT	1999 BARE COSTS MAT.	LABOR	EQUIP.	TOTAL	TOTAL INCL O&P
856	8312	12" diameter	B-21	4	7	Ea.	770	175	34	979	1,150
	8500	Tee weld, standard weight									
	8510	4" diameter	Q-16	4.50	5.333	Ea.	50.50	162	11.05	223.55	315
	8512	5" diameter		3.60	6.667		91	203	13.80	307.80	420
	8514	6" diameter		3	8		91	243	16.55	350.55	490
	8516	8" diameter		2.30	10.435		167	320	21.50	508.50	690
	8518	10" diameter		1.80	13.333		278	405	27.50	710.50	950
	8520	12" diameter		1.50	16		420	485	33	938	1,250
	8522	14" diameter		1.20	20		700	610	41.50	1,351.50	1,725
	8524	16" diameter		1	24		850	730	49.50	1,629.50	2,100
	8526	18" diameter		.90	26.667		1,350	810	55	2,215	2,750
	8528	20" diameter		.80	30		2,075	915	62	3,052	3,725
	8530	24" diameter		.70	34.286		2,575	1,050	71	3,696	4,500
	8810	Malleable, standard weight									
	8812	4" diameter	B-20	8	3	Ea.	75.50	73		148.50	198
	8814	5"	B-21	6	4.667		178	117	22.50	317.50	405
	8816	6"		5.30	5.283		205	132	25.50	362.50	460
	8818	8" diameter		4	7		450	175	34	659	805
	8820	10"		3.30	8.485		935	212	41	1,188	1,400
	8822	12" diameter		2.70	10.370		1,300	260	50.50	1,610.50	1,875
	8900	Extra heavy									
	8902	4" diameter	B-20	8	3	Ea.	113	73		186	240
	8904	5"	B-21	6	4.667		266	117	22.50	405.50	500
	8906	6"		5.30	5.283		310	132	25.50	467.50	575
	8908	8" diameter		4	7		450	175	34	659	805
	8910	10"		3.30	8.485		935	212	41	1,188	1,400
	8912	12" diameter		2.70	10.370		1,300	260	50.50	1,610.50	1,875
858	0010	PIPING, VALVES & METERS, GAS DISTRIBUTION									
	0020	not including excavation or backfill									
	0100	Gas stops, with or without checks									
	0140	1-1/4" size	1 Plum	12	.667	Ea.	30.50	21.50		52	66.50
	0180	1-1/2" size		10	.800		35	26		61	78
	0200	2" size		8	1		56	32.50		88.50	111
	0600	Pressure regulator valves, iron and bronze									
	0640	1-1/2" diameter	1 Plum	5	1.600	Ea.	160	52		212	255
	0680	2" diameter	"	4.50	1.778		180	58		238	286
	0700	3" diameter	Q-22A	3.50	9.143		1,300	250	127	1,677	1,950
	0740	4" diameter	"	2.50	12.800		2,850	350	177	3,377	3,850
	0840	10" diameter	Q-1	1	16		14,000	470		14,470	16,100
	2000	Lubricated semi-steel plug valve									
	2040	3/4" diameter	1 Plum	16	.500	Ea.	75	16.30		91.30	107
	2080	1" diameter		14	.571		90	18.65		108.65	127
	2100	1-1/4" diameter		12	.667		110	21.50		131.50	154
	2140	1-1/2" diameter		11	.727		115	23.50		138.50	163
	2180	2" diameter		8	1		135	32.50		167.50	199
	2300	2-1/2" diameter	Q-1	5	3.200		210	94		304	375
	2340	3" diameter	"	4.50	3.556		255	104		359	440

027 050 | Drainage

	Line	Description	Crew	Daily Output	Labor-Hours	Unit	1999 Bare Costs Mat.	Labor	Equip.	Total	Total Incl O&P
054	0010	**DRAINAGE** Geotextiles									
	0100	Fabric, laid in trench, polypropylene, ideal conditions	2 Clab	2,400	.007	S.Y.	1.25	.14		1.39	1.60
	0110	Adverse conditions		1,600	.010	"	1.30	.21		1.51	1.77
	0170	Fabric ply bonded to 3 dimen. nylon mat, .4″ thk, ideal conditions		2,000	.008	S.F.	.80	.17		.97	1.15
	0180	Adverse conditions		1,200	.013	"	1	.29		1.29	1.55
	0185	Soil drainage mat on vertical wall, 0.44″ thick		265	.060	S.Y.	14	1.30		15.30	17.45
	0188	0.25″ thick		300	.053	"	8.50	1.14		9.64	11.15
	0190	0.8″ thick, ideal conditions		2,400	.007	S.F.	1.25	.14		1.39	1.60
	0200	Adverse conditions	↓	1,600	.010	"	1.65	.21		1.86	2.16
	0300	Drainage material, 3/4″ gravel fill in trench	B-6	260	.092	C.Y.	18.20	2.16	.84	21.20	24.50
	0400	Pea stone	"	260	.092	"	19	2.16	.84	22	25.50

027 100 | Subdrainage Systems

	Line	Description	Crew	Daily Output	Labor-Hours	Unit	1999 Bare Costs Mat.	Labor	Equip.	Total	Total Incl O&P
108	0010	**PIPING, SUBDRAINAGE, CONCRETE**									
	0021	Not including excavation and backfill									
	3000	Porous wall concrete underdrain, std. strength, 4″ diameter	B-20	335	.072	L.F.	1.78	1.74		3.52	4.70
	3020	6″ diameter	"	315	.076		2.32	1.85		4.17	5.45
	3040	8″ diameter	B-21	310	.090		2.86	2.26	.44	5.56	7.15
	3060	12″ diameter		285	.098		6.05	2.46	.48	8.99	11
	3080	15″ diameter		230	.122		6.95	3.05	.59	10.59	13.05
	3100	18″ diameter	↓	165	.170		9.20	4.25	.82	14.27	17.65
	4000	Extra strength, 6″ diameter	B-20	315	.076		2.29	1.85		4.14	5.45
	4020	8″ diameter	B-21	310	.090		3.43	2.26	.44	6.13	7.80
	4040	10″ diameter		285	.098		6.85	2.46	.48	9.79	11.90
	4060	12″ diameter		230	.122		7.45	3.05	.59	11.09	13.60
	4080	15″ diameter		200	.140		8.25	3.51	.68	12.44	15.30
	4100	18″ diameter	↓	165	.170	↓	12	4.25	.82	17.07	21
109	0010	**PIPING, SUBDRAINAGE, PLASTIC**									
	0020	Not including excavation and backfill									
	1110	10″				Ea.	4.29			4.29	4.72
	2100	Perforated PVC, schedule 40, 4″ diameter	B-14	314	.153	L.F.	1.44	3.48	.70	5.62	7.80
	2110	6″ diameter		300	.160		2.66	3.64	.73	7.03	9.40
	2120	8″ diameter		290	.166		3.02	3.76	.75	7.53	10.05
	2130	10″ diameter		280	.171		4.67	3.90	.78	9.35	12.10
	2140	12″ diameter	↓	270	.178	↓	6.45	4.04	.81	11.30	14.30
110	0010	**PIPING, SUBDRAINAGE, CORRUGATED METAL**									
	0021	Not including excavation and backfill									
	2010	Aluminum, perforated									
	2020	6″ diameter, 18 ga.	B-14	380	.126	L.F.	2.57	2.87	.58	6.02	7.95
	2200	8″ diameter, 16 ga.		370	.130		3.73	2.95	.59	7.27	9.35
	2220	10″ diameter, 16 ga.		360	.133		4.67	3.03	.61	8.31	10.55
	2240	12″ diameter, 16 ga.		285	.168		5.25	3.83	.77	9.85	12.60
	2260	18″ diameter, 16 ga.	↓	205	.234	↓	7.85	5.30	1.07	14.22	18.05
	3000	Uncoated galvanized, perforated									
	3020	6″ diameter, 18 ga.	B-20	380	.063	L.F.	4	1.54		5.54	6.80
	3200	8″ diameter, 16 ga.	"	370	.065		5.50	1.58		7.08	8.55
	3220	10″ diameter, 16 ga.	B-21	360	.078		8.25	1.95	.38	10.58	12.50
	3240	12″ diameter, 16 ga.		285	.098		8.65	2.46	.48	11.59	13.85
	3260	18″ diameter, 16 ga.	↓	205	.137	↓	13.20	3.42	.66	17.28	20.50
	4000	Steel, perforated, asphalt coated									
	4020	6″ diameter 18 ga.	B-20	380	.063	L.F.	3.20	1.54		4.74	5.95
	4030	8″ diameter 18 ga	"	370	.065		5	1.58		6.58	8
	4040	10″ diameter 16 ga	B-21	360	.078		5.75	1.95	.38	8.08	9.80
	4050	12″ diameter 16 ga		285	.098		6.60	2.46	.48	9.54	11.60
	4060	18″ diameter 16 ga	↓	205	.137	↓	9	3.42	.66	13.08	16

027 100 | Subdrainage Systems			CREW	DAILY OUTPUT	LABOR-HOURS	UNIT	1999 BARE COSTS MAT.	LABOR	EQUIP.	TOTAL	TOTAL INCL O&P	
111	0010	PIPING, SUBDRAINAGE, CORR. PLASTIC TUBING, PERF. OR PLAIN										111
	0020	In rolls, not including excavation and backfill										
	0030	3" diameter	2 Clab	1,200	.013	L.F.	.42	.29		.71	.91	
	0040	4" diameter		1,200	.013		.52	.29		.81	1.03	
	0041	With silt sock		1,200	.013		.61	.29		.90	1.12	
	0050	5" diameter		900	.018		1.26	.38		1.64	1.99	
	0060	6" diameter		900	.018		1.68	.38		2.06	2.45	
	0080	8" diameter	↓	700	.023	↓	2.34	.49		2.83	3.34	
	0200	Fittings										
	0230	Elbows, 3" diameter	1 Clab	32	.250	Ea.	4.08	5.35		9.43	12.95	
	0240	4" diameter		32	.250		4.49	5.35		9.84	13.40	
	0250	5" diameter		32	.250		5.30	5.35		10.65	14.30	
	0260	6" diameter		32	.250		7.15	5.35		12.50	16.30	
	0280	8" diameter		32	.250		8.75	5.35		14.10	18.10	
	0330	Tees, 3" diameter		27	.296		3.97	6.35		10.32	14.35	
	0340	4" diameter		27	.296		4.17	6.35		10.52	14.60	
	0350	5" diameter		27	.296		5.15	6.35		11.50	15.70	
	0360	6" diameter		27	.296		6.95	6.35		13.30	17.65	
	0370	6" x 6" x 4"		27	.296		7.15	6.35		13.50	17.85	
	0380	8" diameter		27	.296		8.35	6.35		14.70	19.15	
	0390	8" x 8" x 6"		27	.296		14.90	6.35		21.25	26.50	
	0430	End cap, 3" diameter		32	.250		1.53	5.35		6.88	10.15	
	0440	4" diameter		32	.250		1.68	5.35		7.03	10.30	
	0460	6" diameter		32	.250		2.52	5.35		7.87	11.25	
	0480	8" diameter		32	.250		5.30	5.35		10.65	14.25	
	0530	Coupler, 3" diameter		32	.250		1.27	5.35		6.62	9.85	
	0540	4" diameter		32	.250		1.33	5.35		6.68	9.90	
	0550	5" diameter		32	.250		1.65	5.35		7	10.25	
	0560	6" diameter		32	.250		2.22	5.35		7.57	10.90	
	0580	8" diameter	↓	32	.250		4.13	5.35		9.48	13	
	0590	Heavy duty highway type, add					10%					
	0660	Reducer, 6" to 4"	1 Clab	32	.250		2.85	5.35		8.20	11.60	
	0680	8" to 6"		32	.250		4.56	5.35		9.91	13.45	
	0730	"Y" fitting, 3" diameter		27	.296		4.99	6.35		11.34	15.50	
	0740	4" diameter		27	.296		5.75	6.35		12.10	16.30	
	0750	5" diameter		27	.296		7.25	6.35		13.60	17.95	
	0760	6" diameter		27	.296		8.75	6.35		15.10	19.60	
	0780	8" diameter	↓	27	.296	↓	10.75	6.35		17.10	22	
	0860	Silt sock only for above tubing, 6" dia.				L.F.	.56			.56	.62	
	0880	8" diameter				"	.92			.92	1.02	
112	0010	PIPING, SUBDRAINAGE, VITRIFIED CLAY										112
	0020	Not including excavation and backfill										
	3000	Perforated, 5' lengths, C700, 4" diameter	B-14	400	.120	L.F.	1.93	2.73	.55	5.21	7	
	3020	6" diameter		315	.152		3	3.47	.69	7.16	9.45	
	3040	8" diameter		290	.166		4.02	3.76	.75	8.53	11.15	
	3050	10" diameter		320	.150		6	3.41	.68	10.09	12.70	
	3060	12" diameter	↓	275	.175		8.15	3.97	.80	12.92	16	
	4000	Channel pipe, 4" diameter	B-20	430	.056		2	1.36		3.36	4.34	
	4020	6" diameter		335	.072		3	1.74		4.74	6.05	
	4060	8" diameter	↓	295	.081		4.50	1.98		6.48	8.05	
	4080	12" diameter	B-21	280	.100	↓	9.50	2.50	.49	12.49	14.90	
	7998	Standard fittings										
	8000	4" diameter	B-20	35	.686	Ea.	9.75	16.70		26.45	37.50	
	8020	6" diameter		30	.800		15	19.45		34.45	47	
	8040	8" diameter		28	.857		24	21		45	59.50	
	8060	10" diameter	↓	25	.960	↓	38	23.50		61.50	79	

		027 100 | Subdrainage Systems	CREW	DAILY OUTPUT	LABOR-HOURS	UNIT	1999 BARE COSTS MAT.	LABOR	EQUIP.	TOTAL	TOTAL INCL O&P	
112	8080	12″ diameter	B-20	20	1.200	Ea.	55	29		84	107	112

		027 150 | Sewage Systems	CREW	DAILY OUTPUT	LABOR-HOURS	UNIT	MAT.	LABOR	EQUIP.	TOTAL	TOTAL INCL O&P	
152	0010	**CATCH BASINS OR MANHOLES** not including footing, excavation,										152
	0020	backfill, frame and cover										
	0050	Brick, 4′ inside diameter, 4′ deep	D-1	1	16	Ea.	263	395		658	900	
	0100	6′ deep		.70	22.857		370	560		930	1,275	
	0150	8′ deep		.50	32		475	785		1,260	1,750	
	0200	For depths over 8′, add		4	4	V.L.F.	100	98.50		198.50	263	
	0400	Concrete blocks (radial), 4′ I.D., 4′ deep		1.50	10.667	Ea.	300	262		562	740	
	0500	6′ deep		1	16		390	395		785	1,050	
	0600	8′ deep		.70	22.857		510	560		1,070	1,425	
	0700	For depths over 8′, add		5.50	2.909	V.L.F.	76.50	71.50		148	195	
	0800	Concrete, cast in place, 4′ x 4′, 8″ thick, 4′ deep	C-14H	2	24	Ea.	335	650	18.70	1,003.70	1,400	
	0900	6′ deep		1.50	32		485	865	25	1,375	1,925	
	1000	8′ deep		1	48		650	1,300	37.50	1,987.50	2,800	
	1100	For depths over 8′, add		8	6	V.L.F.	82	162	4.68	248.68	355	
	1110	Precast, 4′ I.D., 4′ deep	B-22	4.10	7.317	Ea.	315	185	49.50	549.50	690	
	1120	6′ deep		3	10		425	253	68	746	935	
	1130	8′ deep		2	15		470	380	102	952	1,225	
	1140	For depths over 8′, add		16	1.875	V.L.F.	78.50	47.50	12.75	138.75	175	
	1150	5′ I.D., 4′ deep	B-6	3	8	Ea.	445	187	73	705	860	
	1160	6′ deep		2	12		605	280	109	994	1,225	
	1170	8′ deep		1.50	16		760	375	146	1,281	1,575	
	1180	For depths over 8′, add		12	2	V.L.F.	99	46.50	18.25	163.75	202	
	1190	6′ I.D., 4′ deep		2	12	Ea.	730	280	109	1,119	1,350	
	1200	6′ deep		1.50	16		950	375	146	1,471	1,800	
	1210	8′ deep		1	24		1,175	560	219	1,954	2,375	
	1220	For depths over 8′, add		8	3	V.L.F.	153	70	27.50	250.50	305	
	1250	Slab tops, precast, 8″ thick										
	1300	4′ diameter manhole	B-6	8	3	Ea.	148	70	27.50	245.50	300	
	1400	5′ diameter manhole		7.50	3.200		247	75	29	351	420	
	1500	6′ diameter manhole		7	3.429		287	80	31.50	398.50	475	
	1510	Grates only, for pipe bells, plastic, 4″ diameter pipe	1 Clab	50	.160		4.23	3.43		7.66	10.05	
	1514	6″ diameter pipe		50	.160		5.10	3.43		8.53	11	
	1516	8″ diameter pipe		50	.160		5.85	3.43		9.28	11.85	
	1520	For beehive type grate, add					50%					
	1550	Gray iron, 4″ dia. pipe, light duty	1 Clab	50	.160		14.30	3.43		17.73	21	
	1551	Heavy duty		50	.160		24.50	3.43		27.93	32.50	
	1552	6″ diameter, pipe, light duty		50	.160		22.50	3.43		25.93	30	
	1553	Heavy duty		50	.160		28.50	3.43		31.93	37	
	1554	8″ diamter pipe, light duty		50	.160		30.50	3.43		33.93	39	
	1555	Heavy duty		50	.160		32.50	3.43		35.93	41.50	
	1556	10″ diameter pipe, light duty		40	.200		41	4.29		45.29	52	
	1557	Heavy duty		40	.200		45	4.29		49.29	56.50	
	1558	12″ diamter pipe, light duty		40	.200		51	4.29		55.29	63	
	1559	Heavy duty		40	.200		56	4.29		60.29	68.50	
	1560	15″ diamter pipe, light duty		32	.250		63	5.35		68.35	78	
	1561	Heavy duty		32	.250		67.50	5.35		72.85	82.50	
	1562	18″ diameter pipe, light duty		32	.250		71.50	5.35		76.85	87	
	1563	Heavy duty		32	.250		79.50	5.35		84.85	96	
	1568	For beehive type grate, add					50%					
	1570	Covers only for pipe bells, gray iron, med. duty										
	1571	4″ diameter pipe	1 Clab	50	.160	Ea.	28.50	3.43		31.93	37	
	1572	6″ diameter pipe		50	.160		46	3.43		49.43	56	

027 150 | Sewage Systems

				DAILY	LABOR-		1999 BARE COSTS				TOTAL
			CREW	OUTPUT	HOURS	UNIT	MAT.	LABOR	EQUIP.	TOTAL	INCL O&P
152	1573	8" diameter pipe	1 Clab	50	.160	Ea.	61	3.43		64.43	73
	1574	10" diameter pipe		40	.200		66.50	4.29		70.79	80
	1575	12" diameter pipe		40	.200		71.50	4.29		75.79	85.50
	1576	15" diameter pipe		32	.250		81.50	5.35		86.85	98.50
	1577	18" diameter pipe	↓	32	.250		112	5.35		117.35	131
	1578	24" diameter pipe	2 Clab	24	.667		138	14.30		152.30	174
	1579	36" diameter pipe	"	24	.667	↓	165	14.30		179.30	205
	1580	Curb inlet frame, grate, and curb box									
	1582	Large 24" x 36" heavy duty	B-24	2	12	Ea.	355	300		655	860
	1590	Small 10" x 21" medium duty	"	2	12		281	300		581	775
	1600	Frames & covers, C.I., 24" square, 500 lb.	B-6	7.80	3.077		213	72	28	313	380
	1700	26" D shape, 600 lb.		7	3.429		214	80	31.50	325.50	395
	1800	Light traffic, 18" diameter, 100 lb.		10	2.400		76.50	56	22	154.50	195
	1900	24" diameter, 300 lb.		8.70	2.759		138	64.50	25	227.50	279
	2000	36" diameter, 900 lb.		5.80	4.138		385	96.50	37.50	519	610
	2100	Heavy traffic, 24" diameter, 400 lb.		7.80	3.077		172	72	28	272	330
	2200	36" diameter, 1150 lb.		3	8		510	187	73	770	930
	2300	Mass. State standard, 26" diameter, 475 lb.		7	3.429		214	80	31.50	325.50	395
	2400	30" diameter, 620 lb.		7	3.429		243	80	31.50	354.50	425
	2500	Watertight, 24" diameter, 350 lb.		7.80	3.077		293	72	28	393	465
	2600	26" diameter, 500 lb.		7	3.429		330	80	31.50	441.50	525
	2700	32" diameter, 575 lb.	↓	6	4	↓	390	93.50	36.50	520	610
	2800	3 piece cover & frame, 10" deep,									
	2900	1200 lbs., for heavy equipment	B-6	3	8	Ea.	770	187	73	1,030	1,225
	3000	Raised for paving 1-1/4" to 2" high,									
	3100	4 piece expansion ring									
	3200	20" to 26" diameter	1 Clab	3	2.667	Ea.	102	57		159	202
	3300	30" to 36" diameter	"	3	2.667	"	143	57		200	247
	3320	Frames and covers, existing, raised for paving 2", including									
	3340	row of brick, concrete collar, up to 12" wide frame	B-6	18	1.333	Ea.	31	31	12.15	74.15	96
	3360	20" to 26" wide frame		11	2.182		41	51	19.90	111.90	146
	3380	30" to 36" wide frame	↓	9	2.667		51	62.50	24.50	138	180
	3400	Inverts, single channel brick	D-1	3	5.333		57	131		188	267
	3500	Concrete		5	3.200		45	78.50		123.50	172
	3600	Triple channel, brick		2	8		86.50	197		283.50	400
	3700	Concrete	↓	3	5.333		61	131		192	271
	3800	Steps, heavyweight cast iron, 7" x 9"	1 Bric	40	.200		8.35	5.50		13.85	17.75
	3900	8" x 9"		40	.200		12.50	5.50		18	22.50
	3928	12" x 10-1/2"		40	.200		13	5.50		18.50	23
	4000	Standard sizes, galvanized steel		40	.200		11.75	5.50		17.25	21.50
	4100	Aluminum	↓	40	.200	↓	13	5.50		18.50	23
156	0010	**PIPE REPAIR**									
	0020	Not including excavation or backfill									
	0100	Clamp, stainless steel, lightweight, for steel pipe									
	0110	3" long, 1/2" diameter pipe	1 Plum	34	.235	Ea.	6.60	7.65		14.25	18.85
	0120	3/4" diameter pipe		32	.250		6.80	8.15		14.95	19.85
	0130	1" diameter pipe		30	.267		7.30	8.70		16	21
	0140	1-1/4" diameter pipe		28	.286		7.75	9.30		17.05	22.50
	0150	1-1/2" diameter pipe		26	.308		8	10.05		18.05	24
	0160	2" diameter pipe		24	.333		8.75	10.85		19.60	26
	0170	2-1/2" diameter pipe		23	.348		9.50	11.35		20.85	27.50
	0180	3" diameter pipe		22	.364		10.50	11.85		22.35	29.50
	0190	3-1/2" diameter pipe	↓	21	.381		11.25	12.40		23.65	31
	0200	4" diameter pipe	B-20	56	.429		11.75	10.40		22.15	29.50
	0210	5" diameter pipe	↓	53	.453	↓	13.50	11		24.50	32

027 150 | Sewage Systems

156		CREW	DAILY OUTPUT	LABOR-HOURS	UNIT	1999 BARE COSTS MAT.	LABOR	EQUIP.	TOTAL	TOTAL INCL O&P
0220	6" diameter pipe	B-20	48	.500	Ea.	15	12.15		27.15	35.50
0230	8" diameter pipe		30	.800		17.50	19.45		36.95	50
0240	10" diameter pipe		28	.857		45	21		66	82.50
0250	12" diameter pipe		24	1		48	24.50		72.50	91.50
0260	14" diameter pipe		22	1.091		50	26.50		76.50	97
0270	16" diameter pipe		20	1.200		54	29		83	106
0280	18" diameter pipe		18	1.333		57	32.50		89.50	114
0290	20" diameter pipe		16	1.500		63	36.50		99.50	127
0300	24" diameter pipe		14	1.714		70	41.50		111.50	143
0360	For 6" long, add					100%	40%			
0370	For 9" long, add					200%	100%			
0380	For 12" long, add					300%	150%			
0390	For 18" long, add					500%	200%			
1000	Clamp, stainless steel, with threaded service tap									
1040	Full seal for iron, steel, PVC pipe									
1100	6" long, 2" diameter pipe	1 Plum	17	.471	Ea.	85	15.35		100.35	117
1110	2-1/2" diameter pipe		16	.500		90	16.30		106.30	124
1120	3" diameter pipe		15.60	.513		96	16.70		112.70	132
1130	3-1/2" diameter pipe		15	.533		100	17.40		117.40	137
1140	4" diameter pipe	B-20	40	.600		110	14.60		124.60	144
1150	6" diameter pipe		34	.706		127	17.15		144.15	167
1160	8" diameter pipe		21	1.143		158	28		186	218
1170	10" diameter pipe		20	1.200		175	29		204	239
1180	12" diameter pipe		17	1.412		200	34.50		234.50	274
1205	For 9" long, add					20%	45%			
1210	For 12" long, add					40%	80%			
1220	For 18" long, add					70%	110%			
1600	Clamp, stainless steel, single section									
1640	Full seal for iron, steel, PVC pipe									
1700	6" long, 2" diameter pipe	1 Plum	17	.471	Ea.	44	15.35		59.35	72
1710	2-1/2" diameter pipe		16	.500		48	16.30		64.30	77.50
1720	3" diameter pipe		15.60	.513		50	16.70		66.70	80.50
1730	3-1/2" diameter pipe		15	.533		49.50	17.40		66.90	81
1740	4" diameter pipe	B-20	40	.600		60	14.60		74.60	89
1750	6" diameter pipe		34	.706		75	17.15		92.15	110
1760	8" diameter pipe		21	1.143		100	28		128	154
1770	10" diameter pipe		20	1.200		115	29		144	173
1780	12" diameter pipe		17	1.412		125	34.50		159.50	192
1800	For 9" long, add					40%	45%			
1810	For 12" long, add					60%	80%			
1820	For 18" long, add					120%	110%			
2000	Clamp, stainless steel, two section									
2040	Full seal, for iron, steel, PVC pipe									
2100	6" long, 4" diameter pipe	B-20	24	1	Ea.	85	24.50		109.50	132
2110	6" diameter pipe		20	1.200		100	29		129	156
2120	8" diameter pipe		13	1.846		110	45		155	192
2130	10" diameter pipe		12	2		160	48.50		208.50	253
2140	12" diameter pipe		10	2.400		175	58.50		233.50	285
2200	9" long, 4" diameter pipe		16	1.500		110	36.50		146.50	179
2210	6" diameter pipe		13	1.846		125	45		170	209
2220	8" diameter pipe		9	2.667		130	65		195	245
2230	10" diameter pipe		8	3		180	73		253	315
2240	12" diameter pipe		7	3.429		200	83.50		283.50	350
2250	14" diameter pipe		6.40	3.750		300	91		391	475
2260	16" diameter pipe		6	4		315	97.50		412.50	500
2270	18" diameter pipe		5	4.800		325	117		442	545

2 SITE WORK

		027 150 | Sewage Systems	CREW	DAILY OUTPUT	LABOR-HOURS	UNIT	1999 BARE COSTS MAT.	LABOR	EQUIP.	TOTAL	TOTAL INCL O&P
156	2280	20″ diameter pipe	B-20	4.60	5.217	Ea.	350	127		477	585
	2290	24″ diameter pipe		4	6		600	146		746	890
	2320	For 12″ long, add to 9″					15%	25%			
	2330	For 18″ long, add to 9″					70%	55%			
	8000	For internal cleaning and inspection, see Div. 026-108									
	8100	For pipe testing, see Div. 151-220									
162	0010	**PIPING, DRAINAGE & SEWAGE, CONCRETE** R027-110									
	0020	Not including excavation or backfill									
	0050	Box culvert, cast in place, 6′ x 6′	C-15	16	4.500	L.F.	85	115		200	276
	0060	8′ x 8′		14	5.143		125	132		257	345
	0070	12′ x 12′		10	7.200		245	185		430	560
	0100	Box culvert, precast, base price, 8′ long, 6′ x 3′	B-69	140	.343		180	8.05	7.85	195.90	219
	0150	6′ x 7′		125	.384		250	9.05	8.80	267.85	299
	0200	8′ x 3′		133	.361		210	8.50	8.25	226.75	253
	0250	8′ x 8′		100	.480		310	11.30	11	332.30	370
	0300	10′ x 3′		110	.436		250	10.25	10	270.25	300
	0350	10′ x 8′		80	.600		335	14.10	13.75	362.85	405
	0400	12′ x 3′		100	.480		315	11.30	11	337.30	375
	0401	Piping, precast concrete, box culvert, 8′ long, 12′ x 3′		100	.480			11.30	11	22.30	29.50
	0450	12′ x 8′		67	.716		420	16.85	16.40	453.25	505
	0500	Set up charge at plant, add to base price				Job	2,500			2,500	2,750
	0510	Inserts and keyway, add				Ea.	220			220	242
	0520	Sloped or skewed end, add				″	350			350	385
	1000	Non-reinforced pipe, extra strength, B&S or T&G joints									
	1010	6″ diameter	B-14	265.04	.181	L.F.	3.47	4.12	.83	8.42	11.15
	1020	8″ diameter		224	.214		3.81	4.87	.98	9.66	12.85
	1030	10″ diameter		216	.222		4.23	5.05	1.01	10.29	13.65
	1040	12″ diameter		200	.240		5.20	5.45	1.09	11.74	15.45
	1050	15″ diameter		180	.267		6.05	6.05	1.22	13.32	17.50
	1060	18″ diameter		144	.333		7.45	7.60	1.52	16.57	21.50
	1070	21″ diameter		112	.429		9.20	9.75	1.95	20.90	27.50
	1080	24″ diameter		100	.480		11.25	10.90	2.19	24.34	32
	1560	Reinforced culvert, class 2, no gaskets									
	1590	27″ diameter	B-21	88	.318	L.F.	15.25	7.95	1.54	24.74	31
	1592	30″ diameter	B-13	80	.700		17	16.25	6.85	40.10	52
	1594	36″ diameter	″	72	.778		25	18.10	7.65	50.75	64
	2000	Reinforced culvert, class 3, no gaskets									
	2010	12″ diameter	B-14	210	.229	L.F.	8.10	5.20	1.04	14.34	18.15
	2020	15″ diameter		175	.274		9.90	6.25	1.25	17.40	22
	2030	18″ diameter		130	.369		11.55	8.40	1.68	21.63	27.50
	2035	21″ diameter		120	.400		14.80	9.10	1.82	25.72	32.50
	2040	24″ diameter		100	.480		18.35	10.90	2.19	31.44	39.50
	2045	27″ diameter	B-13	92	.609		22	14.15	5.95	42.10	53
	2050	30″ diameter		88	.636		26.50	14.80	6.25	47.55	59
	2060	36″ diameter		72	.778		36	18.10	7.65	61.75	76
	2070	42″ diameter	B-13B	72	.778		44	18.10	11	73.10	88.50
	2080	48″ diameter		64	.875		57.50	20.50	12.40	90.40	108
	2090	60″ diameter		48	1.167		61.50	27	16.50	105	128
	2100	72″ diameter		40	1.400		95.50	32.50	19.80	147.80	178
	2120	84″ diameter		32	1.750		203	40.50	25	268.50	315
	2140	96″ diameter		24	2.333		243	54	33	330	390
	2200	With gaskets, class 3, 12″ diameter	B-21	168	.167		7.25	4.17	.81	12.23	15.45
	2220	15″ diameter		160	.175		8.70	4.38	.85	13.93	17.35
	2230	18″ diameter		152	.184		10.90	4.61	.89	16.40	20
	2240	24″ diameter		136	.206		16.30	5.15	1	22.45	27
	2260	30″ diameter	B-13	88	.636		22	14.80	6.25	43.05	54

027 150 | Sewage Systems

162	027 150 Sewage Systems	CREW	DAILY OUTPUT	LABOR-HOURS	UNIT	1999 BARE COSTS MAT.	LABOR	EQUIP.	TOTAL	TOTAL INCL O&P
2270	36" diameter (R027-110)	B-13	72	.778	L.F.	32.50	18.10	7.65	58.25	72.50
2290	48" diameter	B-13B	64	.875		53.50	20.50	12.40	86.40	104
2310	72" diameter	"	40	1.400		138	32.50	19.80	190.30	225
2330	Flared ends, 6'-1" long, 12" diameter	B-21	190	.147		27.50	3.69	.71	31.90	36.50
2340	15" diameter		155	.181		30.50	4.52	.88	35.90	41.50
2400	6'-2" long, 18" diameter		122	.230		34	5.75	1.11	40.86	47
2420	24" diameter		88	.318		45	7.95	1.54	54.49	63.50
2440	36" diameter	B-13	60	.933		57	21.50	9.15	87.65	107
2500	Class 4									
2510	12" diameter	B-21	168	.167	L.F.	7	4.17	.81	11.98	15.15
2512	15" diameter		160	.175		12.50	4.38	.85	17.73	21.50
2514	18" diameter		152	.184		13.75	4.61	.89	19.25	23.50
2516	21" diameter		144	.194		14	4.87	.94	19.81	24
2518	24" diameter		136	.206		20	5.15	1	26.15	31
2520	27" diameter		120	.233		22	5.85	1.13	28.98	34.50
2522	30" diameter	B-13	88	.636		28	14.80	6.25	49.05	61
2524	36" diameter	"	72	.778		35	18.10	7.65	60.75	75
2600	Class 5									
2610	12" diameter	B-21	168	.167	L.F.	10	4.17	.81	14.98	18.45
2612	15" diameter		160	.175		13	4.38	.85	18.23	22
2614	18" diameter		152	.184		15	4.61	.89	20.50	24.50
2616	21" diameter		144	.194		20	4.87	.94	25.81	30.50
2618	24" diameter		136	.206		24	5.15	1	30.15	35.50
2620	27" diameter		120	.233		28	5.85	1.13	34.98	41.50
2622	30" diameter	B-13	88	.636		35	14.80	6.25	56.05	68.50
2624	36" diameter	"	72	.778		41	18.10	7.65	66.75	81.50
2800	Add for rubber joints,					12%				
2810	12"					12%				
2812	15"					12%				
2814	18"					12%				
2816	21"					12%				
2818	24"					12%				
2820	27"					12%				
2822	30"					12%				
2824	36"					12%				
3040	Vitrified plate lined, add to above, 30" to 36" diameter				SFCA	3			3	3.30
3050	42" to 54" diameter, add					3.20			3.20	3.52
3060	60" to 72" diameter, add					3.75			3.75	4.13
3070	Over 72" diameter, add					4			4	4.40
3080	Radius pipe, add to pipe prices, 12" to 60" diameter				L.F.	50%				
3090	Over 60" diameter, add				"	20%				
3500	Reinforced elliptical, 8' lengths, C507 class 3									
3520	14" x 23" inside, round equivalent 18" diameter	B-21	82	.341	L.F.	40	8.55	1.66	50.21	59
3530	24" x 38" inside, round equivalent 30" diameter	B-13	58	.966		50	22.50	9.45	81.95	100
3540	29" x 45" inside, round equivalent 36" diameter		52	1.077		65	25	10.55	100.55	122
3550	38" x 60" inside, round equivalent 48" diameter		38	1.474		100	34.50	14.45	148.95	179
3551	Elliptical, 38" x 60" inside, round equivalent 48" diam.		38	1.474		100	34.50	14.45	148.95	179
3560	48" x 76" inside, round equivalent 60" diameter		26	2.154		135	50	21	206	250
3570	58" x 91" inside, round equivalent 72" diameter		22	2.545		200	59	25	284	340
3780	Concrete slotted pipe, class 4 mortar joint									
3800	12" diameter	B-21	168	.167	L.F.	10.50	4.17	.81	15.48	19
3840	18" diameter	"	152	.184	"	16.25	4.61	.89	21.75	26
3900	Class 4 O-ring									
3940	12" diameter	B-21	168	.167	L.F.	12	4.17	.81	16.98	20.50
3960	18" diameter	"	152	.184	"	18	4.61	.89	23.50	28
6200	Gasket, conc. pipe joint, 12"				Ea.	2.15			2.15	2.37

	027 150 | Sewage Systems	CREW	DAILY OUTPUT	LABOR-HOURS	UNIT	MAT.	LABOR	EQUIP.	TOTAL	TOTAL INCL O&P
						1999 BARE COSTS				
162	6220 24" R027-110				Ea.	3.73			3.73	4.10
	6240 36"					5.55			5.55	6.10
	6260 48"					7.95			7.95	8.75
	6270 60"					11			11	12.10
	6280 72"				↓	11.15			11.15	12.25
164	0010 **PIPING, STORM DRAINAGE, CORRUGATED METAL**									
	0020 Not including excavation or backfill									
	2000 Corrugated metal pipe, galvanized and coated									
	2020 Bituminous coated with paved invert, 20' lengths									
	2040 8" diameter, 16 ga.	B-14	330	.145	L.F.	7.10	3.31	.66	11.07	13.75
	2060 10" diameter, 16 ga.		260	.185		8.90	4.20	.84	13.94	17.25
	2080 12" diameter, 16 ga.		210	.229		11.90	5.20	1.04	18.14	22.50
	2100 15" diameter, 16 ga.		200	.240		12.45	5.45	1.09	18.99	23.50
	2120 18" diameter, 16 ga.		190	.253		13.30	5.75	1.15	20.20	25
	2140 24" diameter, 14 ga.	↓	160	.300		26.50	6.80	1.37	34.67	41
	2160 30" diameter, 14 ga.	B-13	120	.467		28.50	10.85	4.58	43.93	53.50
	2180 36" diameter, 12 ga.		120	.467		50	10.85	4.58	65.43	77
	2200 48" diameter, 12 ga.	↓	100	.560		61.50	13	5.50	80	94.50
	2220 60" diameter, 10 ga.	B-13B	75	.747		101	17.35	10.55	128.90	150
	2240 72" diameter, 8 ga.	"	45	1.244	↓	135	29	17.60	181.60	213
	2250 End sections, 8" diameter, 16 ga.	B-14	20	2.400	Ea.	33	54.50	10.95	98.45	134
	2255 10" diameter, 16 ga.		20	2.400		36	54.50	10.95	101.45	137
	2260 12" diameter, 16 ga.		18	2.667		38.50	60.50	12.15	111.15	150
	2265 15" diameter, 16 ga.		18	2.667		45.50	60.50	12.15	118.15	158
	2270 18" diameter, 16 ga.	↓	16	3		67.50	68	13.70	149.20	197
	2275 24" diameter, 16 ga.	B-13	16	3.500		99.50	81.50	34.50	215.50	274
	2280 30" diameter, 16 ga.		14	4		188	93	39	320	395
	2285 36" diameter, 14 ga.	↓	14	4		280	93	39	412	500
	2290 48" diameter, 14 ga.	B-13B	10	5.600		640	130	79.50	849.50	990
	2292 60" diameter, 14 ga.	B-13C	6	9.333		1,075	217	189	1,481	1,725
	2294 72" diameter, 14 ga.	"	5	11.200		1,325	260	226	1,811	2,100
	2300 Bends or elbows, 8" diameter	B-14	28	1.714		56.50	39	7.80	103.30	132
	2320 10" diameter		25	1.920		65	43.50	8.75	117.25	149
	2340 12" diameter, 16 ga.		30	1.600		109	36.50	7.30	152.80	185
	2342 18" diameter, 16 ga.		20	2.400		152	54.50	10.95	217.45	266
	2344 24" diameter, 14 ga.		16	3		220	68	13.70	301.70	365
	2346 30" diameter, 14 ga.	↓	15	3.200		325	73	14.60	412.60	490
	2348 36" diameter, 14 ga.	B-13	15	3.733		435	87	36.50	558.50	655
	2350 48" diameter, 12 ga.		12	4.667		745	108	46	899	1,025
	2352 60" diameter, 10 ga.		10	5.600		1,200	130	55	1,385	1,600
	2354 72" diameter, 10 ga.	↓	6	9.333		1,775	217	91.50	2,083.50	2,400
	2360 Wyes or tees, 8" diameter	B-20	25	.960		56	23.50		79.50	98.50
	2380 10" diameter	"	21	1.143		65	28		93	116
	2400 12" diameter, 16 ga.	B-14	22.48	2.135		122	48.50	9.75	180.25	221
	2410 18" diameter, 16 ga.		16	3		230	68	13.70	311.70	375
	2412 24" diameter, 14 ga.	↓	16	3		325	68	13.70	406.70	475
	2414 30" diameter, 14 ga.	B-13	12	4.667		420	108	46	574	680
	2416 36" diameter, 14 ga.		11	5.091		555	118	50	723	850
	2418 48" diameter, 12 ga.		10	5.600		990	130	55	1,175	1,375
	2420 60" diameter, 10 ga.		8	7		1,425	163	68.50	1,656.50	1,900
	2422 72" diameter, 10 ga.	↓	5	11.200	↓	1,975	260	110	2,345	2,700
	2500 Galvanized, 20' lengths									
	2520 8" diameter, 16 ga.	B-14	355	.135	L.F.	6.35	3.07	.62	10.04	12.45
	2540 10" diameter, 16 ga.		280	.171		6.70	3.90	.78	11.38	14.30
	2560 12" diameter, 16 ga.	↓	220	.218	↓	7.15	4.96	.99	13.10	16.75

027 150 | Sewage Systems

	027 150 Sewage Systems	CREW	DAILY OUTPUT	LABOR-HOURS	UNIT	1999 BARE COSTS MAT.	LABOR	EQUIP.	TOTAL	TOTAL INCL O&P
164 2580	15" diameter, 16 ga.	B-14	220	.218	L.F.	9.50	4.96	.99	15.45	19.30
2600	18" diameter, 16 ga.		205	.234		12.30	5.30	1.07	18.67	23
2610	21" diameter, 16 ga.		285	.168		13.45	3.83	.77	18.05	21.50
2620	24" diameter, 14 ga.		175	.274		21	6.25	1.25	28.50	34
2630	27" diameter, 14 ga.		270	.178		23.50	4.04	.81	28.35	33
2640	30" diameter, 14 ga.		130	.369		26.50	8.40	1.68	36.58	44.50
2660	36" diameter, 12 ga.	B-13	130	.431		34	10	4.23	48.23	58
2680	48" diameter, 12 ga.		110	.509		53.50	11.85	4.99	70.34	82.50
2700	60" diameter, 10 ga.		78	.718		86	16.70	7.05	109.75	128
2710	72" diameter, 10 ga.		60	.933		98	21.50	9.15	128.65	152
2711	Bends or elbows, 12" diameter, 16 ga.	B-14	30	1.600	Ea.	90.50	36.50	7.30	134.30	165
2712	15" diameter, 16 ga.		25.04	1.917		108	43.50	8.75	160.25	197
2714	18" diameter, 16 ga.		20	2.400		130	54.50	10.95	195.45	241
2716	24" diameter, 14 ga.		16	3		188	68	13.70	269.70	330
2718	30" diameter, 14 ga.		15	3.200		280	73	14.60	367.60	440
2720	36" diameter, 14 ga.	B-13	15	3.733		315	87	36.50	438.50	525
2722	48" diameter, 12 ga.		12	4.667		575	108	46	729	850
2724	60" diameter, 10 ga.		10	5.600		985	130	55	1,170	1,350
2726	72" diameter, 10 ga.		6	9.333		1,325	217	91.50	1,633.50	1,875
2728	Wyes or tees, 12" diameter, 16 ga.	B-14	22.48	2.135		137	48.50	9.75	195.25	238
2730	18" diameter, 16 ga.		15	3.200		196	73	14.60	283.60	345
2732	24" diameter, 14 ga.		15	3.200		279	73	14.60	366.60	435
2734	30" diameter, 14 ga.		14	3.429		370	78	15.65	463.65	545
2736	36" diameter, 14 ga.	B-13	14	4		475	93	39	607	710
2738	48" diameter, 12 ga.		12	4.667		735	108	46	889	1,025
2740	60" diameter, 10 ga.		10	5.600		1,250	130	55	1,435	1,650
2742	72" diameter, 10 ga.		6	9.333		1,725	217	91.50	2,033.50	2,325
2780	End sections, 8" diameter	B-14	24	2		52.50	45.50	9.10	107.10	139
2785	10" diameter		22	2.182		54	49.50	9.95	113.45	148
2790	12" diameter		35	1.371		65	31	6.25	102.25	127
2800	18" diameter		30	1.600		72.50	36.50	7.30	116.30	145
2810	24" diameter		25	1.920		105	43.50	8.75	157.25	193
2820	30" diameter		25	1.920		209	43.50	8.75	261.25	310
2825	36" diameter		20	2.400		299	54.50	10.95	364.45	430
2830	48" diameter		10	4.800		670	109	22	801	930
2835	60" diameter		5	9.600		1,175	218	44	1,437	1,700
2840	72" diameter		4	12		1,375	273	54.50	1,702.50	2,000
2850	Couplings, 12" diameter					9.30			9.30	10.20
2855	18" diameter					14.60			14.60	16.10
2860	24" diameter					25			25	27
2865	30" diameter					39			39	42.50
2870	36" diameter					60			60	65.50
2875	48" diameter					178			178	195
2880	60" diameter					198			198	217
2885	72" diameter					340			340	375
3000	Corrugated galvanized or alum. oval arch culverts, coated & paved									
3020	17" x 13", 16 ga., 15" equivalent	B-14	200	.240	L.F.	11.95	5.45	1.09	18.49	23
3040	21" x 15", 16 ga., 18" equivalent		150	.320		30.50	7.30	1.46	39.26	46.50
3060	28" x 20", 14 ga., 24" equivalent		125	.384		24	8.75	1.75	34.50	42
3080	35" x 24", 14 ga., 30" equivalent		100	.480		50	10.90	2.19	63.09	74.50
3100	42" x 29", 12 ga., 36" equivalent	B-13	100	.560		39.50	13	5.50	58	70
3120	49" x 33", 12 ga., 42" equivalent		90	.622		45.50	14.45	6.10	66.05	79
3140	57" x 38", 12 ga., 48" equivalent		75	.747		93	17.35	7.30	117.65	138
3160	Steel, plain oval arch culverts, plain									
3180	17" x 13", 16 ga., 15" equivalent	B-14	225	.213	L.F.	7.85	4.85	.97	13.67	17.30
3200	21" x 15", 16 ga., 18" equivalent		175	.274		15.50	6.25	1.25	23	28

	027 150 | Sewage Systems		CREW	DAILY OUTPUT	LABOR-HOURS	UNIT	1999 BARE COSTS MAT.	LABOR	EQUIP.	TOTAL	TOTAL INCL O&P
164	3220	28″ x 20″, 14 ga., 24″ equivalent	B-14	150	.320	L.F.	14.60	7.30	1.46	23.36	29
	3240	35″ x 24″, 14 ga., 30″ equivalent	B-13	108	.519		30	12.05	5.10	47.15	57.50
	3260	42″ x 29″, 12 ga., 36″ equivalent		108	.519		36	12.05	5.10	53.15	64
	3280	49″ x 33″, 12 ga., 42″ equivalent		92	.609		42.50	14.15	5.95	62.60	75
	3300	57″ x 38″, 12 ga., 48″ equivalent		75	.747		51	17.35	7.30	75.65	91
	3320	End sections, 17″ x 13″		22	2.545	Ea.	34	59	25	118	157
	3340	42″ x 29″		17	3.294	″	172	76.50	32.50	281	345
	3360	Multi-plate arch, steel	B-20	1,690	.014	Lb.	.60	.35		.95	1.20
	8000	Bends or elbows, 6″ diameter		30	.800	Ea.	40	19.45		59.45	74.50
	8020	8″ diameter		28	.857		50	21		71	88
	8040	10″ diameter		25	.960		60	23.50		83.50	103
	8060	12″ diameter	B-21	20	1.400		75	35	6.80	116.80	145
	8100	Wyes or tees, 6″ diameter	B-20	27	.889		45	21.50		66.50	83.50
	8120	8″ diameter		25	.960		52	23.50		75.50	94
	8140	10″ diameter		21	1.143		60	28		88	110
	8160	12″ diameter	B-21	18	1.556		75	39	7.55	121.55	152
166	0010	**PIPING, DRAINAGE & SEWAGE, PLASTIC**									
	0020	Not including excavation & backfill									
	1000	Reinforced plastic pipe, general strength, 4″ diameter	B-20	190	.126	L.F.	10.80	3.07		13.87	16.75
	1010	6″ diameter	″	170	.141		13.50	3.43		16.93	20.50
	1020	8″ diameter	B-21	160	.175		22.50	4.38	.85	27.73	33
	1030	10″ diameter		140	.200		34	5	.97	39.97	46.50
	1040	12″ diameter		100	.280		40.50	7	1.36	48.86	57
	5000	High strength, 4″ diameter	B-20	190	.126		12.95	3.07		16.02	19.10
	5010	6″ diameter	″	170	.141		16.20	3.43		19.63	23
	5020	8″ diameter	B-21	160	.175		27	4.38	.85	32.23	37.50
	5030	10″ diameter		140	.200		41	5	.97	46.97	54
	5040	12″ diameter		100	.280		48.50	7	1.36	56.86	66
	9100	Bends and elbows, general strength, 4″ diameter	B-20	19	1.263	Ea.	104	30.50		134.50	163
	9130	10″ diameter		8	3		360	73		433	510
	9140	12″ diameter		6	4		495	97.50		592.50	700
	9210	High strength, 4″ diameter		19	1.263		108	30.50		138.50	168
	9220	6″ diameter		12	2		158	48.50		206.50	250
	9230	8″ diameter		11	2.182		252	53		305	360
	9240	10″ diameter		8	3		355	73		428	505
	9250	12″ diameter		6	4		485	97.50		582.50	690
	9610	Wyes and tees, general strength, 4″ diameter		12	2		85	48.50		133.50	170
	9620	6″		7	3.429		225	83.50		308.50	380
	9630	8″ diameter		7	3.429		260	83.50		343.50	415
	9640	10″		6	4		410	97.50		507.50	605
	9650	12″ diameter		5	4.800		500	117		617	735
	9710	High strength, 4″ diameter		12	2		85	48.50		133.50	170
	9720	6″ diameter		7	3.429		220	83.50		303.50	375
	9730	8″ diameter		7	3.429		260	83.50		343.50	415
	9740	10″ diameter		6	4		410	97.50		507.50	605
	9750	12″ diameter		5	4.800		500	117		617	735
168	0010	**PIPING, DRAINAGE & SEWAGE, POLYVINYL CHLORIDE**									
	0020	Not including excavation or backfill									
	2000	10′ lengths, S.D.R. 35, B&S, 4″ diameter	B-20	375	.064	L.F.	2.17	1.56		3.73	4.84
	2040	6″ diameter		350	.069		3.95	1.67		5.62	7
	2080	8″ diameter		335	.072		3.94	1.74		5.68	7.10
	2120	10″ diameter	B-21	330	.085		3.81	2.12	.41	6.34	7.95
	2160	12″ diameter		320	.087		5.05	2.19	.42	7.66	9.45
	2200	15″ diameter		190	.147		11	3.69	.71	15.40	18.70
	3040	Fittings, bends or elbows, 4″ diameter	B-20	19	1.263	Ea.	2.75	30.50		33.25	51.50
	3080	6″ diameter		15	1.600		12	39		51	74.50

027 150 | Sewage Systems

				DAILY	LABOR-		1999 BARE COSTS				TOTAL
			CREW	OUTPUT	HOURS	UNIT	MAT.	LABOR	EQUIP.	TOTAL	INCL O&P
168	3120	Tees, 4" diameter	B-20	12	2	Ea.	3.39	48.50		51.89	80
	3160	6" diameter	↓	10	2.400	↓	12.20	58.50		70.70	105
	3200	Wyes, 4" diameter	↓	12	2	↓	3.92	48.50		52.42	81
	3240	6" diameter	↓	10	2.400	↓	11.70	58.50		70.20	105
170	0010	**PIPING, DRAINAGE & SEWAGE, SEWAGE VENT CAST IRON**									
	0020	Not including excavation or backfill									
	2022	Sewage vent cast iron, B & S, 4" diameter	Q-1	44	.364	L.F.	5.95	10.65		16.60	22.50
	2024	5" diameter	Q-2	62	.387	↓	8.40	11.80		20.20	27
	2026	6" diameter	"	59	.407	↓	10.25	12.40		22.65	30
	2028	8" diameter	Q-3	49	.653	↓	16.45	20.50		36.95	49
	2030	10" diameter	↓	45	.711	↓	27.50	22		49.50	63.50
	2032	12" diameter	↓	39	.821	↓	40	25.50		65.50	82
	2034	15" diameter	↓	35	.914	↓	58	28.50		86.50	107
	2048	For push on joint deduct						25%			
	2050	5" diameter						25%			
	2052	6" diameter						25%			
	2054	8" diameter						25%			
	2056	10" diameter						25%			
	2058	12" diameter						25%			
	2060	15" diameter						25%			
	8001	Fittings, bends and elbows									
	8110	4" diameter	Q-1	13	1.231	Ea.	16.55	36		52.55	72.50
	8112	5" diameter	Q-2	18	1.333	↓	23	40.50		63.50	87
	8114	6" diameter	"	17	1.412	↓	29	43		72	96.50
	8116	8" diameter	Q-3	11	2.909	↓	87	90.50		177.50	233
	8118	10" diameter	↓	10	3.200	↓	127	99.50		226.50	291
	8120	12" diameter	↓	9	3.556	↓	172	111		283	355
	8122	15" diameter	↓	7	4.571	↓	550	142		692	820
	8500	Wyes and tees									
	8510	4" diameter	Q-1	8	2	Ea.	21.50	58.50		80	113
	8512	5" diameter	Q-2	12	2	↓	30.50	61		91.50	126
	8514	6" diameter	"	11	2.182	↓	37.50	66.50		104	142
	8516	8" diameter	Q-3	7	4.571	↓	70.50	142		212.50	293
	8518	10" diameter	↓	6	5.333	↓	129	166		295	395
	8520	12" diameter	↓	4	8	↓	214	249		463	610
	8522	15" diameter	↓	3	10.667	↓	455	330		785	1,000
172	0010	**PIPING, DRAINAGE & SEWAGE, VITRIFIED CLAY** C700									
	0020	Not including excavation or backfill,									
	4030	Extra strength, compression joints, C425									
	5000	4" diameter x 4' long	B-20	265	.091	L.F.	1.77	2.20		3.97	5.40
	5020	6" diameter x 5' long	"	200	.120	↓	2.89	2.92		5.81	7.80
	5040	8" diameter x 5' long	B-21	200	.140	↓	4.08	3.51	.68	8.27	10.75
	5060	10" diameter x 5' long	↓	190	.147	↓	6.70	3.69	.71	11.10	13.95
	5080	12" diameter x 6' long	↓	150	.187	↓	8.80	4.67	.91	14.38	18
	5100	15" diameter x 7' long	↓	110	.255	↓	16.05	6.35	1.23	23.63	29
	5120	18" diameter x 7' long	↓	88	.318	↓	23.50	7.95	1.54	32.99	39.50
	5140	24" diameter x 7' long	↓	45	.622	↓	48	15.60	3.02	66.62	80.50
	5160	30" diameter x 7' long	B-22	31	.968	↓	83.50	24.50	6.55	114.55	138
	5180	36" diameter x 7' long	"	20	1.500	↓	123	38	10.20	171.20	207
	6000	For 3' lengths, add					30%	30%			
	6020	For 2' lengths, add					40%	60%			
	6060	For plain joints, deduct					25%				
	7010	Plain joints, 4" diameter x 4' long	B-20	380	.063	↓	1.50	1.54		3.04	4.07
	7014	6" diameter x 5' long	"	340	.071	↓	2.33	1.72		4.05	5.25
	7016	8" diameter x 5' long	B-21	300	.093	↓	3.33	2.34	.45	6.12	7.80
	7018	10" diameter x 5' long	↓	240	.117	↓	5.10	2.92	.57	8.59	10.80

027 150 | Sewage Systems

172		CREW	DAILY OUTPUT	LABOR-HOURS	UNIT	1999 BARE COSTS				TOTAL INCL O&P
						MAT.	LABOR	EQUIP.	TOTAL	
7020	12″ diameter x 6′ long	B-21	190	.147	L.F.	7.20	3.69	.71	11.60	14.55
7022	15″ diameter x 6′ long		155	.181		13.50	4.52	.88	18.90	23
7024	18″ diameter x 7′ long		122	.230		19.35	5.75	1.11	26.21	31.50
7026	21″ diameter x 7′ long		108	.259		31	6.50	1.26	38.76	45.50
7028	24″ diameter x 7′ long		88	.318		40.50	7.95	1.54	49.99	59
7032	30″ diameter x 7′ long	B-22	40	.750		50.50	19	5.10	74.60	90.50
7034	36″ diameter x 7′ long	″	30	1		61	25.50	6.80	93.30	114
7050	3′ lengths, add to above					30%				
7060	2′ lengths, add to above					40%				
8020	1/8 bends, plain joint, 4″ diameter	B-20	52	.462	Ea.	10.70	11.20		21.90	29.50
8022	5″ diameter		50	.480		16.10	11.65		27.75	36
8024	6″ diameter		48	.500		16.10	12.15		28.25	37
8026	8″ diameter		40	.600		26.50	14.60		41.10	52
8030	12″ diameter		30	.800		57.50	19.45		76.95	94
8032	15″ diameter		20	1.200		162	29		191	224
8034	18″ diameter	B-21	10	2.800		232	70	13.60	315.60	380
8036	24″ diameter		8	3.500		490	87.50	17	594.50	695
8038	30″ diameter		6	4.667		505	117	22.50	644.50	765
8040	36″ diameter		4	7		755	175	34	964	1,150
8050	Compression joint, C425, 4″ diameter	B-20	62	.387		13.55	9.40		22.95	30
8052	5″ diameter		60	.400		22.50	9.75		32.25	40.50
8054	6″ diameter		58	.414		22.50	10.05		32.55	41
8056	8″ diameter		50	.480		34.50	11.65		46.15	56.50
8060	12″ diameter		40	.600		69.50	14.60		84.10	99.50
8062	15″ diameter		30	.800		197	19.45		216.45	248
8064	18″ diameter	B-21	20	1.400		293	35	6.80	334.80	385
8066	24″ diameter	″	18	1.556		600	39	7.55	646.55	730
8068	30″ diameter	B-22	16	1.875		660	47.50	12.75	720.25	820
8070	36″ diameter	″	14	2.143		1,000	54.50	14.55	1,069.05	1,200
8092	1/4 bends, plain joint, 4″ diameter	B-20	52	.462		10.70	11.20		21.90	29.50
8094	5″ diameter		50	.480		16.10	11.65		27.75	36
8096	6″ diameter		48	.500		16.10	12.15		28.25	37
8098	8″ diameter		40	.600		26.50	14.60		41.10	52
8102	12″ diameter		30	.800		57.50	19.45		76.95	94
8104	15″ diameter		20	1.200		162	29		191	224
8106	18″ diameter	B-21	10	2.800		232	70	13.60	315.60	380
8108	24″ diameter	″	8	3.500		490	87.50	17	594.50	695
8110	30″ diameter	B-22	6	5		505	127	34	666	790
8112	36″ diameter	″	4	7.500		755	190	51	996	1,175
8150	Compression joint, C425, 4″ diameter	B-20	62	.387		13.55	9.40		22.95	30
8152	5″ diameter		60	.400		22.50	9.75		32.25	40.50
8154	6″ diameter		58	.414		22.50	10.05		32.55	41
8156	8″ diameter		50	.480		34.50	11.65		46.15	56.50
8160	12″ diameter		40	.600		69.50	14.60		84.10	99.50
8162	15″ diameter		30	.800		197	19.45		216.45	248
8164	18″ diameter	B-21	20	1.400		293	35	6.80	334.80	385
8166	24″ diameter	″	18	1.556		600	39	7.55	646.55	730
8168	30″ diameter	B-22	16	1.875		650	47.50	12.75	710.25	805
8170	36″ diameter	″	14	2.143		1,000	54.50	14.55	1,069.05	1,200
8200	Wyes and tees, plain joints, 4″ diameter	B-20	47	.511		10.70	12.40		23.10	31.50
8212	5′ diameter		45	.533		16.10	12.95		29.05	38
8214	6″ diameter		43	.558		16.10	13.55		29.65	39
8216	8″ diameter		35	.686		26.50	16.70		43.20	55.50
8220	12″ diameter		25	.960		57.50	23.50		81	101
8222	15″ diameter		15	1.600		162	39		201	240
8224	18″ diameter	B-21	8	3.500		232	87.50	17	336.50	410

027 150 | Sewage Systems

172		CREW	DAILY OUTPUT	LABOR-HOURS	UNIT	1999 BARE COSTS MAT.	LABOR	EQUIP.	TOTAL	TOTAL INCL O&P
8226	24″ diameter	B-21	5	5.600	Ea.	490	140	27	657	790
8228	30″ diameter	B-22	3	10		505	253	68	826	1,025
8230	36″ diameter	″	1	30		755	760	204	1,719	2,225
8250	Compression joints, C425, 4″ diameter	B-20	57	.421		13.55	10.25		23.80	31
8252	5″ diameter		55	.436		22.50	10.60		33.10	41.50
8254	6″ diameter		53	.453		22.50	11		33.50	42.50
8256	8″ diameter		45	.533		34.50	12.95		47.45	58.50
8260	12″ diameter		35	.686		69.50	16.70		86.20	103
8262	15″ diameter		25	.960		197	23.50		220.50	254
8264	18″ diameter	B-21	15	1.867		293	46.50	9.05	348.55	410
8266	24″ diameter	″	13	2.154		600	54	10.45	664.45	755
8268	30″ diameter	B-22	12	2.500		690	63.50	17	770.50	875
8270	36″ diameter	″	10	3		1,000	76	20.50	1,096.50	1,250
8300	Bends and elbows, extra strength, C700									
8312	Plain joints, 4″ diameter	B-20	50	.480	Ea.	10.70	11.65		22.35	30
8314	5″ diameter		48	.500		16.10	12.15		28.25	37
8316	6″ diameter		46	.522		16.10	12.70		28.80	37.50
8318	8″ diameter		38	.632		26.50	15.35		41.85	53
8320	10″ diameter		32	.750		41	18.25		59.25	73.50
8322	12″ diameter		28	.857		57.50	21		78.50	96.50
8324	15″ diameter		18	1.333		162	32.50		194.50	229
8326	18″ diameter	B-21	8	3.500		232	87.50	17	336.50	410
8328	21″ diameter		7	4		370	100	19.40	489.40	585
8330	24″ diameter		6	4.667		490	117	22.50	629.50	750
8332	27″ diameter		5	5.600		490	140	27	657	790
8334	30″ diameter		4	7		425	175	34	634	780
8336	36″ diameter		3	9.333		755	234	45.50	1,034.50	1,250
8350	Compression joints, C425, 4″ diameter	B-20	60	.400		13.55	9.75		23.30	30
8352	5″ diameter		58	.414		22.50	10.05		32.55	41
8354	6″ diameter		56	.429		22.50	10.40		32.90	41.50
8356	8″ diameter		48	.500		34.50	12.15		46.65	57
8358	10″ diameter		42	.571		54	13.90		67.90	81.50
8360	12″ diameter		38	.632		69.50	15.35		84.85	101
8362	15″ diameter		28	.857		197	21		218	250
8364	18″ diameter	B-21	18	1.556		293	39	7.55	339.55	395
8366	21″ diameter		17	1.647		450	41	8	499	570
8368	24″ diameter		16	1.750		600	44	8.50	652.50	740
8370	27″ diameter	B-20	15	1.600		580	39		619	700
8372	30″ diameter	B-21	14	2		650	50	9.70	709.70	805
8374	36″ diameter	″	13	2.154		995	54	10.45	1,059.45	1,200
8400	Wyes and tees, extra strength, C700									
8410	Plain joints, 4″ diameter	B-20	43	.558	Ea.	10.70	13.55		24.25	33.50
8412	5″ diameter		41	.585		16.10	14.25		30.35	40
8414	6″ diameter		39	.615		16.10	14.95		31.05	41
8416	8″ diameter		31	.774		26.50	18.85		45.35	58.50
8418	10″ diameter		25	.960		41	23.50		64.50	82
8420	12″ diameter		21	1.143		57.50	28		85.50	108
8422	15″ diameter		11	2.182		162	53		215	262
8424	18″ diameter	B-21	4	7		232	175	34	441	565
8426	21″ diameter		2	14		370	350	68	788	1,025
8428	24″ diameter		1	28		490	700	136	1,326	1,800
8430	27″ diameter		.80	35		550	875	170	1,595	2,175
8432	30″ diameter		.50	56		600	1,400	272	2,272	3,150
8434	36″ diameter		.40	70		835	1,750	340	2,925	4,050
8450	Compression joints, C425, 4″ diameter	B-20	53	.453		13.55	11		24.55	32.50
8452	5″ diameter		51	.471		22.50	11.45		33.95	43

2 SITE WORK

027 150 | Sewage Systems

		027 150 | Sewage Systems	CREW	DAILY OUTPUT	LABOR-HOURS	UNIT	1999 BARE COSTS MAT.	LABOR	EQUIP.	TOTAL	TOTAL INCL O&P
172	8454	6″ diameter	B-20	49	.490	Ea.	22.50	11.90		34.40	44
	8456	8″ diameter		41	.585		34.50	14.25		48.75	60.50
	8458	10″ diameter		35	.686		54	16.70		70.70	86
	8460	12″ diameter		31	.774		69.50	18.85		88.35	106
	8462	15″ diameter		21	1.143		197	28		225	261
	8464	18″ diameter	B-21	14	2		293	50	9.70	352.70	415
	8466	21″ diameter		12	2.333		450	58.50	11.30	519.80	600
	8468	24″ diameter		11	2.545		600	63.50	12.35	675.85	775
	8470	27″ diameter		8	3.500		735	87.50	17	839.50	960
	8472	30″ diameter		5	5.600		800	140	27	967	1,125
	8474	36″ diameter		4	7		1,100	175	34	1,309	1,525
174	0010	**SEWAGE PUMPING STATIONS** Prefabricated steel, concrete									
	0020	or fiberglass, 200 GPM	C-17D	.17	494	Total	27,000	14,100	3,225	44,325	55,500
	0200	1,000 GPM		.07	1,200		40,000	34,200	7,850	82,050	106,500
	0500	Add for generator unit, 200 GPM, steel		.34	247		20,500	7,050	1,625	29,175	35,500
	0600	Concrete		.51	164		13,000	4,700	1,075	18,775	22,900
	1000	Add for generator unit, 1,000 GPM, steel		.30	280		21,800	7,975	1,825	31,600	38,600
	1200	Concrete		.38	221		18,400	6,300	1,450	26,150	31,700
	1500	For drilled water well, if required, add	B-23	.50	80		5,300	1,750	3,900	10,950	12,900
176	0010	**SEWAGE TREATMENT** Plant, not incl. fencing or external piping									
	0020	Steel packaged, blown air aeration plants									
	0100	1,000 GPD				Gal.				15	17.25
	0200	5,000 GPD								10	11.50
	0300	15,000 GPD								5.50	6.30
	0400	30,000 GPD								5.20	6
	0500	50,000 GPD								4	4.60
	0600	100,000 GPD								3.50	4
	0700	200,000 GPD								2.50	2.88
	0800	500,000 GPD								2.45	2.80
	1000	Concrete, extended aeration, primary and secondary treatment									
	1010	10,000 GPD				Gal.				11	12.65
	1100	30,000 GPD								5.50	6.35
	1200	50,000 GPD								4.50	5.18
	1400	100,000 GPD								3.50	4.05
	1500	500,000 GPD								2.50	2.90
	1700	Municipal wastewater treatment facility									
	1720	1.0 MGD				Gal.				4.30	4.95
	1740	1.5 MGD								4.25	4.90
	1760	2.0 MGD								3.65	4.20
	1780	3.0 MGD								2.85	3.30
	1800	5.0 MGD								2.60	3
	2000	Holding tank system, not incl. excavation or backfill									
	2010	Recirculating chemical water closet	2 Plum	4	4	Ea.	660	130		790	925
	2100	For voltage converter, add	″	16	1		175	32.50		207.50	243
	2200	For high level alarm, add	1 Plum	7.80	1.026		100	33.50		133.50	161

027 350 | Wastewater System

		027 350 | Wastewater System	CREW	DAILY OUTPUT	LABOR-HOURS	UNIT	MAT.	LABOR	EQUIP.	TOTAL	TOTAL INCL O&P
354	0010	**WASTEWATER TREATMENT SYSTEM** Fiberglass, 1,000 gallon	B-21	1.29	21.705	Ea.	2,650	545	105	3,300	3,900
	0100	1,500 gallon	″	1.03	27.184	″	5,800	680	132	6,612	7,600

027 400 | Septic Systems

		027 400 | Septic Systems	CREW	DAILY OUTPUT	LABOR-HOURS	UNIT	MAT.	LABOR	EQUIP.	TOTAL	TOTAL INCL O&P
404	0010	**SEPTIC TANKS** Not incl. excav. or piping, precast, 1,000 gallon	B-21	8	3.500	Ea.	450	87.50	17	554.50	650
	0020	1,250 gallon		8	3.500		550	87.50	17	654.50	760
	0060	1,500 gallon		7	4		650	100	19.40	769.40	895
	0100	2,000 gallon		5	5.600		870	140	27	1,037	1,200

027 400 | Septic Systems

			CREW	DAILY OUTPUT	LABOR-HOURS	UNIT	1999 BARE COSTS MAT.	LABOR	EQUIP.	TOTAL	TOTAL INCL O&P	
404	0140	2,500 gallon	B-21	5	5.600	Ea.	1,000	140	27	1,167	1,350	404
	0180	4,000 gallon	↓	4	7		3,000	175	34	3,209	3,600	
	0200	5,000 gallon	B-13	3.50	16		4,300	370	157	4,827	5,475	
	0220	5,000 gal., 4 piece	"	3	18.667		5,575	435	183	6,193	7,025	
	0300	15,000 gallon, 4 piece	B-13B	1.70	32.941		11,000	765	465	12,230	13,800	
	0400	25,000 gallon, 4 piece		1.10	50.909		16,500	1,175	720	18,395	20,800	
	0500	40,000 gallon, 4 piece	↓	.80	70		27,000	1,625	990	29,615	33,300	
	0520	50,000 gallon, 5 piece	B-13C	.60	93.333		31,100	2,175	1,875	35,150	39,700	
	0540	75,000 gallon, cast in place	C-14C	.25	448		37,800	11,700	148	49,648	60,500	
	0560	100,000 gallon	"	.15	746		46,800	19,500	246	66,546	83,000	
	0600	High density polyethylene, 1,000 gallon	B-21	6	4.667		800	117	22.50	939.50	1,100	
	0700	1,500 gallon		4	7		1,000	175	34	1,209	1,400	
	0900	Galley, 4′ x 4′ x 4′	↓	16	1.750		175	44	8.50	227.50	271	
	1000	Distribution boxes, concrete, 7 outlets	2 Clab	16	1		75	21.50		96.50	117	
	1100	9 outlets	"	8	2		225	43		268	315	
	1150	Leaching field chambers, 13′ x 3′-7″ x 1′-4″, standard	B-13	16	3.500		665	81.50	34.50	781	895	
	1200	Heavy duty, 8′ x 4′ x 1′-6″		14	4		320	93	39	452	540	
	1300	13′ x 3′-9″ x 1′-6″		12	4.667		910	108	46	1,064	1,225	
	1350	20′ x 4′ x 1′-6″	↓	5	11.200		750	260	110	1,120	1,350	
	1400	Leaching pit, precast concrete, 3′ diameter, 3′ deep	B-21	8	3.500		145	87.50	17	249.50	315	
	1500	6′ diameter, 3′ section		4.70	5.957		375	149	29	553	680	
	1600	Leaching pit, 6′-6″ diameter, 6′ deep		5	5.600		450	140	27	617	745	
	1620	8′ deep		4	7		530	175	34	739	895	
	1700	8′ diameter, H-20 load, 6′ deep		4	7		730	175	34	939	1,125	
	1720	8′ deep		3	9.333		900	234	45.50	1,179.50	1,400	
	2000	Velocity reducing pit, precast conc., 6′ diameter, 3′ deep	↓	4.70	5.957	↓	225	149	29	403	515	
	2200	Excavation for septic tank, 3/4 C.Y. backhoe	B-12F	145	.110	C.Y.		2.95	3.10	6.05	7.90	
	2400	4′ trench for disposal field, 3/4 C.Y. backhoe	"	335	.048	L.F.		1.28	1.34	2.62	3.41	
	2600	Gravel fill, run of bank	B-6	150	.160	C.Y.	5.50	3.74	1.46	10.70	13.45	
	2800	Crushed stone, 3/4″	"	150	.160	"	18.30	3.74	1.46	23.50	27.50	

027 660 | Relining Exist. Pipelines

			CREW	DAILY OUTPUT	LABOR-HOURS	UNIT	MAT.	LABOR	EQUIP.	TOTAL	TOTAL INCL O&P	
664	0010	**LINING PIPE** with cement, incl. bypass and cleaning										664
	0020	Less than 10,000 L.F., urban, 6″ to 10″	C-17E	130	.615	L.F.	5.80	17.50	.47	23.77	34.50	
	0050	10″ to 12″		125	.640		7.15	18.20	.49	25.84	37	
	0070	12″ to 16″		115	.696		7.35	19.80	.53	27.68	39.50	
	0100	16″ to 20″		95	.842		8.60	24	.64	33.24	48	
	0200	24″ to 36″		90	.889		9.25	25.50	.68	35.43	51	
	0300	48″ to 72″		80	1		14.80	28.50	.76	44.06	62	
	0500	Rural, 6″ to 10″		180	.444		5.80	12.65	.34	18.79	26.50	
	0550	10″ to 12″		175	.457		7.15	13	.35	20.50	29	
	0570	12″ to 16″		160	.500		7.45	14.25	.38	22.08	31	
	0600	16″ to 20″		135	.593		7.90	16.85	.45	25.20	35.50	
	0700	24″ to 36″		125	.640		9.40	18.20	.49	28.09	39.50	
	0800	48″ to 72″		100	.800		14.80	23	.61	38.41	53	
	1000	Greater than 10,000 L.F., urban, 6″ to 10″		160	.500		5.80	14.25	.38	20.43	29.50	
	1050	10″ to 12″		155	.516		7.05	14.70	.39	22.14	31	
	1070	12″ to 16″		140	.571		7.35	16.25	.43	24.03	34	
	1100	16″ to 20″		120	.667		7.90	18.95	.51	27.36	39	
	1200	24″ to 36″		115	.696		9.40	19.80	.53	29.73	42	
	1300	48″ to 72″		95	.842		14.80	24	.64	39.44	55	
	1500	Rural, 6″ to 10″		215	.372		5.80	10.60	.28	16.68	23.50	
	1550	10″ to 12″		210	.381		7.15	10.85	.29	18.29	25.50	
	1570	12″ to 16″		185	.432		7.35	12.30	.33	19.98	28	
	1600	16″ to 20″		150	.533		7.90	15.15	.41	23.46	33	
	1700	24″ to 36″	↓	140	.571	↓	9.40	16.25	.43	26.08	36.50	

027 | Sewerage & Drainage

027 660 | Relining Exist. Pipelines

			CREW	DAILY OUTPUT	LABOR-HOURS	UNIT	1999 BARE COSTS MAT.	LABOR	EQUIP.	TOTAL	TOTAL INCL O&P
664	1800	48″ to 72″	C-17E	120	.667	L.F.	14.90	18.95	.51	34.36	47
	2000	Cured in place pipe, non-pressure, flexible felt resin, 400′ runs									
	2100	6″ diameter				L.F.					30
	2200	8″ diameter									40
	2300	10″ diameter									55
	2400	12″ diameter									70
	2500	15″ diameter									95
	2600	18″ diameter									115
	2700	21″ diameter									135
	2800	24″ diameter									150
	2900	30″ diameter									190
	3000	36″ diameter									250
	3100	48″ diameter									335
	3200	60″ diameter									420
	3300	72″ diameter				↓					500
668	0010	**PIPEBURSTING**, 300′ runs, replace with HDPE pipe									
	0020	Not including excavation, backfill, shoring, or dewatering									
	0100	6″ to 15″ diameter, minimum				L.F.					75
	0200	Maximum									150
	0300	18″ to 36″ diameter, minimum									175
	0400	Maximum				↓					300
	0500	Mobilize and demobilize, minimum				Job					2,500
	0600	Maximum				″					25,000

028 | Site Improvements

028 100 | Irrigation Systems

			CREW	DAILY OUTPUT	LABOR-HOURS	UNIT	1999 BARE COSTS MAT.	LABOR	EQUIP.	TOTAL	TOTAL INCL O&P
104	0010	**SPRINKLER IRRIGATION SYSTEM** For lawns									
	0100	Golf course with fully automatic system	C-17	.05	1,600	9 holes	75,000	45,500		120,500	154,500
	0200	24′ diam. head at 15′ O.C incl. piping, auto oper., minimum	B-20	70	.343	Head	16.50	8.35		24.85	31.50
	0300	Maximum		40	.600		38	14.60		52.60	65
	0500	60′ diam. head at 40′ O.C. incl. piping, auto oper., minimum		28	.857		50	21		71	88
	0600	Maximum		23	1.043	↓	140	25.50		165.50	194
	0800	Residential system, custom, 1″ supply		2,000	.012	S.F.	.25	.29		.54	.74
	0900	1-1/2″ supply	↓	1,800	.013	″	.28	.32		.60	.82
	0990	For renovation work, add to above						50%			
	1020	Pop up spray head w/risers, hi-pop, full circle pattern, 4″	2 Skwk	76	.211	Ea.	3.17	5.90		9.07	12.80
	1030	1/2 circle pattern		76	.211		3.17	5.90		9.07	12.80
	1040	6″, full circle pattern		76	.211		7.45	5.90		13.35	17.50
	1050	1/2 circle pattern		76	.211		7.45	5.90		13.35	17.50
	1060	12″, full circle pattern		76	.211		9.25	5.90		15.15	19.50
	1070	1/2 circle pattern		76	.211		9.30	5.90		15.20	19.55
	1080	Pop up bubbler head w/risers, hi-pop bubbler head, 4″		76	.211		3.27	5.90		9.17	12.90
	1090	6″		76	.211		7.45	5.90		13.35	17.50
	1100	12″		76	.211		9.25	5.90		15.15	19.50
	1110	Impact full/part circle sprinklers, 28′-54′ 25-60 PSI		37	.432		8.80	12.15		20.95	29
	1120	Spaced 37′-49′ @ 25-50 PSI		37	.432		20.50	12.15		32.65	41.50
	1130	Spaced 43′-61′ @ 30-60 PSI		37	.432		42.50	12.15		54.65	65.50
	1140	Spaced 54′-78′ @ 40-80 PSI	↓	37	.432	↓	92.50	12.15		104.65	121

Important: See the Reference Section for critical supporting data - Reference Nos., Crews, & City Cost Indexes

028 100 | Irrigation Systems

		Description	Crew	Daily Output	Labor-Hours	Unit	1999 Bare Costs Mat.	Labor	Equip.	Total	Total Incl O&P
104	1145	Impact rotor pop-up full/part commercial circle sprinklers									
	1150	Spaced 42′-65′ 35-80 PSI	2 Skwk	25	.640	Ea.	20.50	17.95		38.45	51
	1160	Spaced 48′-76′ 45-85 PSI	″	25	.640	″	6.20	17.95		24.15	35.50
	1165	Impact rotor pop-up part. circle comm., 53′-75′, 55-100 PSI, w/ acc									
	1170	Plastic case, metal cover	2 Skwk	25	.640	Ea.	134	17.95		151.95	176
	1180	Rubber cover		25	.640		102	17.95		119.95	141
	1190	Iron case, metal cover		22	.727		127	20.50		147.50	171
	1200	Rubber cover		22	.727		133	20.50		153.50	179
	1250	Plastic case, 2 nozzle, metal cover		25	.640		101	17.95		118.95	140
	1260	Rubber cover		25	.640		107	17.95		124.95	146
	1270	Iron case, 2 nozzle, metal cover		22	.727		143	20.50		163.50	190
	1280	Rubber cover		22	.727		152	20.50		172.50	200
	1282	Impact rotor pop-up full circle comm., 39′-99′, 30-100 PSI									
	1284	Plastic case, metal cover	2 Skwk	25	.640	Ea.	101	17.95		118.95	140
	1286	Rubber cover		25	.640		111	17.95		128.95	151
	1288	Iron case, metal cover		22	.727		149	20.50		169.50	196
	1290	Rubber cover		22	.727		154	20.50		174.50	202
	1292	Plastic case, 2 nozzle, metal cover		22	.727		108	20.50		128.50	150
	1294	Rubber cover		22	.727		113	20.50		133.50	156
	1296	Iron case, 2 nozzle, metal cover		20	.800		144	22.50		166.50	194
	1298	Rubber cover		20	.800		152	22.50		174.50	203
	1305	Electric remote control valve, plastic, 3/4″		18	.889		15.55	25		40.55	56.50
	1310	1″		18	.889		23	25		48	65
	1320	1-1/2″		18	.889		49	25		74	93.50
	1330	2″		18	.889		69.50	25		94.50	116
	1335	Quick coupling valves, brass, locking cover									
	1340	Inlet coupling valve, 3/4″	2 Skwk	18.75	.853	Ea.	38.50	24		62.50	80.50
	1350	1″		18.75	.853		47.50	24		71.50	90
	1360	Controller valve boxes, 6″ round boxes		18.75	.853		3.27	24		27.27	41.50
	1370	10″ round boxes		14.25	1.123		7.20	31.50		38.70	57.50
	1380	12″ square box		9.75	1.641		13.90	46		59.90	88
	1388	Electromech. control, 14 day 3-60 min, auto start to 23/day									
	1390	4 station	2 Skwk	1.04	15.385	Ea.	160	430		590	855
	1400	7 station		.64	25		159	700		859	1,275
	1410	12 station		.40	40		243	1,125		1,368	2,050
	1420	Dual programs, 18 station		.24	66.667		1,200	1,875		3,075	4,275
	1430	23 station		.16	100		1,575	2,800		4,375	6,150
	1435	Backflow preventer, bronze, 0-175 PSI, w/valves, test cocks									
	1440	3/4″	2 Skwk	2	8	Ea.	97	224		321	460
	1450	1″		2	8		99	224		323	465
	1460	1-1/2″		2	8		200	224		424	575
	1470	2″		2	8		214	224		438	590
	1475	Pressure vacuum breaker, brass, 15-150 PSI									
	1480	3/4″	2 Skwk	2	8	Ea.	54	224		278	415
	1490	1″		2	8		81.50	224		305.50	445
	1500	1-1/2″		2	8		161	224		385	530
	1510	2″		2	8		149	224		373	520
110	0010	**SUBSURFACE DRIP IRRIGATION**, looped grid, pressure compensating									
	0100	Preinserted emitter, line, hand bury, irregular area, small	3 Skwk	1,200	.020	L.F.	.30	.56		.86	1.21
	0150	Medium		1,800	.013		.30	.37		.67	.92
	0200	Large		2,520	.010		.30	.27		.57	.75
	0250	Rectangular area, small		2,040	.012		.30	.33		.63	.85
	0300	Medium		2,640	.009		.30	.25		.55	.73
	0350	Large		3,600	.007		.30	.19		.49	.62
	0400	Install in trench, irregular area, small		4,050	.006		.30	.17		.47	.59

028 | Site Improvements

		028 100 | Irrigation Systems	CREW	DAILY OUTPUT	LABOR-HOURS	UNIT	1999 BARE COSTS MAT.	LABOR	EQUIP.	TOTAL	TOTAL INCL O&P
110	0450	Medium	3 Skwk	7,488	.003	L.F.	.30	.09		.39	.47
	0500	Large		16,560	.001		.30	.04		.34	.39
	0550	Rectangular area, small		8,100	.003		.30	.08		.38	.46
	0600	Medium		21,960	.001		.30	.03		.33	.38
	0650	Large		33,264	.001		.30	.02		.32	.36
	0700	Trenching and backfill	B-53	500	.016			.44	.19	.63	.87
	0750	For vinyl tubing, 1/4", add to above						10%			
	0800	Vinyl tubing, 1/4", material only					.06			.06	.07
	0850	Supply tubing, 1/2", material only, 100' coil					.10			.10	.11
	0900	500' coil					.09			.09	.10
	0950	Compression fittings	1 Skwk	90	.089	Ea.	.65	2.49		3.14	4.64
	1000	Barbed fittings, 1/4"		360	.022		.15	.62		.77	1.15
	1100	Flush risers		60	.133		.70	3.74		4.44	6.65
	1150	Flush ends, figure eight		180	.044		.12	1.25		1.37	2.10
	1200	Ball valve, 4-1/2"		20	.400		5.50	11.20		16.70	24
	1250	4-3/4"		20	.400		6.50	11.20		17.70	25
	1300	Auto flush, spring loaded		90	.089		1.75	2.49		4.24	5.85
	1350	Volumetric		90	.089		5.50	2.49		7.99	10
	1400	Air relief valve, inline with compensation tee, 1/2"		45	.178		5.90	4.99		10.89	14.35
	1450	1"		30	.267		16.40	7.50		23.90	30
	1500	Round box for flush ends, 6"		30	.267		3.25	7.50		10.75	15.40
	1550	Fertilizer injector, non-proportional		4	2		3.15	56		59.15	92
	1600	Screen filter, 3/4" screen		12	.667		17.50	18.70		36.20	49
	1650	1" disk		8	1		46	28		74	95.50
	1700	1-1/2" disk		4	2		110	56		166	210
	1750	2" disk		3	2.667		220	75		295	360
	1800	Typical installation 18" O.C., small, minimum				S.F.				.70	.80
	1850	Maximum								1.30	1.50
	1900	Large, minimum								.52	.60
	2000	Maximum								.86	1
	2100	For non-pressure compensating systems, deduct								10%	10%
	2150	For supply piping see division 026 678									
		028 200 | Fountains									
204	0010	**FOUNTAINS/AERATORS**									
	0100	Pump w/controls									
	0200	Single phase, 100' chord, 1/2 H.P. pump	2 Skwk	4.40	3.636	Ea.	2,025	102		2,127	2,375
	0300	3/4 H.P. pump		4.30	3.721		2,275	104		2,379	2,675
	0400	1 H.P. pump		4.20	3.810		2,300	107		2,407	2,700
	0500	1-1/2 H.P. pump		4.10	3.902		2,425	109		2,534	2,825
	0600	2 H.P. pump		4	4		2,450	112		2,562	2,875
	0700	Three phase, 200' chord, 5 H.P. pump		3.90	4.103		6,500	115		6,615	7,325
	0800	7-1/2 H.P. pump		3.80	4.211		7,675	118		7,793	8,600
	0900	10 H.P. pump		3.70	4.324		7,725	121		7,846	8,700
	1000	15 H.P. pump		3.60	4.444		10,400	125		10,525	11,600
	1100	Nozzles, minimum		8	2		135	56		191	238
	1200	Maximum		8	2		246	56		302	360
	1300	Lights w/mounting kits, 200 watt		18	.889		295	25		320	365
	1400	300 watt		18	.889		330	25		355	400
	1500	500 watt		18	.889		360	25		385	435
	1600	Color blender		12	1.333		280	37.50		317.50	365
		028 300 | Fences & Gates									
308	0010	**FENCE, CHAIN LINK INDUSTRIAL**, schedule 40									
	0020	3 strands barb wire, 2" post @ 10' O.C., set in concrete, 6' H									

028 300 | Fences & Gates

308		CREW	DAILY OUTPUT	LABOR-HOURS	UNIT	1999 BARE COSTS MAT.	LABOR	EQUIP.	TOTAL	TOTAL INCL O&P
0200	9 ga. wire, galv. steel	B-80	240	.133	L.F.	7	3.13	2.25	12.38	15
0300	Aluminized steel		240	.133		9	3.13	2.25	14.38	17.20
0500	6 ga. wire, galv. steel		240	.133		11.35	3.13	2.25	16.73	19.80
0600	Aluminized steel		240	.133		13	3.13	2.25	18.38	21.50
0800	6 ga. wire, 6′ high but omit barbed wire, galv. steel		250	.128		11	3	2.16	16.16	19.10
0900	Aluminized steel		250	.128		15.40	3	2.16	20.56	24
0920	8′ H, 6 ga. wire, 2-1/2″ line post, galv. steel		180	.178		17.95	4.17	3	25.12	29.50
0940	Aluminized steel		180	.178		22	4.17	3	29.17	34.50
1400	Gate for 6′ high fence, 1-5/8″ frame, 3′ wide, galv. steel		10	3.200	Ea.	80	75	54	209	264
1500	Aluminized steel		10	3.200	″	110	75	54	239	297
2000	5′-0″ high fence, 9 ga., no barbed wire, 2″ line post,									
2010	10′ O.C., 1-5/8″ top rail									
2100	Galvanized steel	B-80	300	.107	L.F.	6	2.50	1.80	10.30	12.45
2200	Aluminized steel		300	.107	″	7.25	2.50	1.80	11.55	13.85
2400	Gate, 4′ wide, 5′ high, 2″ frame, galv. steel		10	3.200	Ea.	100	75	54	229	286
2500	Aluminized steel		10	3.200	″	110	75	54	239	297
3100	Overhead slide gate, chain link, 6′ high, to 18′ wide		38	.842	L.F.	82.50	19.75	14.20	116.45	137
3105	8′ high		30	1.067		84	25	18	127	151
3108	10′ high		24	1.333		75	31.50	22.50	129	156
3110	Cantilever type		48	.667		38	15.65	11.25	64.90	78.50
3120	8′ high		24	1.333		55	31.50	22.50	109	134
3130	10′ high		18	1.778		65	41.50	30	136.50	169
5000	Double swing gates, incl. posts & hardware									
5010	5′ high, 12′ opening	B-80	3.40	9.412	Opng.	263	221	159	643	805
5020	20′ opening		2.80	11.429		340	268	193	801	1,000
5060	6′ high, 12′ opening		3.20	10		475	235	169	879	1,075
5070	20′ opening		2.60	12.308		655	289	208	1,152	1,400
5080	8′ high, 12′ opening		2.13	15.002		700	350	253	1,303	1,600
5090	20′ opening		1.45	22.069		935	520	370	1,825	2,225
5100	10′ high, 12′ opening		1.31	24.427		800	575	410	1,785	2,225
5110	20′ opening		1.03	31.068		1,200	730	525	2,455	3,025
5120	12′ high, 12′ opening		1.05	30.476		1,175	715	515	2,405	2,950
5130	20′ opening		.85	37.647		1,500	885	635	3,020	3,725
5190	For aluminized steel add					20%				
6500	Auger fence post hole, 3′ deep, medium soil, by hand	1 Clab	30	.267	Ea.		5.70		5.70	9
6510	By machine	B-80	175	.183			4.29	3.08	7.37	10.05
6520	Rock, with jackhammer	B-9C	32	1.250			27.50	5.65	33.15	49
6530	With rock drill	B-47C	65	.246			6	9.75	15.75	19.95
6580	Line posts, galvanized, 2-1/2″ OD, set in conc., 4′	B-80	80	.400		14.10	9.40	6.75	30.25	37.50
6585	5′		76	.421		16.90	9.90	7.10	33.90	41.50
6590	6′		74	.432		17	10.15	7.30	34.45	42.50
6595	7′		72	.444		18.55	10.45	7.50	36.50	45
6600	8′		69	.464		20.50	10.90	7.80	39.20	48
6610	H-beam, 1-7/8″, 4′		83	.386		15.75	9.05	6.50	31.30	38.50
6615	5′		81	.395		18	9.25	6.65	33.90	41.50
6620	6′		78	.410		21	9.60	6.90	37.50	45.50
6625	7′		75	.427		23	10	7.20	40.20	49
6630	8′		73	.438		25	10.30	7.40	42.70	51.50
6635	Vinyl coated, 2-1/2″ OD, set in conc., 4′		79	.405		25.50	9.50	6.85	41.85	50
6640	5′		77	.416		30	9.75	7	46.75	56
6645	6′		74	.432		36	10.15	7.30	53.45	63
6650	7′		72	.444		38.50	10.45	7.50	56.45	66.50
6655	8′		69	.464		42.50	10.90	7.80	61.20	72.50
6660	End gate post, steel, 3″ OD, set in conc.,4′		68	.471		30	11.05	7.95	49	59
6665	5′		65	.492		38.50	11.55	8.30	58.35	69.50
6670	6′		63	.508		40.50	11.90	8.55	60.95	72.50

028 300 | Fences & Gates

		DESCRIPTION	CREW	DAILY OUTPUT	LABOR-HOURS	UNIT	1999 BARE COSTS MAT.	LABOR	EQUIP.	TOTAL	TOTAL INCL O&P
308	6675	7′	B-80	61	.525	Ea.	51.50	12.30	8.85	72.65	85.50
	6685	Vinyl, 4′		68	.471		50	11.05	7.95	69	81
	6690	5′		65	.492		55	11.55	8.30	74.85	87.50
	6695	6′		63	.508		65.50	11.90	8.55	85.95	100
	6705	8′		59	.542		78.50	12.70	9.15	100.35	116
	6710	Corner post, galv. steel, 4″ OD, set in conc., 4′		65	.492		50.50	11.55	8.30	70.35	82.50
	6715	6′		63	.508		69	11.90	8.55	89.45	104
	6720	7′		61	.525		80.50	12.30	8.85	101.65	117
	6725	8′		65	.492		89	11.55	8.30	108.85	125
	6730	Vinyl, 5′		65	.492		88	11.55	8.30	107.85	124
	6735	6′		63	.508		113	11.90	8.55	133.45	152
	6740	7′		61	.525		124	12.30	8.85	145.15	165
	6745	8′	↓	59	.542	↓	134	12.70	9.15	155.85	178
	6935	Fabric, 9 ga., galv., 1.2 oz. coat, 2″ chain link, 4′	B-80A	304	.079	L.F.	2	1.69	.59	4.28	5.50
	6940	5′		285	.084		2.43	1.81	.63	4.87	6.20
	6945	6′		266	.090		4.07	1.94	.67	6.68	8.25
	6950	7′		247	.097		3.33	2.08	.73	6.14	7.75
	6955	8′		228	.105		7	2.26	.79	10.05	12.10
	6960	9 ga., fused, 4′		304	.079		2.15	1.69	.59	4.43	5.65
	6965	5′		285	.084		2.69	1.81	.63	5.13	6.50
	6970	6′		266	.090		3.87	1.94	.67	6.48	8.05
	6975	7′		247	.097		3.77	2.08	.73	6.58	8.20
	6980	8′		228	.105		5.40	2.26	.79	8.45	10.30
	6985	Barbed wire, galv., cost per strand		2,280	.011		.07	.23	.08	.38	.52
	6990	Vinyl coated		2,280	.011	↓	.15	.23	.08	.46	.62
	6995	Extension arms, 3 strands		143	.168	Ea.	5.05	3.60	1.26	9.91	12.60
	7000	6 strands, 2-3/8″		119	.202		6	4.33	1.51	11.84	15.05
	7005	Eye tops, 2-3/8″		143	.168	↓	2.03	3.60	1.26	6.89	9.25
	7010	Top rail, incl. tie wires, 1-5/8″, galv.		912	.026	L.F.	1.86	.56	.20	2.62	3.15
	7020	Vinyl coated		912	.026		2.83	.56	.20	3.59	4.22
	7040	Rail, middle/bottom, w/tie wire, 1-5/8″, galv.		912	.026		1.54	.56	.20	2.30	2.80
	7050	Vinyl coated		912	.026		2.14	.56	.20	2.90	3.47
	7055	Braces, galv. steel		960	.025		1.11	.54	.19	1.84	2.27
	7056	Aluminized steel		960	.025		1.33	.54	.19	2.06	2.52
	7060	Reinforcing wire, coiled spring, 7 ga. galv.		2,279	.011		.10	.23	.08	.41	.56
	7070	9 ga., vinyl coated	↓	2,282	.011	↓	.14	.23	.08	.45	.59
	7071	Privacy slats, vertical, vinyl	1 Clab	500	.016	S.F.	.86	.34		1.20	1.49
	7072	Redwood		450	.018		1.50	.38		1.88	2.25
	7073	Diagonal, aluminum	↓	300	.027	↓	1.75	.57		2.32	2.83
	7770	Gates, sliding w/overhead support, 4′ high	B-80B	35	.914	L.F.	390	21	8.10	419.10	470
	7775	5′ high		32	1		425	23	8.85	456.85	510
	7780	6′ high		28	1.143		455	26	10.10	491.10	550
	7785	7′ high		25	1.280		600	29.50	11.30	640.80	720
	7790	8′ high	↓	23	1.391	↓	745	32	12.30	789.30	885
	7795	Cantilever, manual, exp. roller, (pr), 40′ wide x 8′ high	B-22	1	30	Ea.	2,075	760	204	3,039	3,700
	7800	30′ wide x 8′ high		1	30		1,750	760	204	2,714	3,325
	7805	24′ wide x 8′ high	↓	1	30		1,300	760	204	2,264	2,825
	7810	Motor operators for gates, (no elec wiring), 3′ wide swing	2 Skwk	.50	32		765	900		1,665	2,275
	7815	Up to 20′ wide swing		.50	32		2,400	900		3,300	4,050
	7820	Up to 45′ sliding		.50	32	↓	2,425	900		3,325	4,100
	7825	Overhead gate, 6′ to 18′ wide, sliding/cantilever		45	.356	L.F.	88.50	9.95		98.45	113
	7830	Gate operators, digital receiver		7	2.286	Ea.	176	64		240	294
	7835	Two button transmitter		24	.667		34.50	18.70		53.20	67.50
	7840	3 button station		14	1.143		41	32		73	95.50
	7845	Master slave system	↓	4	4	↓	150	112		262	345

028 300 | Fences & Gates

		CREW	DAILY OUTPUT	LABOR-HOURS	UNIT	1999 BARE COSTS MAT.	LABOR	EQUIP.	TOTAL	TOTAL INCL O&P	Ref.
0010	**FENCE, CHAIN LINK RESIDENTIAL**, sch. 20, 11 ga. wire, 1-5/8" post										312
0020	10' O.C., 1-3/8" top rail, 2" corner post, galv. stl. 3' high	B-1	500	.048	L.F.	3.17	1.06		4.23	5.15	
0050	4' high		400	.060		3.79	1.33		5.12	6.25	
0100	6' high		200	.120		4.83	2.65		7.48	9.50	
0150	Add for gate 3' wide, 1-3/8" frame, 3' high		12	2	Ea.	34	44		78	107	
0170	4' high		10	2.400		42	53		95	130	
0190	6' high		10	2.400		64.50	53		117.50	154	
0200	Add for gate 4' wide, 1-3/8" frame, 3' high		9	2.667		42.50	59		101.50	140	
0220	4' high		9	2.667		56	59		115	155	
0240	6' high		8	3		84.50	66.50		151	197	
0350	Aluminized steel, 11 ga. wire, 3' high		500	.048	L.F.	3.79	1.06		4.85	5.85	
0380	4' high		400	.060		4.48	1.33		5.81	7	
0400	6' high		200	.120		5.65	2.65		8.30	10.45	
0450	Add for gate 3' wide, 1-3/8" frame, 3' high		12	2	Ea.	39	44		83	113	
0470	4' high		10	2.400		52.50	53		105.50	142	
0490	6' high		10	2.400		77	53		130	168	
0500	Add for gate 4' wide, 1-3/8" frame, 3' high		10	2.400		52.50	53		105.50	141	
0520	4' high		9	2.667		69.50	59		128.50	170	
0540	6' high		8	3		104	66.50		170.50	218	
0620	Vinyl covered, 9 ga. wire, 3' high		500	.048	L.F.	3.98	1.06		5.04	6.05	
0640	4' high		400	.060		4.92	1.33		6.25	7.50	
0660	6' high		200	.120		6.15	2.65		8.80	11	
0720	Add for gate 3' wide, 1-3/8" frame, 3' high		12	2	Ea.	48	44		92	122	
0740	4' high		10	2.400		63	53		116	153	
0760	6' high		10	2.400		97.50	53		150.50	191	
0780	Add for gate 4' wide, 1-3/8" frame, 3' high		10	2.400		63.50	53		116.50	154	
0800	4' high		9	2.667		85.50	59		144.50	187	
0820	6' high		8	3		132	66.50		198.50	249	
0860	Tennis courts, 11 ga. wire, 2-1/2" post 10' O.C., 1-5/8" top rail										
0900	10' high	B-1	155	.155	L.F.	8.30	3.43		11.73	14.50	
0920	12' high		130	.185	"	9.85	4.08		13.93	17.20	
1000	Add for gate 4' wide, 1-5/8" frame 7' high		10	2.400	Ea.	130	53		183	227	
1040	Aluminized steel, 11 ga. wire 10' high		155	.155	L.F.	15	3.43		18.43	22	
1100	12' high		130	.185	"	15	4.08		19.08	23	
1140	Add for gate 4' wide, 1-5/8" frame, 7' high		10	2.400	Ea.	180	53		233	282	
1250	Vinyl covered, 9 ga. wire, 10' high		155	.155	L.F.	15.90	3.43		19.33	23	
1300	12' high		130	.185	"	17.80	4.08		21.88	26	
1400	Add for gate 4' wide, 1-5/8" frame, 7' high		10	2.400	Ea.	155	53		208	255	
0010	**FENCE, MISC. METAL** Chicken wire, posts @ 4', 1" mesh, 4' high	B-80	410	.078	L.F.	1.10	1.83	1.32	4.25	5.50	320
0100	2" mesh, 6' high		350	.091		1	2.14	1.54	4.68	6.10	
0200	Galv. steel, 12 ga., 2" x 4" mesh, posts 5' O.C., 3' high		300	.107		1.50	2.50	1.80	5.80	7.50	
0300	5' high		300	.107		2	2.50	1.80	6.30	8.05	
0400	14 ga., 1" x 2" mesh, 3' high		300	.107		1.60	2.50	1.80	5.90	7.60	
0500	5' high		300	.107		2.20	2.50	1.80	6.50	8.30	
1000	Kennel fencing, 1-1/2" mesh, 6' long, 3'-6" wide, 6'-2" high	2 Clab	4	4	Ea.	250	86		336	410	
1050	12' long		4	4		300	86		386	465	
1200	Top covers, 1-1/2" mesh, 6' long		15	1.067		50	23		73	91	
1250	12' long		12	1.333		80	28.50		108.50	133	
1300	For kennel doors, see division 083-729										
4500	Security fence, prison grade, set in concrete, 12' high	B-80	25	1.280	L.F.	20	30	21.50	71.50	92	
4600	16' high	"	20	1.600	"	24.50	37.50	27	89	114	
4750	Gate, transom for 10' fence, galv. steel, single, 3' x 7' x 3'	B-80A	52	.462	Ea.	197	9.90	3.45	210.35	236	
4752	4' x 7' x 3'		10	2.400		206	51.50	17.95	275.45	330	
4754	3' x 10'		8	3		199	64.50	22.50	286	345	
4756	4' x 10'		10	2.400		223	51.50	17.95	292.45	345	
4758	Double transom, 10' x 7' x 3'	B-80B	10	3.200		360	73.50	28.50	462	540	

	028 300 | Fences & Gates		Crew	Daily Output	Labor-Hours	Unit	1999 Bare Costs Mat.	Labor	Equip.	Total	Total Incl O&P
320	4760	12′ x 7′ x 3′	B-80B	6	5.333	Ea.	390	122	47	559	670
	4762	14′ x 7′ x 3′		5	6.400		435	147	56.50	638.50	765
	4764	10′ x 10′		4	8		455	183	70.50	708.50	870
	4766	12′ x 10′		7	4.571		490	105	40.50	635.50	750
	4768	14′ x 10′		7	4.571		530	105	40.50	675.50	790
	4780	Vinyl clad, single transom, 3′ x 7′ x 3′		6	5.333		299	122	47	468	570
	4782	4′ x 7′ x 3′		10	3.200		315	73.50	28.50	417	490
	4784	3′ x 10′		8	4		300	91.50	35.50	427	510
	4786	4′ x 10′		10	3.200		340	73.50	28.50	442	520
	4788	Double transom, 10′ x 7′ x 3′		10	3.200		680	73.50	28.50	782	890
	4790	12′ x 7′ x 3′		6	5.333		740	122	47	909	1,050
	4792	14′ x 7′ x 3′		5	6.400		830	147	56.50	1,033.50	1,200
	4794	10′ x 10′		4	8		725	183	70.50	978.50	1,150
	4796	12′ x 12′		7	4.571		780	105	40.50	925.50	1,075
	4798	14′ x 14′		7	4.571		840	105	40.50	985.50	1,125
	5000	Snow fence on steel posts 10′ O.C., 4′ high	B-1	500	.048	L.F.	1.55	1.06		2.61	3.38
	5300	Tubular picket, steel, 6′ sections, 1-9/16″ posts, 4′ high	B-80	300	.107		15.50	2.50	1.80	19.80	23
	5400	2″ posts, 5′ high		240	.133		21.50	3.13	2.25	26.88	31
	5600	2″ posts, 6′ high		200	.160		24.50	3.75	2.70	30.95	35.50
	5700	Staggered picket 1-9/16″ posts, 4′ high		300	.107		14	2.50	1.80	18.30	21.50
	5800	2″ posts, 5′ high		240	.133		23	3.13	2.25	28.38	33
	5900	2″ posts, 6′ high		200	.160		24	3.75	2.70	30.45	35
	6200	Gates, 4′ high, 3′ wide	B-1	10	2.400	Ea.	135	53		188	233
	6300	5′ high, 3′ wide		10	2.400		175	53		228	277
	6400	6′ high, 3′ wide		10	2.400		180	53		233	282
	6500	4′ wide		10	2.400		210	53		263	315
324	0010	**FENCE, RAIL** Picket, No. 2 cedar, Gothic, 2 rail, 3′ high	B-1	160	.150	L.F.	4.40	3.32		7.72	10.05
	0050	Gate, 3′-6″ wide		9	2.667	Ea.	38	59		97	135
	0400	3 rail, 4′ high		150	.160	L.F.	5.05	3.54		8.59	11.10
	0500	Gate, 3′-6″ wide		9	2.667	Ea.	45.50	59		104.50	143
	5000	Fence rail, redwood, 2″ x 4″, merch grade 8′		2,400	.010	L.F.	.86	.22		1.08	1.30
	5050	Select grade, 8′		2,400	.010	″	2.66	.22		2.88	3.28
	6000	Fence post, select redwood, earthpacked & treated, 4″ x 4″ x 6′		96	.250	Ea.	8.55	5.55		14.10	18.10
	6010	4″ x 4″ x 8′		96	.250		10.15	5.55		15.70	19.85
	6020	Set in concrete, 4″ x 4″ x 6′		50	.480		11	10.60		21.60	29
	6030	4″ x 4″ x 8′		50	.480		13.10	10.60		23.70	31
	6040	Wood post, 4′ high, set in concrete, incl. concrete		50	.480		6.40	10.60		17	24
	6050	Earth packed		96	.250		4.37	5.55		9.92	13.50
	6060	6′ high, set in concrete, incl. concrete		50	.480		8.20	10.60		18.80	26
	6070	Earth packed		96	.250		5.95	5.55		11.50	15.25
328	0010	**FENCE, WOOD** Basket weave, 3/8″ x 4″ boards, 2″ x 4″									
	0020	stringers on spreaders, 4″ x 4″ posts									
	0050	No. 1 cedar, 6′ high	B-1	160	.150	L.F.	7	3.32		10.32	12.90
	0070	Treated pine, 6′ high	″	150	.160	″	8.40	3.54		11.94	14.80
	0200	Board fence, 1″ x 4″ boards, 2″ x 4″ rails, 4″ x 4″ post									
	0220	Preservative treated, 2 rail, 3′ high	B-1	145	.166	L.F.	5	3.66		8.66	11.25
	0240	4′ high		135	.178		5.50	3.93		9.43	12.25
	0260	3 rail, 5′ high		130	.185		6.20	4.08		10.28	13.20
	0300	6′ high		125	.192		7.20	4.25		11.45	14.60
	0320	No. 2 grade western cedar, 2 rail, 3′ high		145	.166		5.45	3.66		9.11	11.75
	0340	4′ high		135	.178		6.50	3.93		10.43	13.35
	0360	3 rail, 5′ high		130	.185		7.50	4.08		11.58	14.65
	0400	6′ high		125	.192		8.25	4.25		12.50	15.75
	0420	No. 1 grade cedar, 2 rail, 3′ high		145	.166		7.50	3.66		11.16	14
	0440	4′ high		135	.178		8.55	3.93		12.48	15.60
	0460	3 rail, 5′ high		130	.185		10.20	4.08		14.28	17.60

028 300 | Fences & Gates		CREW	DAILY OUTPUT	LABOR-HOURS	UNIT	1999 BARE COSTS MAT.	LABOR	EQUIP.	TOTAL	TOTAL INCL O&P
0500	6′ high	B-1	125	.192	L.F.	11.40	4.25		15.65	19.25
0540	Shadow box, 1″ x 6″ board, 2″ x 4″ rail, 4″ x 4″post									
0560	Pine, pressure treated, 3 rail, 6′ high	B-1	150	.160	L.F.	9.55	3.54		13.09	16.10
0600	Gate, 3′-6″ wide		8	3	Ea.	50	66.50		116.50	159
0620	No. 1 cedar, 3 rail, 4′ high		130	.185	L.F.	12.10	4.08		16.18	19.70
0640	6′ high		125	.192		14.70	4.25		18.95	23
0860	Open rail fence, split rails, 2 rail 3′ high, no. 1 cedar		160	.150		4.25	3.32		7.57	9.90
0870	No. 2 cedar		160	.150		3.43	3.32		6.75	8.95
0880	3 rail, 4′ high, no. 1 cedar		150	.160		5.75	3.54		9.29	11.90
0890	No. 2 cedar		150	.160		4.04	3.54		7.58	10
0920	Rustic rails, 2 rail 3′ high, no. 1 cedar		160	.150		2.69	3.32		6.01	8.15
0930	No. 2 cedar		160	.150		2.30	3.32		5.62	7.75
0940	3 rail, 4′ high		150	.160		3.45	3.54		6.99	9.35
0950	No. 2 cedar	↓	150	.160	↓	2.67	3.54		6.21	8.50
0960	Picket fence, gothic, pressure treated pine									
1000	2 rail, 3′ high	B-1	140	.171	L.F.	3.60	3.79		7.39	9.90
1020	3 rail, 4′ high		130	.185	″	4.25	4.08		8.33	11.10
1040	Gate, 3′-6″ wide		9	2.667	Ea.	38	59		97	135
1060	No. 2 cedar, 2 rail, 3′ high		140	.171	L.F.	4.50	3.79		8.29	10.90
1100	3 rail, 4′ high		130	.185	″	4.60	4.08		8.68	11.45
1120	Gate, 3′-6″ wide		9	2.667	Ea.	45	59		104	143
1140	No. 1 cedar, 2 rail 3′ high		140	.171	L.F.	9	3.79		12.79	15.85
1160	3 rail, 4′ high		130	.185	″	10.50	4.08		14.58	17.95
1170	Gate, 3′-6″ wide		9	2.667	Ea.	100	59		159	203
1200	Rustic picket, molded pine, 2 rail, 3′ high		140	.171	L.F.	4	3.79		7.79	10.35
1220	No. 1 cedar, 2 rail, 3′ high		140	.171		4.69	3.79		8.48	11.10
1240	Stockade fence, no. 1 cedar, 3-1/4″ rails, 6′ high		160	.150		8.50	3.32		11.82	14.55
1260	8′ high		155	.155	↓	11	3.43		14.43	17.50
1270	Gate, 3′-6″ wide		9	2.667	Ea.	135	59		194	242
1300	No. 2 cedar, treated wood rails, 6′ high		160	.150	L.F.	8.50	3.32		11.82	14.55
1320	Gate, 3′-6″ wide		8	3	Ea.	50	66.50		116.50	159
1360	Treated pine, treated rails, 6′ high		160	.150		8.30	3.32		11.62	14.35
1400	8′ high		150	.160		12.50	3.54		16.04	19.30
1420	Gate, 3′-6″ wide	↓	9	2.667	↓	55	59		114	154

028 400 | Walk/Road/Parkg Appurt		CREW	DAILY OUTPUT	LABOR-HOURS	UNIT	MAT.	LABOR	EQUIP.	TOTAL	TOTAL INCL O&P
0010	**GUIDE/GUARD RAIL** Corrugated steel, galv. steel posts, 6′-3″ O.C.	B-80	850	.038	L.F.	10	.88	.63	11.51	13.05
0100	Double face		570	.056	″	15	1.32	.95	17.27	19.60
0200	End sections, galvanized, flared		50	.640	Ea.	40	15	10.80	65.80	79.50
0300	Wrap around end		50	.640		60	15	10.80	85.80	101
0350	Anchorage units	↓	15	2.133		480	50	36	566	645
0365	End section, flared	B-80A	78	.308		19.15	6.60	2.30	28.05	34
0370	End section, wrap-around, corrugated steel	″	78	.308	↓	23	6.60	2.30	31.90	38
0400	Timber guide rail, 4″ x 8″ with 6″ x 8″ wood posts, treated	B-80	960	.033	L.F.	12	.78	.56	13.34	15.05
0600	Cable guide rail, 3 at 3/4″ cables, steel posts, single face		900	.036		4.71	.83	.60	6.14	7.15
0650	Double face		635	.050		9.90	1.18	.85	11.93	13.65
0700	Wood posts		950	.034		5.90	.79	.57	7.26	8.35
0750	Double face		650	.049	↓	10.90	1.15	.83	12.88	14.70
0760	Breakaway wood posts		195	.164	Ea.	245	3.85	2.77	251.62	279
0800	Anchorage units, breakaway		15	2.133	″	450	50	36	536	610
0900	Guide rail, steel box beam, 6″ x 6″	↓	120	.267	L.F.	15.50	6.25	4.50	26.25	31.50
0950	End assembly	B-80A	48	.500	Ea.	420	10.75	3.74	434.49	480
1100	Median barrier, steel box beam, 6″ x 8″	B-80	215	.149	L.F.	19.50	3.49	2.51	25.50	29.50
1120	Shop curved	B-80A	92	.261	″	25	5.60	1.95	32.55	38.50
1140	End assembly		48	.500	Ea.	190	10.75	3.74	204.49	230
1150	Corrugated beam	↓	400	.060	L.F.	13.35	1.29	.45	15.09	17.15

028 400 | Walk/Road/Parkg Appurt

		028 400 | Walk/Road/Parkg Appurt	CREW	DAILY OUTPUT	LABOR-HOURS	UNIT	1999 BARE COSTS				TOTAL INCL O&P
							MAT.	LABOR	EQUIP.	TOTAL	
404	1200	Impact barrier, UTMCD, barrel type	B-16	30	1.067	Ea.	237	23.50	14.75	275.25	315
	1400	Resilient guide fence and light shield, 6' high	B-2	130	.308	L.F.	17	6.70		23.70	29.50
	1500	Concrete posts, individual, 6'-5", triangular	B-80	110	.291	Ea.	30	6.80	4.90	41.70	49
	1550	Square		110	.291		32.50	6.80	4.90	44.20	52
	1600	Wood guide posts		150	.213		26	5	3.60	34.60	40
	2000	Median barrier, precast concrete, 3'-6" high, 2' wide, single face	B-29	380	.147	L.F.	28	3.42	1.68	33.10	37.50
	2200	Double face	"	340	.165		23.50	3.83	1.87	29.20	34
	2300	Cast in place, steel forms	C-2	170	.282		45.50	7.55		53.05	62
	2320	Slipformed	C-7	352	.205		18.50	4.76	2.54	25.80	30.50
	2400	Speed bumps, thermoplastic, 10-1/2" x 2-1/4" x 48" long	B-2	120	.333	Ea.	80	7.30		87.30	99.50
	3000	Energy absorbing terminal, 10 bay, 3' wide	B-80B	.10	320		23,600	7,325	2,825	33,750	40,500
	3010	7 bay, 2' - 6" wide		.20	160		17,300	3,650	1,425	22,375	26,300
	3020	5 bay, 2' wide		.20	160		13,500	3,650	1,425	18,575	22,100
	3100	Wide hazard protection, foam sandwich, 7 bay, 7' - 6" wide		.14	228		16,700	5,225	2,025	23,950	28,800
	3110	5' wide		.15	213		16,500	4,875	1,875	23,250	27,900
	3120	3' wide		.18	177		15,900	4,075	1,575	21,550	25,600
408	0010	**PARKING BARRIERS** Timber with saddles, treated type									
	0100	4" x 4" for cars	B-2	520	.077	L.F.	2.58	1.68		4.26	5.50
	0200	6" x 6" for trucks		520	.077	"	5.55	1.68		7.23	8.75
	0400	Folding with individual padlocks		50	.800	Ea.	340	17.50		357.50	405
	0600	Flexible fixed stanchion, 2' high, 3" diameter		100	.400		17.40	8.75		26.15	33
	1000	Wheel stops, precast concrete incl. dowels, 6" x 10" x 6'-0"		120	.333		25.50	7.30		32.80	40
	1100	8" x 13" x 6'-0"		120	.333		29.50	7.30		36.80	44
	1200	Thermoplastic, 6" x 10" x 6'-0"		120	.333		50	7.30		57.30	66.50
	1300	Pipe bollards, conc filled/paint, 8' L x 4' D hole, 6" diam.	B-6	20	1.200		165	28	10.95	203.95	238
	1400	8" diam.		15	1.600		250	37.50	14.60	302.10	350
	1500	12" diam.		12	2		325	46.50	18.25	389.75	455
	9000	Parking lot control, see division 111-501									
412	0010	**SIGNS** Stock, 24" x 24", no posts, .080" alum. reflectorized	B-80	70	.457	Ea.	25	10.70	7.70	43.40	52.50
	0100	High intensity		70	.457		45	10.70	7.70	63.40	74.50
	0300	30" x 30", reflectorized		70	.457		48	10.70	7.70	66.40	78
	0400	High intensity		70	.457		64	10.70	7.70	82.40	95.50
	0600	Guide and directional signs, 12" x 18", reflectorized		70	.457		15.50	10.70	7.70	33.90	42
	0700	High intensity		70	.457		15	10.70	7.70	33.40	41.50
	0900	18" x 24", stock signs, reflectorized		70	.457		24	10.70	7.70	42.40	51.50
	1000	High intensity		70	.457		38	10.70	7.70	56.40	67
	1200	24" x 24", stock signs, reflectorized		70	.457		25	10.70	7.70	43.40	52.50
	1300	High intensity		70	.457		45	10.70	7.70	63.40	74.50
	1500	Add to above for steel posts, galvanized, 10'-0" upright, bolted		200	.160		28	3.75	2.70	34.45	40
	1600	12'-0" upright, bolted		140	.229		30	5.35	3.85	39.20	45.50
	1800	Highway road signs, aluminum, over 20 S. F., reflectorized		350	.091	S.F.	12	2.14	1.54	15.68	18.20
	2000	High intensity		350	.091		15	2.14	1.54	18.68	21.50
	2200	Highway, suspended over road, 80 S.F. min., reflectorized		165	.194		15	4.55	3.27	22.82	27
	2300	High intensity		165	.194		20	4.55	3.27	27.82	32.50
	2350	Roadway delineators and reference markers		500	.064	Ea.	12	1.50	1.08	14.58	16.70
	2360	Delineator post only, 6'		500	.064	"	6	1.50	1.08	8.58	10.10
	5000	Removal of signs, including supports									
	5020	To 10 S.F.	B-80B	16	2	Ea.		46	17.70	63.70	91
	5030	11 S.F. to 20 S.F.	"	5	6.400			147	56.50	203.50	290
	5040	21 S.F. to 40 S.F.	B-14	1.80	26.667			605	122	727	1,075
	5050	41 S.F. to 100 S.F.	B-13	1.30	43.077			1,000	425	1,425	2,025
	5200	Remove and relocate signs, including supports									
	5210	Remove and relocate signs, to 10 S.F.	B-80B	5	6.400	Ea.	196	147	56.50	399.50	505
	5220	11 S.F. to 20 S.F.	"	1.70	18.824		430	430	166	1,026	1,325
	5230	21 S.F. to 40 S.F.	B-14	.56	85.714		450	1,950	390	2,790	3,975
	5240	41 S.F. to 100 S.F.	B-13	.32	175		730	4,075	1,725	6,530	9,025

	028 400 | Walk/Road/Parkg Appurt		Crew	Daily Output	Labor-Hours	Unit	1999 Bare Costs Mat.	Labor	Equip.	Total	Total Incl O&P	
412	8000	For temporary barricades and lights, see div. 015-208										412
416	0010	**STEPS** Incl. excav., borrow & concrete base, where applicable										416
	0100	Brick steps	B-24	35	.686	LF Riser	7.60	17.10		24.70	35	
	0200	Railroad ties	2 Clab	25	.640		2.75	13.75		16.50	24.50	
	0300	Bluestone treads, 12" x 2" or 12" x 1-1/2"	B-24	30	.800		16.35	20		36.35	49	
	0500	Concrete, cast in place, see division 033-130										
	0600	Precast concrete, see division 034-804										
424	0010	**TRAFFIC SIGNALS** Mid block pedestrian crosswalk,										424
	0020	with pushbutton and mast arms	R-2	.30	186	Total	11,700	5,300	905	17,905	21,900	
	0100	Intersection, 8 signals w/three sect. (2 each direction), programmed	"	.15	373		24,500	10,600	1,800	36,900	45,100	
	0120	For each additional traffic phase controller, add	L-9	1.20	30		1,500	745		2,245	2,875	
	0200	Semi-actuated, detectors in side street only, add		.81	44.444		2,500	1,100		3,600	4,550	
	0300	Fully-actuated, detectors in all streets, add		.49	73.469		4,500	1,825		6,325	7,925	
	0400	For pedestrian pushbutton, add		.70	51.429		3,500	1,275		4,775	5,925	
	0500	Optically programmed signal only, add per head		1.64	21.951		2,500	545		3,045	3,650	
	0600	School flashing system, programmed		.41	87.805	Signal	6,000	2,175		8,175	10,200	

	028 600 | Playfield Equipment		Crew	Daily Output	Labor-Hours	Unit	Mat.	Labor	Equip.	Total	Total Incl O&P	
604	0010	**BACKSTOPS** Baseball, prefabricated, 30' wide, 12' high & 1 overhang	B-1	1	24	Ea.	1,825	530		2,355	2,850	604
	0100	40' wide, 12' high & 2 overhangs	"	.75	32		1,800	710		2,510	3,100	
	0300	Basketball, steel, single goal	B-13	3.04	18.421		695	430	181	1,306	1,625	
	0400	Double goal	"	1.92	29.167		505	680	286	1,471	1,925	
	0600	Tennis, wire mesh with pair of ends	B-1	2.48	9.677	Set	1,000	214		1,214	1,425	
	0700	Enclosed court	"	1.30	18.462	Ea.	3,125	410		3,535	4,075	
	0900	Handball or squash court, outdoor, wood	2 Carp	.50	32		2,550	875		3,425	4,200	
	1000	Masonry handball/squash court	D-1	.30	53.333		19,600	1,300		20,900	23,600	
608	0010	**GOAL POSTS** Steel, football, double post	B-1	1.50	16	Pr.	1,350	355		1,705	2,025	608
	0100	Deluxe, single post		1.50	16		1,450	355		1,805	2,150	
	0300	Football, convertible to soccer		1.50	16		1,450	355		1,805	2,150	
	0500	Soccer, regulation		2	12		1,475	265		1,740	2,050	
610	0010	**MODULAR PLAYGROUND** Basic components										610
	0100	Deck, square, steel	B-1	1	24	Ea.	560	530		1,090	1,450	
	0110	Polyurethane		1	24		485	530		1,015	1,375	
	0120	Triangular, steel		1	24		320	530		850	1,200	
	0130	Post, steel, 5" square		18	1.333		17.20	29.50		46.70	65.50	
	0140	Aluminum, 2-3/8" square		20	1.200		14.45	26.50		40.95	58	
	0150	5" square		18	1.333	L.F.	19.20	29.50		48.70	67.50	
	0160	Roof, square poly, 54" side		18	1.333	Ea.	1,150	29.50		1,179.50	1,325	
	0170	Transfer module		3	8	"	1,725	177		1,902	2,175	
	0180	Guardrail, pipe		60	.400	L.F.	199	8.85		207.85	233	
	0190	Steps, deck-to-deck		8	3	Ea.	1,150	66.50		1,216.50	1,375	
	0200	Activity panel, minimum		2	12		262	265		527	710	
	0210	Maximum		2	12		600	265		865	1,075	
	0360	With guardrails		3	8		1,675	177		1,852	2,125	
	0370	Crawl tunnel, straight, 56" long		4	6		785	133		918	1,075	
	0380	90°		4	6		945	133		1,078	1,250	
	1200	Slide, tunnel		8	3		1,450	66.50		1,516.50	1,700	
	1210	Straight, poly		8	3		148	66.50		214.50	266	
	1220	Stainless steel, 54" high		6	4		181	88.50		269.50	340	
	1230	Curved, poly, 40" high		6	4		800	88.50		888.50	1,025	
	1240	Spyroslide, 56" - 72" high		5	4.800		2,475	106		2,581	2,900	
	1300	Ladder, vertical		5	4.800		405	106		511	615	
	1310	Horizontal, 8' long		5	4.800		630	106		736	860	
	1320	Corkscrew climber		3	8		430	177		607	750	

028 600 | Playfield Equipment

			CREW	DAILY OUTPUT	LABOR-HOURS	UNIT	1999 BARE COSTS MAT.	LABOR	EQUIP.	TOTAL	TOTAL INCL O&P
610	1330	Fire pole	B-1	6	4	Ea.	188	88.50		276.50	345
	1340	Bridge, ring		4	6		805	133		938	1,100
	1350	Clatter		4	6	L.F.	285	133		418	525
612	0010	**PLAYGROUND EQUIPMENT** See also individual items									
	0200	Bike rack, 10' long, permanent	B-1	12	2	Ea.	299	44		343	400
	0240	Climber, arch, 6' high		4	6		490	133		623	750
	0260	Fitness trail, with signs, 9 to 10 stations, treated pine, minimum		.25	96		2,400	2,125		4,525	6,000
	0270	Maximum		.17	141		5,000	3,125		8,125	10,400
	0280	Metal, minimum		.25	96		3,500	2,125		5,625	7,200
	0285	Maximum		.17	141		2,500	3,125		5,625	7,675
	0300	Redwood, minimum		.25	96		2,700	2,125		4,825	6,325
	0310	Maximum		.17	141		7,000	3,125		10,125	12,600
	0320	16 to 20 station, treated pine, minimum		.17	141		8,100	3,125		11,225	13,900
	0330	Maximum		.13	184		8,700	4,075		12,775	16,000
	0340	Metal, minimum		.17	141		10,000	3,125		13,125	15,900
	0350	Maximum		.13	184		12,800	4,075		16,875	20,400
	0360	Redwood, minimum		.17	141		6,000	3,125		9,125	11,500
	0370	Maximum		.13	184		14,500	4,075		18,575	22,400
	0400	Horizontal monkey ladder, 14' long		4	6		520	133		653	780
	0590	Parallel bars, 10' long		4	6		174	133		307	400
	0600	Posts, tether ball set, 2-3/8" O.D.		12	2		203	44		247	293
	0800	Poles, multiple purpose, 10'-6" long		12	2	Pr.	110	44		154	191
	1000	Ground socket for movable posts, 2-3/8" post		10	2.400		74	53		127	165
	1100	3-1/2" post		10	2.400		77	53		130	169
	1300	See-saw, steel, 2 units		6	4	Ea.	230	88.50		318.50	390
	1400	4 units		4	6		800	133		933	1,100
	1500	6 units		3	8		740	177		917	1,100
	1700	Shelter, fiberglass golf tee, 3 person		4.60	5.217		2,150	115		2,265	2,550
	1900	Slides, stainless steel bed, 12' long, 6' high		3	8		1,550	177		1,727	2,000
	2000	20' long, 10' high		2	12		1,900	265		2,165	2,500
	2200	Swings, plain seats, 8' high, 4 seats		2	12		815	265		1,080	1,325
	2300	8 seats		1.30	18.462		1,500	410		1,910	2,300
	2500	12' high, 4 seats		2	12		1,000	265		1,265	1,525
	2600	8 seats		1.30	18.462		1,650	410		2,060	2,475
	2800	Whirlers, 8' diameter		3	8		1,500	177		1,677	1,925
	2900	10' diameter		3	8		2,175	177		2,352	2,675
614	0010	**PLAYGROUND SURFACING**									
	0100	Resilient rubber surface, poured in place, 4" thick, black	2 Skwk	300	.053	S.F.	4	1.50		5.50	6.75
	0150	2" thick topping, colors	"	2,800	.006		2.25	.16		2.41	2.73
	0200	Wood chip mulch, 6" deep	1 Clab	300	.027		.80	.57		1.37	1.78
620	0010	**PLATFORM/PADDLE TENNIS COURT** Complete with lighting, etc.									
	0100	Aluminum slat deck with aluminum frame	B-1	.08	300	Court	24,000	6,625		30,625	36,800
	0500	Aluminum slat deck and wood frame	C-1	.12	266		24,400	6,900		31,300	37,600
	0800	Aluminum deck heater, add	B-1	1.18	20.339		3,150	450		3,600	4,175
	0900	Douglas fir planking and wood frame 2" x 6" x 30'	C-1	.12	266		22,500	6,900		29,400	35,600
	1000	Plywood deck with steel frame		.12	266		23,500	6,900		30,400	36,700
	1100	Steel slat deck with wood frame		.12	266		24,500	6,900		31,400	37,800

028 700 | Site/Street Furnishings

			CREW	DAILY OUTPUT	LABOR-HOURS	UNIT	MAT.	LABOR	EQUIP.	TOTAL	TOTAL INCL O&P
704	0010	**BENCHES** Park, precast concrete, w/backs,wood rails, 4' long	2 Clab	5	3.200	Ea.	310	68.50		378.50	450
	0100	8' long		4	4		635	86		721	830
	0300	Fiberglass, without back, one piece, 4' long		10	1.600		355	34.50		389.50	445
	0400	8' long		7	2.286		700	49		749	845

028 700 | Site/Street Furnishings

		028 700 | Site/Street Furnishings	CREW	DAILY OUTPUT	LABOR-HOURS	UNIT	1999 BARE COSTS MAT.	LABOR	EQUIP.	TOTAL	TOTAL INCL O&P
704	0500	Steel barstock pedestals w/backs, 2" x 3" wood rails, 4' long	2 Clab	10	1.600	Ea.	630	34.50		664.50	750
	0510	8' long		7	2.286		745	49		794	895
	0520	3" x 8" wood plank, 4' long		10	1.600		635	34.50		669.50	755
	0530	8' long		7	2.286		740	49		789	890
	0540	Backless, 4" x 4" wood plank, 4' square		10	1.600		615	34.50		649.50	730
	0550	8' long		7	2.286		585	49		634	720
	0600	Aluminum pedestals, with backs, aluminum slats, 8' long		8	2		150	43		193	233
	0610	15' long		5	3.200		250	68.50		318.50	385
	0620	Portable, aluminum slats, 8' long		8	2		170	43		213	255
	0630	15' long		5	3.200		275	68.50		343.50	415
	0800	Cast iron pedestals, back & arms, wood slats, 4' long		8	2		460	43		503	575
	0820	8' long		5	3.200		765	68.50		833.50	950
	0840	Backless, wood slats, 4' long		8	2		400	43		443	510
	0860	8' long		5	3.200		675	68.50		743.50	855
	1700	Steel frame, fir seat, 10' long		10	1.600		150	34.50		184.50	219
708	0010	**BLEACHERS** Outdoor, portable, 3 to 5 tiers, to 300' long, min.	2 Sswk	120	.133	Seat	29.50	4.08		33.58	40
	0100	Maximum, less than 15' long, prefabricated		80	.200		45	6.10		51.10	60.50
	0200	6 to 20 tiers, minimum, up to 300' long		120	.133		29.50	4.08		33.58	39.50
	0300	Max., under 15', (highly prefabricated, on wheels)		80	.200		61	6.10		67.10	78
	0500	Permanent grandstands, wood seat, steel frame, 24" row									
	0600	3 to 15 tiers, minimum	2 Sswk	60	.267	Seat	57.50	8.15		65.65	78.50
	0700	Maximum		48	.333		63.50	10.20		73.70	88.50
	0900	16 to 30 tiers, minimum		60	.267		80	8.15		88.15	103
	0950	Average		55	.291		106	8.90		114.90	132
	1000	Maximum		48	.333		113	10.20		123.20	143
	1200	Seat backs only, 30" row, fiberglass		160	.100		17	3.06		20.06	24.50
	1300	Steel and wood		160	.100		21	3.06		24.06	28.50
	1400	NOTE: average seating is 1.5' in width									
712	0010	**PLANTER BLOCKS** Precast concrete, interlocking									
	0020	"V" blocks for retaining soil	D-1	205	.078	S.F.	3.75	1.92		5.67	7.10
716	0010	**PLANTERS** Concrete, sandblasted, precast, 48" diameter, 24" high	2 Clab	15	1.067	Ea.	425	23		448	505
	0100	Fluted, precast, 7' diameter, 36" high		10	1.600		1,150	34.50		1,184.50	1,300
	0300	Fiberglass, circular, 36" diameter, 24" high		15	1.067		315	23		338	380
	0400	60" diameter, 24" high		10	1.600		645	34.50		679.50	765
	0600	Square, 24" side, 36" high		15	1.067		430	23		453	510
	0700	48" side, 36" high		15	1.067		750	23		773	860
	0900	Planter/bench, 72" square, 36" high		5	3.200		2,700	68.50		2,768.50	3,075
	1000	96" square, 27" high		5	3.200		4,400	68.50		4,468.50	4,950
	1200	Wood, square, 48" side, 24" high		15	1.067		755	23		778	865
	1300	Circular, 48" diameter, 30" high		10	1.600		675	34.50		709.50	800
	1500	72" diameter, 30" high		10	1.600		1,075	34.50		1,109.50	1,225
	1600	Planter/bench, 72"		5	3.200		2,100	68.50		2,168.50	2,425
720	0010	**TRASH CLOSURE** Steel with pullover cover									
	0020	2'-3" wide, 4'-7" high, 6'-2" long	2 Clab	5	3.200	Ea.	680	68.50		748.50	860
	0100	10'-1" long		4	4		900	86		986	1,125
	0300	Wood, 10' wide, 6' high, 10' long		1.20	13.333		710	286		996	1,225
724	0010	**TRASH RECEPTACLE** Fiberglass, 2' square, 18" high	2 Clab	30	.533	Ea.	200	11.45		211.45	238
	0100	2' square, 2'-6" high		30	.533		280	11.45		291.45	330
	0300	Circular , 2' diameter, 18" high		30	.533		180	11.45		191.45	216
	0400	2' diameter, 2'-6" high		30	.533		240	11.45		251.45	282
	1000	Alum. frame, hardboard panels, steel drum base,									
	1020	30 gal. capacity, silk screen on plastic finish	2 Clab	25	.640	Ea.	430	13.75		443.75	495
	1040	Aggregate finish		25	.640		345	13.75		358.75	400
	1100	50 gal. capacity, silk screen on plastic finish		20	.800		440	17.15		457.15	510

028 | Site Improvements

028 700 | Site/Street Furnishings

			CREW	DAILY OUTPUT	LABOR-HOURS	UNIT	1999 BARE COSTS MAT.	LABOR	EQUIP.	TOTAL	TOTAL INCL O&P
724	1140	Aggregate finish	2 Clab	20	.800	Ea.	535	17.15		552.15	615
	1200	Formed plastic liner, 14 gal., silk screen on plastic finish		40	.400		190	8.60		198.60	223
	1240	Aggregate finish		40	.400		235	8.60		243.60	273
	1300	30 gal. capacity, silk screen on plastic finish		35	.457		265	9.80		274.80	305
	1340	Aggregate finish		35	.457		320	9.80		329.80	365
	1400	Redwood slats, plastic liner, leg base, 14 gal. capacity,									
	1420	Varnish w/routed message	2 Clab	40	.400	Ea.	245	8.60		253.60	284
	2000	Concrete, precast, 2′ to 2-1/2′ wide, 3′ high, sandblasted	"	15	1.067	"	460	23		483	540
	3000	Galv. steel frame and panels, leg base, poly bag retainer,									
	3020	40 gal. capacity, silk screen on enamel finish	2 Clab	25	.640	Ea.	265	13.75		278.75	315
	3040	Aggregate finish		25	.640		360	13.75		373.75	415
	3200	Formed plastic liner, 50 gal., silk screen on enamel finish		20	.800		400	17.15		417.15	465
	3240	Aggregate finish		20	.800		465	17.15		482.15	535
	4000	Perforated steel, pole mounted, 12″ diam., 10 gal., painted		25	.640		85	13.75		98.75	115
	4040	Redwood slats		25	.640		220	13.75		233.75	264
	4100	22 gal. capacity, painted		25	.640		125	13.75		138.75	160
	4140	Redwood slats		25	.640		265	13.75		278.75	315
	4400	Trash receptacle									
	4500	Galvanized steel street basket, 52 gal. capacity, unpainted	2 Clab	40	.400	Ea.	175	8.60		183.60	207

029 | Landscaping

029 100 | Shrub/Tree Transplanting

			CREW	DAILY OUTPUT	LABOR-HOURS	UNIT	1999 BARE COSTS MAT.	LABOR	EQUIP.	TOTAL	TOTAL INCL O&P
104	0010	**TREE GUYING** Including stakes, guy wire and wrap									
	0100	Less than 3″ caliper, 2 stakes	2 Clab	35	.457	Ea.	15	9.80		24.80	32
	0200	3″ to 4″ caliper, 3 stakes	"	21	.762	"	17.60	16.35		33.95	45
	1000	Including arrowhead anchor, cable, turnbuckles and wrap									
	1100	Less than 3″ caliper, 3″ anchors	2 Clab	20	.800	Ea.	45	17.15		62.15	76.50
	1200	3″ to 6″ caliper, 4″ anchors		15	1.067		65	23		88	108
	1300	6″ caliper, 6″ anchors		12	1.333		80	28.50		108.50	133
	1400	8″ caliper, 8″ anchors		9	1.778		80	38		118	148
	2000	Tree guard, preformed plastic, 36″ high		168	.095		2.50	2.04		4.54	5.95
	2010	Snow fence		140	.114		1.80	2.45		4.25	5.85
108	0010	**TREE REMOVAL**									
	0100	Dig & lace, shrubs, broadleaf evergreen, 18″-24″	B-1	55	.436	Ea.		9.65		9.65	15.20
	0200	2′-3′	"	35	.686			15.15		15.15	24
	0300	3′-4′	B-6	30	.800			18.70	7.30	26	37
	0400	4′-5′	"	20	1.200			28	10.95	38.95	55.50
	1000	Deciduous, 12″-15″	B-1	110	.218			4.83		4.83	7.60
	1100	18″-24″		65	.369			8.15		8.15	12.85
	1200	2′-3′		55	.436			9.65		9.65	15.20
	1300	3′-4′	B-6	50	.480			11.20	4.38	15.58	22
	2000	Evergreeen, 18″-24″	B-1	55	.436			9.65		9.65	15.20
	2100	2′-0″ to 2′-6″		50	.480			10.60		10.60	16.70
	2200	2′-6″ to 3′-0″		35	.686			15.15		15.15	24
	2300	3′-0″ to 3′-6″		20	1.200			26.50		26.50	42
	3000	Trees, deciduous, small, 2′-3′		55	.436			9.65		9.65	15.20
	3100	3′-4′	B-6	50	.480			11.20	4.38	15.58	22
	3200	4′-5′		35	.686			16.05	6.25	22.30	32

029 100 | Shrub/Tree Transplanting

		Description	CREW	DAILY OUTPUT	LABOR-HOURS	UNIT	1999 BARE COSTS MAT.	LABOR	EQUIP.	TOTAL	TOTAL INCL O&P	
108	3300	5′-6′	B-6	30	.800	Ea.		18.70	7.30	26	37	108
	4000	Shade, 5′-6′		50	.480			11.20	4.38	15.58	22	
	4100	6′-8′		35	.686			16.05	6.25	22.30	32	
	4200	8′-10′		25	.960			22.50	8.75	31.25	44.50	
	4300	2″ caliper		12	2			46.50	18.25	64.75	92.50	
	5000	Evergreen, 4′-5′		35	.686			16.05	6.25	22.30	32	
	5100	5′-6′		25	.960			22.50	8.75	31.25	44.50	
	5200	6′-7′		19	1.263			29.50	11.50	41	58.50	
	5300	7′-8′		15	1.600			37.50	14.60	52.10	74	
	5400	8′-10′		11	2.182			51	19.90	70.90	101	

029 200 | Soil Preparation

		Description	CREW	DAILY OUTPUT	LABOR-HOURS	UNIT	1999 BARE COSTS MAT.	LABOR	EQUIP.	TOTAL	TOTAL INCL O&P	
204	0010	LAWN BED PREPARATION										204
	0100	Rake topsoil, site material, harley rock rake, ideal	B-6	33	.727	M.S.F.		17	6.65	23.65	34	
	0200	Adverse	″	7	3.429			80	31.50	111.50	159	
	0250	By hand (raking)	1 Clab	10	.800			17.15		17.15	27	
	0300	Screened loam, york rake and finish, ideal	B-62	24	1			23.50	4.64	28.14	41.50	
	0400	Adverse	″	20	1.200			28	5.55	33.55	49.50	
	0450	By hand (raking)	1 Clab	15	.533			11.45		11.45	18	
	1000	Remove topsoil & stock pile on site, 75 HP dozer, 6″ deep, 50′ haul	B-10L	30	.400			10.45	10.25	20.70	27.50	
	1050	300′ haul		6.10	1.967			51.50	50.50	102	135	
	1100	12″ deep, 50′ haul		15.50	.774			20	19.85	39.85	53	
	1150	300′ haul		3.10	3.871			101	99.50	200.50	264	
	1200	200 HP dozer, 6″ deep, 50′ haul	B-10B	125	.096			2.50	6.70	9.20	11.25	
	1250	300′ haul		30.70	.391			10.20	27.50	37.70	45.50	
	1300	12″ deep, 50′ haul		62	.194			5.05	13.55	18.60	22.50	
	1350	300′ haul		15.40	.779			20.50	54.50	75	91	
	1400	Alternate method, 75 HP dozer, 50′ haul	B-10L	860	.014	C.Y.		.36	.36	.72	.95	
	1450	300′ haul	″	114	.105			2.75	2.70	5.45	7.20	
	1500	200 HP dozer, 50′ haul	B-10B	2,660	.005			.12	.32	.44	.53	
	1600	300′ haul	″	570	.021			.55	1.47	2.02	2.46	
	1800	Rolling topsoil, hand push roller	1 Clab	3,200	.002	S.F.		.05		.05	.08	
	1850	Tractor drawn roller	B-66	10,666	.001	″		.02	.02	.04	.05	
	1900	Remove rocks & debris from grade, by hand	B-62	80	.300	M.S.F.		7	1.39	8.39	12.45	
	1920	With rock picker	B-10S	140	.086			2.24	2.29	4.53	5.95	
	2000	Root raking and loading, residential, no boulders	B-6	53.30	.450			10.50	4.11	14.61	21	
	2100	With boulders		32	.750			17.55	6.85	24.40	34.50	
	2200	Municipal, no boulders		200	.120			2.80	1.09	3.89	5.55	
	2300	With boulders		120	.200			4.67	1.82	6.49	9.25	
	2400	Large commercial, no boulders	B-10B	400	.030			.78	2.10	2.88	3.51	
	2500	With boulders	″	240	.050			1.30	3.50	4.80	5.85	
	2600	Rough grade & scarify subsoil to receive topsoil, common earth										
	2610	200 H.P. dozer with scarifier	B-11A	80	.200	M.S.F.		4.99	10.50	15.49	19.25	
	2620	180 H.P. grader with scarifier	B-11L	110	.145			3.63	5.25	8.88	11.35	
	2700	Clay and till, 200 H.P. dozer with scarifier	B-11A	50	.320			8	16.80	24.80	31	
	2710	180 H.P. grader with scarifier	B-11L	40	.400			9.95	14.40	24.35	31.50	
	3000	Scarify subsoil, residential, skid steer loader w/scarifiers, 50 HP	B-66	32	.250			6.80	6.25	13.05	17.25	
	3050	Municipal, skid steer loader w/scarifiers, 50 HP	″	120	.067			1.81	1.67	3.48	4.60	
	3100	Large commercial, 75 HP, dozer w/scarifier	B-10L	240	.050			1.30	1.28	2.58	3.41	
	3200	Grader with scarifier, 135 H.P.	B-11L	280	.057			1.42	2.06	3.48	4.46	
	3500	Screen topsoil from stockpile, vibrating screen, wet material (organic)	B-10P	200	.060	C.Y.		1.56	4.33	5.89	7.15	
	3550	Dry material	″	300	.040			1.04	2.88	3.92	4.77	
	3600	Mixing with conditioners, manure and peat	B-10R	550	.022			.57	.44	1.01	1.36	
	3650	Mobilization add for 2 days or less operation	B-34K	3	2.667	Job		59	305	364	425	
	3800	Spread conditioned topsoil, 6″ deep, by hand	B-1	360	.067	S.Y.	3.15	1.47		4.62	5.80	
	3850	300 HP dozer	B-10M	27	.444	M.S.F.	340	11.60	42	393.60	440	

029 200 | Soil Preparation

		CREW	DAILY OUTPUT	LABOR-HOURS	UNIT	1999 BARE COSTS MAT.	LABOR	EQUIP.	TOTAL	TOTAL INCL O&P
204	3900 4" deep, by hand	B-1	470	.051	S.Y.	2.83	1.13		3.96	4.90
	3920 300 H.P. dozer	B-10M	34	.353	M.S.F.	255	9.20	33.50	297.70	330
	3940 180 H.P. grader	B-11L	37	.432	"	255	10.80	15.55	281.35	315
	4000 Spread soil conditioners, alum. sulfate, 1#/S.Y., hand push spreader	A-1	17,500	.001	S.Y.	.90	.01		.91	1.01
	4050 Tractor spreader	B-66	700	.011	M.S.F.	100	.31	.29	100.60	111
	4100 Fertilizer, 0.2#/S.Y., push spreader	A-1	17,500	.001	S.Y.	.05	.01		.06	.08
	4150 Tractor spreader	B-66	700	.011	M.S.F.	5.55	.31	.29	6.15	6.90
	4200 Ground limestone, 1#/S.Y., push spreader	A-1	17,500	.001	S.Y.	.08	.01		.09	.11
	4250 Tractor spreader	B-66	700	.011	M.S.F.	11.90	.31	.29	12.50	13.85
	4300 Lusoil, 3#/S.Y., push spreader	A-1	17,500	.001	S.Y.	.40	.01		.41	.46
	4350 Tractor spreader	B-66	700	.011	M.S.F.	44.50	.31	.29	45.10	50
	4400 Manure, 18#/S.Y., push spreader	A-1	2,500	.003	S.Y.	2.35	.07	.03	2.45	2.72
	4450 Tractor spreader	B-66	280	.029	M.S.F.	261	.78	.72	262.50	289
	4500 Perlite, 1" deep, push spreader	A-1	17,500	.001	S.Y.	7.25	.01		7.26	8
	4550 Tractor spreader	B-66	700	.011	M.S.F.	805	.31	.29	805.60	885
	4600 Vermiculite, push spreader	A-1	17,500	.001	S.Y.	2.20	.01		2.21	2.44
	4650 Tractor spreader	B-66	700	.011	M.S.F.	244	.31	.29	244.60	270
	5000 Spread topsoil, skid steer loader and hand dress	B-62	270	.089	C.Y.	14.15	2.08	.41	16.64	19.25
	5100 Articulated loader and hand dress	B-100	320	.038	↓	14.15	.98	1.59	16.72	18.85
	5200 Articulated loader and 75HP dozer	B-10M	500	.024		14.15	.63	2.26	17.04	19.05
	5300 Road grader and hand dress	B-11L	1,000	.016	↓	14.15	.40	.58	15.13	16.85
	6000 Tilling topsoil, 20 HP tractor, disk harrow, 2" deep	B-66	50,000	.001	S.Y.					.01
	6050 4" deep	↓	40,000	.001			.01	.01	.02	.02
	6100 6" deep		3,000	.003			.07	.07	.14	.18
	6101 Tilling topsoil, 20 HP tractor, disk harrow, 6" deep	↓	30,000	.001			.01	.01	.02	.02
	6150 26" rototiller, 2" deep	A-1	1,250	.006			.14	.05	.19	.28
	6200 4" deep		1,000	.008			.17	.07	.24	.35
	6250 6" deep	↓	750	.011	↓		.23	.09	.32	.46
208	0010 **PLANT BED PREPARATION**									
	0100 Backfill planting pit, by hand, on site topsoil	2 Clab	18	.889	C.Y.		19.05		19.05	30
	0200 Prepared planting mix	"	24	.667			14.30		14.30	22.50
	0300 Skid steer loader, on site topsoil	B-62	340	.071			1.65	.33	1.98	2.92
	0400 Prepared planting mix	"	410	.059			1.37	.27	1.64	2.42
	1000 Excavate planting pit, by hand, sandy soil	2 Clab	16	1			21.50		21.50	34
	1100 Heavy soil or clay	"	8	2			43		43	67.50
	1200 1/2 C.Y. backhoe, sandy soil	B-11C	150	.107			2.66	1.46	4.12	5.70
	1300 Heavy soil or clay	"	115	.139			3.47	1.90	5.37	7.45
	2000 Mix planting soil, incl. loam, manure, peat, by hand	2 Clab	60	.267		25	5.70		30.70	36.50
	2100 Skid steer loader	B-62	150	.160	↓	25	3.74	.74	29.48	34
	3000 Pile sod, skid steer loader	"	2,800	.009	S.Y.		.20	.04	.24	.35
	3100 By hand	2 Clab	400	.040			.86		.86	1.35
	4000 Remove sod, F.E. loader	B-10S	2,000	.006			.16	.16	.32	.42
	4100 Sod cutter	B-12K	3,200	.005			.13	.25	.38	.48
	4200 By hand	2 Clab	240	.067	↓		1.43		1.43	2.25
	6000 For planting bed edging, see div. 025-258									

029 300 | Lawns & Grasses

		CREW	DAILY OUTPUT	LABOR-HOURS	UNIT	MAT.	LABOR	EQUIP.	TOTAL	TOTAL INCL O&P
308	0010 **SEEDING** Athletic field mix, 8#/M.S.F., push spreader (R029-310)	1 Clab	8	1	M.S.F.	20	21.50		41.50	56
	0100 Tractor spreader	B-66	52	.154		20	4.18	3.86	28.04	32.50
	0200 Hydro or air seeding, with mulch & fertil.	B-81	80	.300		22	7.20	7.55	36.75	43.50
	0400 Birdsfoot trefoil, .45#/M.S.F., push spreader	1 Clab	8	1		15.60	21.50		37.10	51
	0500 Tractor spreader	B-66	52	.154		15.60	4.18	3.86	23.64	27.50
	0600 Hydro or air seeding,with mulch & fertil.	B-81	80	.300		30	7.20	7.55	44.75	52.50
	0800 Bluegrass, 4#/M.S.F., common, push spreader	1 Clab	8	1		14	21.50		35.50	49.50
	0900 Tractor spreader	B-66	52	.154	↓	14	4.18	3.86	22.04	26

029 300 | Lawns & Grasses

			CREW	DAILY OUTPUT	LABOR-HOURS	UNIT	1999 BARE COSTS MAT.	LABOR	EQUIP.	TOTAL	TOTAL INCL O&P
308	1000	Hydro or air seeding, with mulch & fertil. R029-310	B-81	80	.300	M.S.F.	23	7.20	7.55	37.75	45
	1100	Baron, push spreader	1 Clab	8	1		18.50	21.50		40	54.50
	1200	Tractor spreader	B-66	52	.154		18.50	4.18	3.86	26.54	31
	1300	Hydro or air seeding, with mulch & fertil.	B-81	80	.300		25.50	7.20	7.55	40.25	47.50
	1500	Clover, 0.67#/M.S.F., white, push spreader	1 Clab	8	1		3	21.50		24.50	37.50
	1600	Tractor spreader	B-66	52	.154		3	4.18	3.86	11.04	13.90
	1700	Hydro or air seeding, with mulch and fertil.	B-81	80	.300		16.50	7.20	7.55	31.25	37.50
	1800	Ladino, push spreader	1 Clab	8	1		4.80	21.50		26.30	39.50
	1900	Tractor spreader	B-66	52	.154		4.80	4.18	3.86	12.84	15.90
	2000	Hydro or air seeding, with mulch and fertil.	B-81	80	.300		21	7.20	7.55	35.75	42.50
	2200	Fescue 5.5#/M.S.F., tall, push spreader	1 Clab	8	1		8	21.50		29.50	43
	2300	Tractor spreader	B-66	52	.154		8	4.18	3.86	16.04	19.40
	2400	Hydro or air seeding, with mulch and fertilizer	B-81	80	.300		26.50	7.20	7.55	41.25	48.50
	2500	Chewing, push spreader	1 Clab	8	1		9	21.50		30.50	44
	2600	Tractor spreader	B-66	52	.154		9	4.18	3.86	17.04	20.50
	2700	Hydro or air seeding, with mulch and fertil.	B-81	80	.300		29.50	7.20	7.55	44.25	52
	2800	Creeping, push spreader	1 Clab	8	1		8.50	21.50		30	43.50
	2810	Tractor spreader	B-66	26	.308		8.50	8.35	7.70	24.55	30.50
	2820	Hydro or air seeding, with mulch and fertilizer	B-81	80	.300		28	7.20	7.55	42.75	50.50
	2900	Crown vetch, 4#/M.S.F., push spreader	1 Clab	8	1		35	21.50		56.50	72.50
	3000	Tractor spreader	B-66	52	.154		35	4.18	3.86	43.04	49
	3100	Hydro or air seeding, with mulch and fertilizer	B-81	80	.300		48	7.20	7.55	62.75	72.50
	3300	Rye, 10#/M.S.F., annual, push spreader	1 Clab	8	1		12	21.50		33.50	47
	3400	Tractor spreader	B-66	52	.154		12	4.18	3.86	20.04	24
	3500	Hydro or air seeding, with mulch and fertilizer	B-81	80	.300		26.50	7.20	7.55	41.25	48.50
	3600	Fine textured, push spreader	1 Clab	8	1		14	21.50		35.50	49.50
	3700	Tractor spreader	B-66	52	.154		14	4.18	3.86	22.04	26
	3800	Hydro or air seeding, with mulch and fertilizer	B-81	80	.300		31	7.20	7.55	45.75	53.50
	4000	Shade mix, 6#/M.S.F., push spreader	1 Clab	8	1		10	21.50		31.50	45
	4100	Tractor spreader	B-66	52	.154		10	4.18	3.86	18.04	21.50
	4200	Hydro or air seeding, with mulch and fertilizer	B-81	80	.300		22	7.20	7.55	36.75	43.50
	4400	Slope mix, 6#/M.S.F., push spreader	1 Clab	8	1		10	21.50		31.50	45
	4500	Tractor spreader	B-66	52	.154		10	4.18	3.86	18.04	21.50
	4600	Hydro or air seeding, with mulch and fertilizer	B-81	80	.300		25	7.20	7.55	39.75	47
	4800	Turf mix, 4#/M.S.F., push spreader	1 Clab	8	1		10	21.50		31.50	45
	4900	Tractor spreader	B-66	52	.154		10	4.18	3.86	18.04	21.50
	5000	Hydro or air seeding, with mulch and fertilizer	B-81	80	.300		25	7.20	7.55	39.75	47
	5200	Utility mix, 7#/M.S.F., push spreader	1 Clab	8	1		7.75	21.50		29.25	42.50
	5300	Tractor spreader	B-66	52	.154		7.75	4.18	3.86	15.79	19.15
	5400	Hydro or air seeidng, with mulch and fertilizer	B-81	80	.300		29	7.20	7.55	43.75	51.50
	5600	Wildflower, .10#/M.S.F., push spreader	1 Clab	8	1		3	21.50		24.50	37.50
	5700	Tractor spreader	B-66	52	.154		3	4.18	3.86	11.04	13.90
	5800	Hydro or air seeding, with mulch and fertilizer	B-81	80	.300		16.50	7.20	7.55	31.25	37.50
	7000	Apply fertilizer, 800 lb./acre	B-66	4	2	Ton	330	54.50	50	434.50	505
	7025	Limestone, mechanical spread	A-1	1.75	4.571	Acre	3.25	98	39	140.25	201
	7050	Apply limestone, 800 lb./acre	B-66	4.25	1.882	Ton	140	51	47	238	284
	7060	Limestone, mechanical spread	A-1	1.74	4.598	Acre	3.25	98.50	39.50	141.25	202
	7100	Apply mulch, see div. 029-516									
316	0010	**SODDING** 1" deep, bluegrass sod, on level ground, over 8 M.S.F.	B-63	22	1.818	M.S.F.	200	41	5.05	246.05	290
	0200	4 M.S.F.		17	2.353	"	220	53	6.55	279.55	330
	0300	1000 S.F.		3.50	11.429	Ea.	250	258	32	540	715
	0500	Sloped ground, over 8 M.S.F.		6	6.667	M.S.F.	200	151	18.55	369.55	475
	0600	4 M.S.F.		5	8		220	181	22	423	550
	0700	1000 S.F.		4	10		250	226	28	504	660
	1000	Bent grass sod, on level ground, over 6 M.S.F.		20	2		460	45	5.55	510.55	580
	1100	3 M.S.F.		18	2.222		500	50	6.20	556.20	635

029 | Landscaping

029 300 | Lawns & Grasses

			CREW	DAILY OUTPUT	LABOR-HOURS	UNIT	1999 BARE COSTS MAT.	LABOR	EQUIP.	TOTAL	TOTAL INCL O&P
316	1200	Sodding 1000 S.F. or less	B-63	14	2.857	M.S.F.	550	64.50	7.95	622.45	715
	1500	Sloped ground, over 6 M.S.F.		15	2.667		460	60.50	7.40	527.90	605
	1600	3 M.S.F.		13.50	2.963		500	67	8.25	575.25	665
	1700	1000 S.F.		12	3.333		300	75.50	9.25	384.75	460
320	0010	**STOLENS, SPRIGGING**									
	0100	6" O.C., by hand	1 Clab	4	2	M.S.F.	12.25	43		55.25	81
	0110	Walk behind sprig planter	"	80	.100		12.25	2.15		14.40	16.90
	0120	Towed sprig planter	B-66	350	.023		12.25	.62	.57	13.44	15.10
	0130	9" O.C., by hand	1 Clab	5.20	1.538		9	33		42	62
	0140	Walk behind sprig planter	"	92	.087		9	1.87		10.87	12.85
	0150	Towed sprig planter	B-66	420	.019		9	.52	.48	10	11.20
	0160	12" O.C., by hand	1 Clab	6	1.333		5.75	28.50		34.25	51.50
	0170	Walk behind sprig planter	"	110	.073		5.75	1.56		7.31	8.80
	0180	Towed sprig planter	B-66	500	.016		5.75	.44	.40	6.59	7.45
	0200	Broadcast, by hand, 2 Bu per M.S.F.	1 Clab	15	.533		5	11.45		16.45	23.50
	0210	4 Bu. per M.S.F.		10	.800		10	17.15		27.15	38
	0220	6 Bu. per M.S.F.		6.50	1.231		15	26.50		41.50	58
	0300	Hydro planter, 6 Bu. per M.S.F.	B-64	100	.160		15	3.46	2.87	21.33	25
	0320	Manure spreader planting 6 Bu. per M.S.F.	B-66	200	.040		15	1.09	1	17.09	19.25

029 500 | Trees/Plants/Grnd Cover

			CREW	DAILY OUTPUT	LABOR-HOURS	UNIT	1999 BARE COSTS MAT.	LABOR	EQUIP.	TOTAL	TOTAL INCL O&P
516	0010	**MULCH**									
	0100	Aged barks, 3" deep, hand spread	1 Clab	100	.080	S.Y.	2	1.72		3.72	4.90
	0150	Skid steer loader	B-63	13.50	2.963	M.S.F.	220	67	8.25	295.25	355
	0200	Hay, 1" deep, hand spread	1 Clab	475	.017	S.Y.	.25	.36		.61	.85
	0250	Power mulcher, small	B-64	180	.089	M.S.F.	18.50	1.92	1.59	22.01	25
	0350	Large	B-65	530	.030	"	18.50	.65	.86	20.01	22.50
	0370	Fiber mulch recycled newsprint hand spread	1 Clab	500	.016	S.Y.	.10	.34		.44	.65
	0380	Power mulcher small	B-64	200	.080	M.S.F.	7.75	1.73	1.43	10.91	12.80
	0390	Power mulcher large	B-65	600	.027	"	7.50	.58	.76	8.84	9.95
	0400	Humus peat, 1" deep, hand spread	1 Clab	700	.011	S.Y.	1.11	.25		1.36	1.61
	0450	Push spreader	A-1	2,500	.003	"	1.39	.07	.03	1.49	1.67
	0550	Tractor spreader	B-66	700	.011	M.S.F.	104	.31	.29	104.60	115
	0600	Oat straw, 1" deep, hand spread	1 Clab	475	.017	S.Y.	.28	.36		.64	.88
	0650	Power mulcher, small	B-64	180	.089	M.S.F.	25	1.92	1.59	28.51	32
	0700	Large	B-65	530	.030	"	25	.65	.86	26.51	29.50
	0750	Add for asphaltic emulsion	B-45	1,770	.009	Gal.	1.60	.23	.42	2.25	2.57
	0800	Peat moss, 1" deep, hand spread	1 Clab	900	.009	S.Y.	1.55	.19		1.74	2
	0850	Push spreader	A-1	2,500	.003	"	1.60	.07	.03	1.70	1.90
	0950	Tractor spreader	B-66	700	.011	M.S.F.	175	.31	.29	175.60	194
	1000	Polyethylene film, 6 mil.	2 Clab	2,000	.008	S.Y.	.15	.17		.32	.44
	1010	4 mil		2,300	.007		.12	.15		.27	.36
	1020	1-1/2 mil		2,500	.006		.08	.14		.22	.31
	1050	Filter fabric weed barrier		2,000	.008		.75	.17		.92	1.10
	1100	Redwood nuggets, 3" deep, hand spread	1 Clab	150	.053		6	1.14		7.14	8.40
	1150	Skid steer loader	B-63	13.50	2.963	M.S.F.	600	67	8.25	675.25	775
	1200	Stone mulch, hand spread, ceramic chips, economy	1 Clab	125	.064	S.Y.	5.75	1.37		7.12	8.50
	1250	Deluxe	"	95	.084	"	8.60	1.81		10.41	12.30
	1300	Granite chips	B-1	10	2.400	C.Y.	28	53		81	115
	1400	Marble chips		10	2.400		105	53		158	200
	1500	Onyx gemstone		10	2.400		310	53		363	425
	1600	Pea gravel		28	.857		40.50	18.95		59.45	74.50
	1700	Quartz		10	2.400		135	53		188	233
	1800	Tar paper, 15 Lb. felt	1 Clab	800	.010	S.Y.	.40	.21		.61	.78
	1900	Wood chips, 2" deep, hand spread	"	220	.036	"	1.65	.78		2.43	3.05

Important: See the Reference Section for critical supporting data - Reference Nos., Crews, & City Cost Indexes

		029 500 | Trees/Plants/Grnd Cover	CREW	DAILY OUTPUT	LABOR-HOURS	UNIT	1999 BARE COSTS				TOTAL INCL O&P
							MAT.	LABOR	EQUIP.	TOTAL	
516	1950	Skid steer loader	B-63	20.30	1.970	M.S.F.	108	44.50	5.50	158	195
520	0010	**PLANTING** Moving shrubs on site, 12" ball	B-62	28	.857	Ea.		20	3.98	23.98	35.50
	0100	24" ball	"	22	1.091			25.50	5.05	30.55	45
	0300	Moving trees on site, 36" ball	B-6	3.75	6.400			150	58.50	208.50	296
	0400	60" ball	"	1	24			560	219	779	1,100
521	0010	**PLANTING** Trees, shrubs and ground cover									
	0100	Light soil									
	0110	Bare root seedlings, 3" to 5"	1 Clab	960	.008	Ea.		.18		.18	.28
	0120	6" to 10"		520	.015			.33		.33	.52
	0130	11" to 16"		370	.022			.46		.46	.73
	0140	17" to 24"		210	.038			.82		.82	1.29
	0200	Potted, 2-1/4" diameter		840	.010			.20		.20	.32
	0210	3" diameter		700	.011			.25		.25	.39
	0220	4" diameter		620	.013			.28		.28	.44
	0300	Container, 1 gallon	2 Clab	84	.190			4.09		4.09	6.45
	0310	2 gallon		52	.308			6.60		6.60	10.40
	0320	3 gallon		40	.400			8.60		8.60	13.50
	0330	5 gallon		29	.552			11.85		11.85	18.60
	0400	Bagged and burlapped, 12" diameter ball, by hand		19	.842			18.05		18.05	28.50
	0410	Backhoe/loader, 48 H.P.	B-6	40	.600			14	5.45	19.45	28
	0415	15" diameter, by hand	2 Clab	16	1			21.50		21.50	34
	0416	Backhoe/loader, 48 H.P.	B-6	30	.800			18.70	7.30	26	37
	0420	18" diameter by hand	2 Clab	12	1.333			28.50		28.50	45
	0430	Backhoe/loader, 48 H.P.	B-6	27	.889			21	8.10	29.10	41.50
	0440	24" diameter by hand	2 Clab	9	1.778			38		38	60
	0450	Backhoe/loader 48 H.P.	B-6	21	1.143			26.50	10.40	36.90	53
	0470	36" diameter, backhoe/loader, 48 H.P.		17	1.412			33	12.90	45.90	65
	0500	For other size root balls see R029-540		160	.150	C.F.		3.51	1.37	4.88	6.95
	0550	Medium soil									
	0560	Bare root seedlings, 3" to 5"	1 Clab	672	.012	Ea.		.26		.26	.40
	0561	6" to 10"		364	.022			.47		.47	.74
	0562	11" to 16"		260	.031			.66		.66	1.04
	0563	17" to 24"		145	.055			1.18		1.18	1.86
	0570	Potted, 2-1/4" diameter		590	.014			.29		.29	.46
	0572	3" diameter		490	.016			.35		.35	.55
	0574	4" diameter		435	.018			.39		.39	.62
	0590	Container, 1 gallon	2 Clab	59	.271			5.80		5.80	9.15
	0592	2 gallon		36	.444			9.55		9.55	15
	0594	3 gallon		28	.571			12.25		12.25	19.30
	0595	5 gallon		20	.800			17.15		17.15	27
	0600	Bagged and burlapped, 12" diameter ball, by hand		13	1.231			26.50		26.50	41.50
	0605	Backhoe/loader, 48 H.P.	B-6	28	.857			20	7.80	27.80	39.50
	0607	15" diameter, by hand	2 Clab	11.20	1.429			30.50		30.50	48
	0608	Backhoe/loader, 48 H.P.	B-6	21	1.143			26.50	10.40	36.90	53
	0610	18" diameter, by hand	2 Clab	8.50	1.882			40.50		40.50	63.50
	0615	Backhoe/loader, 48 H.P.	B-6	19	1.263			29.50	11.50	41	58.50
	0620	24" diameter, by hand	2 Clab	6.30	2.540			54.50		54.50	85.50
	0625	Backhoe/loader, 48 H.P.	B-6	14.70	1.633			38	14.90	52.90	75.50
	0630	36" diameter, backhoe/loader, 48 H.P.	"	12	2			46.50	18.25	64.75	92.50
	0650	For other size root balls, see Reference Section									
	0700	Heavy or stoney soil									
	0710	Bare root seedlings, 3" to 5"	1 Clab	470	.017	Ea.		.37		.37	.57
	0711	6" to 10"		255	.031			.67		.67	1.06
	0712	11" to 16"		182	.044			.94		.94	1.48
	0713	17" to 24"		101	.079			1.70		1.70	2.67

2 SITE WORK

029 500 | Trees/Plants/Grnd Cover

		029 500 Trees/Plants/Grnd Cover	CREW	DAILY OUTPUT	LABOR-HOURS	UNIT	1999 BARE COSTS MAT.	LABOR	EQUIP.	TOTAL	TOTAL INCL O&P
521	0720	Potted, 2-1/4" diameter	1 Clab	101	.079	Ea.		1.70		1.70	2.67
	0722	3" diameter		343	.023			.50		.50	.79
	0724	4" diameter		305	.026			.56		.56	.89
	0730	Container, 1 gallon	2 Clab	41.30	.387			8.30		8.30	13.10
	0732	2 gallon		25.20	.635			13.60		13.60	21.50
	0734	3 gallon		19.60	.816			17.50		17.50	27.50
	0735	5 gallon		14	1.143			24.50		24.50	38.50
	0750	Bagged and burlapped, 12" diameter ball, by hand		9.10	1.758			37.50		37.50	59.50
	0751	Backhoe/loader	B-6	19.60	1.224			28.50	11.15	39.65	57
	0752	15" diameter, by hand	2 Clab	7.80	2.051			44		44	69
	0753	Backhoe/loader, 48 H.P.	B-6	14.70	1.633			38	14.90	52.90	75.50
	0754	18" diameter, by hand	2 Clab	5.60	2.857			61.50		61.50	96.50
	0755	Backhoe/loader, 48 H.P.	B-6	13.30	1.805			42	16.45	58.45	83.50
	0756	24" diameter, by hand	2 Clab	4.40	3.636			78		78	123
	0757	Backhoe/loader, 48 H.P.	B-6	10.30	2.330			54.50	21.50	76	108
	0758	36" diameter backhoe/loader, 48 H.P.	"	8.40	2.857			67	26	93	133
	0790	For other size root balls, see Reference Section									
	2000	Stake out tree and shrub locations	2 Clab	220	.073	Ea.		1.56		1.56	2.45
604	0010	**TREES, DECIDUOUS** zones 2 - 6									
	0100	Acer campestre, (Hedge Maple), Z4, B&B									
	0110	4' to 5'				Ea.	43			43	47.50
	0120	5' to 6'					49			49	54
	0130	1-1/2" to 2" Cal.					97			97	107
	0140	2" to 2-1/2" Cal.					111			111	122
	0150	2-1/2" to 3" Cal.					155			155	171
	0200	Acer ginnala, (Amur Maple), Z2, cont/BB									
	0210	8' to 10'				Ea.	94			94	103
	0220	10' to 12'					124			124	136
	0230	12' to 14'					150			150	165
	0300	Acer griseum, (Paperbark Maple), Z5, B&B									
	0310	1-1/2" to 1-3/4" Cal.				Ea.	110			110	121
	0320	1-3/4" to 2" cal.				"	120			120	132
	0400	Acer palmatum, (Japanese Maple), Z6, cont/BB									
	0410	2' to 2-1/2'				Ea.	34			34	37.50
	0420	4' to 5'					72			72	79
	0430	5' to 6'					92			92	101
	0440	6' to 7'					164			164	180
	0450	7' to 8'					230			230	253
	0460	8' to 10'					280			280	310
	0470	10' to 12'					400			400	440
	0500	Acer palmatum atropurpureum, (Bloodgood Japan Maple), Z5, B&B									
	0510	3' to 3-1/2'				Ea.	87			87	95.50
	0520	3-1/2' to 4'					108			108	119
	0530	4' to 4-1/2'					135			135	149
	0540	4-1/2' to 5'					169			169	186
	0550	5' to 6'					237			237	261
	0600	Acer platanoides, (Norway Maple), Z4, B&B									
	0610	8' to 10'				Ea.	70			70	77
	0620	1-1/2" to 2" Cal.					95			95	105
	0630	2" to 2-1/2" Cal.					105			105	116
	0640	2-1/2" to 3" Cal.					140			140	154
	0650	3" to 3-1/2" Cal.					200			200	220
	0660	Bare root, 8' to 10'					30			30	33
	0670	10' to 12'					50			50	55
	0680	12' to 14'					60			60	66
	0700	Acer platanoides columnare, (Column maple), Z4, B&B									

Important: See the Reference Section for critical supporting data - Reference Nos., Crews, & City Cost Indexes

029 500 | Trees/Plants/Grnd Cover

		Description	CREW	DAILY OUTPUT	LABOR-HOURS	UNIT	1999 BARE COSTS MAT.	LABOR	EQUIP.	TOTAL	TOTAL INCL O&P
604	0710	2" to 2-1/2" Cal.				Ea.	95			95	105
	0720	2-1/2" to 3" Cal.					154			154	169
	0730	3" to 3-1/2" Cal.					200			200	220
	0740	3-1/2" to 4" Cal.					250			250	275
	0750	4" to 4-1/2" Cal.					360			360	395
	0760	4-1/2" to 5" Cal.					500			500	550
	0770	5" to 5-1/2" Cal.					675			675	745
	0800	Acer rubrum, (Red Maple), Z4, B&B									
	0810	1-1/2" to 2" Cal.				Ea.	95			95	105
	0820	2" to 2-1/2" Cal.					100			100	110
	0830	2-1/2" to 3" Cal.					130			130	143
	0840	Bare Root, 8′ to 10′					45			45	49.50
	0850	10′ to 12′					55			55	60.50
	0900	Acer saccharum, (Sugar Maple), Z3, B&B									
	0910	1-1/2" to 2" Cal.				Ea.	92			92	101
	0920	2" to 2-1/2" Cal.					100			100	110
	0930	2-1/2" to 3" Cal.					128			128	141
	0940	3" to 3-1/2" Cal.					183			183	201
	0950	3-1/2" to 4" Cal.					240			240	264
	0960	Bare Root, 8′ to 10′					45			45	49.50
	0970	10′ to 12′					55			55	60.50
	0980	12′ to 14′					65			65	71.50
	1000	Aesculus carnea biroti, (Ruby Horsechestnut), Z3, B&B									
	1010	2′ to 3′				Ea.	40			40	44
	1020	3′ to 4′					55			55	60.50
	1030	4′ to 5′					75			75	82.50
	1040	5′ to 6′					110			110	121
	1100	Amelanchier canadensis, (shadblow), Z4, B&B									
	1110	3′ to 4′				Ea.	29			29	32
	1120	4′ to 5′					40			40	44
	1130	5′ to 6′					55			55	60.50
	1140	6′ to 8′					80			80	88
	1150	8′ to 10′					100			100	110
	1200	Betula maximowicziana, (monarch birch), Z5, B&B									
	1210	6′ to 8′				Ea.	55			55	60.50
	1220	1-1/2" to 2" Cal.					105			105	116
	1230	2" to 2-1/2" Cal.					145			145	160
	1300	Betula nigra, (river birch), Z5, B&B									
	1310	1-1/2" to 2" cal.				Ea.	97			97	107
	1320	2" to 2-1/2" Cal.					116			116	128
	1330	2-1/2" to 3" Cal.					150			150	165
	1340	3" to 3-1/2" Cal.					218			218	240
	1400	Betula papyrifera, (canoe or paper birch), Z2, B&B									
	1410	6′ to 8′				Ea.	65			65	71.50
	1420	1-1/2" to 2" Cal.					95			95	105
	1430	2" to 2-1/2" Cal.					108			108	119
	1440	2-1/2" to 3" Cal.					155			155	171
	1450	3" to 3-1/2" Cal.					195			195	215
	1460	3-1/2" to 4" Cal.					240			240	264
	1470	4" to 4-1/2" Cal.					350			350	385
	1480	4-1/2" to 5" Cal.					455			455	500
	1500	Castanea mollissima, (Chinese Chestnut), Z5, B&B									
	1510	2′ to 3′				Ea.	20			20	22
	1520	6′ to 8′					75			75	82.50
	1530	8′ to 10′					100			100	110
	1540	10′ to 12′					140			140	154

029 500 | Trees/Plants/Grnd Cover

		CREW	DAILY OUTPUT	LABOR-HOURS	UNIT	1999 BARE COSTS MAT.	LABOR	EQUIP.	TOTAL	TOTAL INCL O&P
604 1600	Catalpa speciosa, (northern catalpa), Z5, B&B									
1610	1-1/2" to 2" cal.				Ea.	115			115	127
1620	2" to 2-1/2" Cal.				"	140			140	154
1700	Cercidiphyllum japonica, (katsura tree), Z5, B&B									
1710	6' to 8'				Ea.	72			72	79
1720	1-1/2" to 2" Cal.					108			108	119
1730	2" to 2-1/2" Cal.					150			150	165
1740	2-1/2" to 3" Cal.					190			190	209
1750	3-1/2" to 4" Cal.					275			275	305
1760	4" to 4-1/2" Cal.				↓	350			350	385
1800	Cercis canadensis, (eastern redbud), Z5, B&B									
1810	15 gal.				Ea.	68			68	75
1820	20 Gal.					84			84	92.50
1830	4' to 5'					48			48	53
1840	5' to 6'					73			73	80.50
1850	6' to 8'				↓	88			88	97
1900	Chionanthus virginicus, (fringetree), Z4, B&B									
1910	4' to 5'				Ea.	57			57	62.50
1920	5' to 6'					77			77	84.50
1930	6' to 8'				↓	103			103	113
2000	Cladrastis lutea, (yellowood), Z4, B&B									
2010	1-1/2" to 2" Cal.				Ea.	98			98	108
2020	2" to 2-1/2" Cal.					124			124	136
2030	2-1/2" to 3" Cal.					176			176	194
2040	3" to 3-1/2" Cal.					206			206	227
2050	3-1/2" to 4" Cal.				↓	252			252	277
2100	Cornus florida, (white flowering dogwood), Z5, B&B									
2110	4' to 5'				Ea.	39			39	43
2120	5' to 6'					51			51	56
2130	6' to 7'				↓	66			66	72.50
2200	Cornus florida rubra, (pink flowering dogwood), Z6, B&B									
2210	4' to 5'				Ea.	45			45	49.50
2220	5' to 6'					59			59	65
2230	6' to 7'					73			73	80.50
2240	7' to 8'				↓	94			94	103
2300	Cornus kousa, (Japanese Dogwood), Z5, B&B									
2310	3' to 4'				Ea.	34			34	37.50
2320	4' to 5'					37			37	40.50
2330	5' to 6'					53			53	58.50
2340	6' to 8'					75			75	82.50
2350	8' to 10'				↓	110			110	121
2400	Crataegus crus-galli, (Cockspur Hawthorn), Z5, B&B									
2410	55 Gal.				Ea.	140			140	154
2420	5' to 6'				"	75			75	82.50
2500	Crataegus oxyacantha superba, (Crimson Hawthorn), Z5, B&B									
2510	5' to 6'				Ea.	41			41	45
2520	6' to 8'					50			50	55
2530	1-1/2" to 2" Cal.					106			106	117
2540	2" to 2-1/2" Cal.					115			115	127
2550	2-1/2" to 3" Cal.					143			143	157
2560	3" to 3-1/2" Cal.					165			165	182
2570	3-1/2" to 4" Cal.					226			226	249
2580	4" to 5" Cal.				↓	310			310	340
2600	Crataegus phaenopyrum, (Washington Hawthorn), Z5, B&B									
2610	5' to 6'				Ea.	48			48	53
2620	6' to 7'				↓	60			60	66

2 SITE WORK

Important: See the Reference Section for critical supporting data - Reference Nos., Crews, & City Cost Indexe

029 500 | Trees/Plants/Grnd Cover		CREW	DAILY OUTPUT	LABOR-HOURS	UNIT	1999 BARE COSTS MAT.	LABOR	EQUIP.	TOTAL	TOTAL INCL O&P
604	2630 1-3/4" to 2" Cal.				Ea.	80			80	88
	2640 2" to 2-1/2" Cal.					99			99	109
	2650 2-1/2" to 3" Cal.					125			125	138
	2660 3" to 3-1/2" Cal.				↓	163			163	179
	2700 Cydonia oblonga orange, (Orange Quince), Z5, B&B									
	2710 5' to 6'				Ea.	45			45	49.50
	2720 6' to 8'				"	57			57	62.50
	2800 Diospyros virginiana, (Persimmon), Z4, B&B									
	2810 3' to 4'				Ea.	30			30	33
	2820 4' to 5'					35			35	38.50
	2830 1-1/2" to 2" Cal.					97			97	107
	2840 2" to 2-1/2" Cal.					132			132	145
	2850 2-1/2" to 3" Cal.				↓	161			161	177
	2900 Eleagnus angustifolia, (Russian Olive, tree form), Z3, B&B									
	2910 1-1/2" to 2" Cal.				Ea.	96			96	106
	2920 2" to 2-1/2" Cal.					124			124	136
	2930 2-1/2" to 3" Cal.				↓	155			155	171
	3000 Eucommia ulmoides, (Hardy Rubber Tree), Z4, B&B									
	3010 6' to 8'				Ea.	64			64	70.50
	3020 8' to 10'					79			79	87
	3030 1-1/2" to 2" Cal.					102			102	112
	3040 2" to 2-1/2" Cal.					121			121	133
	3050 2-1/2" to 3" Cal.				↓	160			160	176
	3100 Fagus grandiflora, (American Beech), Z3, B&B									
	3110 1-1/2" to 1-3/4" Cal.				Ea.	95			95	105
	3120 1-3/4" to 2" Cal.					118			118	130
	3130 2" to 2-1/2" Cal.				↓	140			140	154
	3200 Fagus sylvatica, (European Beech), Z5, B&B									
	3210 6' to 8'				Ea.	120			120	132
	3220 1-1/2" to 2" Cal.					138			138	152
	3230 2" to 2-1/2" Cal.				↓	182			182	200
	3300 Fagus sylvatica pendula, (Weeping European Beech), Z4, B&B									
	3310 1-3/4" to 2" cal.				Ea.	124			124	136
	3320 2" to 2-1/2" Cal.					150			150	165
	3330 2-1/2" to 3" Cal.					198			198	218
	3340 3" to 3-1/2" Cal.				↓	262			262	288
	3400 Franklinia alatamaha, (Franklin Tree), Z5, B&B									
	3410 3 Gal.				Ea.	20			20	22
	3420 7 Gal.					45			45	49.50
	3430 15 Gal.					76			76	83.50
	3440 5' to 6'					88			88	97
	3450 6' to 8'				↓	104			104	114
	3500 Fraxinus americana, (Autumn Purple Ash), Z3, B&B									
	3510 20 Gal.				Ea.	90			90	99
	3520 1-1/2" to 2" Cal.					110			110	121
	3530 2" to 2-1/2" Cal.				↓	132			132	145
	3600 Fraxinus pensylvanica, (Seedless Green Ash), Z3, B&B									
	3610 15 gal.				Ea.	70			70	77
	3615 20 Gal.					100			100	110
	3620 55 Gal.					138			138	152
	3625 6' to 8'					57			57	62.50
	3630 1-1/2" to 2" Cal.					103			103	113
	3635 2" to 2-1/2" Cal.					125			125	138
	3640 2-1/2" to 3" Cal.					173			173	190
	3645 3" to 3-1/2" Cal.					224			224	246
	3650 3-1/2" to 4" Cal.				↓	300			300	330

029 500 | Trees/Plants/Grnd Cover		CREW	DAILY OUTPUT	LABOR-HOURS	UNIT	1999 BARE COSTS MAT.	LABOR	EQUIP.	TOTAL	TOTAL INCL O&P
604 3655	4" to 4-1/2" Cal.				Ea.	400			400	440
3660	4-1/2" to 5" Cal.					530			530	585
3665	Bare root, 8' to 10'					47			47	51.50
3670	10' to 12'					70			70	77
3675	12' to 14'				↓	77			77	84.50
3700	Ginkgo biloba, (Maidenhair Tree), Z5, B&B									
3710	6' to 8'				Ea.	60			60	66
3720	1-1/2" to 2" Cal.					95			95	105
3730	2" to 2-1/2" Cal.					141			141	155
3740	2-1/2" to 3" Cal.					168			168	185
3750	3" to 3-1/2" Cal.				↓	189			189	208
3800	Gleditsia triacanthos inermis, (Thornless Honeylocust), Z5, B&B									
3810	20 Gal.				Ea.	80			80	88
3815	55 Gal.					100			100	110
3820	8' to 10'					60			60	66
3825	1-1/2" to 2" Cal.					95			95	105
3830	2" to 2-1/2" Cal.					108			108	119
3835	2-1/2" to 3" Cal.					140			140	154
3840	3" to 3-1/2" Cal.					185			185	204
3845	3-1/2" to 4" Cal.					238			238	262
3850	4" to 4-1/2" Cal.					375			375	415
3855	4-1/2" to 5" Cal.					490			490	540
3860	Bare root, 10' to 12'					44			44	48.50
3865	12' to 14'					71			71	78
3870	14' to 16'				↓	110			110	121
3900	Gymnocladus dioicus, (Kentucky Coffee Tree), Z5, B&B									
3910	2" to 2-1/2" Cal.				Ea.	128			128	141
3920	2-1/2" to 3" Cal.					148			148	163
3930	3" to 3-1/2" Cal.				↓	181			181	199
4000	Halesia carolina, (Silverbell), Z6, B&B									
4010	4' to 5'				Ea.	43			43	47.50
4020	5' to 6'					63			63	69.50
4030	6' to 8'					75			75	82.50
4040	8' to 10'					97			97	107
4050	10' to 12'				↓	126			126	139
4100	Hippophae rhamnoides, (Common Sea Buckthorn), Z3, B&B									
4110	3 Gal.				Ea.	12			12	13.20
4120	5' to 6'					70			70	77
4130	6' to 8'				↓	83			83	91.50
4200	Juglans cinera, (Butternut), Z4, B&B									
4210	4' to 5'				Ea.	26			26	28.50
4220	5' to 6'					35			35	38.50
4230	6' to 8'				↓	54			54	59.50
4300	Juglans nigra, (Black Walnut), Z5, B&B									
4310	4' to 5'				Ea.	28			28	31
4320	5' to 6'					36			36	39.50
4330	1-1/2" to 2" Cal.					120			120	132
4340	2" to 2-1/2" Cal.					138			138	152
4350	2-1/2" to 3" Cal.				↓	200			200	220
4400	Koelreuteria paniculata, (Goldenrain Tree), Z6, B&B									
4410	5 Gal.				Ea.	22			22	24
4420	15 Gal.					52			52	57
4430	20 Gal.					104			104	114
4440	55 Gal.					133			133	146
4450	1-1/2" to 2" Cal.					99			99	109
4460	2" to 2-1/2" Cal.				↓	165			165	182 604

029 500 Trees/Plants/Grnd Cover		CREW	DAILY OUTPUT	LABOR-HOURS	UNIT	1999 BARE COSTS MAT.	LABOR	EQUIP.	TOTAL	TOTAL INCL O&P
604 4470	2-1/2" to 3" Cal.				Ea.	150			150	165
4500	Laburnum x watereri, (Goldenchain tree), Z5, B&B									
4510	6' to 8'				Ea.	51			51	56
4520	8' to 9'					70			70	77
4530	1-3/4" to 2" Cal.					98			98	108
4540	2" to 2-1/2" Cal.					120			120	132
4550	2-1/2" to 3" Cal.				↓	160			160	176
4600	Larix decidua pendula, (Weeping European Larch), Z2, B&B									
4610	3' to 4'				Ea.	37			37	40.50
4620	4' to 5'					48			48	53
4630	5' to 6'					62			62	68
4640	6' to 7'				↓	73			73	80.50
4700	Larix kaempferi, (Japanese Larch), Z4, B&B									
4710	5' to 6'				Ea.	74			74	81.50
4720	6' to 8'					95			95	105
4730	8' to 10'				↓	109			109	120
4800	Liquidambar styraciflua, (Sweetgum), Z6, B&B									
4810	6' to 8'				Ea.	58			58	64
4820	8' to 10'					63			63	69.50
4830	1-1/2" to 2" Cal.					95			95	105
4840	2" to 2-1/2" Cal.					116			116	128
4850	2-1/2" to 3" Cal.				↓	150			150	165
4900	Liriodendron tulipifera, (Tuliptree), Z5, B&B									
4910	7' to 8'				Ea.	80			80	88
4920	1-1/2" to 2" Cal.					112			112	123
4930	2" to 2-1/2" Cal.					126			126	139
4940	2-1/2" to 3" Cal.				↓	170			170	187
5000	Maclura pomifera, (Osage Orange), Z5, B&B									
5010	3' to 4'				Ea.	24			24	26.50
5020	4' to 5'					35			35	38.50
5030	6' to 8'				↓	45			45	49.50
5100	Magnolia loebneri merrill, (Merrill Magnolia), Z5, B&B									
5110	4' to 5'				Ea.	40			40	44
5120	5' to 6'					53			53	58.50
5130	6' to 7'					66			66	72.50
5140	7' to 8'				↓	82			82	90
5200	Magnolia soulangena, (Saucer Magnolia), Z6, B&B									
5210	3' to 4'				Ea.	29			29	32
5220	4' to 5'					35			35	38.50
5230	5' to 6'					47			47	51.50
5240	6' to 7'				↓	73			73	80.50
5300	Magnolia stellata, (Star Magnolia), Z6, B&B									
5310	2' to 3'				Ea.	31			31	34
5320	3' to 4'					42			42	46
5330	4' to 5'					81			81	89
5340	5' to 6'				↓	105			105	116
5400	Magnolia virginiana, (Sweetbay Magnolia), Z5, B&B									
5410	2' to 3'				Ea.	26			26	28.50
5420	3' to 4'					40			40	44
5430	4' to 5'					50			50	55
5440	5' to 6'				↓	64			64	70.50
5500	Magnolia x galaxy, (Galaxy Magnolia), Z6, B&B									
5510	1-1/2" to 2" cal.				Ea.	100			100	110
5520	2" to 2-1/2" Cal.					118			118	130
5530	2-1/2" to 3" Cal.				↓	155			155	171
5600	Malus, (Crabapple), Z5, B&B									

029 500 | Trees/Plants/Grnd Cover		CREW	DAILY OUTPUT	LABOR-HOURS	UNIT	1999 BARE COSTS MAT.	LABOR	EQUIP.	TOTAL	TOTAL INCL O&P
604 5610	5 Gal.				Ea.	25			25	27.50
5615	15 Gal.					67			67	73.50
5620	20 Gal.					72			72	79
5625	55 Gal.					91			91	100
5630	6′ to 8′					63			63	69.50
5635	1-1/2″ to 2″ Cal.					71			71	78
5640	2″ to 2-1/2″ Cal.					98			98	108
5645	2-1/2″ to 3″ Cal.					113			113	124
5650	3″ to 3-1/2″ Cal.					145			145	160
5655	3-1/2″ to 4″ Cal.					194			194	213
5660	4″ to 5″ Cal.					270			270	297
5665	Bare root, 5′ to 6′				↓	23			23	25.50
5700	Metasequoia glyptostroboides, (Dawn Redwood), Z5, B&B									
5710	5′ to 6′				Ea.	68			68	75
5720	6′ to 8′					91			91	100
5730	8′ to 10′				↓	110			110	121
5800	Morus alba, (White Mulberry), Z4, B&B									
5810	15 gal.				Ea.	75			75	82.50
5900	Morus alba tatarica, (Russian Mulberry), Z4, B&B									
5910	6′ to 8′				Ea.	78			78	86
5920	10′ to 12′					118			118	130
5930	2-1/2″ to 3″ Cal.				↓	155			155	171
6000	Nyssa sylvatica, (Tupelo), Z4, B&B									
6010	5′ to 6′				Ea.	52			52	57
6020	1-1/2″ to 2″ Cal.					100			100	110
6030	2″ to 2-1/2″ Cal.					120			120	132
6040	2-1/2″ to 3″ Cal.				↓	161			161	177
6100	Ostrya virginiana, (Hop Hornbeam), Z5, B&B									
6110	2″ to 2-1/2″ Cal.				Ea.	126			126	139
6120	2-1/2″ to 3″ Cal.				″	155			155	171
6200	Oxydendron arboreum, (Sourwood), Z4, B&B									
6210	4′ to 5′				Ea.	61			61	67
6220	5′ to 6′					76			76	83.50
6230	6′ to 7′					93			93	102
6240	7′ to 8′				↓	107			107	118
6300	Parrotia persica, (Persian Parrotia), Z6, B&B									
6310	2′ to 3′				Ea.	50			50	55
6320	3′ to 4′				″	63			63	69.50
6400	Phellodendron amurense, (Amur Corktree), Z3, B&B									
6410	6′ to 8′				Ea.	68			68	75
6420	8′ to 10′					85			85	93.50
6430	2″ to 2-1/2″ Cal.					118			118	130
6440	2-1/2″ to 3″ Cal.					152			152	167
6450	3″ to 3-1/2″ Cal.				↓	183			183	201
6500	Platanus acerifolia, (London Plane Tree), Z5, B&B									
6510	15 gal.				Ea.	68			68	75
6520	20 Gal.					84			84	92.50
6530	8′ to 10′					78			78	86
6540	1-1/2″ to 2″ Cal.					100			100	110
6550	2″ to 2-1/2″ Cal.					118			118	130
6560	2-1/2″ to 3″ Cal.					135			135	149
6570	3″ to 3-1/2″ Cal.				↓	150			150	165
6600	Platanus occidentalis, (Buttonwood, Sycamore), Z4, B&B									
6610	30″ box				Ea.	295			295	325
6620	36″ Box					430			430	470
6630	42″ Box				↓	645			645	710

	029 500 | Trees/Plants/Grnd Cover	CREW	DAILY OUTPUT	LABOR-HOURS	UNIT	1999 BARE COSTS MAT.	LABOR	EQUIP.	TOTAL	TOTAL INCL O&P
604	6640	1-1/2" to 2" Cal.				Ea.	98			98	108
	6650	2" to 2-1/2" Cal.					118			118	130
	6660	2-1/2" to 3" Cal.					160			160	176
	6670	3" to 3-1/2" Cal.				↓	204			204	224
	6700	Populus alba pyramidalis, (Bolleana Poplar), Z3, cont/BB									
	6710	5 Gal.				Ea.	22			22	24
	6720	15 Gal.					75			75	82.50
	6730	20 Gal.				↓	90			90	99
	6800	Populus canadensis, (Hybrid Black Poplar), Z5, B&B									
	6810	6' to 8'				Ea.	32			32	35
	6820	1-1/2" to 2" Cal.					108			108	119
	6830	2" to 2-1/2" Cal.				↓	140			140	154
	6900	Populus nigra italica, (Lombardy Poplar), Z4, cont									
	6910	5 Gal.				Ea.	32			32	35
	6920	20 Gal.					88			88	97
	6930	6' to 8' Potted				↓	35			35	38.50
	7000	Prunus cerasifera pissardi, (Flowering Plum), Z5, cont/BB									
	7010	5 Gal.				Ea.	30			30	33
	7020	15 Gal.					74			74	81.50
	7030	20 Gal.					90			90	99
	7040	5' to 6'					35			35	38.50
	7050	6' to 8'					54			54	59.50
	7060	1-1/2" to 2" Cal.					101			101	111
	7070	2" to 2-1/2" Cal.				↓	126			126	139
	7100	Prunus sargenti columnaris, (Columnar Cherry), Z4, cont/BB									
	7110	6' to 8'				Ea.	59			59	65
	7120	1-1/2" to 2" Cal.					84			84	92.50
	7130	2" to 2-1/2" Cal.					113			113	124
	7140	2-1/2" to 3" Cal.				↓	147			147	162
	7200	Prunus serrulata kwanzan, (Flowering Cherry), Z6, cont/BB									
	7210	20 Gal.				Ea.	98			98	108
	7220	1-1/2" to 2" Cal.					100			100	110
	7230	2" to 2-1/2" Cal.					122			122	134
	7240	3" to 3-1/2" Cal.					182			182	200
	7250	3-1/2" to 4" Cal.					243			243	267
	7260	4" to 5" Cal.				↓	375			375	410
	7300	Prunus subhirtella pendula, (Weeping Cherry), Z5, B&B									
	7310	1-1/4" to 1-1/2" Cal.				Ea.	65			65	71.50
	7320	1-1/2" to 1-3/4" Cal.					79			79	87
	7330	1-3/4" to 2" Cal.					95			95	105
	7340	2" to 2-1/2" Cal.					112			112	123
	7350	2-1/2" to 3" Cal.				↓	148			148	163
	7400	Prunus yedoensis, (Yoshino Cherry), Z5, B&B									
	7410	6' to 8'				Ea.	59			59	65
	7420	1-1/2" to 2" Cal.					92			92	101
	7430	2" to 2-1/2" Cal.					106			106	117
	7440	2-1/2" to 3" Cal.				↓	135			135	149
	7500	Pyrus calleryana aristocrat, (Aristocrat Flwrng Pear), Z5, cont/BB									
	7510	15 gal.				Ea.	77			77	84.50
	7520	20 Gal.					99			99	109
	7530	6' to 8'					67			67	73.50
	7540	1-1/2" to 2" Cal.					103			103	113
	7550	2" to 2-1/2" Cal.					122			122	134
	7560	2-1/2" to 3" Cal.					160			160	176
	7570	3" to 3-1/2" Cal.					214			214	235
	7580	3-1/2" to 4" Cal.				↓	278			278	305

029 500 | Trees/Plants/Grnd Cover		CREW	DAILY OUTPUT	LABOR-HOURS	UNIT	1999 BARE COSTS MAT.	LABOR	EQUIP.	TOTAL	TOTAL INCL O&P
604 7600	Pyrus calleryana bradford, (Bradford Flowering Pear), Z5, cont/BB									
7610	15 gal.				Ea.	81			81	89
7620	20 Gal.					103			103	113
7630	1-1/2" to 2" Cal.					76			76	83.50
7640	2" to 2-1/2" Cal.					86			86	94.50
7650	2-1/2" to 3" Cal.					110			110	121
7660	3" to 3-1/2" Cal.					115			115	127
7670	3-1/2" to 4" Cal.				↓	200			200	220
7700	Quercus acutissima, (Sawtooth Oak), Z5, B&B									
7710	6' to 8'				Ea.	53			53	58.50
7720	8' to 10'					78			78	86
7730	10' to 12'					92			92	101
7740	1-1/2" to 2" Cal.					112			112	123
7750	2" to 2-1/2" Cal.				↓	131			131	144
7800	Quercus coccinea, (Scarlet Oak), Z4, B&B									
7810	1-1/2" to 2" cal.				Ea.	95			95	105
7820	2" to 2-1/2" Cal.				"	120			120	132
7900	Quercus palustris, (Pin Oak), Z5, B&B									
7910	6' to 8'				Ea.	53			53	58.50
7920	1-1/2" to 2" Cal.					89			89	98
7930	2" to 2-1/2" Cal.					104			104	114
7940	2-1/2" to 3" Cal.					151			151	166
7950	3" to 3-1/2" Cal.					965			965	1,050
7960	3-1/2" to 4" Cal.					242			242	266
7970	4" to 4-1/2" Cal.					330			330	365
7980	4-1/2" to 5" Cal.					375			375	410
7985	6" to 7" Cal.					790			790	870
7990	8" to 9" Cal.				↓	1,300			1,300	1,425
8000	Quercus prinus, (Chesnut Oak), Z4, B&B									
8010	6' to 8'				Ea.	55			55	60.50
8020	8' to 10'				"	85			85	93.50
8100	Quercus robur fastigiata, (Columnar English Oak), Z4, B&B									
8110	6' to 8'				Ea.	55			55	60.50
8120	8' to 10'					72			72	79
8130	1-1/2" to 2" Cal.					102			102	112
8140	2" to 2-1/2" Cal.					116			116	128
8150	2-1/2" to 3" Cal.				↓	165			165	182
8200	Quercus rubra, (Red Oak), Z4, cont/BB									
8210	6' to 8'				Ea.	55			55	60.50
8220	1-1/2" to 2" Cal.					100			100	110
8230	2" to 2-1/2" Cal.					122			122	134
8240	2-1/2" to 3" Cal.				↓	160			160	176
8300	Robinia pseudoacacia, (Black Locust), Z5, cont									
8310	24" Box				Ea.	170			170	187
8320	30" Box					300			300	330
8330	36" Box				↓	440			440	485
8400	Salix babylonica, (Weeping Willow), Z5, cont/BB									
8410	24" Box				Ea.	170			170	187
8420	36" Box					380			380	420
8430	1-1/2" to 2" Cal.					97			97	107
8440	2" to 2-1/2" Cal.					116			116	128
8450	2-1/2" to 3" Cal.				↓	129			129	142
8500	Sophora japonica "Regent", (Regent Scholartree), Z4, B&B									
8510	1-3/4" to 2" Cal.				Ea.	93			93	102
8520	2" to 2-1/2" Cal.					130			130	143
8530	2-1/2" to 3" Cal.				↓	170			170	187

029 500 | Trees/Plants/Grnd Cover		CREW	DAILY OUTPUT	LABOR-HOURS	UNIT	1999 BARE COSTS				TOTAL INCL O&P
						MAT.	LABOR	EQUIP.	TOTAL	
604 8600	Sorbus alnifolia, (Korean Mountain Ash), Z5, B&B									
8610	8' to 10'				Ea.	75			75	82.50
8620	1-1/2" to 2" Cal.					96			96	106
8630	2" to 2-1/2" Cal.					114			114	125
8640	2-1/2" to 3" Cal.					141			141	155
8700	Sorbus aucuparia, (European Mountain Ash), Z2, cont/BB									
8710	15 Gal.				Ea.	74			74	81.50
8715	20 Gal.					90			90	99
8720	1-1/2" to 2" Cal.					102			102	112
8725	2" to 2-1/2" Cal.					118			118	130
8730	2-1/2" to 3" Cal.					162			162	178
8735	3" to 3-1/2" Cal.					210			210	231
8740	3-1/2" to 4" Cal.					300			300	330
8745	4" to 5" Cal.					390			390	425
8750	5" to 6" Cal.					595			595	655
8755	Bare root, 8' to 10'					50			50	55
8760	10' to 12'					65			65	71.50
8800	Stewartia pseudocamellia, (Japanese Stewartia), Z5, B&B									
8810	5' to 6'				Ea.	114			114	125
8820	6' to 8'					152			152	167
8830	8' to 10'					241			241	265
8900	Styrax japonica, (Japanese Snowbell), Z6, B&B									
8910	4' to 5'				Ea.	52			52	57
8920	5' to 6'					78			78	86
8930	6' to 8'					107			107	118
9000	Syringa japonica, (Japanese Tree, Lilac), Z4, B&B									
9010	5' to 6'				Ea.	55			55	60.50
9020	6' to 8'					70			70	77
9030	8' to 10'					95			95	105
9040	10' to 12'					132			132	145
9100	Taxodium distichum, (Bald Cypress), Z4, B&B									
9110	5' to 6'				Ea.	45			45	49.50
9120	6' to 7'					60			60	66
9130	7' to 8'					75			75	82.50
9140	8' to 10'					95			95	105
9200	Tilia americana "Redmond", (Redmond Linden), Z3, B&B									
9210	8' to 10'				Ea.	64			64	70.50
9220	10' to 12'					88			88	97
9230	1-1/2" to 2" Cal.					93			93	102
9240	2" to 2-1/2" Cal.					108			108	119
9250	2-1/2" to 3" Cal.					132			132	145
9300	Tilia cordata greenspire, (Littleleaf Linden), Z4, B&B									
9310	1-1/2" to 2" Cal.				Ea.	100			100	110
9315	2" to 2-1/2" Cal.					113			113	124
9320	2-1/2" to 3" Cal.					135			135	149
9325	3" to 3-1/2" Cal.					201			201	221
9330	3-1/2" to 4" Cal.					240			240	264
9335	4" to 4-1/2" Cal.					340			340	375
9340	4-1/2" to 5" Cal.					445			445	490
9345	5" to 5-1/2" Cal.					500			500	550
9350	Bare root, 10' to 12'					58			58	64
9355	12' to 14'					74			74	81.50
9400	Tilia tomentosa, (Silver Linden), Z4, B&B									
9410	6' to 8'				Ea.	52			52	57
9420	8' to 10'					62			62	68
9430	10' to 12'					101			101	111

029 500 | Trees/Plants/Grnd Cover		CREW	DAILY OUTPUT	LABOR-HOURS	UNIT	1999 BARE COSTS MAT.	LABOR	EQUIP.	TOTAL	TOTAL INCL O&P
604 9440	2″ to 2-1/2″ Cal.				Ea.	119			119	131
9450	2-1/2″ to 3″ Cal.					157			157	173
9500	Ulmus americana, (American Elm), Z3, B&B									
9510	6′ to 8′				Ea.	55			55	60.50
9600	Zeklova serrata, (Japanese Keaki Tree), Z4, B&B									
9610	1-1/2″ to 2″ Cal.				Ea.	97			97	107
9620	2″ to 2-1/2″ Cal.					116			116	128
9630	2-1/2″ to 3″ Cal.					145			145	160
608 0010	**TREES, CONIFERS** zone 2-7									
1000	Abies balsamea nana, (Dwarf Balsam Fir), Z3, cont									
1001	12″ to 15″				Ea.	33			33	36.50
1002	15″ to 18″					39			39	43
1003	18″ to 24″					47			47	51.50
1050	Abies concolor, (White Fir), Z4, B&B									
1051	2′ to 3′				Ea.	35			35	38.50
1052	3′ to 4′					50			50	55
1053	4′ to 5′					60			60	66
1054	5′ to 6′					89			89	98
1055	6′ to 8′					105			105	116
1100	Abies concolor violacea, Z4, cont/BB									
1101	2′ to 3′				Ea.	42			42	46
1102	3′ to 4′				″	55			55	60.50
1150	Abies fraseri, (Fraser Balsam Fir), Z5, B&B									
1151	5′ to 6′				Ea.	78			78	86
1152	6′ to 7′				″	108			108	119
1200	Abies iasiocarpa arizonica, (Cork Fir), Z5, cont/BB									
1201	2′ to 3′				Ea.	62			62	68
1202	3′ to 4′					88			88	97
1203	4′ to 5′					145			145	160
1204	5′ to 6′					171			171	188
1205	6′ to 7′					210			210	231
1206	8′ to 10′					310			310	345
1207	10′ to 12′					395			395	435
1250	Abies iasiocarpa compacta, (Dwarf Alpine Fir), Z5, cont/BB									
1251	18″ to 24″				Ea.	65			65	71.50
1252	2′ to 3′					135			135	149
1253	3′ to 4′					185			185	204
1254	4′ to 5′					275			275	305
1255	5′ to 6′					335			335	365
1300	Abies koreana, (Korean Fir), Z5, B&B									
1301	10′ to 12′				Ea.	230			230	253
1302	12′ to 14′				″	291			291	320
1350	Abies pinsapo glauca, (Blue Spanish Fir), Z6, cont									
1351	2′ to 3′ Spread				Ea.	73			73	80.50
1352	3′ to 4′					98			98	108
1353	4′ to 5′					188			188	207
1400	Abies procera (nobilis) glauca, (Noble Fir), Z5, B&B									
1401	15″ to 18″ Spread				Ea.	42			42	46
1402	18″ to 24″ Spread					60			60	66
1403	2′ to 3′					103			103	113
1404	3′ to 4′					189			189	208
1450	Abies veitchi, (Veitch Fir), Z3, cont/BB									
1451	4′ to 5′				Ea.	65			65	71.50
1452	5′ to 6′				″	80			80	88
1500	Cedrus atlantica glauca, (Blue Atlas Cedar), Z6, cont/BB									
1501	2′ to 3′				Ea.	25			25	27.50

029 500 | Trees/Plants/Grnd Cover		CREW	DAILY OUTPUT	LABOR-HOURS	UNIT	1999 BARE COSTS MAT.	LABOR	EQUIP.	TOTAL	TOTAL INCL O&P
608	1502	3′ to 4′				Ea.	42			42	46
	1503	4′ to 5′					58			58	64
	1504	5′ to 6′					80			80	88
	1505	7′ to 8′				↓	112			112	123
	1550	Cedrus deodara, (Indian Cedar), Z7, cont/BB									
	1551	5 Gal.				Ea.	16			16	17.60
	1552	6′ to 8′					105			105	116
	1553	8′ to 10′					118			118	130
	1554	10′ to 12′					160			160	176
	1555	12′ to 14′					210			210	231
	1556	14′ to 16′				↓	275			275	305
	1600	Cedrus deodara kashmir, (Hardy Deodar Cedar), Z6, cont/BB									
	1601	3′ to 4′				Ea.	44			44	48.50
	1602	4′ to 5′					52			52	57
	1603	5′ to 6′				↓	66			66	72.50
	1650	Cedrus deodara shalimar, (Hardy Deodar Cedar), Z6, cont/BB									
	1651	2′ to 3′				Ea.	38			38	42
	1652	3′ to 4′				"	45			45	49.50
	1700	Cedrus libani sargenti, (Weeping Cedar of Lebanon), Z6, cont									
	1701	2′ to 3′				Ea.	36			36	39.50
	1702	3′ to 4′				"	52			52	57
	1750	Chamaecyparis lawsoniana, (Dwarf Lawson's Cypress), Z6, cont									
	1751	18″ to 24″				Ea.	26			26	28.50
	1752	2′ to 2-1/2′				"	36			36	39.50
	1800	Chamaecyparis nootkatensis glauca, (Alaska Cedar), Z6, cont/BB									
	1801	2′ to 3′				Ea.	41			41	45
	1802	3′ to 4′					53			53	58.50
	1803	4′ to 5′					90			90	99
	1804	5′ to 6′				↓	117			117	129
	1850	Chamaecyparis nootkatensis lutea, (Gold Cedar), Z6, cont/BB									
	1851	3′ to 4′				Ea.	32			32	35
	1852	4′ to 5′					45			45	49.50
	1853	5′ to 6′				↓	65			65	71.50
	1900	Chamaecyparis nootkatensis pendula, (Alaska Cedar), Z6, cont/BB									
	1901	2′ to 3′				Ea.	43			43	47.50
	1902	4′ to 5′					70			70	77
	1903	5′ to 6′					92			92	101
	1904	6′ to 7′				↓	136			136	150
	1950	Chamaecyparis obtusa coralliformis, (Torulosa), Z5, cont/BB									
	1951	15″ to 18″ Spread				Ea.	26			26	28.50
	1952	18″ to 24″					32			32	35
	1953	2′ to 3′					54			54	59.50
	1954	3′ to 4′					97			97	107
	1955	4′ to 5′					132			132	145
	1956	7′ to 8′				↓	291			291	320
	2000	Chamaecyparis obtusa crippsi, (Gold Hinoki Cypress), Z6, cont/BB									
	2001	3 Gal.				Ea.	28			28	31
	2002	5 Gal.					50			50	55
	2003	2′ to 3′					63			63	69.50
	2004	3′ to 4′					81			81	89
	2005	4′ to 5′				↓	150			150	165
	2050	Chamaecyparis obtusa filiciodes, (Fernspray Cypress), Z5, cont/BB									
	2051	18″ to 24″				Ea.	24			24	26.50
	2052	2′ to 3′					36			36	39.50
	2053	3′ to 4′				↓	60			60	66
	2100	Chamaecyparis obtusa gracilis, (Hinoki Cypress), Z5, cont/BB									

029 500 | Trees/Plants/Grnd Cover		CREW	DAILY OUTPUT	LABOR-HOURS	UNIT	1999 BARE COSTS MAT.	LABOR	EQUIP.	TOTAL	TOTAL INCL O&P
608 2101	2' to 3'				Ea.	60			60	66
2102	3' to 4'				Ea.	83			83	91.50
2103	10' to 12'				Ea.	550			550	605
2150	Chamaecyparis obtusa gracilis compacta, (Cypress), Z5, cont/BB									
2151	15" to 18"				Ea.	36			36	39.50
2152	18" to 24"				Ea.	50			50	55
2153	2' to 3'				Ea.	76			76	83.50
2154	3' to 4'				Ea.	132			132	145
2200	Chamaecyparis obtusa gracilis nana, (Hinoki Cypress), Z5, cont/BB									
2201	10" to 12"				Ea.	22			22	24
2202	12" to 15"				Ea.	32			32	35
2203	15" to 18"				Ea.	41			41	45
2204	18" to 24"				Ea.	60			60	66
2250	Chamaecyparis obtusa lycopodioides, (Cypress), Z5, cont/BB									
2251	18" to 24"				Ea.	36			36	39.50
2252	2' to 3'				Ea.	45			45	49.50
2253	3' to 4'				Ea.	62			62	68
2254	4' to 5'				Ea.	88			88	97
2255	5' to 6'				Ea.	115			115	127
2256	6' to 7'				Ea.	145			145	160
2257	7' to 8'				Ea.	200			200	220
2300	Chamaecyparis obtusa magnifica, (Hinoki Cypress), Z5, cont/BB									
2301	2' to 3'				Ea.	19			19	21
2302	3' to 4'				Ea.	31			31	34
2303	4' to 5'				Ea.	40			40	44
2304	5' to 6'				Ea.	59			59	65
2305	6' to 7'				Ea.	94			94	103
2350	Chamaecyparis pisifera, (Sawara Cypress), Z5, cont/BB									
2351	4' to 5'				Ea.	41			41	45
2352	5' to 6'				Ea.	62			62	68
2353	6' to 7'				Ea.	77			77	84.50
2354	7' to 8'				Ea.	120			120	132
2400	Chamaecyparis pisifera aurea, (Sawara Cypress), Z4, cont/BB									
2401	4' to 5'				Ea.	40			40	44
2402	5' to 6'				Ea.	60			60	66
2403	6' to 7'				Ea.	75			75	82.50
2404	7' to 8'				Ea.	119			119	131
2450	Chamaecyparis pisifera boulevard, (Moss Cypress), Z5, cont/BB									
2451	18" to 24"				Ea.	22			22	24
2452	2' to 3'				Ea.	39			39	43
2453	3' to 4'				Ea.	56			56	61.50
2454	4' to 5'				Ea.	88			88	97
2455	5' to 6'				Ea.	160			160	176
2456	6' to 7'				Ea.	235			235	259
2500	Chamaecyparis pisifera compacta, (Sawara Cypress), Z4, cont									
2501	15" to 18" Spread				Ea.	15			15	16.50
2502	18" to 24" Spread				Ea.	20			20	22
2503	2' to 2-1/2' Spread				Ea.	30			30	33
2504	2-1/2' to 3' Spread				Ea.	45			45	49.50
2550	Chamaecyparis pisifera filifera, (Thread Cypress), Z4, cont/BB									
2551	18" to 24"				Ea.	25			25	27.50
2552	2' to 3'				Ea.	49			49	54
2553	3' to 4'				Ea.	70			70	77
2554	4' to 5'				Ea.	90			90	99
2555	5' to 6'				Ea.	120			120	132
2556	6' to 7'				Ea.	200			200	220

029 500 | Trees/Plants/Grnd Cover		CREW	DAILY OUTPUT	LABOR-HOURS	UNIT	1999 BARE COSTS MAT.	LABOR	EQUIP.	TOTAL	TOTAL INCL O&P
2600	Chamaecyparis pisifera filifera aurea, (Thread Cypress), Z4, cont									
2601	1 Gal.				Ea.	7			7	7.70
2602	2 Gal.				↓	12			12	13.20
2603	5 Gal.				↓	15			15	16.50
2650	Chamaecyparis pisifera filifera nana, (Thread Cypress), Z5, cont									
2651	15″ to 18″				Ea.	24			24	26.50
2652	18″ to 24″				↓	33			33	36.50
2653	2′ to 2-1/2′				↓	43			43	47.50
2654	2-1/2′ to 3′				↓	66			66	72.50
2655	3′ to 4′				↓	110			110	121
2671	12″ to 15″				↓	24			24	26.50
2672	18″ to 24″				↓	41			41	45
2673	Bare root, 2′ to 3′				↓	85			85	93.50
2700	Chamaecyparis pisifera plumosa aurea, (Plumed Cypress), Z6, B&B									
2701	18″ to 24″				Ea.	18			18	19.80
2702	3′ to 4′				↓	32			32	35
2703	4′ to 5′				↓	40			40	44
2750	Chamaecyparis pisifera plumosa compress., Z6, cont/BB									
2751	12″ to 15″				Ea.	14			14	15.40
2752	15″ to 18″				↓	18			18	19.80
2753	18″ to 24″				↓	21			21	23
2754	2′ to 2-1/2′				↓	28			28	31
2755	2-1/2′ to 3′				↓	40			40	44
2800	Chamaecyparis pisifera squarrosa, (Cypress), Z5, cont/BB									
2801	2′ to 3′				Ea.	22			22	24
2802	3′ to 4′				↓	25			25	27.50
2803	4′ to 5′				↓	40			40	44
2804	5′ to 6′				↓	61			61	67
2850	Chamaecyparis pisifera squarrosa pygmaea, (Cypress), Z6, cont									
2851	12″ to 15″				Ea.	15			15	16.50
2852	15″ to 18″				↓	18			18	19.80
2853	18″ to 24″				↓	21			21	23
2900	Chamaecyparis thyoides Andelyensis, (White Cedar), Z5, cont/BB									
2901	2′ to 3′				Ea.	21			21	23
2902	3′ to 4′				↓	30			30	33
2903	4′ to 5′				↓	51			51	56
2904	5′ to 6′				↓	65			65	71.50
2905	6′ to 7′				↓	72			72	79
2950	Cryptomeria japonica cristata, (Japanese Cedar), Z6, B&B									
2951	3′ to 4′				Ea.	30			30	33
2952	4′ to 5′				″	44			44	48.50
3000	Cryptomeria japonica lobbi, (Lobb Cryptomeria), Z6, B&B									
3001	6′ to 8′				Ea.	126			126	139
3002	8′ to 10′				↓	160			160	176
3003	10′ to 12′				↓	191			191	210
3050	Cupressocyparis leylandi blue, (Leylandi Cypress), Z6, B&B									
3051	5′ to 6′				Ea.	91			91	100
3052	6′ to 7′				″	105			105	116
3100	Cupressocyparis leylandii, (Leylandi Cypress), Z6, cont/BB									
3101	5 Gal.				Ea.	20			20	22
3102	15 Gal.				↓	79			79	87
3103	4′ to 5′				↓	43			43	47.50
3104	5′ to 6′				↓	58			58	64
3105	6′ to 7′				↓	77			77	84.50
3106	Box, 24″				↓	158			158	174
3150	Cupressus arizonica, (Arizona Cypress), Z6, cont									

029 500 | Trees/Plants/Grnd Cover

			CREW	DAILY OUTPUT	LABOR-HOURS	UNIT	1999 BARE COSTS MAT.	LABOR	EQUIP.	TOTAL	TOTAL INCL O&P
608	3151	5 Gal.				Ea.	18			18	19.80
	3152	15 Gal.				"	48			48	53
	3200	Cupressus macrocarpa, (Monterey Cypress), Z8,									
	3201	1 Gal.				Ea.	4			4	4.40
	3202	5 Gal.					15			15	16.50
	3203	15 Gal.				↓	48			48	53
	3250	Cupressus sempervirens, (Italian Cypress), Z7, cont									
	3251	5 Gal.				Ea.	18			18	19.80
	3252	15 Gal.				"	50			50	55
	3300	Juniperus chinensis, (Green Clmnr Chinese Juniper), Z4, cont/BB									
	3301	5 Gal.				Ea.	18			18	19.80
	3302	2′ to 3′					25			25	27.50
	3303	4′ to 5′					50			50	55
	3304	5′ to 6′					70			70	77
	3305	12′ to 14′				↓	300			300	330
	3350	Juniperus chinensis "Armstrongii", (Armstrong Juniper), Z4, cont									
	3351	1 Gal.				Ea.	6			6	6.60
	3352	2 Gal.					11			11	12.10
	3353	5 Gal.				↓	16			16	17.60
	3400	Juniperus chinensis "Blue Point", (Blue Point Juniper), Z4, cont/BB									
	3401	1 Gal.				Ea.	8			8	8.80
	3402	5 Gal.					18.50			18.50	20.50
	3403	4′ to 5′					49			49	54
	3404	5′ to 6′					67			67	73.50
	3405	6′ to 8′				↓	84			84	92.50
	3450	Juniperus chinensis "Keteleeri", (Keteleer Juniper), Z5, cont/BB									
	3451	5 Gal.				Ea.	12			12	13.20
	3452	15 Gal.				"	70			70	77
	3500	Juniperus chinensis "Sargenti Glauca", (Sargent Juniper), Z4, cont									
	3501	1 Gal.				Ea.	6			6	6.60
	3502	2 Gal.					12			12	13.20
	3503	3 Gal.					14			14	15.40
	3504	5 Gal.				↓	18			18	19.80
	3550	Juniperus chinensis "Sea Green", (Sea Green Juniper), Z4, cont									
	3555	1 Gal.				Ea.	7			7	7.70
	3556	2 Gal.					12			12	13.20
	3557	3 Gal.					14			14	15.40
	3558	5 Gal.				↓	18			18	19.80
	3600	Juniperus chinensis blaauwi, (Blaauwi Juniper), Z4, B&B									
	3601	18″ to 24″				Ea.	15			15	16.50
	3602	2′ to 3′					28			28	31
	3603	3′ to 4′					41			41	45
	3604	4′ to 5′					55			55	60.50
	3605	5′ to 6′				↓	92			92	101
	3650	Juniperus chinensis columnaris, (Chinese Juniper), Z4, B&B									
	3651	4′ to 5′				Ea.	48			48	53
	3652	5′ to 6′					65			65	71.50
	3653	8′ to 10′				↓	140			140	154
	3700	Juniperus chinensis densaerecta, (Spartan Juniper), Z5, cont									
	3701	1 Gal.				Ea.	7			7	7.70
	3702	5 Gal.					17			17	18.70
	3703	15 Gal.					75			75	82.50
	3704	3′ to 4′					36			36	39.50
	3705	4′ to 5′					50			50	55
	3706	5′ to 6′					79			79	87
	3707	6′ to 8′				↓	88			88	97

029 500 | Trees/Plants/Grnd Cover

029 500 Trees/Plants/Grnd Cover		CREW	DAILY OUTPUT	LABOR-HOURS	UNIT	1999 BARE COSTS MAT.	LABOR	EQUIP.	TOTAL	TOTAL INCL O&P
3750	Juniperus chinensis Hetzii, (Hetz Juniper), Z4, cont/BB									
3751	2 Gal.				Ea.	10			10	11
3752	5 Gal.					15			15	16.50
3753	2' to 2-1/2'					25			25	27.50
3754	2-1/2' to 3'					28			28	31
3755	3' to 4'				↓	36			36	39.50
3800	Juniperus chinensis iowa, (Chinese Juniper), Z4, cont/BB									
3801	2' to 3'				Ea.	28			28	31
3802	3' to 4'					36			36	39.50
3803	4' to 5'					48			48	53
3804	5' to 6'				↓	65			65	71.50
3850	Juniperus chinensis mountbatten, (Chinese Juniper), Z4, cont/BB									
3851	2' to 3'				Ea.	25			25	27.50
3852	3' to 4'					36			36	39.50
3853	4' to 5'					45			45	49.50
3854	5' to 6'					62			62	68
3855	6' to 7'					70			70	77
3856	7' to 8'					90			90	99
3857	8' to 10'				↓	135			135	149
3900	Juniperus chinensis obelisk, (Chinese Juniper), Z5, cont/BB									
3901	2' to 3'				Ea.	25			25	27.50
3902	3' to 4'					36			36	39.50
3903	4' to 5'					45			45	49.50
3904	5' to 6'					66			66	72.50
3905	6' to 7'				↓	75			75	82.50
3950	Juniperus chinensis pfitzeriana, (Pfitzer Juniper), Z4, cont/BB									
3951	1 Gal.				Ea.	5			5	5.50
3952	2 Gal.					12			12	13.20
3953	5 Gal.					15			15	16.50
3954	12" to 15"					8			8	8.80
3955	15" to 18"					10			10	11
3956	18" to 24"					12			12	13.20
3957	2' to 2-1/2'					17			17	18.70
3958	2-1/2' to 3'					20			20	22
3959	3' to 3-1/2'				↓	29			29	32
3970	Juniperus chinensis pfitzeriana aurea, (Pfitzer Juniper), Z4, cont									
3971	1 Gal.				Ea.	4.75			4.75	5.20
3972	2 Gal.					9.35			9.35	10.30
3973	5 Gal.				↓	13.60			13.60	14.95
4000	Juniperus chinensis pfitzeriana compact, (Pfitzer Juniper), Z4, cont									
4001	5 Gal.				Ea.	16			16	17.60
4002	15" to 18"					14.70			14.70	16.15
4003	18" to 24"					17.75			17.75	19.50
4004	2' to 2-1/2'					26.50			26.50	29
4005	2-1/2' to 3'				↓	30.50			30.50	33.50
4050	Juniperus chinensis robusta green, (Green Juniper), Z5, cont/BB									
4051	5 Gal.				Ea.	18			18	19.80
4052	2' to 3'					26			26	28.50
4053	3' to 4'					3			3	3.30
4054	4' to 5'					52			52	57
4055	5' to 6'					62			62	68
4056	6' to 7'					75			75	82.50
4057	7' to 8'					90			90	99
4058	8' to 10'				↓	135			135	149
4100	Juniperus chinensis torulosa, (Hollywood Juniper), Z6, cont/BB									
4101	1 Gal.				Ea.	7			7	7.70

608

029 500 | Trees/Plants/Grnd Cover		CREW	DAILY OUTPUT	LABOR-HOURS	UNIT	1999 BARE COSTS MAT.	LABOR	EQUIP.	TOTAL	TOTAL INCL O&P
608 4102	5 Gal.				Ea.	17			17	18.70
4103	15 Gal.					75			75	82.50
4104	4′ to 5′					50			50	55
4105	5′ to 6′					68			68	75
4150	Juniperus horizontalis "Bar Harbor", (Bar Harbor Juniper), Z4, cont									
4151	1 Gal.				Ea.	6			6	6.60
4152	2 Gal.					12			12	13.20
4153	3 Gal.					14			14	15.40
4154	5 Gal.					18			18	19.80
4200	Juniperus horizontalis "Wiltoni", (Blue Rug Juniper), Z3, cont									
4201	1 Gal.				Ea.	6			6	6.60
4202	2 Gal.					12			12	13.20
4203	3 Gal.					14			14	15.40
4204	5 Gal.					18			18	19.80
4250	Juniperus horizontalis Plumosa Compacta, (Juniper), Z3, cont									
4251	1 Gal.				Ea.	6			6	6.60
4252	2 Gal.					12			12	13.20
4253	3 Gal.					14			14	15.40
4254	5 Gal.					18			18	19.80
4300	Juniperus procumbens "Nana", (Procumbens Juniper), Z4, cont									
4301	1 Gal.				Ea.	6			6	6.60
4302	2 Gal.					12			12	13.20
4303	3 Gal.					14			14	15.40
4304	5 Gal.					18			18	19.80
4350	Juniperus sabina "Broadmoor", (Broadmoor Juniper), Z3, cont									
4351	1 Gal.				Ea.	6			6	6.60
4352	2 Gal.					12			12	13.20
4353	3 Gal.					14			14	15.40
4354	5 Gal.					18			18	19.80
4400	Juniperus sabina Tamariscifolia New Blue, (Tam Juniper), Z4, cont									
4401	1 Gal.				Ea.	6			6	6.60
4402	2 Gal.					12			12	13.20
4403	5 Gal.					18			18	19.80
4450	Juniperus scopulorum "Gray Gleam", (Gray Gleam Juniper), Z3, cont									
4451	1 Gal.				Ea.	8			8	8.80
4452	5 Gal.				"	20.50			20.50	22.50
4500	Juniperus scopulorum blue heaven, (Blue Heaven Juniper), Z4, B&B									
4501	4′ to 5′				Ea.	44			44	48.50
4502	5′ to 6′					60			60	66
4503	7′ to 8′					82			82	90
4504	8′ to 10′					124			124	136
4550	Juniperus scopulorum cologreen, (Cologreen Juniper), Z5, cont									
4551	5 Gal.				Ea.	20.50			20.50	22.50
4600	Juniperus scopulorum hillburn's silver globe, (Juniper), Z5, BB									
4601	3′ to 4′				Ea.	69			69	76
4602	4′ to 5′				"	87			87	95.50
4650	Juniperus scopulorum pathfinder, (Pathfinder Juniper), Z5, B&B									
4651	4′ to 5′				Ea.	49			49	54
4652	5′ to 6′					64			64	70.50
4653	7′ to 8′					92			92	101
4700	Juniperus scopulorum, (Rocky Mountain Juniper), Z5, cont/BB									
4701	20 Gal.				Ea.	102			102	112
4702	3′ to 4′					39			39	43
4703	4′ to 5′					53			53	58.50
4704	5′ to 6′					68			68	75
4750	Juniperus scopulorum welchii, (Welchi Juniper), Z3, cont									

029 500 | Trees/Plants/Grnd Cover

029 500	Trees/Plants/Grnd Cover	CREW	DAILY OUTPUT	LABOR-HOURS	UNIT	1999 BARE COSTS MAT.	LABOR	EQUIP.	TOTAL	TOTAL INCL O&P
608 4751	5 Gal.				Ea.	20			20	22
4752	15 Gal.				″	81			81	89
4800	Juniperus scopulorum wichita blue, (Wichita Blue Juniper), Z5, cont									
4801	5 Gal.				Ea.	19			19	21
4802	3′ to 4′				″	37			37	40.50
4850	Juniperus squamata "Parsonii", (Parson's Juniper), Z4, cont									
4851	1 Gal.				Ea.	5			5	5.50
4852	2 Gal.					11			11	12.10
4853	3 Gal.					13			13	14.30
4854	5 Gal.				↓	16			16	17.60
4900	Juniperus squamata Blue Star, (Blue Juniper), Z4, cont/BB									
4901	1 Gal.				Ea.	7			7	7.70
4902	3 Gal.				″	17			17	18.70
4950	Juniperus squamata meyeri, (Meyer's Juniper), Z5, cont/BB									
4951	3 Gal.				Ea.	13			13	14.30
4952	5 Gal.					16			16	17.60
4953	2′ to 3′				↓	26			26	28.50
5000	Juniperus virginiana, (Eastern Red Cedar), Z2, B&B									
5001	2′ to 3′				Ea.	26			26	28.50
5002	3′ to 4′					38			38	42
5003	4′ to 5′					50			50	55
5004	5′ to 6′				↓	62			62	68
5050	Juniperus virginiana burkii, (Burki Juniper), Z2, cont									
5051	5 Gal.				Ea.	19			19	21
5052	15 Gal.				″	78			78	86
5100	Juniperus virginiana canaerti, (Canaert Red Cedar), Z4, B&B									
5101	2′ to 3′				Ea.	26			26	28.50
5102	3′ to 4′					36			36	39.50
5103	4′ to 5′					50			50	55
5104	5′ to 6′				↓	63			63	69.50
5150	Juniperus virginiana glauca, (Silver Cedar), Z4, cont/BB									
5151	2′ to 3′				Ea.	27			27	29.50
5152	3′ to 4′					40			40	44
5153	4′ to 5′				↓	52			52	57
5200	Juniperus virginiana globusa, (Globe Red Cedar), Z4, cont/BB									
5201	2′ to 3′				Ea.	40			40	44
5250	Juniperus virginiana gray owl, (Gray Owl Juniper), Z4, cont									
5251	2 Gal.				Ea.	12			12	13.20
5252	5 Gal.				″	18			18	19.80
5300	Juniperus virginiana skyrocket, (Skyrocket Juniper), Z5, cont/BB									
5301	5 Gal.				Ea.	19			19	21
5302	15 Gal.					78			78	86
5303	2′ to 3′					28			28	31
5304	3′ to 4′					39			39	43
5305	4′ to 5′					54			54	59.50
5306	5′ to 6′					70			70	77
5307	6′ to 8′				↓	87			87	95.50
5350	Picea abies, (Norway Spruce), Z2, cont/BB									
5351	2′ to 3′				Ea.	26			26	28.50
5352	3′ to 4′					34			34	37.50
5353	4′ to 5′					40			40	44
5354	5′ to 6′					51			51	56
5355	6′ to 7′					80			80	88
5356	7′ to 8′				↓	112			112	123
5400	Picea abies clanbrasiliana, (Barry Spruce), Z4, cont/BB									
5401	2′ to 3′				Ea.	135			135	149

029 500 | Trees/Plants/Grnd Cover		CREW	DAILY OUTPUT	LABOR-HOURS	UNIT	1999 BARE COSTS MAT.	LABOR	EQUIP.	TOTAL	TOTAL INCL O&P
608 5402	3′ to 4′				Ea.	215			215	237
5450	Picea abies gregoryana, (Gregory's Dwarf Spruce), Z4, cont/BB									
5451	18″ to 24″				Ea.	70			70	77
5452	2′ to 2-1/2′				″	100			100	110
5500	Picea abies nidiformis, (Nest Spruce), Z4, cont/BB									
5501	12″ to 15″				Ea.	25			25	27.50
5502	15″ to 18″					25			25	27.50
5503	18″ to 24″					35			35	38.50
5504	2′ to 3′					63			63	69.50
5505	3′ to 4′					89			89	98
5506	4′ to 5′					280			280	310
5550	Picea abies pendula, (Weeping Norway Spruce), Z4, cont/BB									
5551	2′ to 3′				Ea.	64			64	70.50
5552	3′ to 4′					85			85	93.50
5553	4′ to 5′					100			100	110
5554	6′ to 7′					152			152	167
5555	7′ to 8′					182			182	200
5600	Picea abies remonti, (Remont Spruce), Z4, cont/BB									
5601	2′ to 3′				Ea.	101			101	111
5650	Picea glauca, (White or Canadian Spruce), Z3, cont/BB									
5651	3′ to 4′				Ea.	33			33	36.50
5652	4′ to 5′					42			42	46
5653	5′ to 6′					55			55	60.50
5654	6′ to 7′					66			66	72.50
5655	7′ to 8′					82			82	90
5700	Picea glauca conica, (Dwarf Alberta Spruce), Z3, cont/BB									
5701	12″ to 15″				Ea.	22			22	24
5702	15″ to 18″					29			29	32
5703	18″ to 24″					36			36	39.50
5704	2′ to 2-1/2′					45			45	49.50
5705	2-1/2′ to 3′					69			69	76
5750	Picea omorika, (Serbian Spruce), Z4, cont/BB									
5751	2′ to 3′				Ea.	23			23	25.50
5752	3′ to 4′					29			29	32
5753	4′ to 5′					40			40	44
5754	5′ to 6′					54			54	59.50
5755	6′ to 7′					74			74	81.50
5800	Picea omorika nana, (Dwarf Serbian Spruce), Z4, cont									
5801	9″ to 12″				Ea.	31			31	34
5802	12″ to 15″					39			39	43
5803	18″ to 24″					69			69	76
5850	Picea omorika pendula, (Weeping Serbian Spruce), Z4, cont/BB									
5851	3′ to 4′				Ea.	118			118	130
5900	Picea pungens, (Colorado Spruce), Z3, cont/BB									
5901	18″ to 24″				Ea.	19			19	21
5902	2′ to 3′					25			25	27.50
5903	3′ to 4′					35			35	38.50
5904	4′ to 5′					46.50			46.50	51
5905	5′ to 6′					88			88	97
5906	6′ to 7′					110			110	121
5907	7′ to 8′					130			130	143
5950	Picea pungens glauca, (Blue Colorado Spruce), Z3, cont/BB									
5951	18″ to 24″				Ea.	27			27	29.50
5952	2′ to 3′					30			30	33
5953	3′ to 4′					43			43	47.50
5954	4′ to 5′					66			66	72.50

029 500 Trees/Plants/Grnd Cover		CREW	DAILY OUTPUT	LABOR-HOURS	UNIT	1999 BARE COSTS MAT.	LABOR	EQUIP.	TOTAL	TOTAL INCL O&P
5955	5' to 6'				Ea.	93			93	102
5956	6' to 7'				↓	123			123	135
6000	Picea pungens hoopsi, (Hoops Blue Spruce), Z3, cont/BB									
6001	2' to 3'				Ea.	53			53	58.50
6002	3' to 4'					74			74	81.50
6003	4' to 5'					97			97	107
6004	5' to 6'					151			151	166
6005	14' to 16'					960			960	1,050
6006	16' to 18'					280			280	310
6007	18' to 20'				↓	700			700	770
6050	Picea pungens hunnewelliana, (Dwarf Blue Spruce), Z4, cont/BB									
6051	12" to 15"				Ea.	30			30	33
6052	15" to 18"					52			52	57
6053	18" to 24"					75			75	82.50
6054	2' to 2-1/2'				↓	125			125	138
6100	Picea pungens kosteriana, (Koster's Blue Spruce), Z3, cont/BB									
6101	2' to 3'				Ea.	55			55	60.50
6102	3' to 4'					78			78	86
6103	4' to 5'					103			103	113
6104	5' to 6'					148			148	163
6105	6' to 7'					198			198	218
6106	7' to 8'				↓	250			250	275
6150	Picea pungens montgomery, (Dwarf Blue Spruce), Z4, cont/BB									
6151	12" to 15"				Ea.	38			38	42
6152	15" to 18"					44			44	48.50
6153	18" to 24"					73			73	80.50
6154	4' to 5'					400			400	440
6155	5' to 6'				↓	555			555	615
6200	Pinus albicaulis, (White Bark Pine), Z4, cont/BB									
6201	2' to 3'				Ea.	38			38	42
6202	3' to 4'					50			50	55
6203	5' to 6'					108			108	119
6204	6' to 8'					160			160	176
6205	8' to 10'					211			211	232
6206	10' to 12'					280			280	310
6207	12' to 14'					400			400	440
6208	14' to 16'				↓	575			575	635
6250	Pinus aristata, (Bristlecone Pine), Z5, cont/BB									
6251	15" to 18"				Ea.	20			20	22
6252	18" to 24"					31			31	34
6253	2' to 3'					38			38	42
6254	3' to 4'					65			65	71.50
6255	4' to 5'				↓	125			125	138
6300	Pinus banksiana, (Jack Pine), Z3, cont/BB									
6301	5' to 6'				Ea.	51.50			51.50	56.50
6302	6' to 8'					70			70	77
6303	8' to 10'				↓	117			117	129
6350	Pinus bungeana, (Lacebark Pine), Z5, cont/BB									
6351	2' to 3'				Ea.	69			69	76
6352	3' to 4'					102			102	112
6353	4' to 5'					130			130	143
6354	5' to 6'					175			175	193
6355	6' to 7'					240			240	264
6356	7' to 8'				↓	270			270	297
6400	Pinus canariensis, (Canary Island Pine), Z8, cont									
6401	1 Gal.				Ea.	4			4	4.40

029 500 | Trees/Plants/Grnd Cover		CREW	DAILY OUTPUT	LABOR-HOURS	UNIT	1999 BARE COSTS MAT.	LABOR	EQUIP.	TOTAL	TOTAL INCL O&P
608 6402	5 Gal.				Ea.	12			12	13.20
6403	15 Gal.				↓	56			56	61.50
6450	Pinus cembra, (Swiss Stone Pine), Z3, cont/BB									
6451	2' to 2-1/2'				Ea.	53			53	58.50
6452	3' to 4'					87			87	95.50
6453	4' to 5'					118			118	130
6454	5' to 6'				↓	170			170	187
6500	Pinus densiflora umbraculifera, (Tanyosho Pine), Z5, cont/BB									
6501	18" to 24"				Ea.	45			45	49.50
6502	2' to 3'					70			70	77
6503	3' to 4'					110			110	121
6504	4' to 5'					195			195	215
6505	5' to 6'					245			245	270
6506	7' to 8'				↓	350			350	385
6550	Pinus flexilis, (Limber Pine), Z4, B&B									
6551	3' to 4'				Ea.	41			41	45
6552	4' to 5'					60			60	66
6553	5' to 6'				↓	85			85	93.50
6600	Pinus koraiensis, (Korean Pine), Z4, cont/BB									
6601	3' to 4'				Ea.	44			44	48.50
6602	4' to 5'					60			60	66
6603	5' to 6'				↓	85			85	93.50
6650	Pinus lambertiana, (Sugar Pine), Z6, cont/BB									
6651	3' to 4'				Ea.	42			42	46
6652	4' to 5'					55			55	60.50
6653	7' to 8'					135			135	149
6654	8' to 10'				↓	200			200	220
6700	Pinus monticola, (Western White Pine), Z6, B&B									
6701	5' to 6'				Ea.	69			69	76
6702	6' to 7'					101			101	111
6703	7' to 8'					130			130	143
6704	8' to 10'				↓	181			181	199
6750	Pinus mugo var. mugo, (Mugho Pine), Z3, cont/BB									
6751	12" to 15" Spread				Ea.	18			18	19.80
6752	15" to 18" Spread					22			22	24
6753	18" to 24" Spread					31			31	34
6754	2' to 2-1/2' Spread					38			38	42
6755	2-1/2' to 3' Spread					49			49	54
6756	3' to 3-1/2' Spread					60			60	66
6757	3-1/2' to 4' Spread					75			75	82.50
6758	1 Gal.					7			7	7.70
6759	2 Gal.				↓	15			15	16.50
6800	Pinus nigra, (Austrian Pine), Z5, cont/BB									
6801	2' to 3'				Ea.	28			28	31
6802	3' to 4'					41			41	45
6803	4' to 5'					60			60	66
6804	5' to 6'					85			85	93.50
6805	6' to 7'					99			99	109
6806	7' to 8'					114			114	125
6807	8' to 9'					155			155	171
6808	9' to 10'					193			193	212
6809	10' to 12'				↓	237			237	261
6850	Pinus parviflora, (Japanese White Pine), Z5, cont									
6851	12' to 14'				Ea.	500			500	550
6900	Pinus parviflora glauca, (Silver Japanese White Pine), Z5, cont/BB									
6901	2' to 3'				Ea.	38			38	42

029 500 | Trees/Plants/Grnd Cover		CREW	DAILY OUTPUT	LABOR-HOURS	UNIT	1999 BARE COSTS MAT.	LABOR	EQUIP.	TOTAL	TOTAL INCL O&P
608 6902	3′ to 4′				Ea.	45			45	49.50
6903	4′ to 5′					64			64	70.50
6904	5′ to 6′					96			96	106
6905	6′ to 7′				↓	128			128	141
6950	Pinus ponderosa, (Western Yellow Pine), Z6, B&B									
6951	16′ to 20′				Ea.	850			850	935
7000	Pinus radiata, (Monterey Pine), Z8, cont									
7001	1 Gal.				Ea.	4			4	4.40
7002	5 Gal.					12			12	13.20
7003	15 Gal.					55			55	60.50
7004	24″ Box				↓	170			170	187
7050	Pinus resinosa, (Red Pine), Z3, cont/BB									
7051	2′ to 3′				Ea.	25			25	27.50
7052	3′ to 4′					28			28	31
7053	4′ to 5′					44			44	48.50
7054	5′ to 6′					60			60	66
7055	6′ to 8′					78			78	86
7056	8′ to 10′					135			135	149
7057	10′ to 12′					160			160	176
7058	12′ to 14′				↓	240			240	264
7100	Pinus rigida, (Pitch Pine), Z5, cont/BB									
7101	2′ to 3′				Ea.	20			20	22
7150	Pinus strobus, (White Pine), Z3, cont/BB									
7151	2′ to 3′				Ea.	18			18	19.80
7152	3′ to 4′					28			28	31
7153	4′ to 5′					42			42	46
7154	5′ to 6′					52			52	57
7155	6′ to 8′					78			78	86
7156	8′ to 10′					119			119	131
7157	10′ to 12′					152			152	167
7158	12′ to 14′				↓	199			199	219
7200	Pinus strobus fastigiata, (Upright White Pine), Z3, B&B									
7201	5′ to 6′				Ea.	69			69	76
7250	Pinus strobus nana, (Dwarf White Pine), Z3, cont/BB									
7251	12″ to 15″ Spread				Ea.	26			26	28.50
7252	15″ to 18″ Spread					30			30	33
7253	18″ to 24″ Spread					45			45	49.50
7254	2′ to 2-1/2′ Spread					61			61	67
7255	2-1/2′ to 3′ Spread					81			81	89
7256	3′ to 4′ Spread					155			155	171
7257	4′ to 5′ Spread					310			310	345
7258	5′ to 6′ Spread				↓	405			405	445
7300	Pinus strobus pendula, (Weeping White Pine), Z3, cont/BB									
7301	3′ to 4′				Ea.	51			51	56
7302	4′ to 5′					76			76	83.50
7303	6′ to 8′					148			148	163
7304	8′ to 10′					187			187	206
7305	10′ to 12′				↓	237			237	261
7350	Pinus sylvestris, (Scotch Pine), Z3, cont/BB									
7351	Seedlings				Ea.	1			1	1.10
7352	3′ to 4′					35			35	38.50
7353	4′ to 5′					56			56	61.50
7354	5′ to 6′					75			75	82.50
7355	6′ to 7′					112			112	123
7356	7′ to 8′				↓	145			145	160
7400	Pinus sylvestris beauvronensis, (Mini Scotch Pine), Z4, cont/BB									

029 500 | Trees/Plants/Grnd Cover		CREW	DAILY OUTPUT	LABOR-HOURS	UNIT	1999 BARE COSTS MAT.	LABOR	EQUIP.	TOTAL	TOTAL INCL O&P
608 7401	12″ to 15″				Ea.	35			35	38.50
7402	15″ to 18″					44			44	48.50
7403	2′ to 2-1/2′				↓	78			78	86
7450	Pinus sylvestris fastigiata, (Pyramidal Scotch Pine), Z3, cont/BB									
7451	2′ to 3′				Ea.	30			30	33
7452	3′ to 4′				″	41			41	45
7500	Pinus thumbergi, (Japanese Black Pine), Z5, cont/BB									
7501	1 Gal.				Ea.	5			5	5.50
7502	5 Gal.					17			17	18.70
7503	15 Gal.					55			55	60.50
7504	2′ to 3′					22			22	24
7505	3′ to 4′					31			31	34
7506	4′ to 5′					41			41	45
7507	5′ to 6′					64			64	70.50
7508	6′ to 7′					80			80	88
7509	7′ to 8′				↓	101			101	111
7550	Podocarpus macrophyllus "Maki", (Podocarpus), Z7, cont/BB									
7551	1 Gal.				Ea.	6			6	6.60
7552	3 Gal.					8			8	8.80
7553	7 Gal.					22			22	24
7554	15 Gal.					35			35	38.50
7555	4′ to 5′				↓	54			54	59.50
7600	Pseudotsuga menziesii, (Douglas Fir), Z5, cont/BB									
7601	2′ to 3′				Ea.	32			32	35
7602	3′ to 4′					39			39	43
7603	4′ to 5′					69			69	76
7604	5′ to 6′					75			75	82.50
7605	6′ to 7′				↓	87			87	95.50
7650	Sciadopitys verticillata, (Umbrella Pine), Z6, B&B									
7651	18″ to 24″				Ea.	45			45	49.50
7652	2′ to 2-1/2′					69			69	76
7653	6′ to 7′					525			525	580
7654	7′ to 8′				↓	625			625	690
7700	Sequoia sempervirens, (Coast Redwood), Z7, cont									
7701	5 Gal.				Ea.	15			15	16.50
7702	15 Gal.					55			55	60.50
7703	24″ Box				↓	175			175	193
7750	Taxus baccata adpressa fowle, (Midget Boxleaf Yew), Z5, cont/BB									
7751	15″ to 18″				Ea.	18.50			18.50	20.50
7752	18″ to 24″					20.50			20.50	22.50
7753	2′ to 2-1/2′					36			36	39.50
7754	2-1/2′ to 3′					52			52	57
7755	3′ to 4′					55			55	60.50
7756	4′ to 5′				↓	104			104	114
7800	Taxus baccata repandens, (Spreading English Yew), Z5, B&B									
7801	12″ to 15″				Ea.	21			21	23
7802	15″ to 18″					25			25	27.50
7803	2′ to 2-1/2′					51			51	56
7804	2-1/2′ to 3′					59			59	65
7805	3′ to 3-1/2′				↓	64			64	70.50
7850	Taxus cuspidata, (Spreading Japanese Yew), Z4, cont/BB									
7851	15″ to 18″				Ea.	19.50			19.50	21.50
7852	18″ to 24″					24			24	26.50
7853	2′ to 2-1/2′					29			29	32
7854	2-1/2′ to 3′					38			38	42
7855	3′ to 3-1/2′				↓	42			42	46

Important: See the Reference Section for critical supporting data - Reference Nos., Crews, & City Cost Indexe

029 500 | Trees/Plants/Grnd Cover		CREW	DAILY OUTPUT	LABOR-HOURS	UNIT	1999 BARE COSTS MAT.	LABOR	EQUIP.	TOTAL	TOTAL INCL O&P
7900	Taxus cuspidata capitata, (Upright Japanese Yew), Z4, cont/BB									
7901	15" to 18"				Ea.	19.50			19.50	21.50
7902	18" to 24"					23			23	25.50
7903	2' to 2-1/2'					28			28	31
7904	2-1/2' to 3'					37			37	40.50
7905	3' to 4'					53			53	58.50
7906	4' to 5'					85			85	93.50
7907	5' to 6'					99			99	109
7908	6' to 7'					143			143	157
7909	7' to 8'				↓	230			230	253
7950	Taxus cuspidata densiformis, (Dense Spreading Yew), Z5, cont/BB									
7951	15" to 18"				Ea.	16			16	17.60
7952	18" to 24"					20			20	22
7953	2' to 2-1/2'					28			28	31
7954	2-1/2' to 3'					36			36	39.50
7955	3' to 3-1/2'				↓	44			44	48.50
8000	Taxus cuspidata hicksi, (Hick's Yew), Z4, cont/BB									
8001	15" to 18"				Ea.	18			18	19.80
8002	18" to 24"					22			22	24
8003	2' to 2-1/2'					28			28	31
8004	2-1/2' to 3'				↓	37			37	40.50
8050	Taxus cuspidata nana, (Dwarf Yew), Z4, cont/BB									
8051	15" to 18"				Ea.	30			30	33
8052	18" to 24"					38			38	42
8053	3' to 4'					90			90	99
8054	4' to 5'					135			135	149
8055	5' to 6'				↓	190			190	209
8100	Thuja occidentalis "Smaragd", (Emerald Arborvitae), Z3, cont/BB									
8101	1 Gal.				Ea.	5			5	5.50
8102	2 Gal.					8			8	8.80
8103	3' to 4'					31			31	34
8104	4' to 5'					40			40	44
8105	5' to 6'				↓	49			49	54
8150	Thuja occidentalis "Woodwardii", (Globe Arborvitae), Z2, B&B									
8151	15" to 18"				Ea.	12			12	13.20
8152	18" to 24"					14			14	15.40
8153	2' to 2-1/2'					16			16	17.60
8154	2-1/2' to 3'				↓	23			23	25.50
8200	Thuja occidentalis douglasi pyramidalis, (Arborvitae), Z4, cont/BB									
8201	2' to 3'				Ea.	19			19	21
8202	3' to 4'					23			23	25.50
8203	4' to 5'					34			34	37.50
8204	5' to 6'					46			46	50.50
8205	6' to 7'					66			66	72.50
8206	7' to 8'				↓	77			77	84.50
8250	Thuja occidentalis nigra, (Dark American Arborvitae), Z4, cont/BB									
8251	2' to 3'				Ea.	20			20	22
8252	3' to 4'					28			28	31
8253	4' to 5'					32			32	35
8254	5' to 6'					44			44	48.50
8255	6' to 8'					74			74	81.50
8256	8' to 10'				↓	100			100	110
8300	Thuja occidentalis techney, (Mission Arborvitae), Z4, cont/BB									
8301	2' to 3'				Ea.	22			22	24
8302	3' to 4'					29			29	32
8303	4' to 5'				↓	36			36	39.50

608

		029 500 | Trees/Plants/Grnd Cover	CREW	DAILY OUTPUT	LABOR-HOURS	UNIT	1999 BARE COSTS MAT.	LABOR	EQUIP.	TOTAL	TOTAL INCL O&P	
608	8304	5′ to 6′				Ea.	48			48	53	608
	8305	6′ to 7′					75			75	82.50	
	8306	7′ to 8′				↓	85			85	93.50	
	8350	Thuja orientalis "aurea nana", (Gold Arborvitae), Z4, cont										
	8351	1 Gal.				Ea.	5			5	5.50	
	8352	5 Gal.				"	18			18	19.80	
	8400	Thuja orientalis "blue cone", (Blue Cone Arborvitae), Z6, cont										
	8401	1 Gal.				Ea.	5			5	5.50	
	8402	5 Gal.				"	16			16	17.60	
	8450	Tsuga canadensis, (Canadian Hemlock), Z4, cont/BB										
	8451	2′ to 2-1/2′				Ea.	21			21	23	
	8452	2-1/2′ to 3′					27			27	29.50	
	8453	3′ to 3-1/2′					37			37	40.50	
	8454	3-1/2′ to 4′					45			45	49.50	
	8455	4′ to 5′					59			59	65	
	8456	5′ to 5-1/2′					75			75	82.50	
	8457	5-1/2′ to 6′					92			92	101	
	8458	6′ to 7′					126			126	139	
	8459	7′ to 8′				↓	150			150	165	
	8500	Tsuga canadensis sargentii, (Sargents Weeping Hemlock), Z8, cont										
	8501	1 Gal.				Ea.	6			6	6.60	
	8502	3 Gal.					20			20	22	
	8503	30″ to 36″					46			46	50.50	
	8504	3′ to 4′					60			60	66	
	8505	4′ to 5′					75			75	82.50	
	8506	5′ to 6′				↓	90			90	99	
	8550	Tsuga caroliniana, (Caroline Hemlock), Z5, B&B										
	8551	2-1/2′ to 3′				Ea.	28			28	31	
	8552	3′ to 3-1/2′					36			36	39.50	
	8553	3-1/2′ to 4′					44			44	48.50	
	8554	4′ to 5′				↓	60			60	66	
	8600	Tsuga diversifolia, (Northern Japanese Hemlock), Z6, cont/BB										
	8601	3′ to 4′				Ea.	55			55	60.50	
	8602	4′ to 5′				"	85			85	93.50	
610	0010	**TREES & PALMS**, warm temperate and subtropical										610
	0100	Acacia baileyana, (Bailey Acacia), Z9, cont										
	0110	1 Gal.				Ea.	3			3	3.30	
	0120	5 Gal.					10			10	11	
	0130	15 Gal.				↓	52			52	57	
	0200	Acacia latifolia, (Broadleaf Acacia (multistem)), Z10, cont										
	0210	1 Gal.				Ea.	3			3	3.30	
	0220	5 Gal.					10			10	11	
	0230	15 Gal.				↓	44			44	48.50	
	0300	Acacia melanoxylon, (Blackwood Acacia), Z9, cont										
	0310	1 Gal.				Ea.	3			3	3.30	
	0320	5 Gal.					11			11	12.10	
	0330	15 Gal.				↓	55			55	60.50	
	0400	Agonis flexuosa, (Willow Myrtle), Z9, cont										
	0410	5 Gal.				Ea.	14.25			14.25	15.70	
	0420	15 Gal.				"	15			15	16.50	
	0500	Albizzia julibrissin, (Silk Tree), Z7, cont										
	0510	5 Gal.				Ea.	14			14	15.40	
	0520	15 Gal.					59			59	65	
	0530	20 Gal.					77			77	84.50	
	0540	24″ Box				↓	195			195	215	
	0600	Arbutus menziesii, (Pacific Madrone), Z7, cont										

029 500 | Trees/Plants/Grnd Cover

	029 500 Trees/Plants/Grnd Cover	CREW	DAILY OUTPUT	LABOR-HOURS	UNIT	1999 BARE COSTS MAT.	LABOR	EQUIP.	TOTAL	TOTAL INCL O&P
610 0610	1 Gal.				Ea.	6			6	6.60
0620	5 Gal.					17			17	18.70
0630	15 Gal.				↓	63			63	69.50
0700	Arbutus unedo, (Strawberry Tree), Z7, cont									
0710	1 Gal.				Ea.	4			4	4.40
0720	5 Gal.					10			10	11
0730	15 Gal.				↓	45			45	49.50
0800	Archontophoenix cunninghamiana, (King Palm), Z10, cont									
0810	5 Gal.				Ea.	18			18	19.80
0820	15 Gal.					60			60	66
0830	24″ Box				↓	190			190	209
0900	Arecastrum romanzoffianum, (Queen Palm), Z10, cont									
0910	5 Gal.				Ea.	18			18	19.80
0920	15 Gal.					61			61	67
0930	24″ Box				↓	188			188	207
1000	Bauhinia variegata, (Orchid Tree), Z9, cont									
1010	5 Gal.				Ea.	13			13	14.30
1020	15 Gal.					57			57	62.50
1030	24″ Box				↓	180			180	198
1100	Brachychiton populneus, (Kurrajong), Z9, cont									
1110	5 Gal.				Ea.	14			14	15.40
1120	15 Gal.					64			64	70.50
1130	24″ Box				↓	195			195	215
1200	Callistemon citroides, (Lemon Bottlebrush (Tree Form)), Z9, cont									
1210	15 Gal.				Ea.	55			55	60.50
1220	24″ Box					175			175	193
1230	36″ Box				↓	435			435	480
1300	Casuarina stricta, (Mountain She-Oak), Z9, cont									
1310	5 Gal.				Ea.	16			16	17.60
1320	15 Gal.				″	55			55	60.50
1400	Ceratoma siliqua, (Carob), Z10, cont									
1410	5 Gal.				Ea.	14			14	15.40
1420	15 Gal.					48			48	53
1430	24″ Box				↓	166			166	183
1500	Chamaerops humilis, (Mediterranean Fan Palm), Z9, cont									
1510	5 Gal.				Ea.	19			19	21
1520	15 Gal.					60			60	66
1530	24″ Box				↓	189			189	208
1600	Chrysobalanus icaco, (Coco-plum), Z10, cont									
1610	1 Gal.				Ea.	3			3	3.30
1620	3 Gal.					5			5	5.50
1630	10 Gal.					35			35	38.50
1640	20 Gal.				↓	55			55	60.50
1700	Cinnamonum camphora, (Camphor Tree), Z9, cont									
1710	5 Gal.				Ea.	14			14	15.40
1720	15 Gal.					55			55	60.50
1730	24″ Box				↓	180			180	198
1800	Cornus nattali, (California Laurel), Z7, B&B									
1810	5′ to 6′				Ea.	40			40	44
1820	6′ to 8′					50			50	55
1830	8′ to 10′				↓	74			74	81.50
1900	Cupaniopsis anagardioides, (Carrotwood), Z10, cont									
1910	5 Gal.				Ea.	12			12	13.20
1920	15 Gal.					55			55	60.50
1930	24″ Box				↓	180			180	198
2000	Eriobotrya japonica, (Loquat), Z8, cont									

029 500 | Trees/Plants/Grnd Cover

		Crew	Daily Output	Labor-Hours	Unit	1999 Bare Costs Mat.	Labor	Equip.	Total	Total Incl O&P
610	2010 3 Gal.				Ea.	8			8	8.80
	2020 10 Gal.					29			29	32
	2030 15 Gal.					55			55	60.50
	2040 15 Gal. Espalier					75			75	82.50
	2100 Erythrina caffra, (Kaffirboom Coral Tree), Z10, cont									
	2110 5 Gal.				Ea.	11			11	12.10
	2120 15 Gal.					57			57	62.50
	2130 24" Box					195			195	215
	2200 Eucalyptus camaldulensis, (Red Gum), Z9, cont									
	2210 5 Gal.				Ea.	11			11	12.10
	2220 15 Gal.					57			57	62.50
	2230 24" Box					180			180	198
	2300 Eucalyptus ficifolia, (Flaming Gum), Z10, cont									
	2310 5 Gal.				Ea.	13			13	14.30
	2320 15 Gal.					63			63	69.50
	2330 24" Box					185			185	204
	2400 Feijoa sellowiana, (Guava), Z8, cont									
	2410 1 Gal.				Ea.	4			4	4.40
	2420 5 Gal.					11			11	12.10
	2430 15 Gal.					51			51	56
	2500 Ficus benjamina, (Benjamin Fig), Z10, cont									
	2510 5 Gal.				Ea.	17			17	18.70
	2520 15 Gal.					56			56	61.50
	2530 24" Box					185			185	204
	2600 Ficus retusa nitida, (India Laurel Fig), Z10, cont									
	2610 5 Gal.				Ea.	16			16	17.60
	2620 15 Gal.					64			64	70.50
	2630 24" Box					200			200	220
	2700 Fraxinus uhdei, (Shamel Ash), Z9, cont									
	2710 5 Gal.				Ea.	14			14	15.40
	2720 15 Gal.					63			63	69.50
	2730 24" Box					190			190	209
	2740 36" Box					455			455	500
	2800 Fraxinus velutina, (Velvet Ash), Z7, cont									
	2810 5 Gal.				Ea.	12			12	13.20
	2820 15 Gal.					55			55	60.50
	2830 24" Box					175			175	193
	2900 Gordonia lasianthus, (Loblolly Bay Gordonia), Z8, cont									
	2910 5 Gal.				Ea.	18			18	19.80
	2920 7 Gal.				"	29			29	32
	3000 Grevillea robusta, (Silk Oak), Z10, cont									
	3010 5 Gal.				Ea.	14			14	15.40
	3020 15 Gal.				"	62			62	68
	3100 Hakea suaveolens, (Sea Urchin Tree), Z10, cont									
	3110 1 Gal.				Ea.	3			3	3.30
	3120 5 Gal.					13			13	14.30
	3130 15 Gal.					48			48	53
	3200 Harpephyllum caffrum, (Kaffir Plum), Z10, cont									
	3210 5 Gal.				Ea.	12			12	13.20
	3220 15 Gal.					55			55	60.50
	3230 24" Box					172			172	189
	3300 Hymenosporum flavum, (Sweetshade), Z10, cont									
	3310 5 Gal.				Ea.	12			12	13.20
	3320 15 Gal.					56			56	61.50
	3330 24" Box					180			180	198
	3400 Jacaranda mimosifolia, (Jacaranda), Z10, cont									

Important: See the Reference Section for critical supporting data - Reference Nos., Crews, & City Cost Indexes

029 500 | Trees/Plants/Grnd Cover		CREW	DAILY OUTPUT	LABOR-HOURS	UNIT	1999 BARE COSTS MAT.	LABOR	EQUIP.	TOTAL	TOTAL INCL O&P
610	3410	5 Gal.			Ea.	13			13	14.30
	3420	15 Gal.				58			58	64
	3430	24" Box			↓	185			185	204
	3500	Jatropha curcas, (Barbados Nut), Z10, cont/BB								
	3510	3 Gal.			Ea.	5			5	5.50
	3520	10 Gal.				25			25	27.50
	3530	6' to 7'			↓	75			75	82.50
	3600	Laurus nobilis, (Sweet Bay), Z8, cont								
	3610	1 Gal.			Ea.	5			5	5.50
	3620	5 Gal.				15			15	16.50
	3630	15 Gal.				64			64	70.50
	3640	24" Box			↓	182			182	200
	3700	Leptospermum laevigatum, (Australian Tea Tree), Z9, cont								
	3710	1 Gal.			Ea.	4			4	4.40
	3720	5 Gal.				12			12	13.20
	3730	15 Gal.			↓	55			55	60.50
	3800	Leptospermum scoparium Ruby Glow, (Tea Tree), Z9, cont								
	3810	1 Gal.			Ea.	2.75			2.75	3.03
	3820	5 Gal.			"	10.75			10.75	11.80
	3900	Magnolia grandiflora, (Southern Magnolia), Z7, B&B								
	3910	4' to 5'			Ea.	35			35	38.50
	3920	5' to 6'				45.50			45.50	50
	3930	6' to 7'				58			58	64
	3940	7' to 8'			↓	73.50			73.50	80.50
	4000	Malaleuca linariifolia, (Flaxleaf Paperbark), Z9, cont								
	4010	5 Gal.			Ea.	10.65			10.65	11.70
	4020	15 Gal.				53.50			53.50	58.50
	4030	24" Box			↓	172			172	190
	4100	Maytenus boaria, (Mayten Tree), Z9, cont								
	4110	5 Gal.			Ea.	11.35			11.35	12.50
	4120	15 Gal.			"	53.50			53.50	58.50
	4200	Metrosideros excelsus, (New Zealand Christmas Tree), Z10, cont								
	4210	5 Gal.			Ea.	10.65			10.65	11.70
	4220	15 Gal.				53.50			53.50	58.50
	4230	24" Box			↓	172			172	190
	4300	Olea europea, (Olive), Z9, cont								
	4310	1 Gal.			Ea.	5			5	5.50
	4320	5 Gal.				15			15	16.50
	4330	15 Gal.			↓	63			63	69.50
	4400	Parkinsonia aculeata, (Jerusalem Thorn), Z9, cont								
	4410	5 Gal.			Ea.	10.65			10.65	11.70
	4420	15 Gal.				53.50			53.50	58.50
	4430	24" Box			↓	172			172	190
	4500	Pistachia chinensis, (Chinese Pistache), Z9, cont								
	4510	5 Gal.			Ea.	14			14	15.40
	4520	15 Gal.				61			61	67
	4530	24" Box			↓	177			177	195
	4600	Pyrus Kawakamii, (Evergreen Pear), Z8, cont								
	4610	5 Gal.			Ea.	13.50			13.50	14.85
	4620	15 Gal.				53.50			53.50	58.50
	4630	15 Gal. Espalier				67			67	73.50
	4640	24" Box			↓	175			175	193
	4700	Quercus agrifolia, (Coast Live Oak), Z9, cont								
	4710	5 Gal.			Ea.	14			14	15.40
	4720	15 Gal.				59			59	65
	4730	20 Gal.			↓	72			72	79

029 500 | Trees/Plants/Grnd Cover			CREW	DAILY OUTPUT	LABOR-HOURS	UNIT	1999 BARE COSTS MAT.	LABOR	EQUIP.	TOTAL	TOTAL INCL O&P
610	4800	Sapium sebiferum, (Chinese Tallow Tree), Z8, cont									
	4810	5 Gal.				Ea.	14			14	15.40
	4820	15 Gal.					56			56	61.50
	4830	24" Box				↓	180			180	198
	4900	Schinus molle, (Pepper Tree), Z9, cont									
	4910	5 Gal.				Ea.	14			14	15.40
	4920	15 Gal.					59			59	65
	4930	24" Box				↓	180			180	198
	5000	Strelitzia nicolai, (Giant Bird of Paradise), Z10, cont									
	5010	5 Gal.				Ea.	12			12	13.20
	5020	15 Gal.					55			55	60.50
	5030	24" Box				↓	172			172	189
	5100	Syzygium paniculatum, (Brush Cherry), Z9, cont									
	5110	1 Gal.				Ea.	3			3	3.30
	5120	5 Gal.					12			12	13.20
	5130	15 Gal.				↓	53			53	58.50
	5200	Tabebuia chrysotricha, (Golden Trumpet Tree), Z10, cont									
	5210	5 Gal.				Ea.	16			16	17.60
	5220	15 Gal.					72			72	79
	5230	24" Box				↓	180			180	198
	5300	Thevetia peruviana, (Yellow Oleander), Z10, cont									
	5310	1 Gal.				Ea.	3			3	3.30
	5320	5 Gal.					12			12	13.20
	5330	15 Gal.				↓	45			45	49.50
	5400	Tipuana tipi, (Common Tiputree), Z10, cont									
	5410	1 Gal.				Ea.	3.20			3.20	3.52
	5420	5 Gal.					10.65			10.65	11.70
	5430	15 Gal.					53			53	58.50
	5440	24" Box				↓	172			172	189
	5500	Tristania conferta, (Brisbane Box), Z10, cont									
	5510	5 Gal.				Ea.	10.65			10.65	11.70
	5520	15 Gal.				"	53.50			53.50	59
	5600	Umbellularia californica, (California Laurel), Z7, cont									
	5610	1 Gal.				Ea.	10			10	11
	5620	5 Gal.					30			30	33
	5630	15 Gal.				↓	61			61	67
615	0010	**SHRUBS**, temperate zones 2 - 6									
	1000	Abelia grandiflora, (Abelia), Z6, cont									
	1001	1 Gal.				Ea.	4.90			4.90	5.40
	1002	3 Gal.					11.75			11.75	12.95
	1003	5 Gal.				↓	13.90			13.90	15.30
	1050	Acanthopanax sieboldianus, (Five-Leaved Azalia), Z5, cont/BB									
	1051	2' to 3'				Ea.	18.15			18.15	19.95
	1052	3' to 4'					23.50			23.50	26
	1053	4' to 5'				↓	31.50			31.50	34.50
	1100	Aronia arbutifolia brilliantissima, (Brilliant Chokeberry), Z5, B&B									
	1101	2' to 3'				Ea.	13			13	14.30
	1102	3' to 4'					16			16	17.60
	1103	4' to 5'					19			19	21
	1104	5' to 6'				↓	21			21	23
	1150	Aronia melancarpa "Elata", (Elata Black Chokeberry), Z3, B&B									
	1151	2' to 3'				Ea.	11.60			11.60	12.75
	1152	3' to 4'				"	12.95			12.95	14.25
	1200	Azalea arborescens, (Sweet Azalia), Z5, cont									
	1201	15" to 18"				Ea.	17.80			17.80	19.60
	1202	18" to 24"				"	21.50			21.50	23.50

029 500 | Trees/Plants/Grnd Cover

	029 500 Trees/Plants/Grnd Cover	CREW	DAILY OUTPUT	LABOR-HOURS	UNIT	1999 BARE COSTS MAT.	LABOR	EQUIP.	TOTAL	TOTAL INCL O&P
615 1250	Azalea Evergreen Hybrids, (Azalea Evergreen), Z6, cont/BB									
1251	1 Gal.				Ea.	5.30			5.30	5.85
1252	2 Gal.					12			12	13.20
1253	3 Gal.					15			15	16.50
1254	5 Gal.					17			17	18.70
1255	15″ to 18″					15			15	16.50
1256	18″ to 24″					20			20	22
1257	2′ to 2-1/2′					27			27	29.50
1258	2-1/2′ to 3′					43			43	47.50
1300	Azalea exbury, Ilam Hybrids, (Azalea exbury), Z5, cont/BB									
1301	1 Gal.				Ea.	5.95			5.95	6.55
1302	2 Gal.					12			12	13.20
1303	15″ to 18″					18.35			18.35	20
1304	18″ to 24″					23.50			23.50	26
1305	2′ to 3′					34.50			34.50	38
1306	3′ to 4′					50.50			50.50	55.50
1307	4′ to 5′					68			68	74.50
1350	Azalea gandavensis, (Ghent Azalea), Z4, B&B									
1351	15″ to 18″				Ea.	18.15			18.15	19.95
1352	18″ to 24″					22.50			22.50	25
1353	2′ to 3′					36.50			36.50	40.50
1354	3′ to 4′					50.50			50.50	55.50
1355	4′ to 5′					68			68	74.50
1356	5′ to 6′					90.50			90.50	99.50
1400	Azalea kaempferi, (Torch Azalea), Z6, cont/BB									
1401	12″ to 15″				Ea.	15.10			15.10	16.60
1402	15″ to 18″					18.15			18.15	19.95
1403	18″ to 24″					22.50			22.50	25
1404	2′ to 3′					37.50			37.50	41.50
1450	Berberis julianae "Nana", (Dwarf Wintergreen Barberry), Z6, B&B									
1451	18″ to 24″				Ea.	16			16	17.60
1452	2′ to 2-1/2′					20			20	22
1453	2-1/2′ to 3′					24			24	26.50
1500	Berberis mentorensis, (Mentor Barberry), Z6, cont/BB									
1501	1 Gal.				Ea.	4.60			4.60	5.05
1502	5 Gal.					13			13	14.30
1503	2′ to 3′					18.10			18.10	19.90
1504	3′ to 4′					28			28	30.50
1550	Berberis thumbergii (Crimson Pygmy Barberry), Z5, cont									
1551	1 Gal.				Ea.	6			6	6.60
1552	2 Gal.					12			12	13.20
1553	3 Gal.					15			15	16.50
1600	Berberis thunbergi, (Japanese Green Barberry), Z5, cont/BB									
1601	5 Gal.				Ea.	18			18	19.80
1602	12″ to 15″					15			15	16.50
1603	15″ to 18″					15			15	16.50
1604	18″ to 24″					19			19	21
1605	2′ to 3′					21			21	23
1650	Berberis verruculosa, (Barberry), Z5, cont/BB									
1651	15″ to 18″				Ea.	23.50			23.50	26
1652	18″ to 24″				"	31			31	34.50
1700	Buddleia davidii, (Orange-Eye Butterfly Bush), Z5, cont									
1701	1 Gal.				Ea.	7			7	7.70
1702	2 Gal.				"	13			13	14.30
1750	Buxus microphylla japonica, (Japanese Boxwood), Z5, cont/BB									
1751	1 Gal.				Ea.	5.25			5.25	5.80

	029 500 | Trees/Plants/Grnd Cover	CREW	DAILY OUTPUT	LABOR-HOURS	UNIT	1999 BARE COSTS MAT.	LABOR	EQUIP.	TOTAL	TOTAL INCL O&P
615	1752 5 Gal.				Ea.	14.15			14.15	15.55
	1753 12" to 15"					15.05			15.05	16.55
	1754 15" to 18"					19.35			19.35	21.50
	1755 18" to 24"					26			26	28.50
	1756 24" to 30"					38.50			38.50	42.50
	1757 30" to 36"				↓	54			54	59
	1800 Buxus sempervirens, (Common Boxwood), Z5, cont/BB									
	1801 12" to 15"				Ea.	15.05			15.05	16.55
	1802 15" to 18"					19.35			19.35	21.50
	1803 18" to 24"					26			26	28.50
	1804 24" to 30"				↓	38.50			38.50	42.50
	1850 Callicarpa japonica, (Japanese Beautyberry), Z6, B&B									
	1851 2' to 3'				Ea.	12.35			12.35	13.60
	1852 3' to 4'					14			14	15.40
	1853 4' to 5'				↓	17.15			17.15	18.85
	1900 Caragana arborescens, (Siberian Peashrub), Z2, B&B									
	1901 3' to 4'				Ea.	20.50			20.50	22.50
	1902 4' to 5'					28			28	30.50
	1903 5' to 6'					36.50			36.50	40
	1904 6' to 8'				↓	62.50			62.50	68.50
	1950 Caragana arborescens, (Walkers Weeping Peashrub), Z3, B&B									
	1951 1-1/2" to 1-3/4" Cal.				Ea.	88			88	97
	1952 1-3/4" to 2" Cal.					102			102	112
	1953 1-1/2" to 1-3/4" Cal.					88.50			88.50	97
	1954 1-3/4" to 2" Cal.				↓	102			102	113
	2000 Caryopteris x clandonensis, (Bluebeard), Z5, cont									
	2001 1 Qt.				Ea.	5			5	5.50
	2002 1 Gal.					7			7	7.70
	2003 2 Gal.				↓	11			11	12.10
	2050 Chaenomeles speciosa, (Flowering Quince), Z5, cont/BB									
	2051 18" to 24"				Ea.	13.60			13.60	14.95
	2052 2' to 3'					18.15			18.15	19.95
	2053 3' to 4'				↓	28			28	31
	2100 Clethra alnifolia rosea, (Pink Summersweet), Z4, B&B									
	2101 2' to 3'				Ea.	20			20	22
	2102 3' to 4'					26			26	28.50
	2103 4' to 5'				↓	35			35	38.50
	2150 Cornus alba siberica, (Siberian Dogwood), Z3, B&B									
	2151 3' to 4'				Ea.	20.50			20.50	22.50
	2152 4' to 5'					24.50			24.50	26.50
	2153 5' to 6'					34.50			34.50	38
	2154 6' to 7'				↓	41			41	45
	2200 Cornus alternifolia, (Alternate Leaved Dogwood), Z3, B&B									
	2201 4' to 5'				Ea.	27			27	29.50
	2202 5' to 6'					36.50			36.50	40.50
	2203 6' to 8'				↓	50.50			50.50	55.50
	2250 Cornus mas, (Cornelian Cherry), Z5, B&B									
	2251 2' to 3'				Ea.	22			22	24
	2252 3' to 4'					28			28	31
	2253 4' to 5'					36			36	39.50
	2254 5' to 6'				↓	49			49	54
	2300 Cornus racemosa, (Gray Dogwood), Z4, B&B									
	2301 18" to 24"				Ea.	11			11	12.10
	2302 2' to 3'					13			13	14.30
	2303 3' to 4'				↓	17			17	18.70
615	2350 Cornus stolonifera flaviramea, (Goldentwig Dogwood), Z3, cont/BB									

029 500 | Trees/Plants/Grnd Cover			CREW	DAILY OUTPUT	LABOR-HOURS	UNIT	1999 BARE COSTS MAT.	LABOR	EQUIP.	TOTAL	TOTAL INCL O&P
615	2351	2′ to 3′				Ea.	13.60			13.60	14.95
	2352	3′ to 4′					15.85			15.85	17.45
	2353	4′ to 5′					20.50			20.50	22.50
	2354	5′ to 6′					31.50			31.50	34.50
	2355	6′ to 7′				↓	40			40	44
	2400	Corylus americana, (Filbert, American Hazelnut), Z4, B&B									
	2401	2′ to 3′				Ea.	15.85			15.85	17.45
	2402	3′ to 4′					20.50			20.50	22.50
	2403	4′ to 5′				↓	28			28	31
	2450	Cotinus coggygria, (Smoke Tree), Z5, B&B									
	2451	5′ to 6′				Ea.	47			47	51.50
	2452	6′ to 8′					66			66	72.50
	2453	8′ to 10′				↓	99			99	109
	2500	Cotoneaster acutifolius, (Peking Cotoneaster), Z4, B&B									
	2501	2′ to 3′				Ea.	11.75			11.75	12.95
	2502	3′ to 4′					13.75			13.75	15.15
	2503	4′ to 5′				↓	17.15			17.15	18.85
	2550	Cotoneaster adpressa, (Creeping Cotoneaster), Z5, cont									
	2551	12″ to 15″ Spread				Ea.	11.30			11.30	12.45
	2600	Cotoneaster adpressa praecox, (Early Cotoneaster), Z5, cont									
	2601	1 Gal.				Ea.	4.85			4.85	5.35
	2602	2 Gal.					9.55			9.55	10.50
	2603	3 Gal.				↓	11.30			11.30	12.45
	2650	Cotoneaster divaricatus, (Spreading Cotoneaster), Z5, cont/BB									
	2651	1 Gal.				Ea.	7			7	7.70
	2652	2 Gal.					13			13	14.30
	2653	2′ to 3′					15.05			15.05	16.55
	2654	3′ to 4′					20.50			20.50	22.50
	2655	4′ to 5′				↓	34.50			34.50	38
	2700	Cotoneaster horizontalis, (Rock Cotoneaster), Z5, cont									
	2701	1 Gal.				Ea.	7			7	7.70
	2702	2 Gal.					13			13	14.30
	2703	5 Gal.				↓	17			17	18.70
	2750	Cytisus praecox, (Warminster Broom), Z6, cont									
	2751	1 Gal.				Ea.	4.85			4.85	5.35
	2752	2 Gal.					9.55			9.55	10.50
	2753	5 Gal.				↓	13.45			13.45	14.80
	2800	Daphne burkwoodi somerset, (Burkwood Daphne), Z6, cont									
	2801	15″ to 18″				Ea.	15.80			15.80	17.40
	2802	18″ to 24″				″	20.50			20.50	22.50
	2850	Deutzia gracilis, (Slender Deutzia), Z5, cont									
	2851	12″ to 15″				Ea.	11			11	12.10
	2852	18″ to 24″				″	13.90			13.90	15.30
	2900	Diervilla sessilifolia, (Southern Bush Honeysuckle), Z5, B&B									
	2901	18″ to 24″				Ea.	10.25			10.25	11.30
	2902	2′ to 3′					12.35			12.35	13.60
	2903	3′ to 4′				↓	15.85			15.85	17.45
	2950	Elaeagnus angustifolia, (Russian Olive), Z3, B&B									
	2951	3′ to 4′				Ea.	22.50			22.50	25
	3000	Elaeagnus umbellata, (Autumn Olive), Z3, B&B									
	3001	3′ to 4′				Ea.	16			16	17.60
	3002	4′ to 5′					22			22	24
	3003	5′ to 6′				↓	28			28	31
	3050	Enkianthus campanulatus, (Bellflowertree), Z5, cont/BB									
	3051	2′ to 3′				Ea.	24			24	26.50
	3052	3′ to 4′				↓	27			27	29.50

029 500 | Trees/Plants/Grnd Cover		CREW	DAILY OUTPUT	LABOR-HOURS	UNIT	1999 BARE COSTS MAT.	LABOR	EQUIP.	TOTAL	TOTAL INCL O&P
615 3053	4′ to 5′				Ea.	45			45	49.50
3054	5′ to 6′				↓	62			62	68
3100	Euonymus alatus, (Winged Burning Bush), Z4, B&B									
3101	18″ to 24″				Ea.	14			14	15.40
3102	2′ to 2-1/2′					19			19	21
3103	2-1/2′ to 3′					24			24	26.50
3104	3′ to 3-1/2′					29			29	32
3105	3-1/2′ to 4′					34			34	37.50
3106	4′ to 5′					50			50	55
3107	5′ to 6′					63			63	69.50
3108	6′ to 8′				↓	73			73	80.50
3150	Euonymus alatus compacta, (Dwarf Winged Burning Bush), Z4, B&B									
3151	15″ to 18″				Ea.	13			13	14.30
3152	18″ to 24″					15			15	16.50
3153	2′ to 2-1/2′					20			20	22
3154	2-1/2′ to 3′					27			27	29.50
3155	3′ to 3-1/2′					31			31	34
3156	3-1/2′ to 4′					38			38	42
3157	4′ to 5′					53			53	58.50
3158	5′ to 6′				↓	63			63	69.50
3200	Euonymus europaeus aldenhamensis, (Spindletree), Z4, B&B									
3201	3′ to 4′				Ea.	17.80			17.80	19.60
3202	4′ to 5′					19.40			19.40	21.50
3203	5′ to 6′					33.50			33.50	37
3204	6′ to 8′				↓	56			56	61.50
3250	Euonymus fortunei, (Wintercreeper), Z5, cont/BB									
3251	1 Gal.				Ea.	6			6	6.60
3252	2 Gal.					13			13	14.30
3253	5 Gal.					18			18	19.80
3254	24″ to 30″				↓	21			21	23
3300	Euonymus fortunei "Emerald 'n Gold", (Wintercreeper), Z5, cont									
3301	1 Gal.				Ea.	7			7	7.70
3302	2 Gal.					15			15	16.50
3303	3 Gal.					20			20	22
3304	5 Gal.				↓	23			23	25.50
3350	Euonymus fortunei emerald gaiety, (Gaiety Wintercreeper), Z5, cont									
3351	1 Gal.				Ea.	7			7	7.70
3352	2 Gal.					15			15	16.50
3353	3 Gal.					20			20	22
3354	5 Gal.				↓	23			23	25.50
3400	Forsythia intermedia, (Border Goldenbells), Z5, cont/BB									
3401	5 Gal.				Ea.	14			14	15.40
3402	2′ to 3′					12			12	13.20
3403	3′ to 4′					15			15	16.50
3404	4′ to 5′					18			18	19.80
3405	5′ to 6′					21			21	23
3406	6′ to 8′				↓	30			30	33
3450	Forsythia ovata robusta, (Korean Forsythia), Z5, B&B									
3451	2′ to 3′				Ea.	12			12	13.20
3452	3′ to 4′					15			15	16.50
3453	4′ to 5′				↓	18			18	19.80
3500	Hamamelis vernalis, (Vernal Witch-Hazel), Z4, B&B									
3501	2′ to 3′				Ea.	19			19	21
3502	3′ to 4′					23			23	25.50
3503	4′ to 5′				↓	28			28	31
3550	Hibiscus syriacus, (Rose of Sharon), Z6, cont/BB									

029 500 | Trees/Plants/Grnd Cover

029 500	Trees/Plants/Grnd Cover	CREW	DAILY OUTPUT	LABOR-HOURS	UNIT	1999 BARE COSTS MAT.	LABOR	EQUIP.	TOTAL	TOTAL INCL O&P
615 3551	5 Gal.				Ea.	16			16	17.60
3552	2′ to 3′					15			15	16.50
3553	3′ to 4′					20			20	22
3554	4′ to 5′					28			28	31
3555	5′ to 6′					36			36	39.50
3600	Hydrangea arborescens, (Smooth Hydrangea), Z4, cont									
3601	2 Gal.				Ea.	14			14	15.40
3602	3 Gal.				″	15			15	16.50
3650	Hydrangea macrophylla nikko blue, (Blue Hydrangea), Z6, cont									
3651	2 Gal.				Ea.	13			13	14.30
3652	5 Gal.				″	20			20	22
3700	Hydrangea paniculata grandiflora, (Peegee Hydrangea), Z5, B&B									
3701	2′ to 3′				Ea.	18			18	19.80
3702	3′ to 4′				″	24			24	26.50
3750	Hydrangea quercifolia, (Oakleaf Hydrangea), Z6, cont									
3751	2 Gal.				Ea.	11			11	12.10
3800	Hypericum hidcote, (St. John's Wort), Z6, cont									
3801	2 Gal.				Ea.	13			13	14.30
3850	Ilex cornuta "Carissa", (Carissa Holly), Z7, cont									
3851	1 Gal.				Ea.	4.85			4.85	5.35
3852	2 Gal.					11			11	12.10
3853	5 Gal.					14			14	15.40
3900	Ilex cornuta rotunda, (Dwarf Chinese Holly), Z6, cont									
3901	1 Gal.				Ea.	4.85			4.85	5.35
3902	2 Gal.					11			11	12.10
3903	5 Gal.					15			15	16.50
3950	Ilex crenata "Helleri", (Hellers Japanese Holly), Z6, cont									
3951	1 Gal.				Ea.	6			6	6.60
3952	2 Gal.					13			13	14.30
3953	3 Gal.					16			16	17.60
4000	Ilex crenata "Hetzi", (Hetzi Japanese Holly), Z6, cont/BB									
4001	2 Gal.				Ea.	13			13	14.30
4002	3 Gal.					16			16	17.60
4003	2′ to 3′					29			29	32
4004	3′ to 4′					40			40	44
4050	Ilex glabra compacta, (Compact Inkberry), Z4, B&B									
4051	15″ to 18″				Ea.	20			20	22
4052	18″ to 24″					24			24	26.50
4053	2′ to 2-1/2′					30			30	33
4100	Ilex meserve blue holly hybrids, (Blue Holly), Z5, cont/BB									
4101	2 Gal.				Ea.	13			13	14.30
4102	3 Gal.					16			16	17.60
4103	5 Gal.					18			18	19.80
4104	3′ to 3-1/2′					76			76	83.50
4105	3-1/2′ to 4′					99			99	109
4106	4′ to 5′					138			138	152
4107	5′ to 6′					192			192	211
4150	Ilex opaca, (American Holly), Z6, B&B									
4151	2′ to 3′				Ea.	32			32	35
4152	3′ to 4′					53			53	58.50
4153	4′ to 5′					71			71	78
4154	5′ to 6′					85			85	93.50
4155	6′ to 7′					116			116	128
4156	7′ to 8′					164			164	180
4157	8′ to 10′					240			240	264
4158	10′ to 12′					365			365	400

029 500 | Trees/Plants/Grnd Cover

029 500	Trees/Plants/Grnd Cover	CREW	DAILY OUTPUT	LABOR-HOURS	UNIT	1999 BARE COSTS MAT.	LABOR	EQUIP.	TOTAL	TOTAL INCL O&P
615 4159	12′ to 14′				Ea.	445			445	490
4200	Ilex verticillata female, (Winterberry), Z4, B&B									
4201	2′ to 3′				Ea.	30			30	33
4202	3′ to 4′				"	40			40	44
4250	Ilex verticillata male, (Winterberry), Z4, B&B									
4251	2′ to 3′				Ea.	22			22	24
4252	3′ to 4′					32			32	35
4253	4′ to 5′					38			38	42
4300	Ilex x attenuata "Fosteri", (Foster Holly), Z6, B&B									
4301	4′ to 5′				Ea.	62			62	68
4302	5′ to 6′					83			83	91.50
4303	6′ to 7′					108			108	119
4304	7′ to 8′					141			141	155
4350	Itea virginica, (Virginia Sweetspire), Z6, B&B									
4351	2′ to 3′				Ea.	34.50			34.50	38
4400	Kalmia latifolia, (Mountain Laurel), Z5, cont/BB									
4401	15″ to 18″				Ea.	18			18	19.80
4402	18″ to 24″					25			25	27.50
4403	2′ to 3′					38			38	41.50
4404	4′ to 5′					69			69	76
4405	5′ to 6′					106			106	116
4406	6′ to 8′					162			162	178
4450	Kerria japonica, (Kerria), Z6, cont									
4451	1 Gal.				Ea.	5			5	5.50
4452	5 Gal.				"	15			15	16.50
4500	Kolkwitzia amabilis, (Beautybush), Z5, cont/BB									
4501	2 Gal.				Ea.	12			12	13.20
4502	5 Gal.					16			16	17.60
4503	2′ to 3′					14.75			14.75	16.25
4504	3′ to 4′					18.15			18.15	19.95
4505	4′ to 5′					20.50			20.50	22.50
4506	5′ to 6′					34.50			34.50	38
4507	6′ to 8′					45.50			45.50	50
4550	Leucothoe axillaris, (Coast Leucothoe), Z6, cont									
4551	1 Gal.				Ea.	6			6	6.60
4552	2 Gal.					13			13	14.30
4553	3 Gal.					16			16	17.60
4600	Leucothoe fontanesiana, (Drooping Leucothoe), Z6, cont/BB									
4601	2 Gal.				Ea.	10.70			10.70	11.75
4602	3 Gal.				"	14			14	15.40
4650	Ligistrum obtusifolium regelianum, (Regal Privet), Z4, cont/BB									
4651	18″ to 24″				Ea.	17			17	18.70
4652	2′ to 3′					21			21	23
4653	18″ to 24″					4.30			4.30	4.73
4654	2′ to 2-1/2′					5.40			5.40	5.95
4700	Ligustrum amurense, (Amur Privet), Z4, BR									
4701	2′ to 3′				Ea.	4.55			4.55	5
4702	3′ to 4′				"	5.80			5.80	6.40
4750	Lindera benzoin, (Spicebush), Z5, cont/BB									
4751	2′ to 3′				Ea.	33			33	36.50
4752	3′ to 4′				"	40			40	44
4800	Lonicera fragrantissima, (Winter Honeysuckle), Z6, cont/BB									
4801	2′ to 3′				Ea.	26			26	28.50
4802	3′ to 4′				"	31.50			31.50	34.50
4850	Lonicera xylosteum, (Emerald Mound Honeysuckle), Z4, cont									
4851	2 Gal.				Ea.	9.55			9.55	10.50

029 500 | Trees/Plants/Grnd Cover		CREW	DAILY OUTPUT	LABOR-HOURS	UNIT	1999 BARE COSTS MAT.	LABOR	EQUIP.	TOTAL	TOTAL INCL O&P
615 4852	3 Gal.				Ea.	11.70			11.70	12.85
4853	5 Gal.					13.90			13.90	15.30
4900	Mahonia aquifolium, (Oregon Grape Holly), Z6, cont/BB									
4901	1 Gal.				Ea.	5.10			5.10	5.60
4902	2 Gal.					11.70			11.70	12.85
4903	3 Gal.					16.10			16.10	17.70
4950	Myrica pensylvanica, (Northern Bayberry), Z2, B&B									
4951	15″ to 18″				Ea.	14			14	15.40
4952	2′ to 2-1/2′					18			18	19.80
4953	2-1/2′ to 3′					25			25	27.50
5000	Osmanthus heterophyllus "Gulftide", (Gulftide Sweet Holly), Z6, B&B									
5001	18″ to 24″				Ea.	12.90			12.90	14.20
5002	2′ to 2-1/2′					15			15	16.50
5003	2-1/2′ to 3′					19			19	21
5004	3′ to 3-1/2′					21			21	23
5050	Paxistima canbyi, (Canby Paxistima), Z5, B&B									
5051	6″ to 9″				Ea.	7.50			7.50	8.25
5052	9″ to 12″					9.65			9.65	10.60
5053	12″ to 15″					12.90			12.90	14.20
5100	Philadelphus coronarius, (Sweet Mockorange), Z5, cont/BB									
5101	2′ to 3′				Ea.	12.15			12.15	13.35
5102	3′ to 4′					16.90			16.90	18.60
5103	4′ to 5′					19.20			19.20	21
5104	5′ to 6′					31			31	34.50
5150	Philadelphus virginalis, (Hybrid Mockorange), Z6, cont/BB									
5151	5 Gal.				Ea.	15			15	16.50
5152	2′ to 3′					18			18	19.80
5153	3′ to 4′					23			23	25.50
5154	4′ to 5′					29			29	32
5200	Physocarpus opulifolius luteus, (Golden Ninebark), Z3, cont/BB									
5201	2′ to 3′				Ea.	15.80			15.80	17.40
5202	3′ to 4′					20.50			20.50	22.50
5203	4′ to 5′					28			28	30.50
5250	Pieris floribunda, (Mountain Andromeda), Z5, cont									
5251	12″ to 15″ Spread				Ea.	13			13	14.30
5252	15″ to 18″ Spread				″	28			28	30.50
5300	Pieris japonica, (Japanese Andromeda), Z6, cont/BB									
5301	12″ to 15″				Ea.	16.90			16.90	18.60
5302	15″ to 18″					19.35			19.35	21.50
5303	18″ to 24″					24			24	26.50
5304	2′ to 2-1/2′					36			36	39.50
5305	2-1/2′ to 3′					53			53	58.50
5350	Potentilla fruticosa, (Shrubby Cinquefoil), Z2, cont/BB									
5351	1 Gal.				Ea.	4.50			4.50	4.95
5352	5 Gal.					13.45			13.45	14.80
5353	12″ to 15″					9.15			9.15	10.05
5354	15″ to 18″					12			12	13.20
5355	18″ to 24″					14			14	15.40
5356	2′ to 3′					17			17	18.70
5400	Prunus cistena, (Purple-Leaf Sand Cherry), Z3, cont/BB									
5401	5 Gal.				Ea.	16			16	17.60
5402	18″ to 24″					17			17	18.70
5403	2′ to 3′					20			20	22
5404	3′ to 4′					26			26	28.50
5405	4′ to 5′					31			31	34
5406	5′ to 6′					39			39	43

615	029 500 | Trees/Plants/Grnd Cover	CREW	DAILY OUTPUT	LABOR-HOURS	UNIT	1999 BARE COSTS MAT.	LABOR	EQUIP.	TOTAL	TOTAL INCL O&P
5450	Prunus glandulosa albo plena, (Flowering Almond), Z6, cont/BB									
5451	15″ to 18″				Ea.	11.10			11.10	12.20
5452	18″ to 24″					14.20			14.20	15.60
5453	3′ to 4′					24			24	26.50
5454	5′ to 6′					46			46	50.50
5500	Prunus glandulosa sinensis, (Pink Flowering Almond), Z6, cont/BB									
5501	5 Gal.				Ea.	13.65			13.65	15
5502	18″ to 24″					15.30			15.30	16.85
5503	2′ to 3′					17.80			17.80	19.60
5550	Prunus laurocerasus 'Schiphaewsis, (Cherry Laurel), Z5, B&B									
5551	15″ to 18″				Ea.	16			16	17.60
5552	18″ to 24″					21			21	23
5553	24″ to 30″					25			25	27.50
5600	Prunus maritima, (Beach Plum), Z4, cont/BB									
5601	1 Gal.				Ea.	4.90			4.90	5.40
5602	2 Gal.					9.20			9.20	10.10
5603	18″ to 24″					17			17	18.70
5650	Pyracantha coccinea, (LaLande Firethorn), Z6, cont/BB									
5651	1 Gal.				Ea.	6			6	6.60
5652	5 Gal.					21			21	23
5653	5 Gal. Espalier					49			49	54
5654	10 Gal. Espalier					112			112	123
5700	Rhamus frangula columnaris, (Tallhedge), Z2, B&B									
5701	4′ to 5′				Ea.	27			27	29.50
5702	5′ to 6′					37			37	40.50
5703	6′ to 8′					48			48	53
5704	10′ to 12′					89			89	98
5750	Rhododendron carolinianum, (Carolina Rhododendron), Z5, cont/BB									
5751	15″ to 18″				Ea.	15.30			15.30	16.85
5752	18″ to 24″					27			27	29.50
5753	24″ to 30″					40			40	44
5800	Rhododendron catawbiense hybrids, (Rhododendron), Z5, cont/BB									
5801	1 Gal.				Ea.	5.95			5.95	6.55
5802	2 Gal.					14			14	15.40
5803	3 Gal.					19			19	21
5804	5 Gal.					22			22	24
5805	15″ to 18″					17.15			17.15	18.85
5806	18″ to 24″					23.50			23.50	26
5807	24″ to 30″					32			32	35.50
5808	30″ to 36″					37.50			37.50	41.50
5809	36″ to 42″					54			54	59
5850	Rhododendron PJM, (PJM Rhododendron), Z4, cont/BB									
5851	15″ to 18″				Ea.	19.35			19.35	21.50
5852	18″ to 24″					23.50			23.50	26
5853	2′ to 2-1/2′					34.50			34.50	38
5854	2-1/2′ to 3′					44			44	48.50
5855	3′ to 3-1/2′					54			54	59
5900	Rhus aromatica "Gro-Low", (Fragrant Sumac), Z3, cont									
5901	2 Gal.				Ea.	10.65			10.65	11.70
5902	3 Gal.					11.30			11.30	12.45
5903	5 Gal.					12.35			12.35	13.60
5950	Rhus glabra cismontana, (Dwarf Smooth Sumac), Z2, cont									
5951	1 Gal.				Ea.	4.85			4.85	5.35
5952	5 Gal.				″	13.45			13.45	14.80
6000	Rhus typhina lacianata, (Cut-Leaf Staghorn Sumac), Z4, B&B									
6001	4′ to 5′				Ea.	20.50			20.50	22.50

029 500 | Trees/Plants/Grnd Cover		CREW	DAILY OUTPUT	LABOR-HOURS	UNIT	1999 BARE COSTS MAT.	LABOR	EQUIP.	TOTAL	TOTAL INCL O&P
615 6002	5′ to 6′				Ea.	28			28	30.50
6003	6′ to 7′				↓	34.50			34.50	38
6050	Ribes alpinum, (Alpine Currant), Z3, cont/BB									
6051	18″ to 24″				Ea.	16.90			16.90	18.60
6052	2′ to 3′				″	22.50			22.50	25
6100	Rosa hugonis, (Father Hugo Rose), Z6, cont									
6101	1 Gal.				Ea.	6			6	6.60
6102	2 Gal.				″	9			9	9.90
6150	Rosa meidiland hybrids, Z6, cont									
6151	2 Gal.				Ea.	7.40			7.40	8.15
6152	3 Gal.				″	9.60			9.60	10.55
6200	Rosa multiflora, (Japanese Rose), Z6, cont									
6201	1 Gal.				Ea.	4.85			4.85	5.35
6202	2 Gal.				″	7.40			7.40	8.15
6250	Rosa rugosa and hybrids, Z4, cont									
6251	1 Gal.				Ea.	7			7	7.70
6252	2 Gal.				″	10			10	11
6300	Rosa: hybrid teas, grandifloras and climbers, , Zna, cont									
6301	2 Gal. Non Pat.				Ea.	9			9	9.90
6302	2 Gal. Patented				″	11			11	12.10
6350	Salix caprea, (French Pussy), Z5, B&B									
6351	4′ to 5′				Ea.	16.95			16.95	18.65
6352	5′ to 6′				″	20.50			20.50	22.50
6400	Salix discolor, (Pussy Willow), Z2, cont									
6401	5 Gal.				Ea.	13.50			13.50	14.85
6450	Sambucus canadensis, (American Elder), Z4, B&B									
6451	2′ to 3′				Ea.	15.65			15.65	17.20
6500	Skimmia japonica, (Japanese Skimmia), Z6, cont									
6501	1 Gal.				Ea.	4.85			4.85	5.35
6502	2 Gal.					9.70			9.70	10.65
6503	5 Gal.				↓	15.65			15.65	17.20
6550	Spiraea arguta compacta, (Garland Spirea), Z5, B&B									
6551	3′ to 4′				Ea.	22.50			22.50	25
6552	4′ to 5′				″	34.50			34.50	38
6600	Spirea bumalda, (Bumalda Spirea), Z5, cont/BB									
6601	1 Gal.				Ea.	6			6	6.60
6602	5 Gal.					16			16	17.60
6603	2′ to 3′					19			19	21
6604	3′ to 4′				↓	25			25	27.50
6650	Spirea japonica, (Japanese Spirea), Z5, cont									
6651	1 Gal.				Ea.	6			6	6.60
6652	2 Gal.					12			12	13.20
6653	3 Gal.					14			14	15.40
6654	5 Gal.				↓	17			17	18.70
6700	Spirea nipponica snowmound, (Snowmound Spirea), Z5, cont/BB									
6701	5 Gal.				Ea.	14			14	15.40
6702	18″ to 24″					12			12	13.20
6703	2′ to 3′					17			17	18.70
6704	3′ to 4′					20			20	22
6705	4′ to 5′				↓	34			34	37.50
6750	Spirea vanhouttei, (Van Houtte Spirea), Z4, cont/BB									
6751	5 Gal.				Ea.	14			14	15.40
6752	2′ to 3′					17			17	18.70
6753	3′ to 4′					19			19	21
6754	4′ to 5′					25			25	27.50
6755	5′ to 6′				↓	37			37	40.50

615

	029 500 | Trees/Plants/Grnd Cover	CREW	DAILY OUTPUT	LABOR-HOURS	UNIT	1999 BARE COSTS MAT.	LABOR	EQUIP.	TOTAL	TOTAL INCL O&P
615	6800 Symphoricarpos albus, (Common Snowberry), Z4, cont/BB									
	6801 18″ to 24″				Ea.	13.45			13.45	14.80
	6802 2′ to 3′					15.05			15.05	16.55
	6803 3′ to 4′				↓	18.05			18.05	19.85
	6850 Syringa chinensis, (Roven Lilac), Z6, B&B									
	6851 3′ to 4′				Ea.	18.10			18.10	19.90
	6852 4′ to 5′					28			28	30.50
	6853 5′ to 6′				↓	35.50			35.50	39
	6900 Syringa hugo koster, (Hugo Koster Lilac), Z4, B&B									
	6901 2′ to 3′				Ea.	18.10			18.10	19.90
	6902 3′ to 4′					28			28	30.50
	6903 4′ to 5′					34.50			34.50	38
	6904 5′ to 6′				↓	57			57	62.50
	6950 Syringa meyeri, (Meyer Lilac), Z6, cont/BB									
	6951 5 Gal.				Ea.	14			14	15.40
	6952 15″ to 18″					16			16	17.60
	6953 18″ to 24″					18			18	19.80
	6954 2′ to 3′					22			22	24
	6955 3′ to 4′				↓	33			33	36.50
	7000 Syringa persica, (Persian Lilac), Z5, cont									
	7001 5 Gal.				Ea.	16			16	17.60
	7050 Syringa prestonae hybrids, (Preston Lilac), Z3, B&B									
	7051 2′ to 3′				Ea.	20			20	22
	7052 3′ to 4′					28			28	31
	7053 4′ to 5′					49			49	54
	7054 5′ to 6′					68			68	75
	7055 6′ to 7′				↓	79			79	87
	7100 Syringa vulgaris and Mixed Hybrids, (Common Lilac), Z4, cont/BB									
	7101 5 Gal.				Ea.	13.60			13.60	14.95
	7102 2′ to 3′					19.40			19.40	21.50
	7103 3′ to 4′					26			26	28.50
	7104 4′ to 5′					35.50			35.50	39
	7105 5′ to 6′				↓	47.50			47.50	52
	7200 Vaccinium corymbosum, (Highbush Blueberry), Z4, cont/BB									
	7201 18″ to 24″				Ea.	20			20	22
	7202 2′ to 3′					28			28	31
	7203 3′ to 4′					38			38	42
	7204 4′ to 5′				↓	56			56	61.50
	7250 Viburnum bodnantense, (Bodnant Viburnum), Z5, cont/BB									
	7251 18″ to 24″				Ea.	18			18	19.80
	7252 3′ to 4′					34			34	37.50
	7253 4′ to 5′					44			44	48.50
	7254 5′ to 6′				↓	55			55	60.50
	7300 Viburnum carlcephalum, (Fragrant Snowball), Z6, cont/BB									
	7301 2′ to 3′				Ea.	26			26	28.50
	7302 3′ to 4′					33			33	36.50
	7303 4′ to 5′				↓	49			49	54
	7350 Viburnum carlesi, (Korean Spice Viburnum), Z4, B&B									
	7351 18″ to 24″				Ea.	19.40			19.40	21.50
	7352 2′ to 2-1/2′					23.50			23.50	26
	7353 2-1/2′ to 3′					30			30	33
	7354 3′ to 4′				↓	39			39	42.50
	7400 Viburnum dilatatum, (Linden Viburnum), Z6, B&B									
	7401 3′ to 4′				Ea.	23			23	25.50
	7402 4′ to 5′					34			34	37.50
	7403 5′ to 6′				↓	45			45	49.50

		029 500 | Trees/Plants/Grnd Cover	CREW	DAILY OUTPUT	LABOR-HOURS	UNIT	1999 BARE COSTS MAT.	LABOR	EQUIP.	TOTAL	TOTAL INCL O&P	
615	7450	Viburnum juddii, (Judd Viburnum), Z5, B&B										615
	7451	18" to 24"				Ea.	18			18	19.80	
	7452	2' to 2-1/2'					22			22	24	
	7453	2-1/2' to 3'					27			27	29.50	
	7454	3' to 4'				↓	36			36	39.50	
	7500	Viburnum opulus sterile, (Common Snowball), Z4, cont/BB										
	7501	5 Gal.				Ea.	13.50			13.50	14.85	
	7502	18" to 24"					12.95			12.95	14.25	
	7503	2' to 3'					16.95			16.95	18.65	
	7504	3' to 4'				↓	23.50			23.50	26	
	7550	Viburnum plicatum, (Japanese Snowball), Z6, cont/BB										
	7551	5 Gal.				Ea.	15			15	16.50	
	7552	5' to 6'				"	54			54	59.50	
	7600	Viburnum plicatum var. tomentosum, (Viburnum), Z5, cont/BB										
	7601	5 Gal.				Ea.	13.50			13.50	14.85	
	7602	2' to 3'					21			21	23	
	7603	3' to 4'					28			28	31	
	7604	4' to 5'				↓	39			39	42.50	
	7650	Viburnum prunifolium, (Blackhaw), Z4, B&B										
	7651	3' to 4'				Ea.	23.50			23.50	26	
	7652	4' to 5'					36.50			36.50	40.50	
	7653	5' to 6'					45			45	49.50	
	7654	6' to 7'				↓	60.50			60.50	66.50	
	7700	Viburnum rhytidophyllum, (Leatherleaf Viburnum), Z6, B&B										
	7701	2' to 3'				Ea.	23.50			23.50	26	
	7702	3' to 4'					28			28	30.50	
	7703	4' to 5'				↓	39.50			39.50	43	
	7750	Viburnum sieboldi, (Siebold Viburnum), Z5, B&B										
	7751	3' to 4'				Ea.	26			26	28.50	
	7752	4' to 5'					38.50			38.50	42.50	
	7753	5' to 6'					47.50			47.50	52	
	7754	6' to 7'				↓	62.50			62.50	68.50	
	7800	Viburnum trilobum, (American Cranberry Bush), Z2, cont/BB										
	7801	2' to 3'				Ea.	20			20	22	
	7802	3' to 4'					25			25	27.50	
	7803	5' to 6'					45			45	49.50	
	7804	6' to 8'					63			63	69.50	
	7805	8' to 10'				↓	80			80	88	
	7850	Weigela florida and Hybrids, (Old Fashioned Weigela), Z6, cont/BB										
	7851	2' to 3'				Ea.	15			15	16.50	
	7852	3' to 4'					19			19	21	
	7853	4' to 5'					25			25	27.50	
	7854	5' to 6'				↓	30			30	33	
	7900	Xanthorhiza simplicissima, (Yellowroot), Z4, cont										
	7901	4" Pot				Ea.	4.05			4.05	4.46	
	7902	1 Gal.				"	9.70			9.70	10.65	
	7950	Yucca filamentosa, (Adams Needle), Z4, cont										
	7951	1 Gal.				Ea.	3.80			3.80	4.18	
	7952	3 Gal.					7.40			7.40	8.15	
	7953	5 Gal.				↓	11.70			11.70	12.85	
620	0010	**FRUITS AND NUTS**										620
	1000	Apple, Different Varieties, Z4, cont										
	1010	5 Gal.				Ea.	16.10			16.10	17.70	
	1020	Standard, 2-3 yr.					20.50			20.50	22.50	
	1030	Semidwarf, 2-3 yr.				↓	26			26	28.50	
	1500	Apricot, Z5, cont/BB										

		029 500 | Trees/Plants/Grnd Cover	CREW	DAILY OUTPUT	LABOR-HOURS	UNIT	1999 BARE COSTS				TOTAL INCL O&P	
							MAT.	LABOR	EQUIP.	TOTAL		
620	1510	5 Gal.				Ea.	16.10			16.10	17.70	620
	1520	Standard, 2-3 yr.					20.50			20.50	22.50	
	1530	Semidwarf, 2-3 yr.				↓	26			26	28.50	
	2000	Carya ovata, Z4, bare root										
	2010	Bare root, 2 yr.				Ea.	2.50			2.50	2.75	
	2500	Castanea mollissima, (Chinese Chestnut), Z4, bare root										
	2510	Bare root, 2 yr.				Ea.	2.50			2.50	2.75	
	3000	Cherry, Z4, B&B										
	3010	Standard, 2-3 yr.				Ea.	20.50			20.50	22.50	
	3020	Semidwarf, 2-3 yr.				"	26			26	28.50	
	3500	Corylus americana, (Filbert, American Hazelnut), Z5, B&B										
	3510	2′ to 3′				Ea.	15.85			15.85	17.45	
	3520	3′ to 4′				"	19.25			19.25	21	
	4500	Grape, (Different Varieties), Z5, potted										
	4510	Potted				Ea.	4.75			4.75	5.20	
	5000	Juglans cinera, (Butternut), Z4, bare root										
	5010	Bare Root, 2 yr.				Ea.	2.85			2.85	3.14	
	5500	Juglans nigra, (Black Walnut), Z4, bare root										
	5510	Bare Root, 2 yr.				Ea.	3.15			3.15	3.47	
	6000	Peach, Z6, cont/BB										
	6010	5 Gal.				Ea.	16.10			16.10	17.70	
	6020	Standard, 2-3 yr.					20.50			20.50	22.50	
	6030	Semidwarf, 2-3 yr.				↓	27			27	29.50	
	6500	Pear, zone 5, B&B										
	6510	Standard, 2-3 yr.				Ea.	20.50			20.50	22.50	
	6520	Semidwarf, 2-3 yr.				"	28			28	30.50	
	7000	Plum, zone 5, container or B&B										
	7010	5 Gal.				Ea.	16.10			16.10	17.70	
	7020	Standard, 2-3 yr.					20.50			20.50	22.50	
	7030	Semidwarf, 2-3 yr.				↓	27			27	29.50	
	7500	Raspberry, (Everbearing), Zone 4										
	7510	3 per pot				Ea.	6.25			6.25	6.90	
	8000	Strawberry, (Hybrid), zone 5										
	8010	6 per box				Ea.	2.50			2.50	2.75	
	8020	25 per box				"	4.60			4.60	5.05	
630	0010	**SHRUBS**, warm temperate & subtropical, zones 7 - 10										630
	1000	Aspidistra elatior, (Cast Iron Plant), Z7, cont										
	1010	1 Gal.				Ea.	4			4	4.40	
	1020	2 Gal.				"	9			9	9.90	
	1100	Aucuba japonica "Variegata", (Gold Dust Plant), Z7, cont										
	1110	1 Gal.				Ea.	4			4	4.40	
	1120	3 Gal.					9			9	9.90	
	1130	5 Gal.				↓	13			13	14.30	
	1200	Azara microphylla, (Boxleaf Azara), Z8, cont										
	1210	5 Gal.				Ea.	14			14	15.40	
	1220	15 Gal.				"	52			52	57	
	1300	Boronia megastigma, (Sweet Boronia), Z9, cont										
	1310	1 Gal.				Ea.	5			5	5.50	
	1320	5 Gal.				"	13			13	14.30	
	1400	Brunfelsia pauciflora calycina, (Brazil Raisin-tree), Z10, cont										
	1410	1 Gal.				Ea.	4			4	4.40	
	1420	5 Gal.				"	12			12	13.20	
	1500	Calliandra haematocephala, (Pink Powerpuff), Z10, cont										
	1510	1 Gal.				Ea.	4			4	4.40	
	1520	5 Gal.				↓	13			13	14.30	

029 500 | Trees/Plants/Grnd Cover		CREW	DAILY OUTPUT	LABOR-HOURS	UNIT	1999 BARE COSTS MAT.	LABOR	EQUIP.	TOTAL	TOTAL INCL O&P
1530	15 Gal.				Ea.	45			45	49.50
1600	Callistemon citrinus, (Lemon Bottlebrush), Z9, cont									
1610	1 Gal.				Ea.	4			4	4.40
1620	3 Gal.					8			8	8.80
1630	10 Gal.					37			37	40.50
1640	20 Gal.				↓	56			56	61.50
1700	Camellia japonica, (Japanese Camellia), Z7, cont									
1710	1 Gal.				Ea.	5			5	5.50
1720	3 Gal.					9			9	9.90
1730	5 Gal.				↓	15			15	16.50
1800	Camellia sasanqua, (Sasanqua Camellia), Z7, cont									
1810	1 Gal.				Ea.	5			5	5.50
1820	3 Gal.					9			9	9.90
1830	5 Gal.				↓	15			15	16.50
1900	Carissa grandiflora, (Natal Plum), Z9, cont									
1910	1 Gal.				Ea.	4			4	4.40
1920	5 Gal.					13			13	14.30
1930	15 Gal.				↓	47			47	51.50
2000	Carpenteria californica, (Evergreen Mockorange), Z8, cont									
2010	1 Gal.				Ea.	4.15			4.15	4.57
2020	5 Gal.				"	14			14	15.40
2100	Cassia artemisioides, (Feathery Cassia), Z9, cont									
2110	1 Gal.				Ea.	3			3	3.30
2120	5 Gal.					19			19	21
2130	15 Gal.				↓	46			46	50.50
2200	Ceanothus concha, (Concha Wild Lilac), Z8, cont									
2210	1 Gal.				Ea.	5			5	5.50
2220	5 Gal.				"	11			11	12.10
2300	Choisa ternata, (Mexican Orange), Z7, cont									
2310	1 Gal.				Ea.	3.25			3.25	3.58
2320	5 Gal.				"	12			12	13.20
2400	Cistus Hybridus, (White Rockrose), Z7, cont									
2410	1 Gal.				Ea.	4			4	4.40
2420	5 Gal.				"	14			14	15.40
2500	Coprosma repens, (Mirror Plant), Z9, cont									
2510	1 Gal.				Ea.	3.25			3.25	3.58
2520	5 Gal.				"	9.15			9.15	10.05
2600	Correa pulchella "Carminebells", (Australian Fuchsia), Z9, cont									
2610	1 Gal.				Ea.	4			4	4.40
2620	5 Gal.				"	12			12	13.20
2700	Cyrtomium falcatum, (Japanese Holly Fern), Z10, cont									
2710	1 Gal.				Ea.	4.85			4.85	5.35
2720	5 Gal.				"	14.55			14.55	16
2800	Daphne odora "Marginata", (Variegated Winter Daphne), Z7, cont									
2810	1 Gal.				Ea.	5.25			5.25	5.80
2820	5 Gal.					17			17	18.70
2830	15 Gal.				↓	75			75	82.50
2900	Dizygotheca elegantissima, (False Aralia), Z10, cont									
2910	1 Gal.				Ea.	3.15			3.15	3.47
2920	5 Gal.					9.95			9.95	10.95
2930	15 Gal.				↓	47.50			47.50	52
3000	Dodonaea viscosa, (Akeake), Z10, cont									
3010	1 Gal.				Ea.	4			4	4.40
3020	5 Gal.					12			12	13.20
3030	15 Gal.				↓	54			54	59.50
3100	Eleagnus ebbingei "Gilt Edge", (Variegated Silverberry), Z7, cont									

630

029 500	Trees/Plants/Grnd Cover	CREW	DAILY OUTPUT	LABOR-HOURS	UNIT	1999 BARE COSTS MAT.	LABOR	EQUIP.	TOTAL	TOTAL INCL O&P
630 3110	1 Gal.				Ea.	4			4	4.40
3120	3 Gal.				↓	6			6	6.60
3130	15 Gal.				↓	37			37	40.50
3200	Eleagnus pungens, (Silverberry), Z7, cont									
3210	1 Gal.				Ea.	4			4	4.40
3220	3 Gal.				↓	8			8	8.80
3230	5 Gal.				↓	12			12	13.20
3300	Escallonia rubra, (Red Escallonia), Z8, cont									
3310	1 Gal.				Ea.	4			4	4.40
3320	5 Gal.				↓	12			12	13.20
3330	15 Gal.				↓	42			42	46
3400	Euonymus japonica "Aureo-Variegata", (Gold Euonymus), Z7, cont									
3410	1 Gal.				Ea.	4.05			4.05	4.46
3420	2 Gal.				↓	8.55			8.55	9.40
3430	5 Gal.				↓	11.70			11.70	12.85
3500	Fatsia japonica, (Japanese Fatsia), Z8, cont									
3510	1 Gal.				Ea.	4			4	4.40
3520	3 Gal.				↓	7			7	7.70
3530	5 Gal.				↓	12			12	13.20
3540	15 Gal.				↓	48			48	53
3600	Fremontodendron californicum, (Pacific "Fremontia"), Z9, cont									
3610	2 Gal.				Ea.	12.40			12.40	13.65
3620	5 Gal.				↓	18			18	19.80
3630	15 Gal.				↓	60			60	66
3700	Gardenia jasminoides, (Cape Jasmine), Z8, cont									
3710	1 Gal.				Ea.	3.50			3.50	3.85
3720	3 Gal.				↓	13			13	14.30
3730	5 Gal.				↓	17			17	18.70
3800	Garrya elliptica, (Silk Tassel), Z8, cont									
3810	1 Gal.				Ea.	4.85			4.85	5.35
3820	5 Gal.				"	15.05			15.05	16.55
3900	Grewia caffra, (Lavender Starflower), Z9, cont									
3910	5 Gal.				Ea.	14			14	15.40
3920	15 Gal.				"	46			46	50.50
4000	Hebe buxifolia, (Boxleaf Hebe), Z7, cont									
4010	1 Gal.				Ea.	4			4	4.40
4020	5 Gal.				"	12			12	13.20
4100	Heteromeles arbutifolia, (Toyon), Z9, cont									
4110	1 Gal.				Ea.	3.65			3.65	4.02
4120	5 Gal.				↓	11.80			11.80	13
4130	15 Gal.				↓	60			60	66
4200	Hibiscus rosa-sinensis, (Chinese Hibiscus), Z9, cont									
4210	1 Gal.				Ea.	3.50			3.50	3.85
4220	5 Gal.				↓	13			13	14.30
4230	15 Gal.				↓	51			51	56
4300	Ilex vomitoria, (Yaupon Holly), Z7, cont									
4310	1 Gal.				Ea.	4.55			4.55	5
4320	3 Gal.				↓	9.20			9.20	10.10
4330	10 Gal.				↓	32.50			32.50	35.50
4400	Illicium parvifolium, (Florida Anise Tree), Z8, cont									
4410	1 Gal.				Ea.	4			4	4.40
4420	3 Gal.				"	8			8	8.80
4500	Lagerstroemia indica, (Crapemyrtle), Z7, cont									
4510	1 Gal.				Ea.	4			4	4.40
4520	5 Gal.				↓	15			15	16.50
4530	3′ to 4′				↓	17			17	18.70

029 500 | Trees/Plants/Grnd Cover

				CREW	DAILY OUTPUT	LABOR-HOURS	UNIT	1999 BARE COSTS MAT.	LABOR	EQUIP.	TOTAL	TOTAL INCL O&P	
630	4540	4' to 5'					Ea.	25			25	27.50	630
	4550	5' to 6'					↓	38			38	42	
	4600	Ligustrum japonicum texanum, (Wax Leaf Privet), Z7, cont											
	4610	1 Gal.					Ea.	4			4	4.40	
	4620	2 Gal.						7			7	7.70	
	4630	5 Gal.					↓	12			12	13.20	
	4700	Ligustrum lucidum, (Glossy Privet), Z7, cont											
	4710	1 Gal.					Ea.	3			3	3.30	
	4720	5 Gal.						9.15			9.15	10.05	
	4730	15 Gal.					↓	45			45	49.50	
	4800	Ligustrum sinensis "Variegatum", (Variegated Privet), Z7, cont											
	4810	1 Gal.					Ea.	4			4	4.40	
	4820	2 Gal.						8			8	8.80	
	4830	3 Gal.					↓	10			10	11	
	4900	Mahonia lomariifolia, (Burmese Grape), Z8, cont											
	4910	1 Gal.					Ea.	4.30			4.30	4.73	
	4920	5 Gal.						11.25			11.25	12.40	
	4930	15 Gal.					↓	40			40	44	
	5000	Myoporum laetum, (Coast Sandalwood), Z9, cont											
	5010	1 Gal.					Ea.	4			4	4.40	
	5020	5 Gal.						10.60			10.60	11.65	
	5030	15 Gal.					↓	53			53	58.50	
	5100	Myrica californica, (Pacific Wax Myrtle), Z7, cont											
	5110	1 Gal.					Ea.	4			4	4.40	
	5120	5 Gal.					"	12			12	13.20	
	5200	Myrtus communis "Compacta", (Dwarf Roman Myrtle), Z8, cont											
	5210	1 Gal.					Ea.	4			4	4.40	
	5220	5 Gal.					"	12			12	13.20	
	5300	Nandina domestica, (Heavenly Bamboo), Z7, cont											
	5310	1 Gal.					Ea.	4			4	4.40	
	5320	2 Gal.						9			9	9.90	
	5330	5 Gal.					↓	13			13	14.30	
	5400	Nerium oleander, (Oleander), Z8, cont											
	5410	1 Gal.					Ea.	4			4	4.40	
	5420	3 Gal.						9			9	9.90	
	5430	5 Gal.					↓	12			12	13.20	
	5500	Osmanthus fragrans, (Fragrant Tea Olive), Z8, cont											
	5510	1 Gal.					Ea.	4			4	4.40	
	5520	3 Gal.					"	9			9	9.90	
	5600	Philodendron selloum, (Selloum), Z10, cont											
	5610	1 Gal.					Ea.	4			4	4.40	
	5620	5 Gal.						12			12	13.20	
	5630	15 Gal.					↓	46			46	50.50	
	5700	Photinia fraseri, (Christmas Berry), Z7, cont											
	5710	1 Gal.					Ea.	4			4	4.40	
	5720	2 Gal.						9			9	9.90	
	5730	5 Gal.						12			12	13.20	
	5740	15 Gal.					↓	74			74	81.50	
	5800	Phyllostachys aurea, (Golden Bamboo), Z7, cont											
	5810	1 Gal.					Ea.	4.30			4.30	4.73	
	5820	5 Gal.						12			12	13.20	
	5830	15 Gal.					↓	47			47	52	
	5900	Pittosporum tobira, (Japanese Pittosporum), Z8, cont											
	5910	1 Gal.					Ea.	4			4	4.40	
	5920	3 Gal.						7			7	7.70	
	5930	5 Gal.					↓	12			12	13.20	

029 500 | Trees/Plants/Grnd Cover		CREW	DAILY OUTPUT	LABOR-HOURS	UNIT	1999 BARE COSTS MAT.	LABOR	EQUIP.	TOTAL	TOTAL INCL O&P
630 6000	Plumbago auriculata, (Cape Plumbago), Z9, cont									
6010	1 Gal.				Ea.	4			4	4.40
6020	5 Gal.				↓	12			12	13.20
6030	15 Gal.				↓	46			46	50.50
6100	Prunus carolina, (Carolina Cherry Laurel), Z7, cont									
6110	1 Gal.				Ea.	4			4	4.40
6120	3 Gal.				↓	9			9	9.90
6130	15 Gal.				↓	49			49	54
6200	Punica granatum "nana", (Dwarf Pomegranate), Z7, cont									
6210	1 Gal.				Ea.	4			4	4.40
6220	3 Gal.				↓	9.45			9.45	10.40
6230	7 Gal.				↓	19.40			19.40	21.50
6300	Pyracantha koidzumii "Santa Cruz", (Pyracantha), Z8, cont									
6310	1 Gal.				Ea.	4.05			4.05	4.46
6320	5 Gal.				↓	12			12	13.20
6330	5 Gal. Espalier				↓	36			36	39.50
6400	Raphiolepis indica, (India Hawthorne), Z8, cont									
6410	1 Gal.				Ea.	4			4	4.40
6420	2 Gal.				↓	8			8	8.80
6430	5 Gal.				↓	12			12	13.20
6500	Rhamnus alaternus, (Italian Buckthorn), Z7, cont									
6510	1 Gal.				Ea.	6			6	6.60
6520	5 Gal.				↓	12			12	13.20
6530	15 Gal.				↓	50			50	55
6600	Sarcococca hookerana humilis, (Himalayan Sarcococca), Z8, cont									
6610	2 Gal.				Ea.	10			10	11
6620	3 Gal.				"	13			13	14.30
6700	Sarcococca ruscifolia, (Fragrant Sarcococca), Z7, cont									
6710	1 Gal.				Ea.	3.80			3.80	4.18
6720	5 Gal.				"	12			12	13.20
6800	Solanum rantonnetii, (Paraguay Nightshade), Z9, cont									
6810	1 Gal.				Ea.	4			4	4.40
6820	5 Gal.				"	12			12	13.20
6900	Ternstroemia gymnanthera, (Japanese Cleyera), Z7, cont									
6910	1 Gal.				Ea.	4			4	4.40
6920	3 Gal.				↓	9			9	9.90
6930	7 Gal.				↓	23			23	25.50
7000	Viburnum davidii, (David Viburnum), Z8, cont									
7010	1 Gal.				Ea.	4			4	4.40
7020	3 Gal.				"	8.10			8.10	8.90
7100	Viburnum japonicum, (Japanese Viburnum), Z8, cont									
7110	1 Gal.				Ea.	2.75			2.75	3.03
7120	3 Gal.				"	7.50			7.50	8.25
7200	Viburnum tinus, (Spring Bouquet Laurustinis), Z7, cont									
7210	1 Gal.				Ea.	4			4	4.40
7220	5 Gal.				"	12			12	13.20
7300	Vitex agnus castus, (Chastetree), Z7, cont									
7310	5 Gal.				Ea.	13			13	14.30
7400	Xylosma congestum, (Shiny Xylosma), Z8, cont									
7410	1 Gal.				Ea.	3.25			3.25	3.58
7420	5 Gal.				↓	10			10	11
7430	15 Gal.				↓	42			42	46
7500	Yucca aloifolia, (Spanish Bayonet), Z7, cont									
7510	1 Gal.				Ea.	2.75			2.75	3.03
7520	3 Gal.				"	7.25			7.25	8

029 500 | Trees/Plants/Grnd Cover		CREW	DAILY OUTPUT	LABOR-HOURS	UNIT	1999 BARE COSTS MAT.	LABOR	EQUIP.	TOTAL	TOTAL INCL O&P
0010	**GROUND COVERS**, vines and climbing plants									
1000	Achillea tomentosa, (Woolly Yarrow), Z4, cont									
1010	1 Qt.				Ea.	1.50			1.50	1.65
1020	2 Gal.				"	4.75			4.75	5.20
1100	Aegopodium podagaria, (Snow-on-the-Mountain Goutweed), Z3, cont									
1110	1 Qt.				Ea.	1.50			1.50	1.65
1120	1 Gal.					3.65			3.65	4.02
1130	Flat/24 plants				↓	15.65			15.65	17.20
1200	Ajuga reptans, (Carpet Bugle), Z4, cont									
1210	3" Pot				Ea.	.90			.90	.99
1220	1 Qt.					1.50			1.50	1.65
1230	1 Gal.					3.65			3.65	4.02
1240	Flat/24 plants				↓	15.65			15.65	17.20
1300	Akebia quinata, (Five-Leaf Akebia), Z4, cont									
1310	1 Gal.				Ea.	4.75			4.75	5.20
1320	3 Gal.				"	10			10	11
1400	Alchemilla vulgaris, (Common Lady's-Mantle), Z3, cont									
1410	1 Qt.				Ea.	1.50			1.50	1.65
1420	1 Gal.				"	3.65			3.65	4.02
1500	Ampelopsis brevipedunculata elegans, (Ampelopsis), Z4, cont									
1510	1 Gal.				Ea.	4.75			4.75	5.20
1520	3 Gal.				"	9.30			9.30	10.25
1600	Arctostaphylos uva-ursi, (Bearberry), Z2, cont									
1610	3" Pot				Ea.	1.30			1.30	1.43
1620	1 Gal.					4.75			4.75	5.20
1630	5 Gal.				↓	12.65			12.65	13.90
1700	Aristolochia durior, (Dutchman's Pipe), Z5, cont									
1710	2 Gal.				Ea.	12.65			12.65	13.90
1720	5 Gal.				"	23			23	25.50
1800	Armeria maritima, (Thrift, Sea Pink), Z6, cont									
1810	1 Qt.				Ea.	1.50			1.50	1.65
1820	1 Gal.				"	3.65			3.65	4.02
1900	Artemisia schmitdiana, (Silver Mound/Satiny Wormwood), Z3, cont									
1910	1 Qt.				Ea.	1.50			1.50	1.65
1920	1 Gal.				"	3.65			3.65	4.02
2000	Artemisia stellerana, (Dusty Miller), Z2, cont									
2010	1 Qt.				Ea.	1.50			1.50	1.65
2020	1 Gal.				"	3.65			3.65	4.02
2100	Asarum europaeum, (European Wild Ginger), Z4, cont									
2110	4" Pot				Ea.	3.15			3.15	3.47
2120	1 Gal.				"	7.90			7.90	8.70
2200	Baccharis pilularis, (Dwarf Coyote Bush), Z7, cont									
2210	1 Qt.				Ea.	1.50			1.50	1.65
2220	1 Gal.				"	2.65			2.65	2.92
2300	Bergenia cordifolia, (Heartleaf Bergenia), Z3, cont									
2310	1 Qt.				Ea.	2.65			2.65	2.92
2320	1 Gal.				"	4.75			4.75	5.20
2400	Bougainvillea glabra, , Z10, cont									
2410	1 Gal.				Ea.	5			5	5.50
2420	5 Gal.				"	11			11	12.10
2500	Calluna vulgaris cultivars, (Scotch Heather), Z4, cont									
2510	3" Pot				Ea.	1.30			1.30	1.43
2520	1 Qt.					2.75			2.75	3.03
2530	1 Gal.					4.95			4.95	5.45
2540	2 Gal.				↓	10			10	11
2600	Campsis radicans, (Trumpet Creeper), Z5, cont									

	029 500 | Trees/Plants/Grnd Cover		CREW	DAILY OUTPUT	LABOR-HOURS	UNIT	1999 BARE COSTS MAT.	LABOR	EQUIP.	TOTAL	TOTAL INCL O&P
638	2610	1 Gal.				Ea.	5.55			5.55	6.10
	2620	2 Gal.				↓	8.40			8.40	9.25
	2630	5 Gal.				↓	20			20	22
	2700	Celastrus scandens, (American Bittersweet), Z2, cont									
	2710	1 Gal.				Ea.	4.95			4.95	5.45
	2720	2 Gal.				″	8.40			8.40	9.25
	2800	Cerastium tomentosum, (Snow-in-Summer), Z2, cont									
	2810	1 Qt.				Ea.	1.60			1.60	1.76
	2820	1 Gal.				″	3.65			3.65	4.02
	2900	Ceratostigma, (Dwarf Plumbago), Z6, cont									
	2910	1 Qt.				Ea.	1.50			1.50	1.65
	2920	1 Gal.				″	3.65			3.65	4.02
	3000	Cissus antartica, (Kangeroo Vine), Z8, cont									
	3010	1 Gal.				Ea.	2.95			2.95	3.25
	3020	5 Gal.				″	9.25			9.25	10.20
	3100	Cistus crispus, (Wrinkleleaf Rock Rose), Z7, cont									
	3110	1 Gal.				Ea.	2.95			2.95	3.25
	3120	5 Gal.				″	10.05			10.05	11.05
	3200	Clematis, (Hybrids & Clones), Z6, cont									
	3210	1 Gal.				Ea.	7			7	7.70
	3220	2 Gal.				↓	12.70			12.70	13.95
	3230	5 Gal.				↓	21			21	23
	3300	Cocculus laurifolius, (Laurel-leaf Snailseed), Z8, cont									
	3310	1 Gal.				Ea.	2.95			2.95	3.25
	3320	5 Gal.				″	10.55			10.55	11.60
	3400	Conuallaria majalis, (Lily-of-the-Valley), Z3, cont									
	3410	1 Qt.				Ea.	1.50			1.50	1.65
	3420	1 Gal.				″	3.65			3.65	4.02
	3500	Coprosma prostrata, (Prostrate Coprosma), Z7, cont									
	3510	1 Gal.				Ea.	2.55			2.55	2.80
	3520	5 Gal.				″	9.25			9.25	10.20
	3600	Coronilla varia, (Crown Vetch), Z3, cont									
	3610	2-1/4″ Pot				Ea.	.60			.60	.66
	3620	Flat/50 plants				″	20.50			20.50	22.50
	3700	Cotoneaster dammeri, (Bearberry Cotoneaster), Z5, cont									
	3710	1 Gal.				Ea.	4.95			4.95	5.45
	3720	2 Gal.				↓	9.45			9.45	10.40
	3730	5 Gal.				↓	16.65			16.65	18.30
	3800	Cytisus decumbens, (Prostrate Broom), Z5, cont									
	3810	1 Gal.				Ea.	4.95			4.95	5.45
	3820	2 Gal.				″	11.05			11.05	12.15
	3900	Dianthus plumarius, (Grass Pinks), Z4, cont									
	3910	1 Qt.				Ea.	3			3	3.30
	3920	1 Gal.				″	5			5	5.50
	4000	Dicentra formosa, (Western Bleeding Heart), Z3, cont									
	4010	2 Qt.				Ea.	2.65			2.65	2.92
	4020	1 Gal.				″	3.65			3.65	4.02
	4100	Distictis buccinatoria, (Blood Red Trumpet Vine), Z9, cont									
	4110	1 Gal.				Ea.	3.15			3.15	3.47
	4120	5 Gal.				″	11.05			11.05	12.15
	4200	Epimedium grandiflorum, (Barrenwort, Bishopshat), Z3, cont									
	4210	1 Qt.				Ea.	3.15			3.15	3.47
	4220	1 Gal.				″	5.30			5.30	5.85
	4300	Euonymus fortunei colorata, (Purple-Leaf Wintercreeper), Z5, cont									
	4310	1 Qt.				Ea.	2.55			2.55	2.80
	4320	1 Gal.				↓	3.95			3.95	4.35

029 500 | Trees/Plants/Grnd Cover

		CREW	DAILY OUTPUT	LABOR-HOURS	UNIT	1999 BARE COSTS MAT.	LABOR	EQUIP.	TOTAL	TOTAL INCL O&P
4330	Flat/24 plants				Ea.	15.75			15.75	17.35
4400	Euonymus fortunei kewensis, (Euonymus), Z5, cont									
4410	1 Gal.				Ea.	3.95			3.95	4.35
4420	2 Gal.				"	9			9	9.90
4500	Fragaria vesca, (Alpine Strawberry), Z6, cont									
4510	4" Pot				Ea.	1.30			1.30	1.43
4520	1 Qt.				↓	1.85			1.85	2.04
4530	1 Gal.				↓	3.65			3.65	4.02
4600	Galium odoratum, (Sweet Woodruff), Z4, cont									
4610	4" Pot				Ea.	1.30			1.30	1.43
4620	1 Gal.				"	3.65			3.65	4.02
4700	Gardenia jasminoides "Radicans", (Creeping Gardenia), Z8, cont									
4710	1 Gal.				Ea.	3.15			3.15	3.47
4720	5 Gal.				"	11			11	12.10
4800	Gelsemium sempervirens, (Carolina Jessamine), Z7, cont									
4810	1 Gal.				Ea.	5			5	5.50
4820	5 Gal.				"	15			15	16.50
4900	Genista pilosa, (Silky-leaf Woadwaxen), Z5, cont									
4910	1 Gal.				Ea.	4.95			4.95	5.45
4920	2 Gal.				"	11.05			11.05	12.15
5000	Hardenbergia violacea, (Happy Wanderer), Z9, cont									
5010	1 Gal.				Ea.	5			5	5.50
5020	5 Gal.				"	13			13	14.30
5100	Hedera helix varieties, (English Ivy), Z5, cont									
5110	3" Pot				Ea.	.90			.90	.99
5120	1 Gal.				↓	5			5	5.50
5130	Flat/50 plants				↓	21			21	23
5200	Helianthemum nummularium, (Sun-Rose), Z5, cont									
5210	1 Qt.				Ea.	1.60			1.60	1.76
5220	1 Gal.				"	3.65			3.65	4.02
5300	Hemerocallis hybrids, (Assorted Daylillies), Z4, cont									
5310	2 Qt.				Ea.	2.95			2.95	3.25
5320	1 Gal.				↓	4.15			4.15	4.57
5330	2 Gal.				↓	6.85			6.85	7.55
5400	Hosta assorted, (Plantain Lily), Z4, cont									
5410	2 Qt.				Ea.	2.95			2.95	3.25
5420	1 Gal.				↓	4.15			4.15	4.57
5430	2 Gal.				↓	6.85			6.85	7.55
5500	Hydrangea petiolaris, (Climbing Hydrangea), Z4, cont									
5510	1 Gal.				Ea.	8			8	8.80
5520	2 Gal.				↓	12.30			12.30	13.55
5530	7 Gal.				↓	20			20	22
5600	Hypericum calycinum, (St. John's Wort), Z6, cont									
5610	1 Qt.				Ea.	1.50			1.50	1.65
5620	1 Gal.				↓	3.65			3.65	4.02
5630	5 Gal.				↓	9.40			9.40	10.35
5700	Iberis sempervirens, (Candytuft), Z4, cont									
5710	1 Qt.				Ea.	1.50			1.50	1.65
5720	1 Gal.				"	3.65			3.65	4.02
5800	Jasminum nudiflorum, (Winter Jasmine), Z5, cont									
5810	1 Gal.				Ea.	5			5	5.50
5820	2 Gal.				↓	8			8	8.80
5830	5 Gal.				↓	15			15	16.50
5900	Juniperus horizontalis varieties, (Prostrate Juniper), Z2, cont									
5910	1 Gal.				Ea.	4.15			4.15	4.57
5920	3 Gal.				↓	8.90			8.90	9.80

638

029 500 | Trees/Plants/Grnd Cover		CREW	DAILY OUTPUT	LABOR-HOURS	UNIT	1999 BARE COSTS MAT.	LABOR	EQUIP.	TOTAL	TOTAL INCL O&P
5930	5 Gal.				Ea.	14			14	15.40
6000	Lamium maculatum, (Dead Nettle), Z6, cont									
6010	1 Qt.				Ea.	1.50			1.50	1.65
6020	1 Gal.					3.65			3.65	4.02
6030	Flat/24 plants				↓	15.65			15.65	17.20
6100	Lantana camara, (Common Lantana), Z8, cont									
6110	4" Pot				Ea.	.90			.90	.99
6120	1 Gal.				"	2.65			2.65	2.92
6200	Lantana montevidensis, (Trailing Lantana), Z10, cont									
6210	1 Gal.				Ea.	2.65			2.65	2.92
6220	5 Gal.				"	8.40			8.40	9.25
6300	Lavandula angustifolia, (English Lavender), Z5, cont									
6310	4" Pot				Ea.	1.30			1.30	1.43
6320	1 Qt.					2.55			2.55	2.80
6330	1 Gal.				↓	4.15			4.15	4.57
6400	Lirope spicata, (Creeping Lilytuft), Z6, cont									
6410	2 Qt.				Ea.	2.65			2.65	2.92
6420	1 Gal.					4.75			4.75	5.20
6430	5 Gal.				↓	10			10	11
6500	Lonicera heckrotti, (Gold Flame Honeysuckle), Z5, cont									
6510	1 Gal.				Ea.	5			5	5.50
6520	2 Gal.				"	7			7	7.70
6600	Lonicera japonica "Halliana", (Halls Honeysuckle), Z4, cont									
6610	3" Pot				Ea.	.90			.90	.99
6620	1 Gal.					4.15			4.15	4.57
6630	2 Gal.				↓	9			9	9.90
6700	Lysimachia nummularia aurea, (Creeping Jenny), Z3, cont									
6710	2 Qt.				Ea.	1.50			1.50	1.65
6720	1 Gal.				"	2.95			2.95	3.25
6800	Macfadyena unguis-cali, (Cat Claw Vine), Z8, cont									
6810	1 Gal.				Ea.	5			5	5.50
6820	5 Gal.				"	15			15	16.50
6900	Myoporum parvifolium, (Prostrate Myoporum), Z9, cont									
6910	1 Gal.				Ea.	2.65			2.65	2.92
6920	5 Gal.				"	8.90			8.90	9.80
7000	Nepeta, (Persian Catmint), Z5, cont									
7010	1 Qt.				Ea.	1.50			1.50	1.65
7020	1 Gal.				"	3.15			3.15	3.47
7100	Ophiupogon japonicus, (Mondo-Grass), Z7, cont									
7110	4" Pot				Ea.	.90			.90	.99
7120	1 Gal.					3.65			3.65	4.02
7130	5 Gal.				↓	8.40			8.40	9.25
7200	Pachysandra terminalis, (Japanese Spurge), Z5, cont									
7210	2-1/2" Pot				Ea.	.50			.50	.55
7220	1 Gal.					3.65			3.65	4.02
7230	Tray/50 plants					10.55			10.55	11.60
7240	Tray/100 plants				↓	19			19	21
7300	Parthenocissus quinquefolia, (Virginia Creeper), Z3, cont									
7310	3" Pot				Ea.	.90			.90	.99
7320	1 Gal.				"	4.75			4.75	5.20
7400	Parthenocissus tricuspidata veitchi, (Boston Ivy), Z4, cont									
7410	4" Pot				Ea.	3			3	3.30
7420	1 Gal.					5			5	5.50
7430	5 Gal.				↓	15			15	16.50
7500	Phlox subulata, (Moss Phlox, Creeping Phlox), Z4, cont									
7510	1 Qt.				Ea.	1.50			1.50	1.65

029 500 | Trees/Plants/Grnd Cover			CREW	DAILY OUTPUT	LABOR-HOURS	UNIT	1999 BARE COSTS MAT.	LABOR	EQUIP.	TOTAL	TOTAL INCL O&P	
638	7520	1 Gal.				Ea.	3.65			3.65	4.02	638
	7600	Polygonum auberti, (Silver Fleece Vine), Z4, cont										
	7610	1 Qt.				Ea.	1.50			1.50	1.65	
	7620	1 Gal.				"	4.75			4.75	5.20	
	7700	Polygonum cuspidatum compactum, (Fleece Flower), Z4, cont										
	7710	1 Qt.				Ea.	1.50			1.50	1.65	
	7720	1 Gal.				"	4.75			4.75	5.20	
	7800	Potentilla verna, (Spring Cinquefoil), Z6, cont										
	7810	1 Qt.				Ea.	1.50			1.50	1.65	
	7820	1 Gal.				"	3.95			3.95	4.35	
	7900	Rosmarinus officinalis "Prostratus", (Creeping Rosemary), Z8, cont										
	7910	3" Pot				Ea.	.90			.90	.99	
	7920	1 Gal.				"	3.65			3.65	4.02	
	8000	Sagina subulata, (Irish Moss, Scotch Moss), Z6, cont										
	8010	3" Pot				Ea.	.90			.90	.99	
	8020	4" Pot				↓	1.30			1.30	1.43	
	8030	1 Qt.				↓	2.35			2.35	2.58	
	8100	Sedum, (Stonecrop), Z4, cont										
	8110	1 Qt.				Ea.	1.50			1.50	1.65	
	8120	1 Gal.				"	3.65			3.65	4.02	
	8200	Solanum jasminoides, (Jasmine Nightshade), Z9, cont										
	8210	1 Gal.				Ea.	3.65			3.65	4.02	
	8220	5 Gal.				"	12.70			12.70	13.95	
	8300	Stachys byzantina, (Lambs'-ears), Z4, cont										
	8310	1 Qt.				Ea.	1.50			1.50	1.65	
	8320	1 Gal.				"	3.65			3.65	4.02	
	8400	Thymus pseudolanuginosus, (Wooly Thyme), Z5, cont										
	8410	3" Pot				Ea.	.90			.90	.99	
	8420	1 Qt.				↓	1.50			1.50	1.65	
	8430	1 Gal.				↓	3.65			3.65	4.02	
	8500	Vaccinium crassifolium procumbent, (Creeping Blueberry), Z7, cont										
	8510	1 Gal.				Ea.	2.95			2.95	3.25	
	8520	3 Gal.				"	7.90			7.90	8.70	
	8600	Verbena peruviana "St. Paul", , Z8, cont										
	8610	2 Qt.				Ea.	1.95			1.95	2.15	
	8620	1 Gal.				"	2.95			2.95	3.25	
	8700	Veronica repens, (Creeping Speedwell), Z5, cont										
	8710	1 Qt.				Ea.	1.50			1.50	1.65	
	8720	1 Gal.				"	3.65			3.65	4.02	
	8800	Vinca major variegata, (Variegated Greater Periwinkle), Z7, cont										
	8810	2-1/4" Pot				Ea.	.50			.50	.55	
	8820	4" Pot				↓	1.50			1.50	1.65	
	8830	Bare Root				↓	.50			.50	.55	
	8900	Vinca minor, (Periwinkle), Z4, cont										
	8910	2-1/4" Pot				Ea.	.50			.50	.55	
	8920	4" Pot				↓	1.50			1.50	1.65	
	8930	Bare Root				↓	.50			.50	.55	
	9000	Waldsteinia sibirica, (Barren Strawberry), Z4, cont										
	9010	1 Qt.				Ea.	1.50			1.50	1.65	
	9020	1 Gal.				"	3.65			3.65	4.02	
	9100	Wisteria sinensis, (Chinese Wisteria), Z5, cont										
	9110	1 Gal.				Ea.	5			5	5.50	
	9120	3 Gal.				"	12			12	13.20	
644	0010	**ORNAMENTAL GRASSES**										644
	2000	Acorus gramineus, (Japanese Sweet Flag), Z7, cont										

029 500 | Trees/Plants/Grnd Cover		CREW	DAILY OUTPUT	LABOR-HOURS	UNIT	1999 BARE COSTS MAT.	LABOR	EQUIP.	TOTAL	TOTAL INCL O&P
644 2010	1 Gal.				Ea.	3.80			3.80	4.18
2100	Arundo donax, (Giant Reed), Z7, cont									
2110	1 Qt.				Ea.	4.85			4.85	5.35
2120	1 Gal.				″	6.95			6.95	7.65
2200	Calamagrostis acutiflora stricta, (Feather Reed Grass), Z5, cont									
2210	1 Qt.				Ea.	2.75			2.75	3.03
2220	1 Gal.				↓	3.80			3.80	4.18
2230	2 Gal.				↓	5.90			5.90	6.50
2300	Carex elata "Bowles Golden", (Variegated Sedge), Z5, cont									
2310	1 Qt.				Ea.	5.35			5.35	5.90
2320	1 Gal.				″	6.95			6.95	7.65
2400	Carex morrowii "aureo variegata", (Japanese Sedge), Z5, cont									
2410	1 Qt.				Ea.	2.75			2.75	3.03
2420	1 Gal.				↓	3.80			3.80	4.18
2430	2 Gal.				↓	7			7	7.70
2500	Cortaderia selloana, (Pampas Grass), Z8, cont									
2510	1 Gal.				Ea.	4.85			4.85	5.35
2520	2 Gal.				↓	5.90			5.90	6.50
2530	3 Gal.				↓	6.95			6.95	7.65
2540	10 Gal.				↓	27			27	29.50
2600	Deschampsia caespitosa, (Tufted Hair Grass), Z4, cont									
2610	1 Qt.				Ea.	2.40			2.40	2.64
2620	1 Gal.				″	3.80			3.80	4.18
2700	Elymus glaucus, (Blue Wild Rye), Z4, cont									
2710	1 Qt.				Ea.	2.75			2.75	3.03
2720	1 Gal.				″	4.05			4.05	4.46
2800	Erianthus ravennae, (Ravenna Grass), Z6, cont									
2810	1 Gal.				Ea.	4.85			4.85	5.35
2820	3 Gal.				↓	8			8	8.80
2830	7 Gal.				↓	14			14	15.40
2900	Festuca cineria "Sea Urchin", (Sea Urchin Blue Fescue), Z6, cont									
2910	1 Qt.				Ea.	2.10			2.10	2.31
2920	1 Gal.				↓	3.80			3.80	4.18
2930	2 Gal.				↓	5.90			5.90	6.50
3000	Glyceria maxima variegata, (Manna Grass), Z5, cont									
3010	1 Qt.				Ea.	2.35			2.35	2.58
3020	1 Gal.				″	3.80			3.80	4.18
3100	Hakonechloa macra aureola, (Variegated Hakonechloa), Z7, cont									
3110	1 Qt.				Ea.	5.35			5.35	5.90
3120	1 Gal.				″	8.40			8.40	9.25
3200	Helictotrichon sempervirens, (Blue Oat Grass), Z4, cont									
3210	1 Qt.				Ea.	2.85			2.85	3.14
3220	1 Gal.				↓	4.20			4.20	4.62
3230	2 Gal.				↓	6.15			6.15	6.75
3300	Imperata cylindrica rubra, (Japanese Blood Grass), Z7, cont									
3310	1 Qt.				Ea.	3			3	3.30
3320	1 Gal.				↓	5			5	5.50
3330	2 Gal.				↓	8			8	8.80
3400	Juncus effusus spiralis, (Corkscrew Rush), Z4, cont									
3410	1 Qt.				Ea.	2.45			2.45	2.70
3420	2 Gal.				″	6.15			6.15	6.75
3500	Koeleria glauca, (Blue Hair Grass), Z6, cont									
3510	1 Qt.				Ea.	2.20			2.20	2.42
3520	1 Gal.				↓	3.95			3.95	4.35
3530	3 Gal.				↓	7.25			7.25	8
3600	Miscanthus sinensis gracillimus, (Maiden Grass), Z6, cont									

029 500 | Trees/Plants/Grnd Cover		CREW	DAILY OUTPUT	LABOR-HOURS	UNIT	1999 BARE COSTS MAT.	LABOR	EQUIP.	TOTAL	TOTAL INCL O&P	
3610	1 Gal.				Ea.	5			5	5.50	644
3620	3 Gal.					11			11	12.10	
3630	5 Gal.					13			13	14.30	
3700	Miscanthus sinensis zebrinus, (Zebra Grass), Z6, cont										
3710	1 Gal.				Ea.	5			5	5.50	
3720	3 Gal.					11			11	12.10	
3730	5 Gal.					13			13	14.30	
3800	Molinia caerulea arundinacea, (Tall Purple Moor Grass), Z5, cont										
3810	1 Gal.				Ea.	4.30			4.30	4.73	
3820	3 Gal.				″	9.20			9.20	10.10	
3900	Molinia caerulea variegata, (Var. Purple Moor Grass), Z5, cont										
3910	1 Qt.				Ea.	2.45			2.45	2.70	
3920	1 Gal.				″	5			5	5.50	
4000	Pennisetum alopecuroides, (Fountain Grass), Z6, cont										
4010	1 Qt.				Ea.	2.40			2.40	2.64	
4020	1 Gal.					3.80			3.80	4.18	
4030	3 Gal.					7			7	7.70	
4100	Phalaris arundinacea "Picta", (Ribbon Grass), Z4, cont										
4110	1 Qt.				Ea.	2.40			2.40	2.64	
4120	1 Gal.					3.80			3.80	4.18	
4130	2 Gal.					5.40			5.40	5.95	
4200	Sesleria autumnalis, (Autumn Moor Grass), Z5, cont										
4210	1 Qt.				Ea.	2.40			2.40	2.64	
4220	1 Gal.				″	3.80			3.80	4.18	
4300	Spartina pectinata "Aureomarginata", (Cord Grass), Z5, cont										
4310	1 Gal.				Ea.	4.30			4.30	4.73	
4320	2 Gal.				″	7			7	7.70	
4400	Stipa gigantea, (Giant Feather Grass), Z7, cont										
4410	1 Qt.				Ea.	5.40			5.40	5.95	
4420	1 Gal.				″	8.10			8.10	8.90	
0010	**PERENNIALS**										648
0100	Aclillea filipendulina, (Fernleaf Yarrow), Z2, cont										
0110	4" Pot				Ea.	.95			.95	1.05	
0120	1 Qt.					1.60			1.60	1.76	
0130	1 Gal.					3.95			3.95	4.35	
0200	Agapanthus africanus, (Lily-of-the-Nile), Z9, cont										
0210	1 Gal.				Ea.	3.60			3.60	3.96	
0220	5 Gal.				″	14			14	15.40	
0300	Amsonia ciliata, (Blue Star), Z7, cont										
0310	1 Qt.				Ea.	2.85			2.85	3.14	
0320	1 Gal.				″	4.50			4.50	4.95	
0400	Amsonia tabernaemontana, (Willow Amsonia), Z3, cont										
0410	1 Qt.				Ea.	2.85			2.85	3.14	
0420	1 Gal.				″	4.50			4.50	4.95	
0500	Anchusa azurea, (Italian Bugloss), Z3, cont										
0510	4" Pot				Ea.	.95			.95	1.05	
0520	1 Qt.					1.60			1.60	1.76	
0530	1 Gal.					3.95			3.95	4.35	
0600	Anemone x hybrida, (Japanese Anemone), Z5, cont										
0610	4" Pot				Ea.	.95			.95	1.05	
0620	1 Qt.					2.85			2.85	3.14	
0630	2 Gal.					5.10			5.10	5.60	
0700	Anthemis tinctoria, (Golden Marguerite), Z3, cont										
0710	4" Pot				Ea.	.95			.95	1.05	

	029 500 | Trees/Plants/Grnd Cover	CREW	DAILY OUTPUT	LABOR-HOURS	UNIT	1999 BARE COSTS MAT.	LABOR	EQUIP.	TOTAL	TOTAL INCL O&P
648	0720 1 Qt.				Ea.	1.60			1.60	1.76
	0730 1 Gal.				↓	3.95			3.95	4.35
	0800 Aonitum napellus, (Aconite Monkshood), Z2, cont									
	0810 4" Pot				Ea.	.95			.95	1.05
	0820 1 Qt.					1.60			1.60	1.76
	0830 1 Gal.				↓	3.95			3.95	4.35
	0900 Aquilegia species and hybrids, (Columbine), Z2, cont									
	0910 4" Pot				Ea.	.95			.95	1.05
	0920 1 Qt.					1.60			1.60	1.76
	0930 1 Gal.				↓	3.95			3.95	4.35
	1000 Aruncus dioicus, (Goatsbeard), Z4, cont									
	1010 1 Qt.				Ea.	1.60			1.60	1.76
	1020 1 Gal.					3.60			3.60	3.96
	1030 2 Gal.				↓	5.10			5.10	5.60
	1100 Asclepias tuberosa, (Butterfly Flower), Z4, cont									
	1110 1 Qt.				Ea.	1.60			1.60	1.76
	1120 1 Gal.				"	3.95			3.95	4.35
	1200 Aster species and hybrids, (Aster, Michaelemas Daisy), Z4, cont									
	1210 4" Pot				Ea.	.95			.95	1.05
	1220 1 Qt.					1.60			1.60	1.76
	1230 1 Gal.				↓	3.95			3.95	4.35
	1300 Astilbe species and hybrids, (False Spirea), Z4, cont									
	1310 1 Qt.				Ea.	4			4	4.40
	1320 1 Gal.					6			6	6.60
	1330 3 Gal.				↓	8			8	8.80
	1400 Aubrieta deltoidea, (Purple Rock-cress), Z4, cont									
	1410 4" Pot				Ea.	.95			.95	1.05
	1420 1 Qt.					1.60			1.60	1.76
	1430 1 Gal.				↓	3.95			3.95	4.35
	1500 Aurinia saxatilis, (Basket of Gold), Z3, cont									
	1510 4" Pot				Ea.	.95			.95	1.05
	1520 1 Qt.					1.60			1.60	1.76
	1530 1 Gal.				↓	3.95			3.95	4.35
	1600 Baptisia australis, (False Indigo), Z3, cont									
	1610 1 Qt.				Ea.	1.60			1.60	1.76
	1620 1 Gal.				"	3.95			3.95	4.35
	1700 Begonia grandis, (Evans Begonia), Z6, cont									
	1710 1 Qt.				Ea.	2.85			2.85	3.14
	1720 1 Gal.				"	5.10			5.10	5.60
	1800 Belamcanda chinensis, (Blackberry Lily), Z5, cont									
	1810 1 Qt.				Ea.	1.60			1.60	1.76
	1820 1 Gal.				"	3.95			3.95	4.35
	1900 Boltonia asteroides "Snowbank", (White Boltonia), Z3, cont									
	1910 1 Qt.				Ea.	2.85			2.85	3.14
	1920 1 Gal.				"	5.10			5.10	5.60
	2000 Brunnera macrophylla, (Siberian Bugloss), Z3, cont									
	2010 1 Qt.				Ea.	2.75			2.75	3.03
	2020 1 Gal.				"	4.90			4.90	5.40
	2100 Caltha palustris, (Marsh Marigold), Z3, cont									
	2110 1 Qt.				Ea.	1.50			1.50	1.65
	2120 1 Gal.				"	3.80			3.80	4.18
	2200 Campanula carpatica "Blue Chips", (Bellflower), Z3, cont									
	2210 1 Qt.				Ea.	1.50			1.50	1.65
	2220 1 Gal.				"	3.80			3.80	4.18
	2300 Ceratostigma plumbaginoides, (Blue Plumbago), Z6, cont									
	2310 1 Qt.				Ea.	1.50			1.50	1.65

029 500 | Trees/Plants/Grnd Cover		CREW	DAILY OUTPUT	LABOR-HOURS	UNIT	1999 BARE COSTS MAT.	LABOR	EQUIP.	TOTAL	TOTAL INCL O&P
648 2320	1 Gal.				Ea.	3.80			3.80	4.18
2400	Chrysanthemum hybrids, (Hardy Chrysanthemum), Z5, cont									
2410	1 Qt.				Ea.	1.30			1.30	1.43
2420	6" to 8" Pot				↓	2.20			2.20	2.42
2430	1 Gal.				↓	3.50			3.50	3.85
2500	Chrysogonum virginianum, (Golden Star), Z6, cont									
2510	1 Qt.				Ea.	1.50			1.50	1.65
2520	1 Gal.				"	3.80			3.80	4.18
2600	Cimicifuga racemosa, (Bugbane), Z5, cont									
2610	4" Pot				Ea.	2.40			2.40	2.64
2620	1 Qt.				↓	3.80			3.80	4.18
2630	1 Gal.				↓	6.55			6.55	7.20
2700	Coreopsis verticillata, (Threadleaf Coreopsis), Z5, cont									
2710	1 Qt.				Ea.	1.50			1.50	1.65
2720	1 Gal.				"	3.80			3.80	4.18
2800	Delphinium elatum and hybrids, (Delphinium), Z2, cont									
2810	1 Qt.				Ea.	1.50			1.50	1.65
2820	1 Gal.				↓	3.80			3.80	4.18
2830	2 Gal.				↓	4.90			4.90	5.40
2900	Dianthus species and hybrids, (Cottage Pink/Scotch Pink), Z3, cont									
2910	1 Qt.				Ea.	1.50			1.50	1.65
2920	1 Gal.				"	3.80			3.80	4.18
3000	Dicentra spectabilis, (Bleeding Heart), Z3, cont									
3010	2 Qt.				Ea.	2.75			2.75	3.03
3020	1 Gal.				↓	3.80			3.80	4.18
3030	3 Gal.				↓	6.80			6.80	7.50
3100	Dictamnus, (Gas Plant), Z3, cont									
3110	1 Gal.				Ea.	4.35			4.35	4.79
3120	2 Gal.				↓	7.50			7.50	8.25
3130	3 Gal.				↓	9.30			9.30	10.25
3200	Digitalis purpurea, (Foxglove), Z4, cont									
3210	1 Gal.				Ea.	3.50			3.50	3.85
3220	2 Gal.				↓	4.90			4.90	5.40
3230	3 Gal.				↓	7.10			7.10	7.80
3300	Doronicum cordatun, (Leopards Bane), Z4, cont									
3310	2 Qt.				Ea.	2.75			2.75	3.03
3320	1 Gal.				"	3.80			3.80	4.18
3400	Echinacea purpurea, (Purple Coneflower), Z3, cont									
3410	2 Qt.				Ea.	2.75			2.75	3.03
3420	1 Gal.				↓	3.80			3.80	4.18
3430	2 Gal.				↓	4.90			4.90	5.40
3500	Echinops exaltatus, (Globe Thistle), Z3, cont									
3510	1 Qt.				Ea.	1.50			1.50	1.65
3520	1 Gal.				"	3.80			3.80	4.18
3600	Erigeron speciosus, (Oregon Fleabane), Z3, cont									
3610	1 Qt.				Ea.	1.50			1.50	1.65
3620	1 Gal.				"	3.80			3.80	4.18
3700	Euphorbia epithymoides, (Cushion Spurge), Z4, cont									
3710	1 Qt.				Ea.	1.60			1.60	1.76
3720	1 Gal.				"	3.95			3.95	4.35
3800	Filipendula hexapetala, (Meadowsweet), Z5, cont									
3810	1 Qt.				Ea.	1.60			1.60	1.76
3820	1 Gal.				"	3.95			3.95	4.35
3900	Gaillardia aristata, (Blanket Flower), Z2, cont									
3910	1 Qt.				Ea.	1.60			1.60	1.76
3920	1 Gal.				↓	3.60			3.60	3.96

029 500 | Trees/Plants/Grnd Cover

648

No.	Description	CREW	DAILY OUTPUT	LABOR-HOURS	UNIT	1999 BARE COSTS MAT.	LABOR	EQUIP.	TOTAL	TOTAL INCL O&P
3930	2 Gal.				Ea.	5.10			5.10	5.60
4000	Geranium sanguineum, (Bloodred Geranium), Z3, cont									
4010	1 Qt.				Ea.	2.70			2.70	2.97
4020	1 Gal.				"	4.50			4.50	4.95
4100	Geum x borisii, (Boris Avens), Z3, cont									
4110	1 Qt.				Ea.	1.60			1.60	1.76
4120	1 Gal.				"	4			4	4.40
4200	Gypsophila paniculata, (Baby's Breath), Z3, cont									
4210	2 Qt.				Ea.	4			4	4.40
4220	1 Gal.				↓	3.95			3.95	4.35
4230	3 Gal.				↓	8			8	8.80
4300	Helenium autumnale, (Sneezeweed), Z3, cont									
4310	2 Qt.				Ea.	4			4	4.40
4320	1 Gal.				↓	3.95			3.95	4.35
4330	2 Gal.				↓	8			8	8.80
4400	Helianthus salicifolius, (Willow Leaf Sunflower), Z5, cont									
4410	1 Qt.				Ea.	4			4	4.40
4420	1 Gal.				↓	5.10			5.10	5.60
4430	2 Gal.				↓	8			8	8.80
4500	Heliopsis helianthoides, (False Sunflower), Z3, cont									
4510	2 Qt.				Ea.	4			4	4.40
4520	1 Gal.				↓	3.95			3.95	4.35
4530	2 Gal.				↓	5.10			5.10	5.60
4600	Helleborus niger, (Christmas Rose), Z3, cont									
4610	2 Qt.				Ea.	5.60			5.60	6.15
4620	1 Gal.				"	8.45			8.45	9.30
4700	Hemerocallis hybrids (Diploid), (Day Lilly), Z3, cont									
4710	1 Qt.				Ea.	2			2	2.20
4720	1 Gal.				↓	4			4	4.40
4730	3 Gal.				↓	7.75			7.75	8.55
4800	Hemerocallis hybrids (Tetraploid), , Z3, cont									
4810	1 Qt.				Ea.	2			2	2.20
4820	1 Gal.				↓	6			6	6.60
4830	3 Gal.				↓	8			8	8.80
4900	Hibiscus moscheutos, (Rose Mallow), Z5, cont									
4910	1 Qt.				Ea.	3			3	3.30
4920	2 Gal.				↓	6			6	6.60
4930	3 Gal.				↓	8			8	8.80
5000	Hosta species and hybrids, (Plantain Lily), Z3, cont									
5010	1 Gal.				Ea.	8			8	8.80
5020	2 Gal.				↓	10			10	11
5030	3 Gal.				↓	15			15	16.50
5100	Iris hybrids, (Pacific Coast Iris), Z6, cont									
5110	1 Gal.				Ea.	4			4	4.40
5120	3 Gal.				"	7			7	7.70
5200	Iris sibiricu and hybrids, (Siberian Iris), Z3, cont									
5210	1 Qt.				Ea.	4			4	4.40
5220	2 Gal.				↓	5.10			5.10	5.60
5230	3 Gal.				↓	8			8	8.80
5300	Kniphofia uvaria and hybrids, (Torch Lily), Z7, cont									
5310	1 Qt.				Ea.	3			3	3.30
5320	1 Gal.				"	4			4	4.40
5400	Liatris spicata, (Blazing Star, Gayfeather), Z3, cont									
5410	1 Qt.				Ea.	3			3	3.30
5420	1 Gal.				"	5			5	5.50
5500	Ligularia dentata, (Golden Groundsel), Z5, cont									

Important: See the Reference Section for critical supporting data - Reference Nos., Crews, & City Cost Indexe

029 500 | Trees/Plants/Grnd Cover

648		CREW	DAILY OUTPUT	LABOR-HOURS	UNIT	1999 BARE COSTS MAT.	LABOR	EQUIP.	TOTAL	TOTAL INCL O&P
5510	2 Qt.				Ea.	5			5	5.50
5520	2 Gal.				″	8			8	8.80
5600	Linum perenne, (Perennial Flax), Z4, cont									
5610	1 Qt.				Ea.	3			3	3.30
5620	1 Gal.				″	5			5	5.50
5700	Lobelia cardinalis, (Cardinal Flower), Z3, cont									
5710	1 Qt.				Ea.	1.55			1.55	1.70
5720	1 Gal.				″	3.80			3.80	4.18
5800	Lupinus "Rusell Hybrids", (Russell Lupines), Z3, cont									
5810	1 Gal.				Ea.	2.75			2.75	3.03
5820	2 Gal.				↓	4.95			4.95	5.45
5830	3 Gal.				↓	6.55			6.55	7.20
5900	Lychnis chalcedonica, (Maltese Cross), Z3, cont									
5910	1 Qt.				Ea.	1.55			1.55	1.70
5920	1 Gal.				″	3.80			3.80	4.18
6000	Lysimachia clethroides, (Gooseneck Loosestrife), Z3, cont									
6010	1 Qt.				Ea.	1.55			1.55	1.70
6020	1 Gal.				″	3.80			3.80	4.18
6100	Lythrum salicaria, (Purple Loosestrife), Z3, cont									
6110	1 Qt.				Ea.	1.55			1.55	1.70
6120	1 Gal.				″	3.80			3.80	4.18
6200	Macleaya cordata, (Plume Poppy), Z3, cont									
6210	1 Qt.				Ea.	1.55			1.55	1.70
6220	1 Gal.				″	3.80			3.80	4.18
6300	Mertensia virginica, (Virginia Bluebells), Z3, cont									
6310	1 Qt.				Ea.	1.55			1.55	1.70
6320	1 Gal.				″	3.80			3.80	4.18
6400	Paeonia hybrids, (Herbaceous Peony), Z5, cont									
6410	1 Gal.				Ea.	6			6	6.60
6420	2 Gal.				↓	10			10	11
6430	3 Gal.				↓	13			13	14.30
6500	Papaver orientalis, (Oriental Poppy), Z3, cont									
6510	1 Qt.				Ea.	2.20			2.20	2.42
6520	1 Gal.				″	4.35			4.35	4.79
6600	Penstemon azureus, (Azure Penstemon), Z8, cont									
6610	1 Qt.				Ea.	1.55			1.55	1.70
6620	1 Gal.				″	3.80			3.80	4.18
6700	Phlox paniculata, (Garden Phlox), Z4, cont									
6710	1 Qt.				Ea.	1.55			1.55	1.70
6720	1 Gal.				″	3.80			3.80	4.18
6800	Physostegia virginiana, (False Dragonhead), Z3, cont									
6810	1 Qt.				Ea.	1.55			1.55	1.70
6820	1 Gal.				″	3.80			3.80	4.18
6900	Platycodon grandiflorus, (Balloon Flower), Z3, cont									
6910	1 Qt.				Ea.	1.55			1.55	1.70
6920	1 Gal.				″	3.80			3.80	4.18
7000	Pulmonaria saccharata, (Bethlehem Sage), Z4, cont									
7010	4″ Pot				Ea.	.90			.90	.99
7020	1 Qt.				↓	1.55			1.55	1.70
7030	1 Gal.				↓	3.80			3.80	4.18
7100	Rudbeckia fulgida "Goldstrum", (Black-eyed Susan), Z4, cont									
7110	1 Qt.				Ea.	1.55			1.55	1.70
7120	1 Gal.				↓	3.80			3.80	4.18
7130	2 Gal.				↓	4.95			4.95	5.45
7200	Salvia x superba, (Perennial Salvia), Z5, cont									
7210	4″ Pot				Ea.	.90			.90	.99

2 SITE WORK

029 500 | Trees/Plants/Grnd Cover

			CREW	DAILY OUTPUT	LABOR-HOURS	UNIT	1999 BARE COSTS MAT.	LABOR	EQUIP.	TOTAL	TOTAL INCL O&P
648	7220	1 Qt.				Ea.	1.55			1.55	1.70
	7230	1 Gal.					3.80			3.80	4.18
	7300	Scabiosa caucasica, (Pincushion Flower), Z3, cont									
	7310	1 Qt.				Ea.	1.55			1.55	1.70
	7320	1 Gal.				"	3.80			3.80	4.18
	7400	Sedum spectabile, (Showy Sedum), Z3, cont									
	7410	4" Pot				Ea.	.90			.90	.99
	7420	1 Qt.					1.55			1.55	1.70
	7430	1 Gal.					3.80			3.80	4.18
	7500	Solidago hybrids, (Goldenrod), Z3, cont									
	7510	1 Qt.				Ea.	1.55			1.55	1.70
	7520	1 Gal.				"	3.80			3.80	4.18
	7700	Trollius europaeus, (Common Globeflower), Z4, cont									
	7710	1 Qt.				Ea.	1.55			1.55	1.70
	7720	1 Gal.				"	3.80			3.80	4.18
	7800	Veronica hybrids, (Speedwell), Z4, cont									
	7810	1 Qt.				Ea.	1.55			1.55	1.70
	7820	1 Gal.				"	3.80			3.80	4.18

029 700 | Site Maintenance

			CREW	DAILY OUTPUT	LABOR-HOURS	UNIT	MAT.	LABOR	EQUIP.	TOTAL	TOTAL INCL O&P
710	0010	**SITE MAINTENANCE**									
	0800	Flower bed maintenance									
	0810	Cultivate bed-no mulch	1 Clab	14	.571	M.S.F.		12.25		12.25	19.30
	0830	Fall clean-up of flower bed, including pick-up mulch for re-use		1	8			172		172	270
	0840	Fertilize flower bed, dry granular 3 lb./c.s.f.		85	.094		9	2.02		11.02	13.10
	1130	Police, hand pickup		30	.267			5.70		5.70	9
	1140	Vacuum (outside)	A-1	48	.167			3.58	1.43	5.01	7.20
	1200	Spring prepare	1 Clab	2	4			86		86	135
	1300	Weed mulched bed		20	.400			8.60		8.60	13.50
	1310	Unmulched bed		8	1			21.50		21.50	34
	3000	Lawn maintenance									
	3010	Aerate lawn, 18" cultivating width, walk behind	A-1	95	.084	M.S.F.		1.81	.72	2.53	3.63
	3040	48" cultivating width	B-66	750	.011			.29	.27	.56	.73
	3060	72" cultivating width	"	1,100	.007			.20	.18	.38	.50
	3100	Edge lawn, by hand at walks	1 Clab	16	.500	C.L.F.		10.75		10.75	16.90
	3150	At planting beds	"	7	1.143			24.50		24.50	38.50
	3200	Using gas powered edger at walks	A-1	88	.091			1.95	.78	2.73	3.93
	3250	At planting beds		24	.333			7.15	2.86	10.01	14.40
	3260	Vacuum, 30" gas, outdoors with hose		96	.083	M.L.F.		1.79	.72	2.51	3.60
	3400	Weed lawn, by hand	1 Clab	3	2.667	M.S.F.		57		57	90
	4510	Power rake	A-1	45	.178	"		3.81	1.53	5.34	7.70
	4700	Seeding lawn, see Division 029-308									
	4750	Sodding, see Division 029-316									
	5900	Road & walk maintenance									
	5910	Asphaltic concrete paving, cold patch, 2" thick	B-37	350	.137	S.Y.	3.80	3.12	.44	7.36	9.55
	5913	3" thick	"	260	.185	"	5.50	4.20	.59	10.29	13.25
	5915	De-icing roads and walks									
	5920	Calcium Chloride in truckload lots see Division 025-112									
	6000	Ice melting comp., 90% Calc. Chlor., effec. to -30°F									
	6010	50-80 lb. poly bags, med. applic. 19 lbs./M.S.F., by hand	1 Clab	60	.133	M.S.F.	14	2.86		16.86	19.90
	6050	With hand operated rotary spreader		110	.073		14	1.56		15.56	17.85
	6100	Rock salt, med. applic. on road & walkway, by hand		60	.133		3.50	2.86		6.36	8.35
	6110	With hand operated rotary spreader		110	.073		3.50	1.56		5.06	6.30
	6130	Hosing, sidewalks & other paved areas		30	.267			5.70		5.70	9
	6260	Sidewalk, brick pavers, steam cleaning	A-1	950	.008	S.F.	.05	.18	.07	.30	.42
	6400	Sweep walk by hand	1 Clab	15	.533	M.S.F.		11.45		11.45	18

Important: See the Reference Section for critical supporting data - Reference Nos., Crews, & City Cost Indexe

029 700 | Site Maintenance

	Line	Description	CREW	DAILY OUTPUT	LABOR-HOURS	UNIT	1999 BARE COSTS MAT.	LABOR	EQUIP.	TOTAL	TOTAL INCL O&P
710	6410	Power vacuum	A-1	100	.080	M.S.F.		1.72	.69	2.41	3.46
	6420	Drives & parking areas with power vacuum	1 Clab	120	.067			1.43		1.43	2.25
	6600	Shrub maintenance									
	6640	Shrub bed fertilize dry granular 3 lbs./M.S.F.	1 Clab	85	.094	M.S.F.	.85	2.02		2.87	4.12
	6800	Weed, by handhoe		8	1			21.50		21.50	34
	6810	Spray out		32	.250			5.35		5.35	8.45
	6820	Spray after mulch		48	.167			3.58		3.58	5.65
	7100	Tree maintenance									
	7200	Fertilize, tablets, slow release, 30 gram/tree	1 Clab	100	.080	Ea.	.30	1.72		2.02	3.03
	7420	Pest control, spray		24	.333		14	7.15		21.15	26.50
	7430	Systemic		48	.167		13	3.58		16.58	19.95
715	0010	**SNOW REMOVAL**									
	0020	Plowing, 12 ton truck, 2"-4" deep	A-3	250	.032	M.S.F.		.71	1.46	2.17	2.68
	0040	4"-10" deep		200	.040			.88	1.82	2.70	3.35
	0060	10"-15" deep		150	.053			1.18	2.43	3.61	4.48
	0080	Pickup truck, 2"-4" deep	A-3A	175	.046			.99	.83	1.82	2.43
	0100	4"-10" deep		130	.062			1.34	1.12	2.46	3.28
	0120	10"-15" deep		75	.107			2.32	1.94	4.26	5.70
	0140	Load and haul snow, 1 mile round trip	A-3B	230	.070	C.Y.		1.76	1.92	3.68	4.79
	0160	2 mile round trip		175	.091			2.31	2.53	4.84	6.30
	0180	3 mile round trip		150	.107			2.69	2.95	5.64	7.35
	0200	4 mile round trip		130	.123			3.11	3.40	6.51	8.50
	0220	5 mile round trip		120	.133			3.37	3.69	7.06	9.20
	0240	Clearing with wheeled skid steer loader, 1 C.Y.	A-3C	80	.100			2.72	2.88	5.60	7.30
	0260	Spread sand and salt mix	A-3	375	.021	M.S.F.	3.59	.47	.97	5.03	5.75
	0280	Sidewalks and drives, by hand	1 Clab	780	.010	C.F.		.22		.22	.35
	0300	Power, 24" blower	A-1	1,200	.007	"		.14	.06	.20	.28
	0320	2"-4" deep, single driveway (10' x 50')		8	1	Ea.		21.50	8.60	30.10	43.50
	0340	Double driveway (20' x 50')		4.50	1.778			38	15.25	53.25	77
	0360	4"-10" deep, single driveway		6	1.333			28.50	11.45	39.95	57.50
	0380	Double driveway		3.25	2.462			53	21	74	106
	0400	10"-15" deep, single driveway		4	2			43	17.15	60.15	86.50
	0420	Double driveway		2.25	3.556			76.50	30.50	107	154
	0440	For heavy wet snow, add								20%	
720	0010	**FERTILIZE**									
	0100	Dry granular, 4#/M.S.F., hand spread	1 Clab	24	.333	M.S.F.	1.85	7.15		9	13.30
	0110	Push rotary	"	140	.057		1.70	1.23		2.93	3.80
	0120	Tractor towed spreader, 8'	B-66	500	.016		1.70	.44	.40	2.54	2.97
	0130	12' spread		800	.010		1.70	.27	.25	2.22	2.56
	0140	Truck whirlwind spreader		1,200	.007		1.70	.18	.17	2.05	2.33
	0180	Water soluable, hydro spread, 1.5 # /MSF	B-64	600	.027		.70	.58	.48	1.76	2.19
	0190	Add for weed control					.28			.28	.31
730	0010	**MOWING**									
	1650	Mowing brush, tractor with rotary mower									
	1660	Light density	B-84	22	.364	M.S.F.		10.35	9.50	19.85	26
	1670	Medium density		13	.615			17.50	16.05	33.55	44
	1680	Heavy density		9	.889			25	23	48	64
	4000	Lawn mowing, improved areas, 16" hand push	1 Clab	48	.167			3.58		3.58	5.65
	4050	Power mower, 18" - 22"		65	.123			2.64		2.64	4.15
	4100	22" - 30"		110	.073			1.56		1.56	2.45
	4150	30" - 32"	A-1	140	.057			1.23	.49	1.72	2.47
	4160	Riding mower, 36" - 44"	B-66	300	.027			.73	.67	1.40	1.84
	4170	48" - 58"	"	480	.017			.45	.42	.87	1.15
	4175	Mowing with tractor & attachments									

029 700 | Site Maintenance

		029 700 Site Maintenance	CREW	DAILY OUTPUT	LABOR-HOURS	UNIT	1999 BARE COSTS MAT.	LABOR	EQUIP.	TOTAL	TOTAL INCL O&P
730	4180	3 gang reel, 7′	B-66	930	.009	M.S.F.		.23	.22	.45	.60
	4190	5 gang reel, 12′		1,200	.007			.18	.17	.35	.46
	4200	Cutter or sickle-bar, 5′, rough terrain		210	.038			1.04	.95	1.99	2.63
	4210	Cutter or sickle-bar, 5′, smooth terrain		340	.024			.64	.59	1.23	1.62
	4220	Drainage channel, 5′ sickle bar		5	1.600	Mile		43.50	40	83.50	110
	4250	Lawnmower, rotary type, sharpen (all sizes)	1 Clab	10	.800	Ea.		17.15		17.15	27
	4260	Repair or replace part	"	7	1.143	"		24.50		24.50	38.50
	5000	Edge trimming with weed whacker	A-1	5,760	.001	L.F.		.03	.01	.04	.06
731	0010	**TREE PRUNING**									
	0020	1-1/2″ caliper	1 Clab	84	.095	Ea.		2.04		2.04	3.21
	0030	2″ caliper		70	.114			2.45		2.45	3.86
	0040	2-1/2″ caliper		50	.160			3.43		3.43	5.40
	0050	3″ caliper		30	.267			5.70		5.70	9
	0060	4″ caliper, by hand	2 Clab	21	.762			16.35		16.35	25.50
	0070	Aerial lift equipment	B-85	38	1.053			24	21.50	45.50	61
	0100	6″ caliper, by hand	2 Clab	12	1.333			28.50		28.50	45
	0110	Aerial lift equipment	B-85	20	2			46	40.50	86.50	116
	0200	9″ caliper, by hand	2 Clab	7.50	2.133			46		46	72
	0210	Aerial lift equipment	B-85	12.50	3.200			73.50	64.50	138	185
	0300	12″ caliper, by hand	2 Clab	6.50	2.462			53		53	83
	0310	Aerial lift equipment	B-85	10.80	3.704			85	75	160	215
	0400	18″ caliper by hand	2 Clab	5.60	2.857			61.50		61.50	96.50
	0410	Aerial lift equipment	B-85	9.30	4.301			99	87	186	249
	0500	24″ caliper, by hand	2 Clab	4.60	3.478			74.50		74.50	117
	0510	Aerial lift equipment	B-85	7.70	5.195			119	105	224	300
	0600	30″ caliper, by hand	2 Clab	3.70	4.324			93		93	146
	0610	Aerial lift equipment	B-85	6.20	6.452			148	130	278	375
	0700	36″ caliper, by hand	2 Clab	2.70	5.926			127		127	200
	0710	Aerial lift equipment	B-85	4.50	8.889			204	179	383	510
	0800	48″ caliper, by hand	2 Clab	1.70	9.412			202		202	320
	0810	Aerial lift equipment	B-85	2.80	14.286			330	288	618	825
740	0010	**SHRUB PRUNING**									
	6700	Prune, shrub bed	1 Clab	7	1.143	M.S.F.		24.50		24.50	38.50
	6710	Shrub under 3′ height		190	.042	Ea.		.90		.90	1.42
	6720	4′ height		90	.089			1.91		1.91	3
	6730	Over 6′		50	.160			3.43		3.43	5.40
	7350	Prune trees from ground		20	.400			8.60		8.60	13.50
760	0010	**WATERING**									
	4900	Water lawn or planting bed with hose, 1″ of water	1 Clab	16	.500	M.S.F.		10.75		10.75	16.90
	4910	50′ soaker hoses, in place		82	.098			2.09		2.09	3.29
	4920	60′ soaker hoses, in place		89	.090			1.93		1.93	3.03
	7500	Water trees or shrubs, under 1″ caliper		32	.250	Ea.		5.35		5.35	8.45
	7550	1″ - 3″ caliper		17	.471			10.10		10.10	15.90
	7600	3″ - 4″ caliper		12	.667			14.30		14.30	22.50
	7650	Over 4″ caliper		10	.800			17.15		17.15	27
	9000	For sprinkler irrigation systems, see Div. 028-100									
780	0010	**WEEDING**									
	0100	Weed planting bed	1 Clab	800	.010	S.Y.		.21		.21	.34

Division 3 Concrete

Estimating Tips

General

- Carefully check all the plans and specifications. Concrete often appears on drawings other than structural drawings, including mechanical and electrical drawings for equipment pads. The cost of cutting and patching is often difficult to estimate. See Division 020 for demolition costs.
- Always obtain concrete prices from suppliers near the job site. A volume discount can often be negotiated depending upon competition in the area. Remember to add for waste, particularly for slabs and footings on grade.

031 Concrete Formwork

- A primary cost for concrete construction is forming. Most jobs today are constructed with prefabricated forms. The selection of the forms best suited for the job and the total square feet of forms required for efficient concrete forming and placing are key elements in estimating concrete construction. Enough forms must be available for erection to make efficient use of the concrete placing equipment and crew.
- Concrete accessories for forming and placing depend upon the systems used. Study the plans and specifications to assure that all special accessory requirements have been included in the cost estimate such as anchor bolts, inserts and hangers.

032 Concrete Reinforcement

- Ascertain that the reinforcing steel supplier has included all accessories, cutting, bending and an allowance for lapping, splicing and waste. A good rule of thumb is 10% for lapping, splicing and waste. Also, 10% waste should be allowed for welded wire fabric.

033 Cast-in-Place Concrete

- When estimating structural concrete, pay particular attention to requirements for concrete additives, curing methods and surface treatments. Special consideration for climate, hot or cold, must be included in your estimate. Be sure to include requirements for concrete placing equipment and concrete finishing.

034 Precast Concrete
035 Cementitious Decks & Toppings

- The cost of hauling precast concrete structural members is often an important factor. For this reason, it is important to get a quote from the nearest supplier. It may become economically feasible to set up precasting beds on the site if the hauling costs are prohibitive.

Reference Numbers

Reference numbers are shown in bold squares at the beginning of some major classifications. These numbers refer to related items in the Reference Section. The reference information may be an estimating procedure, an alternate pricing method or technical information.

Note: Not all subdivisions listed here necessarily appear in this publication.

031 100 | Struct C.I.P. Formwork

		031 100 Struct C.I.P. Formwork	CREW	DAILY OUTPUT	LABOR-HOURS	UNIT	1999 BARE COSTS MAT.	LABOR	EQUIP.	TOTAL	TOTAL INCL O&P
110	0010	**ACCESSORIES, ANCHOR BOLTS** J-type, incl. nut and washer									
	0020	1/2" diameter, 6" long	1 Carp	90	.089	Ea.	1.11	2.43		3.54	5.05
	0050	10" long		85	.094		1.22	2.57		3.79	5.40
	0100	12" long		85	.094		1.27	2.57		3.84	5.45
	0200	5/8" diameter, 12" long		80	.100		1.40	2.73		4.13	5.85
	0250	18" long		70	.114		1.65	3.12		4.77	6.75
	0300	24" long		60	.133		1.90	3.64		5.54	7.85
	0350	3/4" diameter, 8" long		80	.100		1.70	2.73		4.43	6.20
	0400	12" long		70	.114		2.11	3.12		5.23	7.25
	0450	18" long		60	.133		2.58	3.64		6.22	8.60
	0500	24" long		50	.160		3.42	4.37		7.79	10.60
	0600	7/8" diameter, 12" long		60	.133		2.32	3.64		5.96	8.30
	0650	18" long		50	.160		2.96	4.37		7.33	10.10
	0700	24" long		40	.200		3.09	5.45		8.54	12
	0800	1" diameter, 12" long		55	.145		3.20	3.97		7.17	9.75
	0850	18" long		45	.178		3.61	4.85		8.46	11.60
	0900	24" long		35	.229		4.25	6.25		10.50	14.50
	0950	36" long		25	.320		5.60	8.75		14.35	19.90
	1200	1-1/2" diameter, 18" long		22	.364		8.95	9.95		18.90	25.50
	1250	24" long		18	.444		10.15	12.15		22.30	30.50
	1300	36" long		12	.667	▼	11.95	18.20		30.15	41.50
	1350	Larger sizes	▼	200	.040	Lb.	.88	1.09		1.97	2.69
	2000	Galvanized, incl. nut and washer									
	2100	1/2" diameter, 6" long	1 Carp	90	.089	Ea.	1.39	2.43		3.82	5.35
	2150	10" long		85	.094		1.33	2.57		3.90	5.50
	2200	12" long		85	.094		1.59	2.57		4.16	5.80
	2500	5/8" diameter, 12" long		80	.100		1.40	2.73		4.13	5.85
	2550	18" long		70	.114		1.46	3.12		4.58	6.50
	2600	24" long		60	.133		1.52	3.64		5.16	7.45
	2700	3/4" diameter, 8" long		80	.100		2.13	2.73		4.86	6.65
	2750	12" long		70	.114		2.64	3.12		5.76	7.80
	2800	18" long		60	.133		3.22	3.64		6.86	9.30
	2850	24" long		50	.160		3.60	4.37		7.97	10.80
	3000	7/8" diameter, 12" long		60	.133		3.52	3.64		7.16	9.60
	3050	18" long		50	.160		3.97	4.37		8.34	11.20
	3100	24" long		40	.200		4.68	5.45		10.13	13.75
	3200	1" diameter, 12" long		55	.145		4	3.97		7.97	10.65
	3250	18" long		45	.178		4.51	4.85		9.36	12.60
	3300	24" long		35	.229		5.30	6.25		11.55	15.65
	3350	36" long		25	.320		7	8.75		15.75	21.50
	3500	1-1/2" diameter, 18" long		22	.364		11.20	9.95		21.15	28
	3550	24" long		18	.444		12.70	12.15		24.85	33
	3600	36" long	▼	12	.667	▼	16.75	18.20		34.95	47
	3700	Larger sizes, galvanized	2 Carp	200	.080	Lb.	1.24	2.18		3.42	4.80
	8000	Sleeves, see Division 031-126									
	8800	Templates, steel, 8" bolt spacing	2 Carp	16	1	Ea.	6.40	27.50		33.90	50
	8850	12" bolt spacing		15	1.067		10.90	29		39.90	58
	8900	16" bolt spacing		14	1.143		16.15	31		47.15	67
	8950	24" bolt spacing		12	1.333		30	36.50		66.50	90.50
	9100	Wood, 8" bolt spacing		16	1		6.15	27.50		33.65	50
	9150	12" bolt spacing		15	1.067		6.45	29		35.45	53
	9200	16" bolt spacing		14	1.143		6.75	31		37.75	56.50
	9250	24" bolt spacing	▼	16	1	▼	7.05	27.50		34.55	51
112	0010	**ACCESSORIES, CHAMFER STRIPS**									
	2000	Polyvinyl chloride, 1/2" wide with leg	1 Carp	535	.015	L.F.	.48	.41		.89	1.17

031 100 | Struct C.I.P. Formwork

	031 100 Struct C.I.P. Formwork	CREW	DAILY OUTPUT	LABOR-HOURS	UNIT	1999 BARE COSTS MAT.	LABOR	EQUIP.	TOTAL	TOTAL INCL O&P	
2200	3/4″ wide with leg	1 Carp	525	.015	L.F.	.55	.42		.97	1.26	112
2400	1″ radius with leg		515	.016		.80	.42		1.22	1.55	
2800	1-1/2″ radius with leg		500	.016		1.75	.44		2.19	2.62	
5000	Wood, 1/2″ wide		535	.015		.12	.41		.53	.77	
5200	3/4″ wide		525	.015		.30	.42		.72	.98	
5400	1″ wide		515	.016		.39	.42		.81	1.09	
0010	**ACCESSORIES, DOVETAIL ANCHOR SYSTEM**										116
0500	Anchor slot, galv., filled, 24 ga.	1 Carp	425	.019	L.F.	.56	.51		1.07	1.43	
0600	20 ga.		400	.020		.40	.55		.95	1.30	
0800	16 oz. copper, foam filled		375	.021		1.55	.58		2.13	2.63	
0900	26 ga. stainless steel, foam filled		375	.021		1.10	.58		1.68	2.13	
1200	Brick anchor, corr., galv., 3-1/2″ long, 16 ga.	1 Bric	10.50	.762	C	22	21		43	56.50	
1300	12 ga.		10.50	.762		32	21		53	67.50	
1500	Flat, galv., 3-1/2″ long, 16 ga.		10.50	.762		23.50	21		44.50	58.50	
1600	12 ga.		10.50	.762		45.50	21		66.50	82.50	
6000	Stone anchors, 3-1/2″ long, galv., 1/8″ x 1″ wide		10.50	.762		73	21		94	113	
6100	1/4″ x 1″ wide		10.50	.762		109	21		130	152	
0010	**ACCESSORIES, HANGERS**										120
8500	Wire, black annealed, 9 ga				Cwt.	83			83	91.50	
8600	16 ga				″	88			88	97	
0010	**ACCESSORIES, INSERTS**										122
6000	Thin slab										
6100	1/2″ diameter bolt	1 Carp	60	.133	Ea.	.95	3.64		4.59	6.80	
6150	3/4″ diameter bolt		60	.133		1.35	3.64		4.99	7.25	
6200	1″ diameter bolt		60	.133		2.55	3.64		6.19	8.55	
6250	1-1/4″ diameter bolt		60	.133		5.30	3.64		8.94	11.60	
0010	**SLAB TEXTURE STAMPING,** buy										125
0020	Approx. 3 S.F.- 5 S.F. each, minimum				Ea.	40			40	44	
0030	Average				″	44			44	48.50	
0120	Per S.F. of tool, average				S.F.	48			48	53	
0200	Commonly used chemicals for texture systems										
0210	Hardeners w/colors average				S.F.	.40			.40	.44	
0220	Release agents w/colors, average					.15			.15	.17	
0225	Clear, average					.10			.10	.11	
0230	Sealers, clear, average					.10			.10	.11	
0240	Colors, average					.12			.12	.13	
0010	**ACCESSORIES, SLEEVES AND CHASES**										126
0100	Plastic, 1 use, 9″ long, 2″ diameter	1 Carp	100	.080	Ea.	.55	2.18		2.73	4.05	
0150	4″ diameter		90	.089		1.62	2.43		4.05	5.60	
0200	6″ diameter		75	.107		2.84	2.91		5.75	7.70	
0250	12″ diameter		60	.133		6.65	3.64		10.29	13.05	
5000	Sheet metal, 2″ diameter		100	.080		.85	2.18		3.03	4.38	
5100	4″ diameter		90	.089		.85	2.43		3.28	4.76	
5150	6″ diameter		75	.107		1.23	2.91		4.14	5.95	
5200	12″ diameter		60	.133		2.46	3.64		6.10	8.45	
6000	Steel pipe, 2″ diameter		100	.080		1.80	2.18		3.98	5.40	
6100	4″ diameter		90	.089		5.30	2.43		7.73	9.65	
6150	6″ diameter		75	.107		14.30	2.91		17.21	20.50	
6200	12″ diameter		60	.133		42.50	3.64		46.14	52.50	
0010	**ACCESSORIES, SNAP TIES, FLAT WASHER**										128
0100	3000 lb., to 8″				C	70			70	77	
0150	9″ & 10″					78.50			78.50	86	
0200	11″ & 12″					83			83	91	

	031 100 | Struct C.I.P. Formwork		CREW	DAILY OUTPUT	LABOR-HOURS	UNIT	1999 BARE COSTS MAT.	LABOR	EQUIP.	TOTAL	TOTAL INCL O&P
128	0250	16″				C	88			88	97
	0300	18″					92			92	101
	0500	With plastic cone, to 8″					62.50			62.50	68.50
	0550	9″ & 10″					69.50			69.50	76.50
	0600	11″ & 12″					73.50			73.50	81
	0650	16″					78.50			78.50	86
	0700	18″					81.50			81.50	90
	1000	5000 lb., to 8″					100			100	110
	1100	9″ & 10″					106			106	117
	1150	11″ & 12″					114			114	125
	1200	16″					126			126	138
	1250	18″					131			131	144
	1500	With plastic cone, to 8″					107			107	118
	1550	9″ & 10″					117			117	128
	1600	11″ & 12″					126			126	139
	1650	16″					140			140	154
	1700	18″					146			146	160
130	0010	**ACCESSORIES, WALL AND FOUNDATION**									
	0020	Coil tie system									
	0050	Coil ties 1/2″, 6000 lb., to 8″				C	155			155	171
	0150	24″					271			271	298
	0300	3/4″, 12,000 lb., to 8″					222			222	245
	0400	24″					365			365	405
	0900	Coil bolts, 1/2″ diameter x 3″ long					106			106	117
	0940	12″ long					305			305	335
	1000	3/4″ diameter x 3″ long					234			234	258
	1040	12″ long					640			640	705
	1300	Adjustable coil tie, 3/4″ diameter, 20″ long					900			900	985
	1350	3/4″ diameter, 36″ long					1,150			1,150	1,275
	1400	Tie cones, plastic, 1″ long, 1/2″ diameter					96			96	106
	1420	3/4″ diameter					131			131	144
	1500	Coil tie ends, 1/2″ diameter x 4″					74			74	81.50
	1550	3/4″ diameter x 6″					105			105	116
	1700	Waler holders, 1/2″ diameter					119			119	131
	1750	3/4″ diameter					131			131	144
	1900	Flat washers, 4″ x 5″ x 1/4″ for 3/4″ diameter					225			225	247
	2000	Footings, form braces, solid steel, adjustable				Ea.	6.60			6.60	7.30
	2050	Spreaders for footer, adjustable				"	3.80			3.80	4.18
	2100	Lagstud, threaded, 1/2″ diameter				C.L.F.	46.50			46.50	51
	2200	1″ diameter				"	220			220	242
	2300	Lagnuts, 1/2″ diameter				C	21			21	23
	2400	1″ diameter					152			152	167
	2600	Lagnuts with handle, 1/2″ diameter					83			83	91.50
	2700	1″ diameter					263			263	289
	2750	Plastic set back plugs for 1/2″ & 3/4″ diameter					27			27	29.50
	2800	Rock anchors, 1/2″ diameter					415			415	455
	2900	1″ diameter					575			575	635
	2950	Batter washer, 1/2″ diameter					490			490	540
	3000	Form oil, coverage varies greatly, minimum				Gal.	4.50			4.50	4.95
	3050	Maximum				"	6.70			6.70	7.35
	3500	Form patches, 1-3/4″ diameter				C	8.80			8.80	9.70
	3550	2-3/4″ diameter				"	15.20			15.20	16.70
	4000	Nail stakes, 3/4″ diameter, 18″ long				Ea.	.53			.53	.58
	4050	24″ long					.65			.65	.71
	4200	30″ long					.75			.75	.83

	031 100 | Struct C.I.P. Formwork		CREW	DAILY OUTPUT	LABOR-HOURS	UNIT	1999 BARE COSTS MAT.	LABOR	EQUIP.	TOTAL	TOTAL INCL O&P
130	4250	36″ long				Ea.	.80			.80	.88
	5000	Rods, 1/4″ diameter				C.L.F.	17			17	18.70
	5100	1/2″ diameter				″	74			74	81.50
	5200	Clamps for 1/4″ and 3/8″ rods				Ea.	5.50			5.50	6.05
	5250	For 1/2″ and 5/8″ rods					10.50			10.50	11.55
	5300	Clamping jacks for 1/4″ and 3/8″ rods					100			100	110
	5350	For 1/2″ and 5/8″ rods					126			126	139
	6000	She-bolts, 7/8″ x 20″				C	1,325			1,325	1,450
	6150	1-1/4″ x 20″					2,075			2,075	2,300
	6300	Inside rods, threaded, 1/2″ diameter x 6″					64			64	70.50
	6350	1/2″ diameter x 12″					88			88	97
	6500	3/4″ diameter x 6″					135			135	149
	6550	3/4″ diameter x 12″					200			200	220
	6700	For wing nuts, see taper ties									
	7000	Taper tie system									
	7100	Taper ties, 3/4″ to 5/8″ diameter x 30″				Ea.	23			23	25.50
	7200	1-1/4″ to 1″ diameter x 30″				″	29.50			29.50	32.50
	7300	Wing nuts, 5/8″ diameter				C	660			660	725
	7550	1-1/4″ diameter					870			870	960
	7700	Flat washers, 3/8″ x 5″ sq. 5/8″ end					281			281	310
	7800	5/8″ x 7″ sq. 1″ end					700			700	770
	8000	Fiberglass form ties									
	8010	Fiberglass ties, 4500 lb., 1/4″ diameter				L.F.	.45			.45	.50
	8020	6000 lb., 5/16″ diameter					.45			.45	.50
	8030	15000 lb., 1/2″ diameter					1.10			1.10	1.21
	8100	Fiberglass form tie hardware, 6000 lb. ties									
	8110	Tie holder				Ea.	5.25			5.25	5.80
	8120	Tie wedge					1.15			1.15	1.27
	8130	Tie wedge at waler					3.25			3.25	3.58
	8140	Spreader hardware					.15			.15	.17
	8150	Water stop					.15			.15	.17
	8160	Rock anchors					4.50			4.50	4.95
	8200	Fiberglass form tie hardware, 15,000 lb. ties									
	8210	Tie holder				Ea.	10.60			10.60	11.65
	8220	Tie wedge					7.40			7.40	8.15
	8250	Water stop					.15			.15	.17
	8260	Rock anchors					5.40			5.40	5.95
	9000	Wood form accessories									
	9100	Brace plate				C	540			540	595
	9200	Corner washer					265			265	292
	9300	Outside corner clip					330			330	360
	9400	Panel wedge					110			110	121
	9500	Stud clamp					330			330	365
132	0010	**EXPANSION JOINT** Keyed, cold, 24 ga, incl. stakes, 3-1/2″ high	1 Carp	200	.040	L.F.	.80	1.09		1.89	2.60
	0050	4-1/2″ high		200	.040		.90	1.09		1.99	2.71
	0100	5-1/2″ high		195	.041		1.12	1.12		2.24	2.99
	0150	7-1/2″ high		190	.042		1.35	1.15		2.50	3.30
	0200	9-1/2″ high		185	.043		1.82	1.18		3	3.86
	0300	Poured asphalt, plain, 1/2″ x 1″	1 Clab	450	.018		.31	.38		.69	.94
	0350	1″ x 2″		400	.020		1.13	.43		1.56	1.92
	0500	Neoprene, liquid, cold applied, 1/2″ x 1″		450	.018		1.50	.38		1.88	2.25
	0550	1″ x 2″		400	.020		5.70	.43		6.13	6.95
	0700	Polyurethane, poured, 2 part, 1/2″ x 1″		400	.020		1.75	.43		2.18	2.61
	0750	1″ x 2″		350	.023		6.90	.49		7.39	8.35
	0900	Rubberized asphalt, hot or cold applied, 1/2″ x 1″		450	.018		.58	.38		.96	1.24

	031 100 | Struct C.I.P. Formwork		CREW	DAILY OUTPUT	LABOR-HOURS	UNIT	1999 BARE COSTS MAT.	LABOR	EQUIP.	TOTAL	TOTAL INCL O&P	
132	0950	1″ x 2″	1 Clab	400	.020	L.F.	1.13	.43		1.56	1.92	132
	1100	Hot applied, fuel resistant, 1/2″ x 1″		450	.018		1	.38		1.38	1.70	
	1150	1″ x 2″		400	.020		1.38	.43		1.81	2.20	
	2000	Premolded, bituminous fiber, 1/2″ x 6″	1 Carp	375	.021		.36	.58		.94	1.31	
	2050	1″ x 12″		300	.027		1.71	.73		2.44	3.03	
	2250	Cork with resin binder, 1/2″ x 6″		375	.021		1.75	.58		2.33	2.85	
	2300	1″ x 12″		300	.027		5.15	.73		5.88	6.80	
	2500	Neoprene sponge, closed cell, 1/2″ x 6″		375	.021		1.73	.58		2.31	2.82	
	2550	1″ x 12″		300	.027		6.40	.73		7.13	8.15	
	2750	Polyethylene foam, 1/2″ x 6″		375	.021		.38	.58		.96	1.34	
	2800	1″ x 12″		300	.027		2.10	.73		2.83	3.46	
	3000	Polyethylene backer rod, 3/8″ diameter		460	.017		.02	.47		.49	.77	
	3050	3/4″ diameter		460	.017		.07	.47		.54	.82	
	3100	1″ diameter		460	.017		.13	.47		.60	.89	
	3500	Polyurethane foam, with polybutylene, 1/2″ x 1/2″		475	.017		.80	.46		1.26	1.60	
	3550	1″ x 1″		450	.018		1.40	.49		1.89	2.30	
	3750	Polyurethane foam, regular, closed cell, 1/2″ x 6″		375	.021		.65	.58		1.23	1.63	
	3800	1″ x 12″		300	.027		2.40	.73		3.13	3.79	
	4000	Polyvinyl chloride foam, closed cell, 1/2″ x 6″		375	.021		1.62	.58		2.20	2.70	
	4050	1″ x 12″		300	.027		5.05	.73		5.78	6.70	
	4100	Rod, 3/8″ polyethylene/rubberized asphalt sealer, 1/4″ x 3/4″	2 Clab	600	.027		.21	.57		.78	1.13	
	4250	Rubber, gray sponge, 1/2″ x 6″	1 Carp	375	.021		2.58	.58		3.16	3.76	
	4300	1″ x 12″	″	300	.027		10.35	.73		11.08	12.55	
	4500	Lead wool for joints, 1 ton lots				Lb.	2			2	2.20	
	4550	Retail				″	2.20			2.20	2.42	
	5000	For installation in walls, add						75%				
	5250	For installation in boxouts, add						25%				
138	0010	FORMS IN PLACE, BEAMS AND GIRDERS										138
	0020	See also Elevated Slabs, division 031-150										
	3500	Beam bottoms only, to 30″ wide, plywood, 1 use	C-2	230	.209	SFCA	3.32	5.55		8.87	12.40	
	3550	2 use		265	.181		1.86	4.83		6.69	9.65	
	3600	3 use		280	.171		1.33	4.57		5.90	8.65	
	3650	4 use		290	.166		1.08	4.41		5.49	8.15	
	4000	Beam sides only, vertical, 36″ high, plywood, 1 use		335	.143		3.56	3.82		7.38	9.90	
	4050	2 use		405	.119		1.96	3.16		5.12	7.10	
	4100	3 use		430	.112		1.42	2.98		4.40	6.25	
	4150	4 use		445	.108		1.16	2.88		4.04	5.80	
	4500	Sloped sides, 36″ high, plywood, 1 use		305	.157		3.50	4.20		7.70	10.45	
	4550	2 use		370	.130		1.95	3.46		5.41	7.60	
	4600	3 use		405	.119		1.40	3.16		4.56	6.50	
	4650	4 use		425	.113		1.14	3.01		4.15	6	
	5000	Upstanding beams, 36″ high, plywood, 1 use		225	.213		4.21	5.70		9.91	13.60	
	5050	2 use		255	.188		2.35	5		7.35	10.50	
	5100	3 use		275	.175		1.70	4.65		6.35	9.15	
	5150	4 use		280	.171		1.38	4.57		5.95	8.70	
142	0010	FORMS IN PLACE, COLUMNS (R031-040)										142
	0500	Round fiberglass, 4 use per mo., rent, 12″ diameter	C-1	160	.200	L.F.	2.10	5.15		7.25	10.45	
	0550	16″ diameter (R031-050)		150	.213		2.25	5.50		7.75	11.15	
	0600	18″ diameter		140	.229		2.50	5.90		8.40	12.05	
	0650	24″ diameter (R031-060)		135	.237		3.20	6.15		9.35	13.15	
	0700	28″ diameter		130	.246		3.50	6.35		9.85	13.85	
	0800	30″ diameter		125	.256		4.10	6.60		10.70	14.90	
	0850	36″ diameter		120	.267		4.70	6.90		11.60	16	
	1500	Round fiber tube, 1 use, 8″ diameter		155	.206		1.84	5.35		7.19	10.45	
	1550	10″ diameter		155	.206		2.57	5.35		7.92	11.25	

031 100 | Struct C.I.P. Formwork

		CREW	DAILY OUTPUT	LABOR-HOURS	UNIT	1999 BARE COSTS MAT.	LABOR	EQUIP.	TOTAL	TOTAL INCL O&P	
1600	12″ diameter (R031-040)	C-1	150	.213	L.F.	3.10	5.50		8.60	12.05	142
1650	14″ diameter		145	.221		3.97	5.70		9.67	13.30	
1700	16″ diameter (R031-050)		140	.229		5.40	5.90		11.30	15.25	
1750	20″ diameter		135	.237		8.05	6.15		14.20	18.50	
1800	24″ diameter (R031-060)		130	.246		9.80	6.35		16.15	21	
1850	30″ diameter		125	.256		14.25	6.60		20.85	26	
1900	36″ diameter		115	.278		19.05	7.20		26.25	32.50	
1950	42″ diameter		100	.320		39	8.25		47.25	55.50	
2000	48″ diameter		85	.376		48.50	9.75		58.25	69	
2200	For seamless type, add					15%					
5000	Plywood, 8″ x 8″ columns, 1 use	C-1	165	.194	SFCA	1.56	5		6.56	9.60	
5050	2 use		195	.164		.90	4.24		5.14	7.65	
5100	3 use		210	.152		.62	3.94		4.56	6.90	
5150	4 use		215	.149		.51	3.85		4.36	6.60	
5500	12″ x 12″ columns, 1 use		180	.178		1.58	4.59		6.17	9	
5550	2 use		210	.152		.87	3.94		4.81	7.15	
5600	3 use		220	.145		.63	3.76		4.39	6.60	
5650	4 use		225	.142		.51	3.68		4.19	6.35	
6000	16″ x 16″ columns, 1 use		185	.173		1.63	4.47		6.10	8.85	
6050	2 use		215	.149		.87	3.85		4.72	7	
6100	3 use		230	.139		.65	3.60		4.25	6.35	
6150	4 use		235	.136		.53	3.52		4.05	6.15	
6500	24″ x 24″ columns, 1 use		190	.168		1.75	4.35		6.10	8.80	
6550	2 use		216	.148		.96	3.83		4.79	7.05	
6600	3 use		230	.139		.66	3.60		4.26	6.40	
6650	4 use		238	.134		.57	3.47		4.04	6.10	
7000	36″ x 36″ columns, 1 use		200	.160		1.73	4.13		5.86	8.40	
7050	2 use		230	.139		.97	3.60		4.57	6.70	
7100	3 use		245	.131		.69	3.38		4.07	6.05	
7150	4 use		250	.128		.56	3.31		3.87	5.80	
7500	Steel framed plywood, 4 use per mo., rent, 8″ x 8″		340	.094		2.50	2.43		4.93	6.60	
7550	10″ x 10″		350	.091		2.25	2.36		4.61	6.20	
7600	12″ x 12″		370	.086		2.13	2.23		4.36	5.85	
7650	16″ x 16″		400	.080		2	2.07		4.07	5.45	
7700	20″ x 20″		420	.076		1.90	1.97		3.87	5.20	
7750	24″ x 24″		440	.073		1.80	1.88		3.68	4.94	
7755	30″ x 30″		440	.073		1.80	1.88		3.68	4.94	
0010	**FORMS IN PLACE, CULVERT** 5′ to 8′ square or rectangular, 1 use	C-1	170	.188	SFCA	2.63	4.86		7.49	10.55	146
0050	2 use		180	.178		1.39	4.59		5.98	8.80	
0100	3 use (R031-050)		190	.168		1.09	4.35		5.44	8.05	
0150	4 use		200	.160		.94	4.13		5.07	7.55	
0010	**FORMS IN PLACE, ELEVATED SLABS** (R031-060)										150
1000	Flat plate plywood to 15′ high, 1 use	C-2	470	.102	S.F.	2.49	2.72		5.21	7	
1050	2 use		520	.092		1.37	2.46		3.83	5.40	
1100	3 use		545	.088		1	2.35		3.35	4.79	
1150	4 use		560	.086		.81	2.29		3.10	4.48	
1500	15′ to 20′ high ceilings, 4 use		495	.097		1.01	2.59		3.60	5.20	
1600	21′ to 35′ high ceilings, 4 use		450	.107		1.20	2.84		4.04	5.80	
2000	Flat slab with drop panels, to 15′ high, 1 use		449	.107		2.57	2.85		5.42	7.30	
2050	2 use		509	.094		1.58	2.51		4.09	5.70	
2100	3 use		532	.090		1.29	2.41		3.70	5.20	
2150	4 use		544	.088		1.13	2.35		3.48	4.95	
2250	15′ to 20′ high ceilings, 4 use		480	.100		1.37	2.67		4.04	5.70	
2350	20′ to 35′ high ceilings, 4 use		435	.110		1.51	2.94		4.45	6.30	
3000	Floor slab hung from steel beams, 1 use		485	.099		1.58	2.64		4.22	5.90	

For expanded coverage of these items see *Means Concrete & Masonry Cost Data 1999*

031 100	Struct C.I.P. Formwork		CREW	DAILY OUTPUT	LABOR-HOURS	UNIT	1999 BARE COSTS MAT.	LABOR	EQUIP.	TOTAL	TOTAL INCL O&P
150	3050	2 use (R031-060)	C-2	535	.090	S.F.	1.13	2.39		3.52	5
	3100	3 use		550	.087		.98	2.33		3.31	4.74
	3150	4 use		565	.085		.91	2.26		3.17	4.56
	5000	Box out for slab openings, over 16" deep, 1 use		190	.253	SFCA	1.94	6.75		8.69	12.75
	5050	2 use		240	.200	"	1.07	5.35		6.42	9.55
	5500	Shallow slab box outs, to 10 S.F.		42	1.143	Ea.	9.75	30.50		40.25	59
	5550	Over 10 S.F. (use perimeter)		600	.080	L.F.	1.30	2.13		3.43	4.79
	6000	Bulkhead forms for slab, with keyway, 1 use, 2 piece		500	.096		1.56	2.56		4.12	5.75
	6100	3 piece (see also edge forms)		460	.104		2.09	2.78		4.87	6.70
	6200	Bulkhead forms for slab, w/keyway expanded metal									
	6210	In lieu of 2 piece form	C-1	1,100	.029	L.F.	1.12	.75		1.87	2.41
	6215	In lieu of 3 piece form		960	.033		1.12	.86		1.98	2.59
	6220	6" high, 4 uses		1,100	.029		1.35	.75		2.10	2.67
	6500	Curb forms, wood, 6" to 12" high, on elevated slabs, 1 use		180	.178	SFCA	1.44	4.59		6.03	8.85
	6550	2 use		205	.156		1.14	4.03		5.17	7.60
	6600	3 use		220	.145		.83	3.76		4.59	6.80
	6650	4 use		225	.142		.68	3.68		4.36	6.55
	7000	Edge forms to 6" high, on elevated slab, 4 use		500	.064	L.F.	.36	1.65		2.01	3
	7500	Depressed area forms to 12" high, 4 use		300	.107		.43	2.76		3.19	4.81
	7550	12" to 24" high, 4 use		175	.183		.58	4.73		5.31	8.10
	8000	Perimeter deck and rail for elevated slabs, straight		90	.356		9.10	9.20		18.30	24.50
	8050	Curved		65	.492		12.50	12.70		25.20	34
	8500	Void forms, round fiber, 3" diameter		450	.071		1.08	1.84		2.92	4.08
	8550	4" diameter		425	.075		1.40	1.95		3.35	4.61
	8600	6" diameter		400	.080		2.13	2.07		4.20	5.60
	8650	8" diameter		375	.085		3.46	2.21		5.67	7.25
	8700	10" diameter		350	.091		4.11	2.36		6.47	8.25
	8750	12" diameter		300	.107		4.95	2.76		7.71	9.80
	8800	Metal end closures, loose, minimum				C	30			30	33
	8850	Maximum				"	154			154	169
154	0010	**FORMS IN PLACE, EQUIPMENT FOUNDATIONS** 1 use	C-2	160	.300	SFCA	1.69	8		9.69	14.45
	0050	2 use		190	.253		.93	6.75		7.68	11.60
	0100	3 use		200	.240		.68	6.40		7.08	10.80
	0150	4 use		205	.234		.55	6.25		6.80	10.40
158	0010	**FORMS IN PLACE, FOOTINGS** Continuous wall, plywood, 1 use (R031-050)	C-1	375	.085	SFCA	1.93	2.21		4.14	5.60
	0050	2 use		440	.073		1.06	1.88		2.94	4.13
	0100	3 use (R031-060)		470	.068		.77	1.76		2.53	3.62
	0150	4 use		485	.066		.63	1.70		2.33	3.37
	0500	Dowel supports for footings or beams, 1 use		500	.064	L.F.	.64	1.65		2.29	3.30
	1000	Integral starter wall, to 4" high, 1 use		400	.080		1.20	2.07		3.27	4.57
	1500	Keyway, 4 use, tapered wood, 2" x 4"	1 Carp	530	.015		.19	.41		.60	.86
	1550	2" x 6"		500	.016		.27	.44		.71	.99
	2000	Tapered plastic, 2" x 3"		530	.015		.50	.41		.91	1.20
	2050	2" x 4"		500	.016		.65	.44		1.09	1.40
	2250	For keyway hung from supports, add		150	.053		.82	1.46		2.28	3.19
	3000	Pile cap, square or rectangular, plywood, 1 use	C-1	290	.110	SFCA	1.85	2.85		4.70	6.55
	3050	2 use		346	.092		1.02	2.39		3.41	4.88
	3100	3 use		371	.086		.74	2.23		2.97	4.33
	3150	4 use		383	.084		.60	2.16		2.76	4.06
	4000	Triangular or hexagonal caps, plywood, 1 use		225	.142		2.19	3.68		5.87	8.20
	4050	2 use		280	.114		1.20	2.95		4.15	5.95
	4100	3 use		305	.105		.87	2.71		3.58	5.20
	4150	4 use		315	.102		.71	2.63		3.34	4.91
	5000	Spread footings, plywood, 1 use		305	.105		1.49	2.71		4.20	5.90
	5050	2 use		371	.086		.82	2.23		3.05	4.41
	5100	3 use		401	.080		.60	2.06		2.66	3.90

031 100 | Struct C.I.P. Formwork

	Line	Description	Ref	Crew	Daily Output	Labor-Hours	Unit	1999 Bare Costs Mat.	Labor	Equip.	Total	Total Incl O&P	
158	5150	4 use		C-1	414	.077	SFCA	.49	2		2.49	3.68	158
	6000	Supports for dowels, plinths or templates, 2′ x 2′	R031-050		25	1.280	Ea.	3.10	33		36.10	55.50	
	6050	4′ x 4′ footing	R031-060		22	1.455		6.50	37.50		44	66	
	6100	8′ x 8′ footing			20	1.600		13	41.50		54.50	79.50	
	6150	12′ x 12′ footing			17	1.882		21	48.50		69.50	99.50	
	7000	Plinths, 1 use			250	.128	SFCA	2.40	3.31		5.71	7.85	
	7100	4 use			270	.119	″	.74	3.06		3.80	5.65	
162	0010	**FORMS IN PLACE, GRADE BEAM** Plywood, 1 use	R031-060	C-2	530	.091	SFCA	1.36	2.41		3.77	5.30	162
	0050	2 use			580	.083		.75	2.21		2.96	4.29	
	0100	3 use			600	.080		.54	2.13		2.67	3.96	
	0150	4 use			605	.079		.44	2.12		2.56	3.82	
166	0010	**FORMS IN PLACE, MAT FOUNDATION** 1 use		C-2	290	.166	SFCA	1.49	4.41		5.90	8.60	166
	0050	2 use			310	.155		.82	4.13		4.95	7.40	
	0100	3 use			330	.145		.58	3.88		4.46	6.75	
	0120	4 use			350	.137		.48	3.66		4.14	6.30	
170	0010	**FORMS IN PLACE, SLAB ON GRADE**											170
	1000	Bulkhead forms with keyway, wood, 1 use, 2 piece		C-1	510	.063	L.F.	.65	1.62		2.27	3.26	
	1050	3 piece			400	.080		.80	2.07		2.87	4.13	
	1100	4 piece			350	.091		1.05	2.36		3.41	4.87	
	1400	Bulkhead forms w/keyway, 1 piece expanded metal, left in place											
	1410	In lieu of 2 piece form		C-1	1,375	.023	L.F.	1.12	.60		1.72	2.18	
	1420	In lieu of 3 piece form			1,200	.027		1.12	.69		1.81	2.31	
	1430	In lieu of 4 piece form			1,050	.030		1.12	.79		1.91	2.47	
	2000	Curb forms, wood, 6″ to 12″ high, on grade, 1 use			215	.149	SFCA	1.17	3.85		5.02	7.35	
	2050	2 use			250	.128		.65	3.31		3.96	5.90	
	2100	3 use			265	.121		.47	3.12		3.59	5.45	
	2150	4 use			275	.116		.38	3.01		3.39	5.15	
	3000	Edge forms, wood, 4 use, on grade, to 6″ high			600	.053	L.F.	.24	1.38		1.62	2.43	
	3050	7″ to 12″ high			435	.074	SFCA	.63	1.90		2.53	3.68	
	3060	Over 12″			350	.091	″	.74	2.36		3.10	4.53	
	3500	For depressed slabs, 4 use, to 12″ high			300	.107	L.F.	.61	2.76		3.37	5	
	3550	To 24″ high			175	.183		.79	4.73		5.52	8.30	
	4000	For slab blockouts, to 12″ high, 1 use			200	.160		.62	4.13		4.75	7.20	
	4050	To 24″ high, 1 use			120	.267		.78	6.90		7.68	11.70	
	4100	Plastic (extruded), to 6″ high, multiple use, on grade			800	.040		.22	1.03		1.25	1.87	
	5020	Wood, incl. wood stakes, 1″ x 3″			900	.036		.40	.92		1.32	1.89	
	5050	2″ x 4″			900	.036		1.15	.92		2.07	2.72	
	6000	Trench forms in floor, wood, 1 use			160	.200	SFCA	1.90	5.15		7.05	10.25	
	6050	2 use			175	.183		1.13	4.73		5.86	8.70	
	6100	3 use			180	.178		.82	4.59		5.41	8.15	
	6150	4 use			185	.173		.67	4.47		5.14	7.80	
174	0010	**FORMS IN PLACE, STAIRS** (Slant length x width), 1 use	R031-060	C-2	165	.291	S.F.	3.58	7.75		11.33	16.15	174
	0050	2 use			170	.282		1.97	7.55		9.52	14	
	0100	3 use			180	.267		1.43	7.10		8.53	12.75	
	0150	4 use			190	.253		1	6.75		7.75	11.70	
	1000	Alternate pricing method (0.7 L.F./S.F.), 1 use			100	.480	LF Rsr	4.40	12.80		17.20	25	
	1050	2 use			105	.457		2.60	12.20		14.80	22	
	1100	3 use			110	.436		2.05	11.65		13.70	20.50	
	1150	4 use			115	.417		1.75	11.15		12.90	19.45	
	2000	Stairs, cast on sloping ground (length x width), 1 use			220	.218	S.F.	2.25	5.80		8.05	11.65	
	2100	4 use			240	.200	″	1.25	5.35		6.60	9.80	
182	0010	**FORMS IN PLACE, WALLS**	R031-010										182
	0100	Box out for wall openings, to 16″ thick, to 10 S.F.		C-2	24	2	Ea.	19.60	53.50		73.10	106	
	0150	Over 10 S.F. (use perimeter)		″	280	.171	L.F.	1.76	4.57		6.33	9.15	
	0250	Brick shelf, 4″ w, add to wall forms, use wall area abv shelf	R031-050										

CONCRETE 3

or expanded coverage of these items see *Means Concrete & Masonry Cost Data 1999*

031 100 | Struct C.I.P. Formwork

182

		CREW	DAILY OUTPUT	LABOR-HOURS	UNIT	1999 BARE COSTS MAT.	LABOR	EQUIP.	TOTAL	TOTAL INCL O&P
0260	1 use (R031-010)	C-2	240	.200	SFCA	1.98	5.35		7.33	10.60
0300	2 use		275	.175		1.09	4.65		5.74	8.50
0350	4 use (R031-050)		300	.160		.79	4.27		5.06	7.55
0500	Bulkhead forms, with keyway, 1 use, 2 piece		265	.181	L.F.	2.50	4.83		7.33	10.35
0550	3 piece		175	.274	"	3.10	7.30		10.40	14.90
0600	Bulkhead forms w/keyway, 1 piece expanded metal, left in place									
0610	In lieu of 2 piece form (R031-060)	C-1	800	.040	L.F.	1.05	1.03		2.08	2.78
0620	In lieu of 3 piece form	"	525	.061	"	1.05	1.58		2.63	3.63
0700	Buttress forms, to 8′ high, 1 use	C-2	350	.137	SFCA	2.97	3.66		6.63	9
0750	2 use		430	.112		1.63	2.98		4.61	6.50
0800	3 use		460	.104		1.19	2.78		3.97	5.70
0850	4 use		480	.100		.98	2.67		3.65	5.25
1100	3 use		175	.274	L.F.	.64	7.30		7.94	12.20
1150	4 use		180	.267	"	.52	7.10		7.62	11.75
2000	Job built plyform wall forms, to 8′ high, 1 use, below grade		300	.160	SFCA	1.89	4.27		6.16	8.75
2050	2 use, below grade		365	.132		1.05	3.51		4.56	6.65
2100	3 use, below grade		425	.113		.75	3.01		3.76	5.55
2150	4 use, below grade		435	.110		.62	2.94		3.56	5.30
2400	Over 8′ to 16′ high, 1 use		280	.171		3.60	4.57		8.17	11.15
2450	2 use		345	.139		1.16	3.71		4.87	7.15
2500	3 use		375	.128		.83	3.41		4.24	6.25
2550	4 use		395	.122		.68	3.24		3.92	5.85
2700	Over 16′ high, 1 use		235	.204		2.34	5.45		7.79	11.10
2750	2 use		290	.166		1.29	4.41		5.70	8.35
2800	3 use		315	.152		.93	4.06		4.99	7.45
2850	4 use		330	.145		.76	3.88		4.64	6.95
3000	For architectural finish, add		1,820	.026		.58	.70		1.28	1.75
3500	Polystyrene (expanded) wall forms									
3510	To 8′ high, 1 use, left in place	1 Carp	295	.027	SFCA	1.55	.74		2.29	2.86
4000	Radial wall forms, smooth curved, 1 use	C-2	245	.196		2.30	5.20		7.50	10.75
4050	2 use		300	.160		1.26	4.27		5.53	8.10
4100	3 use		325	.148		.92	3.94		4.86	7.20
4150	4 use		335	.143		.75	3.82		4.57	6.80
4200	Wall forms, smooth curved, below grade, job built plyform, 1 use		225	.213		2.78	5.70		8.48	12
4210	2 use		225	.213		1.54	5.70		7.24	10.65
4220	3 use		225	.213		1.25	5.70		6.95	10.30
4230	4 use		225	.213		.90	5.70		6.60	9.95
4300	Curved, with 2′ chords, 1 use		290	.166		1.89	4.41		6.30	9
4350	2 use		355	.135		1.04	3.60		4.64	6.80
4400	3 use		385	.125		.75	3.32		4.07	6.10
4450	4 use		400	.120		.61	3.20		3.81	5.70
4500	Over 8′ high, 1 use		290	.166		.86	4.41		5.27	7.90
4525	2 use		355	.135		.48	3.60		4.08	6.15
4550	3 use		385	.125		.35	3.32		3.67	5.65
4575	4 use		400	.120		.29	3.20		3.49	5.35
4600	Retaining wall forms, battered, to 8′ high, 1 use		300	.160		1.78	4.27		6.05	8.65
4650	2 use		355	.135		.98	3.60		4.58	6.75
4700	3 use		375	.128		.71	3.41		4.12	6.15
4750	4 use		390	.123		.54	3.28		3.82	5.75
4900	Over 8′ to 16′ high, 1 use		240	.200		1.94	5.35		7.29	10.55
4950	2 use		295	.163		1.07	4.34		5.41	7.95
5000	3 use		305	.157		.78	4.20		4.98	7.45
5050	4 use		320	.150		.63	4		4.63	7
5100	Retaining wall form, smooth curve, 1 use		200	.240		2.89	6.40		9.29	13.25
5120	2 use		235	.204		1.59	5.45		7.04	10.30
5130	3 use		250	.192		1.16	5.10		6.26	9.30

031 100 | Struct C.I.P. Formwork

			CREW	DAILY OUTPUT	LABOR-HOURS	UNIT	1999 BARE COSTS MAT.	LABOR	EQUIP.	TOTAL	TOTAL INCL O&P
182	5140	4 use	C-2	260	.185	SFCA	.95	4.92		5.87	8.80
	5500	For gang wall forming, 192 S.F. sections, deduct (R031-010)					10%	10%			
	5550	384 S.F. sections, deduct					20%	20%			
	5750	Liners for forms (add to wall forms), A.B.S. plastic (R031-050)									
	5800	Aged wood, 4" wide, 1 use	1 Carp	250	.032	SFCA	4.75	.87		5.62	6.55
	5820	2 use (R031-060)		400	.020		2.63	.55		3.18	3.75
	5840	4 use		750	.011		1.60	.29		1.89	2.22
	5900	Fractured rope rib, 1 use		250	.032		7.40	.87		8.27	9.50
	6000	4 use		750	.011		2.45	.29		2.74	3.16
	6100	Ribbed look, 1/2" & 3/4" deep, 1 use		300	.027		5.15	.73		5.88	6.80
	6200	4 use		800	.010		1.65	.27		1.92	2.25
	6300	Rustic brick pattern, 1 use		250	.032		4.80	.87		5.67	6.65
	6400	4 use		750	.011		1.60	.29		1.89	2.22
	6500	Striated, random, 3/8" x 3/8" deep, 1 use		300	.027		5.10	.73		5.83	6.75
	6600	4 use		800	.010		1.65	.27		1.92	2.25
	6800	Rustication strips, A.B.S. plastic, 2 piece snap-on									
	6850	1" deep x 1-3/8" wide, 1 use	C-2	400	.120	L.F.	3.75	3.20		6.95	9.20
	6900	2 use		600	.080		2.10	2.13		4.23	5.65
	6950	4 use		800	.060		1.25	1.60		2.85	3.90
	7050	Wood, beveled edge, 3/4" deep, 1 use		600	.080		.22	2.13		2.35	3.60
	7100	1" deep, 1 use		450	.107		.35	2.84		3.19	4.86
	7200	For solid board finish, uniform, 1 use, add to wall forms		300	.160	SFCA	.74	4.27		5.01	7.50
	7300	Non-uniform finish		250	.192		.65	5.10		5.75	8.75
	7500	Lintel or sill forms, 1 use	1 Carp	30	.267		2.30	7.30		9.60	14
	7520	2 use		34	.235		1.26	6.40		7.66	11.50
	7540	3 use		36	.222		.92	6.05		6.97	10.55
	7560	4 use		37	.216		.75	5.90		6.65	10.10
	7800	Modular prefabricated plywood, to 8' high, 1 use per month	C-2	910	.053		.97	1.41		2.38	3.27
	7820	2 use per month		930	.052		.53	1.38		1.91	2.75
	7840	3 use per month		950	.051		.39	1.35		1.74	2.55
	7860	4 use per month		970	.049		.31	1.32		1.63	2.42
	8000	To 16' high, 1 use per month		550	.087		1.23	2.33		3.56	5
	8020	2 use per month		570	.084		.68	2.25		2.93	4.28
	8040	3 use per month		590	.081		.49	2.17		2.66	3.95
	8060	4 use per month		610	.079		.40	2.10		2.50	3.74
	8100	Over 16' high, 1 use per month		550	.087		1.48	2.33		3.81	5.30
	8120	2 use per month		570	.084		.82	2.25		3.07	4.43
	8140	3 use per month		590	.081		.59	2.17		2.76	4.06
	8160	4 use per month		610	.079		.48	2.10		2.58	3.83
	8600	Pilasters, 1 use		270	.178		2.20	4.74		6.94	9.85
	8620	2 use		330	.145		1.21	3.88		5.09	7.45
	8640	3 use		370	.130		.88	3.46		4.34	6.40
	8660	4 use		385	.125		.72	3.32		4.04	6.05
	9000	Steel framed plywood, to 8' high, 1 use per month		600	.080		1.65	2.13		3.78	5.20
	9020	2 use per month		640	.075		.91	2		2.91	4.15
	9040	3 use per month		655	.073		.66	1.95		2.61	3.80
	9060	4 use per month		665	.072		.54	1.92		2.46	3.62
	9200	Over 8' to 16' high, 1 use per month		455	.105		1.80	2.81		4.61	6.40
	9220	2 use per month		505	.095		.99	2.53		3.52	5.10
	9240	3 use per month		525	.091		.72	2.44		3.16	4.62
	9260	4 use per month		530	.091		.59	2.41		3	4.45
	9400	Over 16' to 20' high, 1 use per month		425	.113		1.65	3.01		4.66	6.55
	9420	2 use per month		435	.110		.91	2.94		3.85	5.65
	9440	3 use per month		455	.105		.66	2.81		3.47	5.15
	9460	4 use per month		465	.103	SFCA	.54	2.75		3.29	4.92
	9480	For battered walls, 1 side battered, add					10%	10%			

031 100 | Struct C.I.P. Formwork

			CREW	DAILY OUTPUT	LABOR-HOURS	UNIT	1999 BARE COSTS MAT.	LABOR	EQUIP.	TOTAL	TOTAL INCL O&P
182	9485	For battered walls, 2 sides battered, add (R031-010)					15%	15%			
186	0010	**GAS STATION FORMS** Curb fascia, with template,									
	0050	12 ga. steel, left in place, 9″ high	1 Carp	50	.160	L.F.	6.80	4.37		11.17	14.35
	1000	Sign or light bases, 18″ diameter, 9″ high		9	.889	Ea.	38	24.50		62.50	80
	1050	30″ diameter, 13″ high		8	1		64.50	27.50		92	114
	2000	Island forms, 10′ long, 9″ high, 3′- 6″ wide	C-1	10	3.200		175	82.50		257.50	325
	2050	4′ wide		9	3.556		190	92		282	355
	2500	20′ long, 9″ high, 4′ wide		6	5.333		310	138		448	555
	2550	5′ wide		5	6.400		325	165		490	620
190	0011	**SCAFFOLDING** See division 015-254									
192	0010	**SHORES** Erect and strip, by hand, horizontal members									
	0500	Aluminum joists and stringers	2 Carp	60	.267	Ea.		7.30		7.30	11.45
	0600	Steel, adjustable beams		45	.356			9.70		9.70	15.25
	0700	Wood joists		50	.320			8.75		8.75	13.75
	0800	Wood stringers		30	.533			14.55		14.55	23
	1000	Vertical members to 10′ high		55	.291			7.95		7.95	12.50
	1050	To 13′ high		50	.320			8.75		8.75	13.75
	1100	To 16′ high		45	.356			9.70		9.70	15.25
	1500	Reshoring		1,400	.011	S.F.	.17	.31		.48	.68
	1600	Flying truss system	C-17D	9,600	.009	SFCA		.25	.06	.31	.45
	1760	Horizontal, aluminum joists, 6′ to 30′ spans, buy				L.F.	10.75			10.75	11.80
	1770	Aluminum stringers, 12′ & 16′ spans				″	16.50			16.50	18.15
	1810	Horizontal, steel beam, adjustable, 4′ to 7′ span				Ea.	110			110	121
	1830	6′ to 10′ span					142			142	156
	1920	9′ to 15′ span					260			260	286
	1940	12′ to 20′ span					300			300	330
	1970	Steel stringer, 6′ to 15′ span				L.F.	8.10			8.10	8.90
	3000	Rent for job duration, aluminum, first month				SF Flr.	.25			.25	.28
	3050	Steel				″	.20			.20	.22
	3500	Vertical, adjustable steel, 5′-7″ to 9′-6″ high, 10,000# cap., buy				Ea.	59			59	65
	3550	7′-3″ to 12′-10″ high, 7800# capacity					71			71	78
	3600	8′-10″ to 12′-4″ high, 10,000# capacity					79			79	87
	3650	8′-10″ to 16′-1″ high, 3800# capacity					86			86	94.50
	4000	Frame shoring systems, aluminum, 10,000# per leg,									
	4050	6′ wide, 5′ & 6′ high				Ea.	235			235	259
	4100	5′ to 7′ post with base, jack screw & top plate				Set	52			52	57
	5010	Steel, 10,000# per leg									
	5040	2′ & 4′ wide, 3′, 4′, 5′ & 6′ high				Ea.	97			97	107
	5250	6′ extension tube with adjusting collar					91			91	100
	5550	Base plate					9.75			9.75	10.75
	5600	12″ adjustable leg					39			39	43
	5650	Top plate					21			21	23
	5750	Flying truss system				SFCA	9			9	9.90
196	0010	**SLIPFORMS** Silos, minimum	C-17E	3,885	.021	″	1	.59	.02	1.61	2.04
	0050	Maximum		1,095	.073	SFCA	1.40	2.08	.06	3.54	4.88
	1000	Buildings, minimum		3,660	.022		1.20	.62	.02	1.84	2.32
	1050	Maximum		875	.091		2.10	2.60	.07	4.77	6.50
198	0010	**WATERSTOP** PVC, ribbed 3/16″ thick, 4″ wide	1 Carp	155	.052	L.F.	.74	1.41		2.15	3.03
	0050	6″ wide		145	.055		1.28	1.51		2.79	3.78
	0500	Ribbed, PVC, with center bulb, 9″ wide, 3/16″ thick		135	.059		1.76	1.62		3.38	4.49
	0550	3/8″ thick		130	.062		2.20	1.68		3.88	5.05
	0800	Dumbbell type, PVC, 6″ wide, 3/16″ thick		150	.053		.92	1.46		2.38	3.30
	0850	3/8″ thick		145	.055		1.80	1.51		3.31	4.36

031 | Concrete Formwork

031 100 | Struct C.I.P. Formwork

		CREW	DAILY OUTPUT	LABOR-HOURS	UNIT	1999 BARE COSTS				TOTAL INCL O&P
						MAT.	LABOR	EQUIP.	TOTAL	
198 1000	9" wide, 3/8" thick, PVC, plain	1 Carp	130	.062	L.F.	3.11	1.68		4.79	6.05
1050	Center bulb		130	.062		5.60	1.68		7.28	8.80
1250	Split PVC, 3/8" thick, 6" wide		145	.055		4.65	1.51		6.16	7.45
1300	9" wide		130	.062		6.85	1.68		8.53	10.15
2000	Rubber, flat dumbbell, 3/8" thick, 6" wide		145	.055		4.70	1.51		6.21	7.50
2050	9" wide		135	.059		7.75	1.62		9.37	11.10
2500	Flat dumbbell split, 3/8" thick, 6" wide		145	.055		6.75	1.51		8.26	9.80
2550	9" wide		135	.059		10	1.62		11.62	13.55
3000	Center bulb, 1/4" thick, 6" wide		145	.055		5.20	1.51		6.71	8.05
3050	9" wide		135	.059		9.50	1.62		11.12	13
3500	Center bulb split, 3/8" thick, 6" wide		145	.055		6.15	1.51		7.66	9.10
3550	9" wide	↓	135	.059	↓	10.90	1.62		12.52	14.55
5000	Waterstop fittings, rubber, flat									
5010	Dumbbell or center bulb, 3/8" thick,									
5200	Field union, 6" wide	1 Carp	50	.160	Ea.	9.25	4.37		13.62	17.05
5250	9" wide		50	.160		13.25	4.37		17.62	21.50
5500	Flat cross, 6" wide		30	.267		34	7.30		41.30	49
5550	9" wide		30	.267		57	7.30		64.30	74
6000	Flat tee, 6" wide		30	.267		33.50	7.30		40.80	48.50
6050	9" wide		30	.267		47.50	7.30		54.80	64
6500	Flat ell, 6" wide		40	.200		32	5.45		37.45	43.50
6550	9" wide		40	.200		46.50	5.45		51.95	59.50
7000	Vertical tee, 6" wide		25	.320		28.50	8.75		37.25	45.50
7050	9" wide		25	.320		43.50	8.75		52.25	62
7500	Vertical ell, 6" wide		35	.229		25	6.25		31.25	37.50
7550	9" wide	↓	35	.229	↓	43	6.25		49.25	57.50

032 | Concrete Reinforcement

032 100 | Reinforcing Steel

		CREW	DAILY OUTPUT	LABOR-HOURS	UNIT	1999 BARE COSTS				TOTAL INCL O&P
						MAT.	LABOR	EQUIP.	TOTAL	
102 0010	**ACCESSORIES** Materials only R032-070									
0700	Clip or bar ties, 16 ga., plain, 3" long				C	8.50			8.50	9.35
0710	4" long					9			9	9.90
0720	6" long					10.50			10.50	11.55
0730	8" long					11			11	12.10
1200	High chairs, individual, no plates (1 HC), to 3" high, plain					425			425	470
1202	Galvanized					57			57	62.50
1204	Stainless					118			118	130
1206	Plastic					55			55	60.50
1210	5" high, plain					62.50			62.50	69
1212	Galvanized					71			71	78
1214	Stainless					140			140	154
1216	Plastic					76			76	83.50
1220	8" high, plain					136			136	150
1222	Galvanized					174			174	191
1224	Stainless					288			288	315
1226	Plastic					171			171	188
1230	12" high, plain					284			284	310
1232	Galvanized					345			345	380
1234	Stainless ↓				↓	520			520	575

032 100 | Reinforcing Steel

			CREW	DAILY OUTPUT	LABOR-HOURS	UNIT	1999 BARE COSTS MAT.	LABOR	EQUIP.	TOTAL	TOTAL INCL O&P
102	1236	Plastic				C	320			320	350
	1240	15″ high, plain R032 -070					555			555	610
	1242	Galvanized					590			590	650
	1244	Stainless					875			875	960
	1246	Plastic					575			575	635
	1250	For each added 1″ up to 24″ high, plain, add					33			33	36.50
	1252	Galvanized, add					38			38	42
	1254	Stainless, add					45			45	49.50
	1256	Plastic, add					38			38	42
	1400	Individual high chairs, with plates, (HCP), to 5″ high, add					110			110	121
	1410	Over 5″ high, add					121			121	133
	1500	Bar chair (BC) for up to 1-3/4″ high, plain					23			23	25.50
	1520	Galvanized					26			26	28.50
	1530	Stainless					69			69	76
	1540	Plastic					45			45	49.50
	1550	Joist chair (JC) for up to 1-1/2″ high, plain					28			28	31
	1580	Galvanized					34			34	37.50
	1600	Stainless					49			49	54
	1620	Plastic					49			49	54
	1630	Epoxy				▼	113			113	124
	1700	Continuous high chairs, legs 8″ O.C. (CHC) to 4″ high, plain				C.L.F.	55			55	60.50
	1705	Galvanized					64			64	70.50
	1710	Stainless					136			136	150
	1715	Plastic					69.50			69.50	76.50
	1718	Epoxy					98			98	108
	1720	6″ high, plain					84			84	92.50
	1725	Galvanized					108			108	119
	1730	Stainless					159			159	175
	1735	Plastic					108			108	119
	1738	Epoxy					131			131	144
	1740	8″ high, plain					115			115	127
	1745	Galvanized					151			151	166
	1750	Stainless					189			189	208
	1755	Plastic					150			150	165
	1758	Epoxy					198			198	218
	1760	12″ high, plain					281			281	310
	1765	Galvanized					325			325	355
	1770	Stainless					450			450	495
	1775	Plastic					320			320	355
	1778	Epoxy					470			470	515
	1780	15″ high, plain					315			315	345
	1785	Galvanized					365			365	400
	1790	Stainless					360			360	395
	1795	Plastic					360			360	395
	1798	Epoxy					615			615	680
	1800	For each added 1″ up to 24″ high, plain, add					26			26	28.50
	1820	Galvanized, add					31			31	34
	1840	Stainless, add					28			28	31
	1860	Plastic, add					28			28	31
	1900	For continuous bottom plate, (CHCP), add					123			123	135
	1940	For upper continuous high chairs, (CHCU), add					123			123	135
	1960	For galvanized wire runners, add				▼	123			123	135
	2200	Screed base, 1/2″ diameter, 2-1/2″ high, plain				C	133			133	146
	2210	Galvanized					138			138	152
	2220	5-1/2″ high, plain					160			160	176
	2250	Galvanized				▼	168			168	185

032 100 | Reinforcing Steel

	032 100 | Reinforcing Steel	CREW	DAILY OUTPUT	LABOR-HOURS	UNIT	1999 BARE COSTS MAT.	LABOR	EQUIP.	TOTAL	TOTAL INCL O&P	
300	3/4" diameter, 2-1/2" high, plain (R032-070)				C	167			167	184	102
310	Galvanized					175			175	193	
320	5-1/2" high, plain					206			206	227	
350	Galvanized					220			220	242	
400	Screed holder, 1/2" diam. for 1" I.D. pipe, plain, 6" long					139			139	153	
420	12" long					230			230	253	
600	3/4" diameter, for 1-1/2" I.D. pipe, 6" long					250			250	275	
620	12" long					410			410	455	
700	Screw anchor for bolts, plain, 1/2" diameter					90			90	99	
720	1" diameter					271			271	298	
740	1-1/2" diameter					450			450	495	
800	Screw eye bolts, 1/2" x 5" long					1,100			1,100	1,200	
820	1" x 9" long					4,000			4,000	4,425	
840	1-1/2" x 14" long					10,200			10,200	11,200	
900	Screw anchor bolts, 1/2" x up to 7" long					420			420	460	
920	1" x up to 12" long					1,375			1,375	1,500	
300	Subgrade chairs, 1/2" diameter, 3-1/2" high					270			270	297	
350	12" high					770			770	845	
900	3/4" diameter, 3-1/2" high					350			350	385	
950	12" high					840			840	925	
200	Subgrade stakes, 3/4" diameter, 12" long					277			277	305	
250	24" long					375			375	415	
300	1" diameter, 12" long					420			420	465	
350	24" long					630			630	690	
500	Tie wire, 16 ga. annealed steel, under 500 lbs.				Cwt.	80			80	88	
520	2,000 to 4,000 lbs.				"	75			75	82.50	
550	Tie wire holder, plastic case				Ea.	31			31	34	
600	Aluminum case				"	36			36	39.50	
010	**COATED REINFORCING** Add to material										104
100	Epoxy coated, A775				Cwt.	23.50			23.50	26	
150	Galvanized, #3					31.50			31.50	34.50	
200	#4					31.50			31.50	34.50	
250	#5					31			31	34	
300	#6 or over					31			31	34	
000	For over 20 tons, #6 or larger, minimum					28.50			28.50	31.50	
500	Maximum					34.50			34.50	38	
010	**REINFORCING** A615 Grade 40, incl. freight from mill										105
200	Average price, cut, bent, and delivered				Ton	480			480	530	
500	Grade 60, incl. freight from mill										
700	Average price, cut, bent, and delivered				Ton	460			460	505	
010	**REINFORCING IN PLACE** A615 Grade 60 (R032-060)										107
100	Beams & Girders, #3 to #7	4 Rodm	1.60	20	Ton	535	610		1,145	1,650	
150	#8 to #18 (R032-080)		2.70	11.852		525	360		885	1,200	
200	Columns, #3 to #7		1.50	21.333		535	650		1,185	1,725	
250	#8 to #18		2.30	13.913		525	425		950	1,300	
300	Spirals, hot rolled, 8" to 15" diameter		2.20	14.545		1,075	440		1,515	1,950	
330	24" to 36" diameter		2.30	13.913		1,000	425		1,425	1,825	
340	36" to 48" diameter		2.40	13.333		1,025	405		1,430	1,825	
400	Elevated slabs, #4 to #7		2.90	11.034		570	335		905	1,200	
500	Footings, #4 to #7		2.10	15.238		510	465		975	1,350	
550	#8 to #18		3.60	8.889		480	270		750	1,000	
600	Slab on grade, #3 to #7		2.30	13.913		510	425		935	1,300	
700	Walls, #3 to #7		3	10.667		510	325		835	1,125	
750	#8 to #18		4	8		510	243		753	980	
000	Typical in place, 10 ton lots, average		1.70	18.824		535	570		1,105	1,575	
100	Over 50 ton lots, average		2.30	13.913		515	425		940	1,300	

expanded coverage of these items see *Means Concrete & Masonry Cost Data 1999*

032 100 | Reinforcing Steel

				CREW	DAILY OUTPUT	LABOR-HOURS	UNIT	1999 BARE COSTS MAT.	LABOR	EQUIP.	TOTAL	TOTAL INCL O&P
107	1200	High strength steel, Grade 75, #14 bars only, add	R032-060				Ton	60			60	66
	2000	Unloading & sorting, add to above		C-5	100	.560			16.60	5.50	22.10	34
	2200	Crane cost for handling, add to above, minimum	R032-080		135	.415			12.30	4.07	16.37	25
	2210	Average			92	.609			18.05	5.95	24	36.50
	2220	Maximum		↓	35	1.600	↓		47.50	15.70	63.20	97
	2400	Dowels, 2 feet long, deformed, #3		2 Rodm	520	.031	Ea.	.22	.94		1.16	1.86
	2410	#4			480	.033		.40	1.01		1.41	2.18
	2420	#5			435	.037		.62	1.12		1.74	2.61
	2430	#6			360	.044	↓	.89	1.35		2.24	3.32
	2450	Longer and heavier dowels			725	.022	Lb.	.44	.67		1.11	1.64
	2500	Smooth dowels, 12" long, 1/4" or 3/8" diameter			140	.114	Ea.	.67	3.47		4.14	6.75
	2520	5/8" diameter			125	.128		1.18	3.89		5.07	8.05
	2530	3/4" diameter			110	.145		1.46	4.42		5.88	9.25
	2700	Dowel caps, 5" long, 1/2" to 3/4" diameter			800	.020		.21	.61		.82	1.28
	2720	1-1/4" diameter		↓	750	.021	↓	.26	.65		.91	1.41

032 200 | Welded Wire Fabric

			CREW	DAILY OUTPUT	LABOR-HOURS	UNIT	MAT.	LABOR	EQUIP.	TOTAL	TOTAL INCL O&P
207	0010	**WELDED WIRE FABRIC** ASTM A185									
	0050	Sheets									
	0100	6 x 6 - W1.4 x W1.4 (10 x 10) 21 lb. per C.S.F.	2 Rodm	35	.457	C.S.F.	6.90	13.90		20.80	31.50
	0200	6 x 6 - W2.1 x W2.1 (8 x 8) 30 lb. per C.S.F.		31	.516		10.40	15.70		26.10	38.50
	0300	6 x 6 - W2.9 x W2.9 (6 x 6) 42 lb. per C.S.F.		29	.552		14.70	16.75		31.45	45
	0400	6 x 6 - W4 x W4 (4 x 4) 58 lb. per C.S.F.		27	.593		21.50	18		39.50	54.50
	0500	4 x 4 - W1.4 x W1.4 (10 x 10) 31 lb. per C.S.F.		31	.516		10.45	15.70		26.15	38.50
	0600	4 x 4 - W2.1 x W2.1 (8 x 8) 44 lb. per C.S.F.		29	.552		12.80	16.75		29.55	43
	0650	4 x 4 - W2.9 x W2.9 (6 x 6) 61 lb. per C.S.F.		27	.593		18.75	18		36.75	51.50
	0700	4 x 4 - W4 x W4 (4 x 4) 85 lb. per C.S.F.	↓	25	.640	↓	28.50	19.45		47.95	64.50
	0750	Rolls									
	0800	2 x 2 - #14 galv. @ 21 lb., beam & column wrap	2 Rodm	6.50	2.462	C.S.F.	14.80	75		89.80	145
	0900	2 x 2 - #12 galv. for gunite reinforcing	"	6.50	2.462	"	21	75		96	152
	0950	Material prices for above include 10% lap									
	1000	Specially fabricated heavier gauges in sheets	4 Rodm	50	.640	C.S.F.		19.45		19.45	33.50
	1010	Material only, minimum				Ton	520			520	570
	1020	Average					725			725	800
	1030	Maximum				↓	930			930	1,025
240	0010	**FIBROUS REINFORCING**									
	0100	Synthetic fibers				Lb.	3.78			3.78	4.15
	0110	1-1/2 lb. per C.Y., add to concrete				C.Y.	5.85			5.85	6.40
	0150	Steel fibers				Lb.	.46			.46	.51
	0155	25 lb. per C.Y., add to concrete				C.Y.	11.50			11.50	12.65
	0160	50 lb. per C.Y., add to concrete					23			23	25.50
	0170	75 lb. per C.Y., add to concrete					35.50			35.50	39
	0180	100 lb. per C.Y., add to concrete				↓	46			46	50.50

032 400 | Fiberglass Reinforcement

			CREW	DAILY OUTPUT	LABOR-HOURS	UNIT	MAT.	LABOR	EQUIP.	TOTAL	TOTAL INCL O&P
407	0010	**GLASS FIBER REINFORCED POLYMER** reinforcing bars									
	0050	#2 bar, .043 lbs/ ft.				L.F.	.25			.25	.28
	0100	#3 bar, .092 lbs/ ft.					.40			.40	.44
	0150	#4 bar, .160 lbs/ ft.					.60			.60	.66
	0200	#5 bar, .258 lbs/ ft.					.78			.78	.86
	0250	#6 bar, .372 lbs/ ft.					1			1	1.10
	0300	#7 bar, .497 lbs/ ft.					1.40			1.40	1.54
	0350	#8 bar, .620 lbs/ ft.					1.70			1.70	1.87
	0400	#9 bar, .800 lbs/ ft.					2			2	2.20
	0450	#10 bar, 1.08 lbs/ ft.				↓	2.50			2.50	2.75

	033 100 | Structural Concrete	CREW	DAILY OUTPUT	LABOR-HOURS	UNIT	1999 BARE COSTS MAT.	LABOR	EQUIP.	TOTAL	TOTAL INCL O&P	
0010	**AGGREGATE** Expanded shale, C.L. lots, 52 lb. per C.F., minimum (R033-020)				Ton	35			35	38.50	102
0050	Maximum				"	45			45	49.50	
0100	Lightweight vermiculite or perlite, 4 C.F. bag, C.L. lots (R033-060)				Bag	8.70			8.70	9.60	
0150	L.C.L. lots				"	9.60			9.60	10.55	
0250	Sand & stone, loaded at pit, crushed bank gravel				Ton	12.40			12.40	13.65	
0350	Sand, washed, for concrete					8.25			8.25	9.05	
0400	For plaster or brick					16.15			16.15	17.80	
0450	Stone, 3/4″ to 1-1/2″					13			13	14.30	
0470	Round, river stone					11.75			11.75	12.95	
0500	3/8″ roofing stone & 1/2″ pea stone					13			13	14.30	
0550	For trucking 10 miles, add to the above					3.50			3.50	3.85	
0600	30 miles, add to the above					7.75			7.75	8.55	
0850	Sand & stone, loaded at pit, crushed bank gravel				C.Y.	16.10			16.10	17.70	
0950	Sand, washed, for concrete					11.15			11.15	12.25	
1000	For plaster or brick					23			23	25.50	
1050	Stone, 3/4″ to 1-1/2″					18.40			18.40	20	
1055	Round, river stone					19.50			19.50	21.50	
1100	3/8″ roofing stone & 1/2″ pea stone					23			23	25.50	
1150	For trucking 10 miles, add to the above					4.90			4.90	5.40	
1200	30 miles, add to the above					10.85			10.85	11.95	
1310	Quartz chips, 50 lb. bags				Cwt.	15.25			15.25	16.80	
1330	Silica chips, 50 lb. bags				"	8.35			8.35	9.20	
1350	Slag				Ton	7.75			7.75	8.55	
1410	White marble, 3/8″ to 1/2″, 50 lb. bags				Cwt.	12.60			12.60	13.85	
1430	3/4″ F.O.B. plant				Ton	25.50			25.50	28	
0010	**CEMENT** Material only (R033-060)										114
0050	Masonry, gray, T.L. or C.L. lots				Bag	5.80			5.80	6.40	
0060	L.T.L. or L.C.L. lots					5.60			5.60	6.15	
0100	Masonry, white, T.L. or C.L. lots					15.85			15.85	17.45	
0120	L.T.L. or L.C.L. lots					16.40			16.40	18.05	
0240	Portland, type I, plain/air entrained, TL lots, 94 lb bags					7			7	7.70	
0300	Trucked in bulk, per cwt				Cwt.	4.05			4.05	4.45	
0400	Type III, high early strength, TL lots, 94 lb bags				Bag	7.90			7.90	8.70	
0420	L.T.L. or L.C.L. lots					8.90			8.90	9.80	
0500	White high early strength, T.L. or C.L. lots, bags					16.90			16.90	18.60	
0520	L.T.L. or L.C.L. lots					21.50			21.50	23.50	
0600	White, T.L. or C.L. lots, bags					16.50			16.50	18.15	
0620	L.T.L. or L.C.L. lots					20			20	22	
0010	**CONCRETE ADMIXTURES & SURFACE TREATMENTS**										118
0040	Abrasives, aluminum oxide, over 20 tons				Lb.	.88			.88	.97	
0070	Under 1 ton					.92			.92	1.01	
0100	Silicon carbide, black, over 20 tons					1.20			1.20	1.32	
0120	Under 1 ton					1.25			1.25	1.38	
0200	Air entraining agent, .7 to 1.5 oz. per bag, 55 gallon lots				Gal.	7.40			7.40	8.10	
0220	5 gallon lots					8.45			8.45	9.25	
0300	Bonding agent, acrylic latex (200-250 S.F. per gallon)					26			26	28.50	
0320	Epoxy resin (70-80 S.F. per gallon)					50			50	55	
0400	Calcium chloride, 100 lb. bags, FOB plant, truckload lots				Ton	330			330	365	
0420	Less than truckload lots				Bag	19.75			19.75	21.50	
0500	Carbon black, liquid, 2 to 8 lbs. per bag of cement				Lb.	2.85			2.85	3.14	
0600	Colors, integral, 2 to 10 lb. per bag of cement, minimum					1.30			1.30	1.43	
0610	Average					2.07			2.07	2.28	
0620	Maximum					3.69			3.69	4.06	
0700	Curing compound, (200 to 400 S.F. per gallon), 55 gal. lots				Gal.	9.90			9.90	10.90	

expanded coverage of these items see *Means Concrete & Masonry Cost Data 1999*

	033 100 | Structural Concrete	CREW	DAILY OUTPUT	LABOR-HOURS	UNIT	1999 BARE COSTS MAT.	LABOR	EQUIP.	TOTAL	TOTAL INCL O&P	
118	0720 5 gallon lots				Gal.	8.40			8.40	9.25	118
	0800 Premium grade, (450 S.F. per gallon), 55 gallon lots					12.80			12.80	14.10	
	0820 5 gallon lots					14.25			14.25	15.70	
	0900 Dustproofing compound, (200-600 S.F./gal.), 55 gallon lots					7.75			7.75	8.55	
	0920 5 gallon lots					8.90			8.90	9.80	
	1000 Epoxy dustproof coating, colors, (300-400 S.F. per coat),										
	1010 or transparent, (400-600 S.F. per coat)				Gal.	56			56	61.50	
	1100 Hardeners, metallic, 55 lb. bags, natural (grey)				Lb.	.76			.76	.84	
	1200 Colors, average					.95			.95	1.05	
	1300 Non-metallic, 55 lb. bags, natural (grey), minimum					.31			.31	.34	
	1310 Maximum					.35			.35	.39	
	1320 Non-metallic, colors, mininum					.43			.43	.47	
	1340 Maximum					.69			.69	.76	
	1400 Non-metallic, non-slip, 100 lb. bags, minimum					.43			.43	.47	
	1420 Maximum					.85			.85	.94	
	1500 Solution type, (300 to 400 S.F. per gallon)				Gal.	5.80			5.80	6.40	
	1550 Release agent, for tilt slabs					6.40			6.40	7.05	
	1570 For forms, average					4.40			4.40	4.84	
	1600 Sealer, hardener and dustproofer, clear, 450 S.F., minimum					6.30			6.30	6.90	
	1620 Maximum					14			14	15.40	
	1700 Colors (300-400 S.F. per gallon)					36			36	39.50	
	1800 Set accelerator for below freezing, 1 to 1-1/2 gal. per C.Y.					5.25			5.25	5.80	
	1900 Set retarder, 2 to 4 fl. oz. per bag of cement					10.75			10.75	11.80	
	2000 Waterproofing, integral 1 lb. per bag of cement				Lb.	.72			.72	.79	
	2100 Powdered metallic, 40 lbs. per 100 S.F., minimum					.95			.95	1.05	
	2120 Maximum					1.95			1.95	2.15	
	2200 Water reducing admixture, average				Gal.	8.75			8.75	9.65	
122	0010 **CONCRETE, FIELD MIX** FOB forms 2250 psi R033-080				C.Y.	62.50			62.50	69	122
	0020 3000 psi				"	65			65	71.50	
126	0010 **CONCRETE, READY MIX** Regular weight R033-060										126
	0020 2000 psi				C.Y.	56.50			56.50	62	
	0100 2500 psi R033-070					57			57	63	
	0150 3000 psi					61			61	67	
	0200 3500 psi					61			61	67	
	0300 4000 psi					62.50			62.50	68.50	
	0350 4500 psi					63.50			63.50	69.50	
	0400 5000 psi					67.50			67.50	74	
	0411 6000 psi					77			77	84.50	
	0412 8000 psi					125			125	138	
	0413 10,000 psi					178			178	196	
	0414 12,000 psi					215			215	236	
	1000 For high early strength cement, add					10%					
	1010 For structural lightweight with regular sand, add					25%					
	2000 For all lightweight aggregate, add					45%					
	3000 For integral colors, 2500 psi, 5 bag mix										
	3100 Red, yellow or brown, 1.8 lb. per bag, add				C.Y.	13.05			13.05	14.35	
	3200 9.4 lb. per bag, add					70.50			70.50	77.50	
	3400 Black, 1.8 lb. per bag, add					15.50			15.50	17.05	
	3500 7.5 lb. per bag, add					65			65	71.50	
	3700 Green, 1.8 lb. per bag, add					31			31	34	
	3800 7.5 lb. per bag, add					149			149	163	
	4000 Flowable fill, ash cement aggregate water										
	4100 40 - 80 psi				C.Y.	25.50			25.50	28	
	4150 Structural: ash cement aggregate water & sand										
	4200 50 psi				C.Y.	49			49	54	

Important: See the Reference Section for critical supporting data - Reference Nos., Crews, & City Cost Indexe

033 100 | Structural Concrete

	033 100 Structural Concrete	CREW	DAILY OUTPUT	LABOR-HOURS	UNIT	1999 BARE COSTS MAT.	LABOR	EQUIP.	TOTAL	TOTAL INCL O&P
126										
250	140 psi R033-060				C.Y.	51			51	56
300	500 psi					53			53	58.50
350	1000 psi R033-070				↓	56			56	62
130										
010	**CONCRETE IN PLACE** Including forms (4 uses), reinforcing R033-100									
300	Beams, 5 kip per L.F., 10′ span	C-14A	15.62	12.804	C.Y.	207	350	44	601	835
350	25′ span R033-130	"	18.55	10.782		188	296	37	521	720
500	Chimney foundations, industrial, minimum	C-14C	32.22	3.476		126	91	1.15	218.15	285
510	Maximum	"	23.71	4.724		147	123	1.56	271.56	360
700	Columns, square, 12″ x 12″, minimum reinforcing	C-14A	11.96	16.722		210	460	57	727	1,025
300	16″ x 16″, minimum reinforcing		16.22	12.330		178	340	42	560	785
900	24″ x 24″, minimum reinforcing		23.66	8.453		149	232	29	410	565
000	36″ x 36″, minimum reinforcing		33.69	5.936		143	163	20.50	326.50	440
200	16″ diameter, minimum reinforcing		31.49	6.351		211	174	21.50	406.50	535
300	20″ diameter, minimum reinforcing		41.04	4.873		200	134	16.65	350.65	450
400	24″ diameter, minimum reinforcing		51.85	3.857		188	106	13.20	307.20	390
500	36″ diameter, minimum reinforcing	↓	75.04	2.665	↓	176	73	9.10	258.10	320
100	Elevated slabs including finish, not									
110	including forms or reinforcing									
150	Regular concrete, 4″ slab	C-8	2,613	.021	S.F.	.77	.52	.25	1.54	1.91
200	6″ slab		2,585	.022		1.21	.52	.25	1.98	2.40
250	2-1/2″ thick floor fill		2,685	.021		.52	.50	.24	1.26	1.61
300	Lightweight, 110# per C.F., 2-1/2″ thick floor fill		2,585	.022		.64	.52	.25	1.41	1.78
400	Cellular concrete, 1-5/8″ fill, under 5000 S.F.		2,000	.028		.45	.67	.32	1.44	1.90
450	Over 10,000 S.F.	↓	2,200	.025	↓	.35	.61	.29	1.25	1.65
800	Footings, spread under 1 C.Y.	C-14C	38.07	2.942	C.Y.	89.50	77	.97	167.47	223
850	Over 5 C.Y.		81.04	1.382		82.50	36	.46	118.96	149
900	Footings, strip, 18″ x 9″, plain		41.04	2.729		80.50	71.50	.90	152.90	204
950	36″ x 12″, reinforced		61.55	1.820		83	47.50	.60	131.10	168
000	Foundation mat, under 10 C.Y.		38.67	2.896		113	75.50	.96	189.46	247
050	Over 20 C.Y.	↓	56.40	1.986		101	52	.66	153.66	194
200	Grade walls, 8″ thick, 8′ high	C-14D	45.83	4.364		97.50	119	14.90	231.40	310
250	14′ high		27.26	7.337		124	199	25	348	480
260	12″ thick, 8′ high		64.32	3.109		89.50	84.50	10.65	184.65	244
270	14′ high		40.01	4.999		99.50	136	17.10	252.60	345
300	15″ thick, 8′ high		80.02	2.499		85.50	68	8.55	162.05	210
350	12′ high		51.26	3.902		89	106	13.35	208.35	281
500	18′ high	↓	48.85	4.094	↓	99	111	14	224	300
520	Handicap access ramp, railing both sides, 3′ wide	C-14H	14.58	3.292	L.F.	94.50	89	2.57	186.07	248
525	5′ wide		12.22	3.928		108	106	3.06	217.06	290
530	With cheek walls and rails both sides, 3′ wide		8.55	5.614		96.50	151	4.38	251.88	350
535	5′ wide	↓	7.31	6.566	↓	97	177	5.10	279.10	395
650	Slab on grade, not including finish, 4″ thick	C-14E	60.75	1.449	C.Y.	73.50	39	.62	113.12	145
700	6″ thick	"	92	.957	"	70.50	25.50	.41	96.41	120
751	Slab on grade, incl. troweled finish, not incl. forms									
760	or reinforcing, over 10,000 S.F., 4″ thick slab	C-14F	3,425	.021	S.F.	.79	.52	.01	1.32	1.67
820	6″ thick slab		3,350	.021		1.16	.53	.01	1.70	2.09
840	8″ thick slab		3,184	.023		1.59	.56	.01	2.16	2.60
900	12″ thick slab		2,734	.026		2.38	.65	.01	3.04	3.63
950	15″ thick slab	↓	2,505	.029	↓	2.99	.71	.01	3.71	4.39
000	Slab on grade, incl. textured finish, not incl. forms									
001	or reinforcing, 4″ thick slab	C-14G	2,873	.019	S.F.	.79	.48	.01	1.28	1.61
010	6″ thick		2,590	.022		1.24	.53	.01	1.78	2.18
020	8″ thick	↓	2,320	.024	↓	1.62	.59	.02	2.23	2.70
200	Lift slab in place above the foundation, incl. forms,									
210	reinforcing, concrete and columns, minimum	C-14B	2,113	.098	S.F.	4.74	2.69	.32	7.75	9.85

033 | Cast-In-Place Concrete

033 100 | Structural Concrete			CREW	DAILY OUTPUT	LABOR-HOURS	UNIT	1999 BARE COSTS MAT.	LABOR	EQUIP.	TOTAL	TOTAL INCL O&P
130	5900	Pile caps, incl. forms and reinf., sq. or rect., under 5 C.Y. R033-100	C-14C	54.14	2.069	C.Y.	86	54	.68	140.68	182
	5950	Over 10 C.Y.		75	1.493		83.50	39	.49	122.99	155
	6000	Triangular or hexagonal, under 5 C.Y. R033-130		53	2.113		79	55	.70	134.70	176
	6050	Over 10 C.Y.		85	1.318		83.50	34.50	.43	118.43	147
	6200	Retaining walls, gravity, 4′ high see division 022-708	C-14D	66.20	3.021		81	82	10.35	173.35	230
	6250	10′ high		125	1.600		72	43.50	5.45	120.95	154
	6300	Cantilever, level backfill loading, 8′ high		70	2.857		88	77.50	9.75	175.25	231
	6350	16′ high		91	2.198		85.50	59.50	7.50	152.50	197
	6800	Stairs, not including safety treads, free standing, 3′-6″ wide	C-14H	83	.578	LF Nose	5.75	15.60	.45	21.80	32
	6850	Cast on ground		125	.384	″	4.05	10.35	.30	14.70	21
	7000	Stair landings, free standing		200	.240	S.F.	2.25	6.50	.19	8.94	13
	7050	Cast on ground		475	.101	″	1.30	2.73	.08	4.11	5.85
134	0010	**CURING** Burlap, 4 uses assumed, 7.5 oz.	2 Clab	55	.291	C.S.F.	2.55	6.25		8.80	12.60
	0100	12 oz.		55	.291		3.74	6.25		9.99	13.90
	0200	Waterproof curing paper, 2 ply, reinforced		70	.229		4.71	4.90		9.61	12.90
	0300	Sprayed membrane curing compound		95	.168		2.42	3.61		6.03	8.35
	0400	Curing blankets, 1″ to 2″ thick, buy, minimum				S.F.	1.10			1.10	1.21
	0450	Maximum					2.75			2.75	3.03
	0500	Electrically heated pads, 110 volts, 15 watts per S.F., buy					4.50			4.50	4.95
	0600	20 watts per S.F., buy					6			6	6.60
	0710	Electrically, heated pads, 15 watts/S.F., 20 uses, minimum					.16			.16	.18
	0800	Maximum					.27			.27	.29
156	0010	**GROUT** Column & machine bases, non-shrink, metallic, 1″ deep	1 Cefi	35	.229	S.F.	4.25	6		10.25	13.60
	0050	2″ deep		25	.320		8.50	8.35		16.85	22
	0300	Non-shrink, non-metallic, 1″ deep		35	.229		7.50	6		13.50	17.15
	0350	2″ deep		25	.320		14.80	8.35		23.15	29
160	0010	**GUNITE**									
	0020	Applied in 1″ layers, no mesh included	C-8	2,000	.028	S.F.	.92	.67	.32	1.91	2.41
	0100	Mesh for gunite 2 x 2, #12, to 3″ thick	2 Rodm	800	.020		.17	.61		.78	1.24
	0150	Over 3″ thick	″	500	.032		.21	.97		1.18	1.92
	0300	Typical in place, including mesh, 2″ thick, minimum	C-16	1,000	.072		1.68	1.83	.65	4.16	5.50
	0350	Maximum		500	.144		2.60	3.67	1.29	7.56	10.15
	0500	4″ thick, minimum		750	.096		2.49	2.45	.86	5.80	7.60
	0550	Maximum		350	.206		3.86	5.25	1.85	10.96	14.65
	0900	Prepare old walls, no scaffolding, minimum	C-10	1,000	.024		.57	.59		1.16	1.52
	0950	Maximum	″	275	.087		1.80	2.15		3.95	5.25
	1100	For high finish requirement or close tolerance, add, minimum						50%			
	1150	Maximum						110%			
162	0010	**SHOTCRETE** Placed @ 10 C.Y. per hour, 3000 psi	C-8C	80	.600	C.Y.	75	14.25	4.70	93.95	110
	0100	2nd pour	C-6	15	3.200	″	67.50	72	4.99	144.49	192
	1010	Fiber reinforced, 1″ thick	C-8E	1,160	.028	S.F.	.56	.68	.67	1.91	2.40
	1020	2″ thick		600	.053		1.12	1.31	1.30	3.73	4.67
	1030	3″ thick		550	.058		1.68	1.43	1.42	4.53	5.60
	1040	4″ thick		500	.064		2.24	1.57	1.56	5.37	6.60
168	0010	**PATCHING CONCRETE**									
	0100	Floors, 1/4″ thick, small areas, regular grout	1 Cefi	170	.047	S.F.	.07	1.23		1.30	1.91
	0150	Epoxy grout	″	100	.080	″	.52	2.09		2.61	3.70
	0300	Slab on Grade, cut outs, up to 50 C.F.	2 Cefi	50	.320	C.F.	4.52	8.35		12.87	17.45
	2000	Walls, including chipping, cleaning and epoxy grout									
	2100	Minimum	1 Cefi	65	.123	S.F.	.12	3.22		3.34	4.94
	2150	Average		50	.160		.19	4.18		4.37	6.45
	2200	Maximum		40	.200		.35	5.25		5.60	8.20
	2400	Walls, including chipping or sand blasting,									
	2410	Priming, and two part polymer mix, min.	1 Cefi	80	.100	S.F.	.08	2.61		2.69	3.99
	2420	Average		60	.133		.12	3.49		3.61	5.35
	2430	Maximum		40	.200		.16	5.25		5.41	8

033 100 | Structural Concrete		CREW	DAILY OUTPUT	LABOR-HOURS	UNIT	1999 BARE COSTS MAT.	LABOR	EQUIP.	TOTAL	TOTAL INCL O&P
2510	Underlayment, P.C based self-leveling, 4100 psi, pumped, 1/4"	C-8	20,000	.003	S.F.	1.16	.07	.03	1.26	1.42
2520	1/2"		19,000	.003		2.09	.07	.03	2.19	2.45
2530	3/4"		18,000	.003		3.25	.07	.04	3.36	3.73
2540	1"		17,000	.003		4.41	.08	.04	4.53	5
2550	1-1/2"		15,000	.004		6.75	.09	.04	6.88	7.60
2560	Hand mix, 1/2"	C-18	4,000	.002		2.09	.05	.01	2.15	2.39
2610	Topping, P.C. based self-level/dry 6100 psi, pumped, 1/4"	C-8	20,000	.003		1.76	.07	.03	1.86	2.08
2620	1/2"		19,000	.003		3.17	.07	.03	3.27	3.63
2630	3/4"		18,000	.003		4.93	.07	.04	5.04	5.55
2660	1"		17,000	.003		6.70	.08	.04	6.82	7.50
2670	1-1/2"		15,000	.004		10.20	.09	.04	10.33	11.45
2680	Hand mix, 1/2"	C-18	4,000	.002		3.17	.05	.01	3.23	3.57
0010	**PLACING CONCRETE** and vibrating, including labor & equipment R033-090									
0050	Beams, elevated, small beams, pumped	C-20	60	1.067	C.Y.		24.50	12	36.50	51.50
0100	With crane and bucket	C-7	45	1.600			37	19.85	56.85	79.50
0200	Large beams, pumped	C-20	90	.711			16.45	8	24.45	34.50
0250	With crane and bucket	C-7	65	1.108			26	13.75	39.75	55
0400	Columns, square or round, 12" thick, pumped	C-20	60	1.067			24.50	12	36.50	51.50
0450	With crane and bucket	C-7	40	1.800			42	22.50	64.50	89.50
0600	18" thick, pumped	C-20	90	.711			16.45	8	24.45	34.50
0650	With crane and bucket	C-7	55	1.309			30.50	16.25	46.75	65
0800	24" thick, pumped	C-20	92	.696			16.10	7.85	23.95	33.50
0850	With crane and bucket	C-7	70	1.029			24	12.75	36.75	51
1000	36" thick, pumped	C-20	140	.457			10.60	5.15	15.75	22
1050	With crane and bucket	C-7	100	.720			16.75	8.95	25.70	36
1400	Elevated slabs, less than 6" thick, pumped	C-20	140	.457			10.60	5.15	15.75	22
1450	With crane and bucket	C-7	95	.758			17.65	9.40	27.05	38
1500	6" to 10" thick, pumped	C-20	160	.400			9.25	4.50	13.75	19.35
1550	With crane and bucket	C-7	110	.655			15.20	8.10	23.30	32.50
1600	Slabs over 10" thick, pumped	C-20	180	.356			8.25	4	12.25	17.20
1650	With crane and bucket	C-7	130	.554			12.90	6.85	19.75	27.50
1900	Footings, continuous, shallow, direct chute	C-6	120	.400			9.05	.62	9.67	14.75
1950	Pumped	C-20	150	.427			9.90	4.80	14.70	20.50
2000	With crane and bucket	C-7	90	.800			18.60	9.90	28.50	40
2100	Footings, continuous, deep, direct chute	C-6	140	.343			7.75	.53	8.28	12.65
2150	Pumped	C-20	160	.400			9.25	4.50	13.75	19.35
2200	With crane and bucket	C-7	110	.655			15.20	8.10	23.30	32.50
2400	Footings, spread, under 1 C.Y., direct chute	C-6	55	.873			19.70	1.36	21.06	32
2450	Pumped	C-20	65	.985			23	11.10	34.10	47.50
2500	With crane and bucket	C-7	45	1.600			37	19.85	56.85	79.50
2600	Footings, spread, over 5 C.Y., direct chute	C-6	120	.400			9.05	.62	9.67	14.75
2650	Pumped	C-20	150	.427			9.90	4.80	14.70	20.50
2700	With crane and bucket	C-7	100	.720			16.75	8.95	25.70	36
2900	Foundation mats, over 20 C.Y., direct chute	C-6	350	.137			3.10	.21	3.31	5.05
2950	Pumped	C-20	400	.160			3.71	1.80	5.51	7.75
3000	With crane and bucket	C-7	300	.240			5.60	2.98	8.58	11.90
3200	Grade beams, direct chute	C-6	150	.320			7.20	.50	7.70	11.80
3250	Pumped	C-20	180	.356			8.25	4	12.25	17.20
3300	With crane and bucket	C-7	120	.600			13.95	7.45	21.40	29.50
3700	Pile caps, under 5 C.Y., direct chute	C-6	90	.533			12.05	.83	12.88	19.65
3750	Pumped	C-20	110	.582			13.45	6.55	20	28
3800	With crane and bucket	C-7	80	.900			21	11.15	32.15	45
000	Over 10 C.Y., direct chute	C-6	215	.223			5.05	.35	5.40	8.25
050	Pumped	C-20	240	.267			6.20	3	9.20	12.90
100	With crane and bucket	C-7	185	.389			9.05	4.83	13.88	19.35
300	Slab on grade, 4" thick, direct chute	C-6	110	.436			9.85	.68	10.53	16.10

168

172

3 CONCRETE

033 100 | Structural Concrete

Ref	Line	Description	Crew	Daily Output	Labor-Hours	Unit	1999 Bare Costs Mat.	Labor	Equip.	Total	Total Incl O&P
172	4350	Pumped (R033-090)	C-20	130	.492	C.Y.		11.40	5.55	16.95	24
	4400	With crane and bucket	C-7	110	.655			15.20	8.10	23.30	32.50
	4600	Over 6" thick, direct chute	C-6	165	.291			6.55	.45	7	10.75
	4650	Pumped	C-20	185	.346			8	3.90	11.90	16.75
	4700	With crane and bucket	C-7	145	.497			11.55	6.15	17.70	24.50
	4900	Walls, 8" thick, direct chute	C-6	90	.533			12.05	.83	12.88	19.65
	4950	Pumped	C-20	100	.640			14.80	7.20	22	31
	5000	With crane and bucket	C-7	80	.900			21	11.15	32.15	45
	5050	12" thick, direct chute	C-6	100	.480			10.85	.75	11.60	17.70
	5100	Pumped	C-20	110	.582			13.45	6.55	20	28
	5200	With crane and bucket	C-7	90	.800			18.60	9.90	28.50	40
	5300	15" thick, direct chute	C-6	105	.457			10.30	.71	11.01	16.85
	5350	Pumped	C-20	120	.533			12.35	6	18.35	26
	5400	With crane and bucket	C-7	95	.758			17.65	9.40	27.05	38
	5600	Wheeled concrete dumping, add to placing costs above									
	5610	Walking cart, 50' haul, add	C-18	32	.281	C.Y.		6.10	1.56	7.66	11.30
	5620	150' haul, add		24	.375			8.15	2.08	10.23	15.10
	5700	250' haul, add		18	.500			10.85	2.77	13.62	20
	5800	Riding cart, 50' haul, add	C-19	80	.112			2.44	1.04	3.48	4.99
	5810	150' haul, add		60	.150			3.25	1.39	4.64	6.65
	5900	250' haul, add		45	.200			4.33	1.85	6.18	8.85
184	0010	**STAIR TREAD INSERTS** Cast iron, abrasive, 3" wide	1 Carp	90	.089	L.F.	5.60	2.43		8.03	9.95
	0020	4" wide		80	.100		6.85	2.73		9.58	11.80
	0040	6" wide		75	.107		8.40	2.91		11.31	13.85
	0050	9" wide		70	.114		12.45	3.12		15.57	18.60
	0100	12" wide		65	.123		18.65	3.36		22.01	26
	0300	Cast aluminum, compared to cast iron, deduct					10%				
	0500	Extruded aluminum safety tread, 3" wide	1 Carp	75	.107		4.75	2.91		7.66	9.80
	0550	4" wide		75	.107		6.40	2.91		9.31	11.65
	0600	6" wide		75	.107		10.75	2.91		13.66	16.40
	0650	9" wide to resurface stairs		70	.114		15.15	3.12		18.27	21.50
	1700	Cement filled pan type, plain	1 Cefi	115	.070	S.F.	2.15	1.82		3.97	5.10
	1750	Non-slip	"	100	.080	"	3.23	2.09		5.32	6.65
192	0010	**WATERPROOFING AND DAMPPROOFING** See division 071									
	0050	Integral waterproofing, add to cost of regular concrete				C.Y.	4.20			4.20	4.62
196	0010	**WINTER PROTECTION** For heated ready mix, add, minimum					3.75			3.75	4.13
	0050	Maximum					4.75			4.75	5.20
	0100	Protecting concrete and temporary heat, add, minimum	2 Clab	6,000	.003	S.F.	.15	.06		.21	.26
	0150	Maximum, see also division 010-094	"	2,000	.008	"	.65	.17		.82	.98
	0200	Temporary shelter for slab on grade, wood frame and polyethylene									
	0201	sheeting, minimum	2 Carp	10	1.600	M.S.F.	275	43.50		318.50	375
	0210	Maximum	"	3	5.333	"	330	146		476	595
	0300	See also Division 033-134									

033 450 | Concrete Finishing

Ref	Line	Description	Crew	Daily Output	Labor-Hours	Unit	1999 Bare Costs Mat.	Labor	Equip.	Total	Total Incl O&P
454	0010	**FINISHING FLOORS** Monolithic, screed finish	1 Cefi	900	.009	S.F.		.23		.23	.35
	0050	Darby finish		750	.011			.28		.28	.42
	0100	Screed and float finish		725	.011			.29		.29	.43
	0150	Screed, float, and broom finish		630	.013			.33		.33	.50
	0200	Screed, float, and hand trowel		600	.013			.35		.35	.52
	0250	Machine trowel		550	.015			.38		.38	.57
	0400	Integral topping and finish, using 1:1:2 mix, 3/16" thick	C-10	1,000	.024		.06	.59		.65	.95
	0450	1/2" thick		950	.025		.15	.62		.77	1.10
	0500	3/4" thick		850	.028		.23	.69		.92	1.30
	0600	1" thick		750	.032		.30	.79		1.09	1.52

Important: See the Reference Section for critical supporting data - Reference Nos., Crews, & City Cost Indexe

033 450 | Concrete Finishing

		CREW	DAILY OUTPUT	LABOR-HOURS	UNIT	1999 BARE COSTS MAT.	LABOR	EQUIP.	TOTAL	TOTAL INCL O&P	
800	Granolithic topping, laid after, 1:1:1-1/2 mix, 1/2" thick	C-10	590	.041	S.F.	.16	1		1.16	1.70	454
820	3/4" thick		580	.041		.24	1.02		1.26	1.81	
850	1" thick		575	.042		.32	1.03		1.35	1.91	
950	2" thick		500	.048		.65	1.18		1.83	2.50	
200	Heavy duty, 1:1:2, 3/4" thick, preshrunk, gray, 20 MSF		320	.075		.21	1.84		2.05	3.02	
300	100 MSF		380	.063		.21	1.55		1.76	2.58	
350	For colors, .50 psf, add, minimum		1,650	.015		.41	.36		.77	1	
400	Maximum		1,500	.016		1.86	.39		2.25	2.65	
600	Exposed local aggregate finish, minimum	1 Cefi	625	.013		.36	.33		.69	.90	
650	Maximum		465	.017		1.14	.45		1.59	1.92	
800	Floor abrasives, .25 psf, add to above, aluminum oxide		850	.009		.15	.25		.40	.54	
850	Silicon carbide		850	.009		.18	.25		.43	.57	
000	Floor hardeners, metallic, light service, .50 psf, add		850	.009		.27	.25		.52	.67	
050	Medium service, .75 psf, add		750	.011		.39	.28		.67	.85	
100	Heavy service, 1.0 psf, add		650	.012		.54	.32		.86	1.07	
150	Extra heavy, 1.5 psf, add		575	.014		.62	.36		.98	1.22	
300	Non-metallic, light service, .50 psf, add		850	.009		.17	.25		.42	.56	
350	Medium service, .75 psf, add		750	.011		.24	.28		.52	.68	
400	Heavy service, 1.00 psf, add		650	.012		.31	.32		.63	.82	
450	Extra heavy, 1.50 psf, add		575	.014		.48	.36		.84	1.06	
600	Add for colored hardeners, metallic					50%					
650	Non-metallic					25%					
800	Trap rock wearing surface for monolithic floors										
810	2.0 psf, add to above	C-10	1,250	.019	S.F.	.78	.47		1.25	1.58	
000	Floor coloring, dusted on, 0.5 psf per S.F., add to above, min.	1 Cefi	1,300	.006		.57	.16		.73	.87	
050	Maximum	"	625	.013		2	.33		2.33	2.69	
100	Colors only, minimum				Lb.	.96			.96	1.06	
120	Maximum				"	3.36			3.36	3.70	
200	Integral colors, see division 033-126										
600	1/2" topping using 5 lb. per bag, regular colors	C-10	590	.041	S.F.	.18	1		1.18	1.72	
650	Blue or green	"	590	.041		.18	1		1.18	1.72	
800	Dustproofing, silicate liquids, 1 coat	1 Cefi	1,900	.004		.07	.11		.18	.24	
850	2 coats		1,300	.006		.13	.16		.29	.38	
000	Epoxy coating, 1 coat, clear		1,500	.005		.36	.14		.50	.61	
050	Colors		1,500	.005		.36	.14		.50	.61	
400	Stair finish, float		275	.029			.76		.76	1.13	
500	Steel trowel finish		200	.040			1.05		1.05	1.56	
600	Silicon carbide finish, .25 psf		150	.053		.50	1.39		1.89	2.63	
010	**FINISHING WALLS** Break ties and patch voids	1 Cefi	540	.015	S.F.	.03	.39		.42	.61	458
050	Burlap rub with grout		450	.018		.03	.46		.49	.72	
100	Carborundum rub, dry		270	.030		.03	.77		.80	1.19	
150	Wet rub		175	.046		.03	1.20		1.23	1.81	
300	Bush hammer, green concrete	B-39	1,000	.048		.03	1.09	.18	1.30	1.94	
350	Cured concrete	"	650	.074		.03	1.68	.28	1.99	2.95	
500	Acid etch	1 Cefi	575	.014		.15	.36		.51	.71	
600	Float finish, 1/16" thick	"	300	.027		.02	.70		.72	1.06	
700	Sandblast, light penetration	C-10	1,100	.022		.20	.54		.74	1.03	
750	Heavy penetration	"	375	.064		.40	1.57		1.97	2.82	
010	**ROLLER COMPACTED CONCRETE** No material, spread and compact										460
100	Mass placement, 1' lift, 1' layer	B-10C	1,280	.009	C.Y.		.24	.74	.98	1.20	
200	2' lift, 6" layer	"	1,600	.008			.20	.59	.79	.95	
210	Vertical face, formed, 1' lift	B-11V	400	.060			1.29	.23	1.52	2.29	
220	6" lift	"	200	.120			2.57	.47	3.04	4.56	
300	Sloped face, nonformed, 1' lift	B-11L	384	.042			1.04	1.50	2.54	3.25	
360	6" lift	"	192	.083			2.08	3	5.08	6.50	
400	Surface preparation, vacuum truck	B-6A	3,280	.006	S.Y.		.15	.10	.25	.34	

expanded coverage of these items see ***Means Concrete & Masonry Cost Data 1999***

033 | Cast-In-Place Concrete

033 450 | Concrete Finishing

Ref.	Line	033 450 Concrete Finishing	Crew	Daily Output	Labor-Hours	Unit	Mat. (1999 Bare Costs)	Labor (1999 Bare Costs)	Equip. (1999 Bare Costs)	Total (1999 Bare Costs)	Total Incl O&P
460	0450	Water clean	B-9A	3,000	.008	S.Y.		.17	.18	.35	.47
	0460	Water blast	B-9B	800	.030			.65	.74	1.39	1.83
	0500	Joint bedding placement, 1" thick	B-11C	975	.016	↓		.41	.22	.63	.88
	0510	Conveyance of materials, 18 C.Y. truck, 5 min. cycle	B-34F	2,048	.004	C.Y.		.09	.51	.60	.69
	0520	10 min. cycle		1,024	.008			.17	1.01	1.18	1.38
	0540	15 min. cycle	↓	680	.012			.26	1.53	1.79	2.08
	0550	With crane and bucket	C-23A	1,600	.025			.60	.85	1.45	1.87
	0560	With 4 C.Y. loader, 4 min. cycle	B-10U	480	.025			.65	1.93	2.58	3.12
	0570	8 min. cycle		240	.050			1.30	3.86	5.16	6.25
	0580	12 min. cycle	↓	160	.075			1.96	5.80	7.76	9.35
	0590	With belt conveyor	C-7D	600	.093			2.12	.30	2.42	3.65
	0600	With 17 C.Y. scraper, 5 min. cycle	B-33J	1,440	.006			.16	1.12	1.28	1.48
	0610	10 min. cycle		720	.011			.32	2.25	2.57	2.95
	0620	15 min. cycle		480	.017			.47	3.37	3.84	4.43
	0630	20 min. cycle	↓	360	.022	↓		.63	4.49	5.12	5.90
	0640	Water cure, small job, < 500 CY	B-94C	8	1	Hr.		21.50	7.30	28.80	42
	0650	Large job, over 500 CY	B-59A	8	3	"		65	66	131	174
	0660	RCC paving, with asphalt paver including material	B-25C	1,000	.048	C.Y.	40	1.16	1.54	42.70	47.50
	0670	8" thick layers		4,200	.011	S.Y.	10.50	.28	.37	11.15	12.40
	0680	12" thick layers	↓	2,800	.017	"	15.60	.41	.55	16.56	18.40

034 | Precast Concrete

034 100 | Structural Precast

Ref.	Line	034 100 Structural Precast	Crew	Daily Output	Labor-Hours	Unit	Mat. (1999 Bare Costs)	Labor (1999 Bare Costs)	Equip. (1999 Bare Costs)	Total (1999 Bare Costs)	Total Incl O&P
104	0011	**BEAMS,** "L" shaped, 20' span, 12" x 20"	C-11	32	2.250	Ea.	1,350	67.50	48.50	1,466	1,675
	1000	Inverted tee beams, add to above, small beams				L.F.	15%				
	1050	Large beams				"	5.55			5.55	6.10
	1200	Rectangular, 20' span, 12" x 20"	C-11	32	2.250	Ea.	925	67.50	48.50	1,041	1,200
	1250	18" x 36"		24	3		1,700	90	64.50	1,854.50	2,100
	1300	24" x 44"		22	3.273		2,450	98	70.50	2,618.50	2,950
	1400	30' span, 12" x 36"		24	3		2,175	90	64.50	2,329.50	2,600
	1450	18" x 44"		20	3.600		3,050	108	77.50	3,235.50	3,625
	1500	24" x 52"		16	4.500		4,325	135	97	4,557	5,100
	1600	40' span, 12" x 52"		20	3.600		4,025	108	77.50	4,210.50	4,700
	1650	18" x 52"		16	4.500		4,900	135	97	5,132	5,750
	1700	24" x 52"		12	6		6,000	180	129	6,309	7,050
	2000	"T" shaped, 20' span, 12" x 20"		32	2.250		1,600	67.50	48.50	1,716	1,950
	2050	18" x 36"		24	3		2,550	90	64.50	2,704.50	3,050
	2100	24" x 44"		22	3.273		3,600	98	70.50	3,768.50	4,200
	2200	30' span, 12" x 36"		24	3		3,650	90	64.50	3,804.50	4,225
	2250	18" x 44"		20	3.600		4,975	108	77.50	5,160.50	5,750
	2300	24" x 52"		16	4.500		5,150	135	97	5,382	6,000
	2500	40' span, 12" x 52"		20	3.600		6,850	108	77.50	7,035.50	7,825
	2550	18" x 52"		16	4.500		7,500	135	97	7,732	8,600
	2600	24" x 52"	↓	12	6	↓	9,150	180	129	9,459	10,600
136	0010	**SLABS** Prestressed roof/floor members, solid, grouted, 4" thick	C-11	3,600	.020	S.F.	4.65	.60	.43	5.68	6.65
	0050	6" thick		4,500	.016		4.48	.48	.35	5.31	6.15
	0100	8" thick, hollow		5,600	.013		4.83	.39	.28	5.50	6.30
	0150	10" thick	↓	8,800	.008	↓	5.70	.25	.18	6.13	6.85

034 | Precast Concrete

034 100 | Structural Precast		CREW	DAILY OUTPUT	LABOR-HOURS	UNIT	1999 BARE COSTS MAT.	LABOR	EQUIP.	TOTAL	TOTAL INCL O&P	
0200	12″ thick	C-11	8,000	.009	S.F.	4.78	.27	.19	5.24	5.95	136

034 500 | Architectural Precast		CREW	DAILY OUTPUT	LABOR-HOURS	UNIT	MAT.	LABOR	EQUIP.	TOTAL	TOTAL INCL O&P	
0011	**WALL PANELS** Material only										504
0050	Uninsulated 4″ thick, smooth gray										
0150	Low rise, 4′ x 8′x 4″ thick	C-11	320	.225	S.F.	8.70	6.75	4.85	20.30	27	
0200	8′x 8′ x 4″ thick		576	.125		8.60	3.74	2.70	15.04	19.05	
0250	8′x 16′x 4″ thick		1,024	.070		8.55	2.11	1.52	12.18	14.80	
0600	High rise, 4′ x 8′ x 4″ thick		288	.250		8.70	7.50	5.40	21.60	29	
0650	8′ x 8′x 4″ thick		512	.141		8.60	4.21	3.03	15.84	20	
0700	8′ x 16′ x 4″ thick		768	.094		8.55	2.81	2.02	13.38	16.60	
0800	Insulated panel, 2″ polystyrene, add					1.60			1.60	1.76	
0850	2″ urethane, add					2.10			2.10	2.31	
1200	Finishes, white, add					1.70			1.70	1.87	
1250	Exposed aggregate, add					.80			.80	.88	
1300	Granite faced, domestic, add					27			27	29.50	
2200	Fiberglass reinforced cement with urethane core										
2210	R20, 8′ x 8′, minimum	E-2	750	.075	S.F.	8.30	2.22	1.35	11.87	14.55	
2220	Maximum	″	600	.093	″	15.50	2.78	1.69	19.97	24	

034 700 | Tilt-Up Precast		CREW	DAILY OUTPUT	LABOR-HOURS	UNIT	MAT.	LABOR	EQUIP.	TOTAL	TOTAL INCL O&P	
0010	**TILT-UP** Wall panel construction, walls only, 5-1/2″ thick	C-14	1,600	.090	S.F.	2.44	2.39	.69	5.52	7.30	704
0100	7-1/2″ thick	″	1,550	.093	″	2.80	2.47	.71	5.98	7.80	

034 800 | Precast Specialties		CREW	DAILY OUTPUT	LABOR-HOURS	UNIT	MAT.	LABOR	EQUIP.	TOTAL	TOTAL INCL O&P	
0010	**LINTELS**										802
0800	Precast concrete, 4″ wide, 8″ high, to 5′ long	D-1	175	.091	L.F.	5.60	2.25		7.85	9.65	
0850	5′-12′ long	D-4	190	.168		5.85	4.13	.58	10.56	13.50	
1000	6″ wide, 8″ high, to 5′ long		185	.173		8.35	4.24	.59	13.18	16.40	
1050	5′-12′ long		190	.168		6.95	4.13	.58	11.66	14.70	
1200	8″ wide, 8″ high, to 5′ long		185	.173		9.60	4.24	.59	14.43	17.75	
1250	5′-12′ long		190	.168		8.85	4.13	.58	13.56	16.80	
1400	10″ wide, 8″ high, to 14′ long		180	.178		31.50	4.36	.61	36.47	42.50	
1450	12″ wide, 8″ high, to 19′ long		185	.173		33.50	4.24	.59	38.33	44	
0010	**STAIRS**, Precast concrete treads on steel stringers, 3′ wide	C-12	75	.640	Riser	53.50	17.30	5.90	76.70	92.50	804
0200	Front entrance 4′ wide with 48″ platform, 2 risers		17	2.824	Flight	265	76	26	367	440	
0250	5 risers		13	3.692		315	99.50	34	448.50	540	
0301	5′ wide with 48″ platform, 2 risers		16	3		278	81	27.50	386.50	465	
1200	Basement entrance stairs, steel bulkhead doors, minimum	B-51	22	2.182		460	47.50	7.80	515.30	590	
1250	Maximum	″	11	4.364		680	95.50	15.60	791.10	915	

037 | Concrete Restoration & Cleaning

037 300 | Concrete Rehabilitation		CREW	DAILY OUTPUT	LABOR-HOURS	UNIT	1999 BARE COSTS MAT.	LABOR	EQUIP.	TOTAL	TOTAL INCL O&P	
0010	**SLABJACKING**										355
0100	4″ thick slab	D-4	1,500	.021	S.F.	.20	.52	.07	.79	1.11	
0150	6″ thick slab		1,200	.027		.28	.65	.09	1.02	1.41	
0200	8″ thick slab		1,000	.032		.33	.78	.11	1.22	1.69	
0250	10″ thick slab		900	.036		.35	.87	.12	1.34	1.87	
0300	12″ thick slab		850	.038		.39	.92	.13	1.44	1.99	

Division Notes

	CREW	DAILY OUTPUT	LABOR-HOURS	UNIT	1999 BARE COSTS MAT.	LABOR	EQUIP.	TOTAL	TOTAL INCL O&P

)ivision 4
1asonry

'mating Tips

1ortar & Masonry Accessories

terms *mortar* and *grout* are often used
rchangeably, and incorrectly. Mortar is
l to bed masonry units, seal the entry of
nd moisture, provide architectural
earance, and allow for size variations in
units. Grout is used primarily in
forced masonry construction and is used
ond the masonry to the reinforcing
l. Common mortar types are M(2500
, S(1800 psi), N(750 psi), and O(350 psi),
conform to ASTM C270. Grout is either
or coarse, conforms to ASTM C476, and
lace strengths generally exceed 2500 psi.
tar and grout are different components
asonry construction and are placed by
rely different methods. An estimator
uld be aware of their unique uses and
s.

042 Unit Masonry

- The most common types of unit masonry are brick and concrete masonry. The major classifications of brick are building brick (ASTM C62), facing brick (ASTM C216) and glazed brick, fire brick and pavers. Many varieties of texture and appearance can exist within these classifications, and the estimator would be wise to check local custom and availability within the project area. On repair and remodeling jobs, matching the existing brick may be the most important criteria.
- Brick and concrete block are priced by the piece and then converted into a price per square foot of wall. Openings less than two square feet are generally ignored by the estimator because any savings in units used is offset by the cutting and trimming required.
- All masonry walls, whether interior or exterior, require bracing. The cost of bracing walls during construction should be included by the estimator and this bracing must remain in place until permanent bracing is complete. Permanent bracing of masonry walls is accomplished by masonry itself, in the form of pilasters or abutting wall corners, or by anchoring the walls to the structural frame. Accessories in the form of anchors, anchor slots and ties are used, but their supply and installation can be by different trades. For instance, anchor slots on spandrel beams and columns are supplied and welded in place by the steel fabricator, but the ties from the slots into the masonry are installed by the bricklayer. Regardless of the installation method the estimator must be certain that these accessories are accounted for in pricing.

Reference Numbers

Reference numbers are shown in bold squares at the beginning of some major classifications. These numbers refer to related items in the Reference Section. The reference information may be an estimating procedure, an alternate pricing method or technical information.

Note: Not all subdivisions listed here necessarily appear in this publication.

041 000 | Mortar				CREW	DAILY OUTPUT	LABOR-HOURS	UNIT	1999 BARE COSTS MAT.	LABOR	EQUIP.	TOTAL	TOTAL INCL O&P
008	0010	**CEMENT** Gypsum 80 lb. bag, T.L. lots					Bag	11.25			11.25	12.40
	0050	L.T.L. lots						11.75			11.75	12.95
	0100	Masonry, 70 lb. bag, T.L. lots						5.80			5.80	6.40
	0150	L.T.L. lots						5.60			5.60	6.15
	0200	White, 70 lb. bag, T.L. lots						15.85			15.85	17.45
	0250	L.T.L. lots						16.40			16.40	18.05
012	0010	**COLORS** 50 lb. bags (2 bags per M bricks),	R041-100									
	0020	range 2 to 10 lb. per bag of cement, minimum					Lb.	1.50			1.50	1.65
	0050	Average						2.50			2.50	2.75
	0100	Maximum						5			5	5.50
016	0010	**GROUTING** Bond bms. & lintels, 8″ dp., pumped, not incl. block	R041-100									
	0020	8″ thick, 0.2 C.F. per L.F.		D-4	1,400	.023	L.F.	.66	.56	.08	1.30	1.69
	0050	10″ thick, 0.25 C.F. per L.F.			1,200	.027		.83	.65	.09	1.57	2.02
	0060	12″ thick, 0.3 C.F. per L.F.			1,040	.031		.99	.75	.11	1.85	2.37
	0200	Concrete block cores, solid, 4″ thk., by hand, 0.067 C.F./S.F.		D-8	1,100	.036	S.F.	.22	.92		1.14	1.66
	0210	6″ thick, pumped, 0.175 C.F. per S.F.		D-4	720	.044		.58	1.09	.15	1.82	2.49
	0250	8″ thick, pumped, 0.258 C.F. per S.F.			680	.047		.85	1.15	.16	2.16	2.90
	0300	10″ thick, pumped, 0.340 C.F. per S.F.			660	.048		1.13	1.19	.17	2.49	3.26
	0350	12″ thick, pumped, 0.422 C.F. per S.F.			640	.050		1.40	1.23	.17	2.80	3.62
	0500	Cavity walls, 2″ space, pumped, no shoring, 0.167 C.F./S.F.			1,700	.019		.55	.46	.06	1.07	1.39
	0550	3″ space, 0.250 C.F./S.F.			1,200	.027		.83	.65	.09	1.57	2.02
	0600	4″ space, 0.333 C.F. per S.F.			1,150	.028		1.10	.68	.10	1.88	2.36
	0700	6″ space, 0.500 C.F. per S.F.			800	.040		1.65	.98	.14	2.77	3.48
	0800	Door frames, 3′ x 7′ opening, 2.5 C.F. per opening			60	.533	Opng.	8.25	13.05	1.82	23.12	31
	0850	6′ x 7′ opening, 3.5 C.F. per opening			45	.711	"	11.60	17.40	2.43	31.43	42.50
	2000	Grout, C476, for bond beams, lintels and CMU cores			350	.091	C.F.	3.31	2.24	.31	5.86	7.45
020	0010	**LIME** Masons, hydrated, 50 lb. bag, T.L. lots					Bag	5.40			5.40	5.95
	0050	L.T.L. lots						5.65			5.65	6.20
	0200	Finish, double hydrated, 50 lb. bag, T.L. lots						6.50			6.50	7.15
	0250	L.T.L. lots						7.40			7.40	8.15
024	0010	**MORTAR**	R041-100									
	0020	With masonry cement										
	0100	Type M, 1:1:6 mix	R041-200	1 Brhe	143	.056	C.F.	3.17	1.21		4.38	5.35
	0200	Type N, 1:3 mix			143	.056		2.96	1.21		4.17	5.15
	0300	Type O, 1:3 mix			143	.056		2.97	1.21		4.18	5.15
	0400	Type PM, 1:1:6 mix, 2500 psi			143	.056		3.17	1.21		4.38	5.35
	0500	Type S, 1/2:1:4 mix			143	.056		3.36	1.21		4.57	5.60
	2000	With portland cement and lime										
	2100	Type M, 1:1/4:3 mix		1 Brhe	143	.056	C.F.	3.82	1.21		5.03	6.10
	2200	Type N, 1:1:6 mix, 750 psi			143	.056		3.11	1.21		4.32	5.30
	2300	Type O, 1:2:9 mix (Pointing Mortar)			143	.056		3.02	1.21		4.23	5.20
	2400	Type PL, 1:1/2:4 mix, 2500 psi			143	.056		3.47	1.21		4.68	5.70
	2500	Type K, 1:3:12 mix, 75 psi			143	.056		2.98	1.21		4.19	5.15
	2600	Type S, 1:1/2:4 mix, 1800 psi			143	.056		3.47	1.21		4.68	5.70
	2700	Mortar for glass block			143	.056		7.25	1.21		8.46	9.85
	2800	Gypsum cement mortar						5.60			5.60	6.15
	2900	Mortar for Fire Brick, 80 lb. bag, T.L. Lots					Bag	12.60			12.60	13.85
028	0010	**POINTING MORTAR** See also Division 041-024					Lb.	.86			.86	.95
	0050	White					"	1.02			1.02	1.12
032	0010	**SAND** For mortar, screened and washed, at the pit					Ton	14.45			14.45	15.90
	0050	With 10 mile haul						15.75			15.75	17.30
	0100	With 30 mile haul						17			17	18.70
	0200	Screened and washed, at the pit					C.Y.	20			20	22
	0250	With 10 mile haul						22			22	24
	0300	With 30 mile haul						23.50			23.50	26

041 000 | Mortar

			CREW	DAILY OUTPUT	LABOR-HOURS	UNIT	1999 BARE COSTS MAT.	LABOR	EQUIP.	TOTAL	TOTAL INCL O&P
036	0010	**SURFACE BONDING** CMU walls with fiberglass mortar,									
	0020	gray or white colors, not incl. block work	1 Bric	540	.015	S.F.	.50	.41		.91	1.19
040	0010	**WATERPROOFING** Admixture, 1 qt. to 2 bags of masonry cement				Qt.	6			6	6.60

041 500 | Masonry Accessories

			CREW	DAILY OUTPUT	LABOR-HOURS	UNIT	1999 BARE COSTS MAT.	LABOR	EQUIP.	TOTAL	TOTAL INCL O&P
508	0010	**CONTROL JOINT** Rubber, 4" and wider wall	1 Bric	400	.020	L.F.	2.40	.55		2.95	3.50
	0050	PVC, 4" wall		400	.020		1.36	.55		1.91	2.36
	0100	Rubber, 6" wall		320	.025		4.30	.69		4.99	5.80
	0120	PVC, 6" wall		320	.025		1.62	.69		2.31	2.85
	0140	Rubber, 8" and wider wall		280	.029		4.40	.79		5.19	6.05
	0160	PVC, 8" wall		280	.029		1.67	.79		2.46	3.07
	0180	12" wall		240	.033		2.79	.92		3.71	4.50
512	0010	**REINFORCING** Steel bars A615, placed horiz., #3 & #4 bars (R041-500)	1 Bric	450	.018	Lb.	.27	.49		.76	1.05
	0020	#5 & #6 bars		800	.010		.27	.28		.55	.72
	0050	Placed vertical, #3 & #4 bars		350	.023		.27	.63		.90	1.27
	0060	#5 & #6 bars		650	.012		.27	.34		.61	.82
	0500	Joint reinforcing, ladder type, mill std galvanized									
	0600	9 ga. sides, 9 ga. ties, 4" wall	1 Bric	30	.267	C.L.F.	6.90	7.35		14.25	19.05
	0650	6" wall		30	.267		7.30	7.35		14.65	19.45
	0700	8" wall		25	.320		7.65	8.85		16.50	22
	0750	10" wall		20	.400		8.15	11.05		19.20	26
	0800	12" wall		20	.400		8.60	11.05		19.65	26.50
	1000	Truss type									
	1100	9 ga. sides, 9 ga. ties, 4" wall	1 Bric	30	.267	C.L.F.	9.85	7.35		17.20	22.50
	1150	6" wall		30	.267		9.95	7.35		17.30	22.50
	1200	8" wall		25	.320		10.40	8.85		19.25	25
	1250	10" wall		20	.400		10.70	11.05		21.75	29
	1300	12" wall		20	.400		11.25	11.05		22.30	29.50
	1500	3/16" sides, 9 ga. ties, 4" wall		30	.267		12.85	7.35		20.20	25.50
	1550	6" wall		30	.267		13.10	7.35		20.45	26
	1600	8" wall		25	.320		13.45	8.85		22.30	28.50
	1650	10" wall		20	.400		13.95	11.05		25	32.50
	1700	12" wall		20	.400		14.50	11.05		25.55	33
	2000	3/16" sides, 3/16" ties, 4" wall		30	.267		18.40	7.35		25.75	32
	2050	6" wall		30	.267		18.65	7.35		26	32
	2100	8" wall		25	.320		19.30	8.85		28.15	35.50
	2150	10" wall		20	.400		20	11.05		31.05	39.50
	2200	12" wall		20	.400		21.50	11.05		32.55	40.50
	2500	Cavity truss type, galvanized									
	2600	9 ga. sides, 9 ga. ties, 4" wall	1 Bric	25	.320	C.L.F.	9.80	8.85		18.65	24.50
	2650	6" wall		25	.320		10.35	8.85		19.20	25
	2700	8" wall		20	.400		10.40	11.05		21.45	28.50
	2750	10" wall		15	.533		10.75	14.70		25.45	35
	2800	12" wall		15	.533		11.35	14.70		26.05	35.50
	3000	3/16" sides, 9 ga. ties, 4" wall		25	.320		13.45	8.85		22.30	28.50
	3050	6" wall		25	.320		13.70	8.85		22.55	29
	3100	8" wall		20	.400		14.05	11.05		25.10	32.50
	3150	10" wall		15	.533		14.50	14.70		29.20	39
	3200	12" wall		15	.533		15.15	14.70		29.85	39.50
	3500	For hot dip galvanizing, add					50%				
518	0010	**WALL PLUGS** For nailing to brickwork, 26 ga., galvanized, plain	1 Bric	10.50	.762	C	23.50	21		44.50	58.50
	0050	Wood filled	"	10.50	.762	"	80	21		101	121
520	0010	**WALL TIES** To brick veneer, galv., corrugated, 7/8" x 7", 22 Ga.	1 Bric	10.50	.762	C	4.47	21		25.47	37.50
	0100	24 Ga.		10.50	.762		3.97	21		24.97	37

or expanded coverage of these items see *Means Concrete & Masonry Cost Data 1999*

041 500 | Masonry Accessories

520		CREW	DAILY OUTPUT	LABOR-HOURS	UNIT	1999 BARE COSTS MAT.	LABOR	EQUIP.	TOTAL	TOTAL INCL O&P
0150	16 Ga.	1 Bric	10.50	.762	C	13.45	21		34.45	47.50
0200	Buck anchors, galv., corrugated, 16 gauge, 2" bend, 8" x 2"		10.50	.762		93	21		114	135
0250	8" x 3"		10.50	.762		99	21		120	142
0300	Adjustable, rectangular, 4-1/8" wide									
0350	Anchor and tie, 3/16" wire, mill galv.									
0400	2-3/4" eye, 3-1/4" tie	1 Bric	1.05	7.619	M	315	210		525	670
0500	2-3/4" eye, 4-3/4" tie		1.05	7.619		345	210		555	705
0520	2-3/4" eye, 5-1/2" tie		1.05	7.619		355	210		565	715
0550	4-3/4" eye, 3-1/4" tie		1.05	7.619		350	210		560	710
0570	4-3/4" eye, 4-3/4" tie		1.05	7.619		380	210		590	740
0580	4-3/4" eye, 5-1/2" tie		1.05	7.619		390	210		600	755
0600	Cavity wall, Z type, galvanized, 6" long, 1/4" diameter		10.50	.762	C	17.45	21		38.45	51.50
0650	3/16" diameter		10.50	.762		9.50	21		30.50	43
0800	8" long, 1/4" diameter		10.50	.762		23.50	21		44.50	58.50
0850	3/16" diameter		10.50	.762		10.45	21		31.45	44
1000	Rectangular type, galvanized, 1/4" diameter, 2" x 6"		10.50	.762		38.50	21		59.50	75
1050	2" x 8" or 4" x 6"		10.50	.762		29	21		50	64
1100	3/16" diameter, 2" x 6"		10.50	.762		17.45	21		38.45	51.50
1150	2" x 8" or 4" x 6"		10.50	.762		19.40	21		40.40	54
1200	Mesh wall tie, 1/2" mesh, hot dip galvanized									
1400	16 gauge, 12" long, 3" wide	1 Bric	9	.889	C	68.50	24.50		93	114
1420	6" wide		9	.889		105	24.50		129.50	153
1440	12" wide		8.50	.941		167	26		193	224
1500	Rigid partition anchors, plain, 8" long, 1" x 1/8"		10.50	.762		48	21		69	85.50
1550	1" x 1/4"		10.50	.762		94	21		115	136
1580	1-1/2" x 1/8"		10.50	.762		66.50	21		87.50	106
1600	1-1/2" x 1/4"		10.50	.762		158	21		179	206
1650	2" x 1/8"		10.50	.762		83	21		104	124
1700	2" x 1/4"		10.50	.762		225	21		246	281
2000	Column flange ties, wire, galvanized									
2300	3/16" diameter, up to 3" wide	1 Bric	10.50	.762	C	54	21		75	92
2350	To 5" wide		10.50	.762		59	21		80	97.50
2400	To 7" wide		10.50	.762		62	21		83	101
2600	To 9" wide		10.50	.762		67	21		88	106
2650	1/4" diameter, up to 3" wide		10.50	.762		85	21		106	126
2700	To 5" wide		10.50	.762		94	21		115	136
2800	To 7" wide		10.50	.762		98	21		119	141
2850	To 9" wide		10.50	.762		107	21		128	151
2900	For hot dip galvanized, add					35%				
4000	Channel slots, 1-3/8" x 1/2" x 8"									
4100	12 gauge, plain	1 Bric	10.50	.762	C	120	21		141	165
4150	16 gauge, galvanized	"	10.50	.762	"	63	21		84	102
4200	Channel slot anchors									
4300	16 gauge, galvanized, 1-1/4" x 3-1/2"				C	28			28	31
4350	1-1/4" x 5-1/2"					33			33	36.50
4400	1-1/4" x 7-1/2"					38			38	42
4500	1/8" plain, 1-1/4" x 3-1/2"					47			47	51.50
4550	1-1/4" x 5-1/2"					57.50			57.50	63.50
4600	1-1/4" x 7-1/2"					68			68	75
4700	For corrugation, add					32			32	35
4750	For hot dip galvanized, add					35%				
5000	Dowels									
5100	Plain, 1/4" diameter, 3" long				C	17			17	18.70
5150	4" long					19.20			19.20	21
5200	6" long					23.50			23.50	26
5300	3/8" diameter, 3" long					23			23	25.50

Important: See the Reference Section for critical supporting data - Reference Nos., Crews, & City Cost Indexes

041 500 | Masonry Accessories		CREW	DAILY OUTPUT	LABOR-HOURS	UNIT	MAT.	LABOR	EQUIP.	TOTAL	TOTAL INCL O&P	
350	4″ long				C	28.50			28.50	31.50	520
400	6″ long					39.50			39.50	43.50	
500	1/2″ diameter, 3″ long					34			34	37.50	
550	4″ long					42			42	46	
600	6″ long					56			56	61.50	
700	5/8″ diameter, 3″ long					47			47	51.50	
750	4″ long					59			59	65	
800	6″ long					83			83	91.50	
000	3/4″ diameter, 3″ long					60			60	66	
100	4″ long					78			78	86	
150	6″ long					114			114	125	
300	For hot dip galvanized, add				↓	35%					

042 | Unit Masonry

042 100 | Brick Masonry		CREW	DAILY OUTPUT	LABOR-HOURS	UNIT	MAT.	LABOR	EQUIP.	TOTAL	TOTAL INCL O&P	
010	**COLUMNS** Brick, scaffolding not included										108
050	8″ x 8″, 9 brick	D-1	56	.286	V.L.F.	3.32	7.05		10.37	14.60	
100	12″ x 8″, 13.5 brick		37	.432		4.99	10.65		15.64	22	
200	12″ x 12″, 20 brick		25	.640		7.40	15.75		23.15	32.50	
300	16″ x 12″, 27 brick		19	.842		9.95	20.50		30.45	43	
400	16″ x 16″, 36 brick		14	1.143		13.30	28		41.30	58	
500	20″ x 16″, 45 brick		11	1.455		16.60	36		52.60	74	
600	20″ x 20″, 56 brick		9	1.778		20.50	43.50		64	91	
700	24″ x 20″, 68 brick		7	2.286		25	56		81	115	
800	24″ x 24″, 81 brick		6	2.667		30	65.50		95.50	135	
000	36″ x 36″, 182 brick	↓	3	5.333	↓	67	131		198	278	
010	**COMMON BUILDING BRICK** C62, TL lots, material only										110
020	Standard, minimum				M	250			250	275	
050	Average (select)				″	310			310	340	
010	**COPING** Stock units										116
050	Precast concrete, 10″ wide, 4″ tapers to 3-1/2″, 8″ wall	D-1	75	.213	L.F.	9.60	5.25		14.85	18.70	
100	12″ wide, 3-1/2″ tapers to 3″, 10″ wall		70	.229		10.10	5.60		15.70	19.85	
110	14″ wide, 4″ tapers to 3-1/2″, 14″ wall		65	.246		9.10	6.05		15.15	19.40	
150	16″ wide, 4″ tapers to 3-1/2″, 14″ wall		60	.267	↓	13.90	6.55		20.45	25.50	
250	Precast concrete corners		40	.400	Ea.	23.50	9.85		33.35	41.50	
300	Limestone for 12″ wall, 4″ thick		90	.178	L.F.	12.90	4.37		17.27	21	
350	6″ thick		80	.200		15.15	4.92		20.07	24.50	
500	Marble, to 4″ thick, no wash, 9″ wide		90	.178		18.70	4.37		23.07	27.50	
550	12″ wide		80	.200		27.50	4.92		32.42	38	
700	Terra cotta, 9″ wide		90	.178		4.55	4.37		8.92	11.80	
750	12″ wide		80	.200		7.60	4.92		12.52	16	
800	Aluminum, for 12″ wall	↓	80	.200	↓	10.75	4.92		15.67	19.45	
010	**FACE BRICK** C216, TL lots, material only R042-550										124
300	Standard modular, 4″ x 2-2/3″ x 8″, minimum				M	330			330	365	
350	Maximum					450			450	495	
450	Economy, 4″ x 4″ x 8″, minimum				↓	425			425	470	

	042 100 | Brick Masonry	CREW	DAILY OUTPUT	LABOR-HOURS	UNIT	1999 BARE COSTS MAT.	LABOR	EQUIP.	TOTAL	TOTAL INCL O&P
124	0500 Maximum R042-550				M	545			545	600
	0510 Economy, 4" x 4" x 12", minimum					425			425	470
	0520 Maximum					545			545	600
	0550 Jumbo, 6" x 4" x 12", minimum					1,150			1,150	1,250
	0600 Maximum					1,425			1,425	1,575
	0610 Jumbo, 8" x 4" x 12", minimum					1,150			1,150	1,250
	0620 Maximum					1,425			1,425	1,575
	0650 Norwegian, 4" x 3-1/5" x 12", minimum					520			520	570
	0700 Maximum					730			730	805
	0710 Norwegian, 6" x 3-1/5" x 12", minimum					520			520	570
	0720 Maximum					730			730	805
	0850 Standard glazed, plain colors, 4" x 2-2/3" x 8", minimum					985			985	1,075
	0900 Maximum					1,150			1,150	1,275
	1000 Deep trim shades, 4" x 2-2/3" x 8", minimum					1,150			1,150	1,250
	1050 Maximum					1,300			1,300	1,425
	1080 Jumbo utility, 4" x 4" x 12"					750			750	825
	1120 4" x 8" x 8"					1,100			1,100	1,200
	1140 4" x 8" x 16"					2,500			2,500	2,750
	1260 Engineer, 4" x 3-1/5" x 8", minimum					295			295	325
	1270 Maximum					485			485	535
	1350 King, 4" x 2-3/4" x 10", minimum					415			415	455
	1360 Maximum					440			440	485
	1400 Norman, 4" x 2-3/4" x 12"					415			415	455
	1450 Roman, 4" x 2" x 12"					415			415	455
	1500 SCR, 6" x 2-2/3" x 12"					415			415	455
	1550 Double, 4" x 5-1/3" x 8"					415			415	455
	1600 Triple, 4" x 5-1/3" x 12"					415			415	455
	1770 Standard modular, double glazed, 4" x 2-2/3" x 8"					1,000			1,000	1,100
	1850 Jumbo, colored glazed ceramic, 6" x 4" x 12"					1,500			1,500	1,650
	2050 Jumbo utility, glazed, 4" x 4" x 12"					1,075			1,075	1,175
	2100 4" x 8" x 8"					1,500			1,500	1,650
	2150 4" x 16" x 8"				↓	3,000			3,000	3,300
	2160 Fire Brick, 2" x 2-2/3" x 9", minimum									
	2165 Maximum				M	910			910	1,000
	2170 For less than truck load lots, add					10				
	2180 For buff or gray brick, add					15				
	3050 Used brick, minimum					280			280	310
	3100 Maximum				↓	305			305	335
	3150 Add for brick to match existing work, minimum					5%				
	3200 Maximum					50%				
162	0010 **SIMULATED BRICK** Aluminum, baked on colors	1 Carp	200	.040	S.F.	2.15	1.09		3.24	4.09
	0050 Fiberglass panels		200	.040		2.25	1.09		3.34	4.20
	0100 Urethane pieces cemented in mastic		150	.053		4.60	1.46		6.06	7.35
	0150 Vinyl siding panels	↓	200	.040		1.90	1.09		2.99	3.81
	0160 Cement base, brick, incl. mastic	D-1	100	.160	↓	3.10	3.94		7.04	9.50
	0170 Corner		50	.320	V.L.F.	7.35	7.85		15.20	20.50
	0180 Stone face, incl. mastic		100	.160	S.F.	6.95	3.94		10.89	13.75
	0190 Corner	↓	50	.320	V.L.F.	8	7.85		15.85	21
170	0010 **STEPS** With select common	D-1	.30	53.333	M	310	1,300		1,610	2,400
184	0010 **WALLS** Building brick, including mortar R042-120									
	0060 Includes 3% brick waste and 25% mortar waste									
	0140 4" thick, facing, 4" x 2-2/3" x 8" R042-500	D-8	1.45	27.586	M	290	695		985	1,400
	0150 4" thick, as back-up, 6.75 bricks per S.F.		1.60	25		290	630		920	1,300
	0204 8" thick, 13.50 bricks per S.F. R042-550		1.80	22.222		297	560		857	1,200
	0250 12" thick, 20.25 bricks per S.F.	↓	1.90	21.053	↓	299	530		829	1,150

042 100 | Brick Masonry

			CREW	DAILY OUTPUT	LABOR-HOURS	UNIT	1999 BARE COSTS MAT.	LABOR	EQUIP.	TOTAL	TOTAL INCL O&P	
0304	16″ thick, 27.00 bricks per S.F.	R042 -120	D-8	2	20	M	300	505		805	1,125	184
0500	Reinforced, 4″ wall, 4″ x 2-2/3″ x 8″			1.40	28.571		300	720		1,020	1,450	
0550	8″ thick, 13.50 bricks per S.F.	R042 -500		1.75	22.857		370	575		945	1,300	
0600	12″ thick, 20.25 bricks per S.F.			1.85	21.622		375	545		920	1,250	
0650	16″ thick, 27.00 bricks per S.F.	R042 -550	↓	1.95	20.513	↓	375	515		890	1,225	
0790	Alternate method of figuring by square foot											
0800	4″ wall, face, 4″ x 2-2/3″ x 8″		D-8	215	.186	S.F.	2.51	4.69		7.20	10.05	
0850	4″ thick, as back up, 6.75 bricks per S.F.			240	.167		1.96	4.20		6.16	8.70	
0900	8″ thick wall, 13.50 brick per S.F.			135	.296		4.01	7.45		11.46	16	
1000	12″ thick wall, 20.25 bricks per S.F.			95	.421		6.05	10.60		16.65	23	
1050	16″ thick wall, 27.00 bricks per S.F.			75	.533		8.15	13.45		21.60	30	
1200	Reinforced, 4″ x 2-2/3″ x 8″ , 4″ wall			205	.195		1.97	4.92		6.89	9.80	
1250	8″ thick wall, 13.50 brick per S.F.			130	.308		4.02	7.75		11.77	16.50	
1300	12″ thick wall, 20.25 bricks per S.F.			90	.444		6.05	11.20		17.25	24	
1350	16″ thick wall, 27.00 bricks per S.F.	↓	↓	70	.571	↓	8.15	14.40		22.55	31.50	

042 200 | Concrete Unit Masonry

		CREW	DAILY OUTPUT	LABOR-HOURS	UNIT	MAT.	LABOR	EQUIP.	TOTAL	TOTAL INCL O&P	
0010	**CONCRETE BLOCK** material only										204
0020	2″ x 8″ x 16″ solid, normal-weight, 2,000 psi				Ea.	.48			.48	.52	
0050	3,500 psi					.59			.59	.65	
0100	5,000 psi					.92			.92	1.01	
0150	Lightweight, std.					.52			.52	.58	
0300	3″ x 8″ x 16″ solid, normal-weight, 2000 psi					.65			.65	.71	
0350	3,500 psi					.79			.79	.87	
0400	5,000 psi					1.15			1.15	1.27	
0450	Lightweight, std.					.65			.65	.71	
0600	4″ x 8″ x 16″ hollow, normal-weight, 2000 psi					.56			.56	.62	
0650	3,500 psi					.63			.63	.69	
0700	5000 psi					.93			.93	1.02	
0750	Lightweight, std.					.64			.64	.70	
1000	75% solid, normal-weight, 2,000 psi					.74			.74	.81	
1050	3,500 psi					.92			.92	1.01	
1100	5,000 psi					1.20			1.20	1.32	
1150	Lightweight, std.					.81			.81	.89	
1300	Solid, normal-weight, 2,000 psi					.95			.95	1.05	
1350	3,500 psi					.78			.78	.86	
1400	5,000 psi					.95			.95	1.05	
1450	Lightweight, std.					.86			.86	.95	
1600	6″ x 8″ x 16″ hollow, normal-weight, 2,000 psi					.84			.84	.92	
1650	3,500 psi					.84			.84	.92	
1700	5,000 psi					1.14			1.14	1.25	
1750	Lightweight, std.					.91			.91	1	
2000	75% solid, normal-weight, 2,000 psi					1.25			1.25	1.38	
2050	3,500 psi					.91			.91	1	
2100	5,000 psi					1.55			1.55	1.70	
2150	Lightweight, std.					1.65			1.65	1.82	
2300	Solid, normal-weight, 2,000 psi					1			1	1.10	
2350	3,500 psi					1.20			1.20	1.32	
2400	5,000 psi					1.50			1.50	1.65	
2450	Lightweight, std.					1.08			1.08	1.19	
2600	8″ x 8″ x 16″ hollow, normal-weight, 2000 psi					.90			.90	.99	
2650	3500 psi					.88			.88	.97	
2700	5,000 psi					1.30			1.30	1.43	
2750	Lightweight, std.					1.03			1.03	1.13	
3000	75% solid, normal-weight, 2,000 psi				↓	1.90			1.90	2.09	

4 MASONRY

042 200 | Concrete Unit Masonry

		Description	CREW	DAILY OUTPUT	LABOR-HOURS	UNIT	1999 BARE COSTS MAT.	LABOR	EQUIP.	TOTAL	TOTAL INCL O&P
204	3050	3,500 psi				Ea.	1.34			1.34	1.47
	3100	5,000 psi					2.35			2.35	2.58
	3150	Lightweight, std.					2.45			2.45	2.70
	3200	Solid, normal-weight, 2,000 psi					1.46			1.46	1.61
	3250	3,500 psi					1.80			1.80	1.98
	3300	5,000 psi					2.05			2.05	2.26
	3350	Lightweight, std.					1.53			1.53	1.68
	3400	10" x 8" x 16" hollow, normal-weight, 2000 psi					.90			.90	.99
	3410	3500 psi					.88			.88	.97
	3420	5,000 psi					1.30			1.30	1.43
	3430	Lightweight, std.					1.03			1.03	1.13
	3440	75% solid, normal-weight, 2,000 psi					1.90			1.90	2.09
	3450	3,500 psi					1.34			1.34	1.47
	3460	5,000 psi					2.35			2.35	2.58
	3470	Lightweight, std.					2.45			2.45	2.70
	3480	Solid, normal-weight, 2,000 psi					1.46			1.46	1.61
	3490	3,500 psi					1.80			1.80	1.98
	3500	5,000 psi					2.05			2.05	2.26
	3510	Lightweight, std.					1.53			1.53	1.68
	3600	12" x 8" x 16" hollow, normal-weight, 2000 psi					1.47			1.47	1.62
	3650	3,500 psi					1.36			1.36	1.50
	3700	5,000 psi					1.66			1.66	1.83
	3750	Lightweight, std.					1.60			1.60	1.75
	4000	75% solid, normal-weight, 2,000 psi					2			2	2.20
	4050	3500 psi					1.82			1.82	2
	4100	5,000 psi					2.50			2.50	2.75
	4150	Lightweight, std.					2.60			2.60	2.86
	4300	Solid, normal-weight, 2,000 psi					2.40			2.40	2.64
	4350	3,500 psi					2.10			2.10	2.31
	4400	5,000 psi					3			3	3.30
	4500	Lightweight, std.				↓	3.10			3.10	3.41
220	0010	**CONCRETE BLOCK, DECORATIVE** Scaffolding not included									
	0020	Embossed, simulated brick face									
	0100	8" x 16" units, 4" thick	D-8	400	.100	S.F.	2.11	2.52		4.63	6.25
	0200	8" thick		340	.118		2.90	2.96		5.86	7.80
	0250	12" thick	↓	300	.133	↓	3.73	3.36		7.09	9.30
	0400	Embossed both sides									
	0500	8" thick	D-8	300	.133	S.F.	3.22	3.36		6.58	8.75
	0550	12" thick	"	275	.145	"	4.27	3.67		7.94	10.40
	1000	Fluted high strength									
	1100	Flutes 1 side, 8" x 16" x 4" thick	D-8	345	.116	S.F.	2.59	2.92		5.51	7.40
	1150	Flutes 2 sides, 8" x 16" x 4" thick		335	.119		3.11	3.01		6.12	8.10
	1200	8" thick	↓	300	.133		4.01	3.36		7.37	9.60
	1250	For special colors, add				↓	.23			.23	.25
	1400	Deep grooved, smooth face									
	1450	8" x 16" x 4" thick	D-8	345	.116	S.F.	1.63	2.92		4.55	6.35
	1500	8" thick	"	300	.133	"	2.80	3.36		6.16	8.30
	2000	Formblock, incl. inserts & reinforcing									
	2100	8" x 16" x 8" thick	D-8	345	.116	S.F.	2.90	2.92		5.82	7.75
	2150	12" thick	"	310	.129	"	3.56	3.25		6.81	8.95
	2500	Ground face									
	2600	8" x 16" x 4" thick	D-8	345	.116	S.F.	3.89	2.92		6.81	8.80
	2650	6" thick		310	.129		4.32	3.25		7.57	9.80
	2700	8" thick	↓	290	.138		4.99	3.48		8.47	10.90
	2750	12" thick	D-9	265	.181	↓	5.85	4.46		10.31	13.40

042 200 | Concrete Unit Masonry

	042 200 | Concrete Unit Masonry	CREW	DAILY OUTPUT	LABOR-HOURS	UNIT	1999 BARE COSTS MAT.	LABOR	EQUIP.	TOTAL	TOTAL INCL O&P	
2900	For special colors, add, minimum					15%					220
2950	For special colors, add, maximum					45%					
4000	Slump block										
4100	4" face height x 16" x 4" thick	D-1	165	.097	S.F.	3.72	2.39		6.11	7.80	
4150	6" thick		160	.100		4.56	2.46		7.02	8.85	
4200	8" thick		155	.103		6.20	2.54		8.74	10.75	
4250	10" thick		140	.114		9.25	2.81		12.06	14.55	
4300	12" thick		130	.123		9.85	3.03		12.88	15.55	
4400	6" face height x 16" x 6" thick		155	.103		3.73	2.54		6.27	8.05	
4450	8" thick		150	.107		4.86	2.62		7.48	9.45	
4500	10" thick		130	.123		7.40	3.03		10.43	12.85	
4550	12" thick		120	.133		7.75	3.28		11.03	13.60	
5000	Split rib profile units, 1" deep ribs, 8 ribs										
5100	8" x 16" x 4" thick	D-8	345	.116	S.F.	1.58	2.92		4.50	6.30	
5150	6" thick		325	.123		1.91	3.10		5.01	6.95	
5200	8" thick		305	.131		2.35	3.30		5.65	7.75	
5250	12" thick	D-9	275	.175		3.05	4.29		7.34	10.05	
5400	For special deeper colors, 4" thick, add					.17			.17	.19	
5450	12" thick, add					.32			.32	.35	
5600	For white, 4" thick, add					.69			.69	.76	
5650	6" thick, add					.92			.92	1.01	
5700	8" thick, add					1.13			1.13	1.24	
5750	12" thick, add					1.58			1.58	1.74	
6000	Split face or scored split face										
6100	8" x 16" x 4" thick	D-8	350	.114	S.F.	1.87	2.88		4.75	6.55	
6150	6" thick		315	.127		2.29	3.20		5.49	7.50	
6200	8" thick		295	.136		3.61	3.42		7.03	9.25	
6250	12" thick	D-9	270	.178		4.84	4.37		9.21	12.15	
6400	For special deeper colors, 4" thick, add					.21			.21	.24	
6450	6" thick, add					.21			.21	.24	
6500	8" thick, add					.23			.23	.25	
6550	12" thick, add					.40			.40	.44	
6650	For white, 4" thick, add					.75			.75	.82	
6700	6" thick, add					.90			.90	.99	
6750	8" thick, add					1.15			1.15	1.27	
6800	12" thick, add					1.62			1.62	1.78	
7000	Scored ground face, 2 to 5 scores										
7100	8" x 16" x 4" thick	D-8	345	.116	S.F.	3.49	2.92		6.41	8.35	
7150	6" thick		310	.129		4.20	3.25		7.45	9.65	
7200	8" thick		290	.138		4.71	3.48		8.19	10.60	
7250	12" thick	D-9	265	.181		5.95	4.46		10.41	13.50	
8000	Hexagonal face profile units, 8" x 16" units										
8100	4" thick, hollow	D-8	345	.116	S.F.	1.97	2.92		4.89	6.70	
8200	Solid		345	.116		2.80	2.92		5.72	7.60	
8300	6" thick, hollow		310	.129		2.28	3.25		5.53	7.55	
8350	8" thick, hollow		290	.138		3.12	3.48		6.60	8.85	
8500	For stacked bond, add						26%				
8550	For high rise construction, add per story	D-8	67.80	.590	M.S.F.		14.85		14.85	23	
8600	For scored block, add					10%					
8650	For honed or ground face, per face, add				Ea.	.24			.24	.26	
8700	For honed or ground end, per end, add				"	2.43			2.43	2.67	
8750	For bullnose block, add					10%					
8800	For special color, add					13%					
0010	**CONCRETE BLOCK, PARTITIONS** Scaffolding not included R042-200										232
1000	Lightweight block, tooled joints, 2 sides, hollow										

MASONRY 4

042 | Unit Masonry

4 MASONRY

042 200 | Concrete Unit Masonry

		042 200 Concrete Unit Masonry	CREW	DAILY OUTPUT	LABOR-HOURS	UNIT	1999 BARE COSTS MAT.	LABOR	EQUIP.	TOTAL	TOTAL INCL O&P
232	1100	Not reinforced, 8" x 16" x 4" thick (R042-200)	D-8	440	.091	S.F.	.86	2.29		3.15	4.50
	1200	8" thick	"	385	.104		3.01	2.62		5.63	7.40
	1300	12" thick	D-9	350	.137		3.28	3.37		6.65	8.85
	1500	Reinforced alternate courses, 4" thick	D-8	435	.092		1.16	2.32		3.48	4.88
	1650	8" thick	"	380	.105		1.46	2.65		4.11	5.70
	1750	12" thick	D-9	345	.139	↓	2.21	3.42		5.63	7.75
	4000	Regular block, tooled joints, 2 sides, hollow									
	4100	Not reinforced, 8" x 16" x 4" thick	D-8	430	.093	S.F.	.77	2.34		3.11	4.49
	4200	8" thick	"	375	.107		1.26	2.69		3.95	5.55
	4300	12" thick	D-9	340	.141		2	3.47		5.47	7.60
	4500	Reinforced alternate courses, 8" x 16" x 4" thick	D-8	425	.094		.84	2.37		3.21	4.62
	4600	8" thick		370	.108		1.34	2.72		4.06	5.70
	4650	10" thick	↓	355	.113		2.03	2.84		4.87	6.65
	4700	12" thick	D-9	335	.143	↓	2.09	3.52		5.61	7.80
252	0010	**INSULATION** See also division 072-108									
	0100	Inserts, styrofoam, plant installed, add to block prices									
	0200	8" x 16" units, 6" thick				S.F.	.74			.74	.81
	0250	8" thick					.74			.74	.81
	0300	10" thick					.87			.87	.96
	0350	12" thick					.91			.91	1
	0500	8" x 8" units, 8" thick					.60			.60	.66
	0550	12" thick				↓	.72			.72	.79
256	0010	**PARGETING** Regular Portland cement, 1/2" thick	D-1	2.50	6.400	C.S.F.	12.85	157		169.85	259
	5100	Waterproof Portland cement	"	2.50	6.400	"	14.05	157		171.05	260
258	0010	**PATIO TILES**									
	0050	Marble, non-slip, 1/2" thick, flame finish	D-1	75	.213	S.F.	11.25	5.25		16.50	20.50

042 300 | Reinforced Unit Masonry

		042 300 Reinforced Unit Masonry	CREW	DAILY OUTPUT	LABOR-HOURS	UNIT	MAT.	LABOR	EQUIP.	TOTAL	TOTAL INCL O&P
304	0010	**CONCRETE BLOCK BOND BEAM** Scaffolding not included									
	0020	Not including grout or reinforcing									
	0100	Regular block, 8" high, 8" thick	D-8	565	.071	L.F.	1.77	1.78		3.55	4.71
	0150	12" thick	D-9	510	.094		2.36	2.32		4.68	6.20
	0500	Lightweight, 8" high, 8" thick	D-8	575	.070		1.57	1.75		3.32	4.46
	0550	12" thick	D-9	520	.092	↓	2.74	2.27		5.01	6.55
	2000	Including grout and 2 #5 bars									
	2100	Regular block, 8" high, 8" thick	D-8	300	.133	L.F.	3.12	3.36		6.48	8.65
	2150	12" thick	D-9	250	.192		4.14	4.72		8.86	11.90
	2500	Lightweight, 8" high, 8" thick	D-8	305	.131		3.29	3.30		6.59	8.75
	2550	12" thick	D-9	255	.188	↓	4.52	4.63		9.15	12.20
310	0010	**CONCRETE BLOCK, EXTERIOR** Not including scaffolding (R042-200)									
	0020	Reinforced alt courses, tooled joints 2 sides, foam inserts									
	0100	Regular, 8" x 16" x 6" thick	D-8	390	.103	S.F.	1.27	2.58		3.85	5.40
	0200	8" thick		365	.110		1.81	2.76		4.57	6.30
	0250	10" thick	↓	355	.113		2.27	2.84		5.11	6.90
	0300	12" thick	D-9	330	.145		2.46	3.58		6.04	8.25
	0500	Lightweight, 8" x 16" x 6" thick	D-8	410	.098		2.11	2.46		4.57	6.15
	0600	8" thick		385	.104		3.07	2.62		5.69	7.45
	0650	10" thick	↓	370	.108		2.67	2.72		5.39	7.20
	0700	12" thick	D-9	350	.137	↓	3.34	3.37		6.71	8.95
320	0010	**CONCRETE BLOCK FOUNDATION WALL** Scaffolding not included									
	0050	Normal-weight, trowel cut joints, parged 1/2" thick, no reinforcing									

042 300 | Reinforced Unit Masonry

		Line	Description	CREW	DAILY OUTPUT	LABOR-HOURS	UNIT	1999 BARE COSTS MAT.	LABOR	EQUIP.	TOTAL	TOTAL INCL O&P	
320	0200	Hollow, 8" x 16" x 6" thick	D-8	450	.089	S.F.	1.36	2.24		3.60	4.98	320
	0250	8" thick		430	.093		1.50	2.34		3.84	5.30	
	0300	10" thick	↓	420	.095		2.08	2.40		4.48	6	
	0350	12" thick	D-9	395	.122		2.25	2.99		5.24	7.15	
	0500	Solid, 8" x 16" block, 6" thick	D-8	440	.091		1.55	2.29		3.84	5.25	
	0550	8" thick	"	415	.096		2.13	2.43		4.56	6.10	
	0600	12" thick	D-9	380	.126	↓	3.30	3.11		6.41	8.45	
	1000	Reinforced										
	1100	Hollow, 8" x 16" block, 4" thick	D-8	455	.088	S.F.	1.38	2.22		3.60	4.96	
	1125	6" thick		445	.090		2	2.27		4.27	5.70	
	1150	8" thick		425	.094		2.40	2.37		4.77	6.35	
	1200	10" thick	↓	415	.096		3.26	2.43		5.69	7.35	
	1250	12" thick	D-9	390	.123		3.70	3.03		6.73	8.80	
	1500	Solid, 8" x 16" block, 6" thick	D-8	435	.092		1.55	2.32		3.87	5.30	
	1600	8" thick	"	410	.098		2.13	2.46		4.59	6.15	
	1650	12" thick	D-9	375	.128	↓	3.30	3.15		6.45	8.55	
330	0010	**CONCRETE BLOCK, LINTELS** Scaffolding not included										330
	0100	Including grout and horizontal reinforcing										
	0200	8" x 8" x 8", 1 #4 bar	D-4	300	.107	L.F.	3.31	2.61	.36	6.28	8.10	
	0250	2 #4 bars		295	.108		3.43	2.66	.37	6.46	8.30	
	0400	8" x 16" x 8", 1 #4 bar		275	.116		5.80	2.85	.40	9.05	11.25	
	0450	2 #4 bars		270	.119		5.95	2.90	.41	9.26	11.45	
	1000	12" x 8" x 8", 1 #4 bar		275	.116		4.68	2.85	.40	7.93	10	
	1100	2 #4 bars		270	.119		4.79	2.90	.41	8.10	10.20	
	1150	2 #5 bars		270	.119		4.92	2.90	.41	8.23	10.35	
	1200	2 #6 bars		265	.121		5.10	2.96	.41	8.47	10.60	
	1500	12" x 16" x 8", 1 #4 bar		250	.128		7.70	3.14	.44	11.28	13.85	
	1600	2 #3 bars		245	.131		7.70	3.20	.45	11.35	13.95	
	1650	2 #4 bars		245	.131		7.80	3.20	.45	11.45	14.05	
	1700	2 #5 bars	↓	240	.133	↓	7.95	3.27	.46	11.68	14.30	

042 550 | Masonry Veneer

	Line	Description	CREW	DAILY OUTPUT	LABOR-HOURS	UNIT	MAT.	LABOR	EQUIP.	TOTAL	TOTAL INCL O&P	
554	0010	**BRICK VENEER** Scaffolding not included, truck load lots R042-120										554
	0015	Material costs incl. 3% brick and 25% mortar waste										
	0020	Standard, select common, 4" x 2-2/3" x 8" (6.75/S.F.) R042-180	D-8	1.50	26.667	M	355	670		1,025	1,450	
	0050	Red, 4" x 2-2/3" x 8", running bond		1.50	26.667		380	670		1,050	1,475	
	0100	Full header every 6th course (7.88/S.F.) R042-500		1.45	27.586		375	695		1,070	1,500	
	0150	English, full header every 2nd course (10.13/S.F.)		1.40	28.571		375	720		1,095	1,550	
	0200	Flemish, alternate header every course (9.00/S.F.) R042-550		1.40	28.571		375	720		1,095	1,550	
	0300	Full headers throughout (13.50/S.F.)		1.40	28.571		375	720		1,095	1,550	
	0450	Soldier course (6.75/S.F.)		1.40	28.571		380	720		1,100	1,550	
	0500	Sailor course (4.50/S.F.)		1.30	30.769		380	775		1,155	1,625	
	0601	Buff or gray face, running bond, (6.75/S.F.)		1.50	26.667		380	670		1,050	1,475	
	0700	Glazed face, 4" x 2-2/3" x 8", running bond		1.40	28.571		1,100	720		1,820	2,325	
	1000	Jumbo, 6" x 4" x 12",(3.00/S.F.)		1.30	30.769		1,300	775		2,075	2,625	
	1051	Norman, 4" x 2-2/3" x 12" (4.50/S.F.)		1.45	27.586		910	695		1,605	2,075	
	1100	Norwegian, 4" x 3-1/5" x 12" (3.75/S.F.)		1.40	28.571		590	720		1,310	1,775	
	1150	Economy, 4" x 4" x 8" (4.50 per S.F.)		1.40	28.571		480	720		1,200	1,650	
	1201	Engineer, 4" x 3-1/5" x 8", (5.63/S.F.)		1.45	27.586		345	695		1,040	1,450	
	1251	Roman, 4" x 2" x 12", (6.00/S.F.)		1.50	26.667		790	670		1,460	1,925	
	1300	S.C.R. 6" x 2-2/3" x 12" (4.50/S.F.)		1.40	28.571		910	720		1,630	2,125	
	1350	Utility, 4" x 4" x 12" (3.00/S.F.)	↓	1.35	29.630		835	745		1,580	2,075	
	1360	For less than truck load lots, add				↓	10					
	1999	Alternate method of figuring by square foot										
	2000	Standard, sel. common, 4" x 2-2/3" x 8", (6.75/S.F.)	D-8	230	.174	S.F.	2.55	4.38		6.93	9.60	
	2020	Standard, red, 4" x 2-2/3" x 8", running bond (6.75/SF)	↓	220	.182	↓	2.55	4.58		7.13	9.90	

042 550 | Masonry Veneer

		Description	Ref	CREW	DAILY OUTPUT	LABOR-HOURS	UNIT	1999 BARE COSTS MAT.	LABOR	EQUIP.	TOTAL	TOTAL INCL O&P
554	2050	Full header every 6th course (7.88/S.F.)	R042-120	D-8	185	.216	S.F.	2.97	5.45		8.42	11.70
	2100	English, full header every 2nd course (10.13/S.F.)			140	.286		3.81	7.20		11.01	15.40
	2150	Flemish, alternate header every course (9.00/S.F.)	R042-180		150	.267		3.39	6.70		10.09	14.20
	2250	Full headers throughout (13.50/S.F.)			105	.381		5.05	9.60		14.65	20.50
	2400	Soldier course (6.75/S.F.)	R042-500		200	.200		2.55	5.05		7.60	10.65
	2600	Buff or gray face, running bond, (6.75/S.F.)			220	.182		2.70	4.58		7.28	10.05
	2700	Glazed face brick, running bond	R042-550		210	.190		7.10	4.80		11.90	15.25
	3000	Jumbo, 6″ x 4″ x 12″ running bond (3.00/S.F.)			435	.092		3.65	2.32		5.97	7.60
	3050	Norman, 4″ x 2-2/3″ x 12″ running bond, (4.5/S.F.)			320	.125		3.36	3.15		6.51	8.60
	3100	Norwegian, 4″ x 3-1/5″ x 12″ (3.75/S.F.)			375	.107		2.15	2.69		4.84	6.55
	3150	Economy, 4″ x 4″ x 8″ (4.50/S.F.)			310	.129		2.15	3.25		5.40	7.40
	3200	Engineer, 4″ x 3-1/5″ x 8″ (5.63/S.F.)			260	.154		1.93	3.88		5.81	8.15
	3250	Roman, 4″ x 2″ x 12″ (6.00/S.F.)			250	.160		4.68	4.03		8.71	11.40
	3300	SCR, 6″ x 2-2/3″ x 12″ (4.50/S.F.)			310	.129		4.19	3.25		7.44	9.65
	3350	Utility, 4″ x 4″ x 12″ (3.00/S.F.)		↓	450	.089	↓	2.45	2.24		4.69	6.15
	3400	For cavity wall construction, add							15%			
	3450	For stacked bond, add							10%			
	3500	For interior veneer construction, add							15%			
	3550	For curved walls, add							30%			

044 | Stone

044 100 | Rough Stone

		Description	CREW	DAILY OUTPUT	LABOR-HOURS	UNIT	1999 BARE COSTS MAT.	LABOR	EQUIP.	TOTAL	TOTAL INCL O&P
104	0011	**ROUGH STONE WALL**, Dry									
	0100	Random fieldstone, under 18″ thick	D-12	60	.533	C.F.	9.75	13.40		23.15	32
	0150	Over 18″ thick	″	63	.508	″	12.50	12.75		25.25	33.50

044 200 | Cut Stone

		Description	CREW	DAILY OUTPUT	LABOR-HOURS	UNIT	1999 BARE COSTS MAT.	LABOR	EQUIP.	TOTAL	TOTAL INCL O&P
204	0010	**BLUESTONE** Cut to size									
	0100	Paving, natural cleft, to 4′, 1″ thick	D-8	150	.267	S.F.	3.83	6.70		10.53	14.65
	0150	1-1/2″ thick		145	.276		4.45	6.95		11.40	15.70
	0200	Smooth finish, 1″ thick		150	.267		5.60	6.70		12.30	16.65
	0250	1-1/2″ thick		145	.276		5.85	6.95		12.80	17.25
	0300	Thermal finish, 1″ thick		150	.267		6.20	6.70		12.90	17.30
	0350	1-1/2″ thick	↓	145	.276	↓	6.50	6.95		13.45	17.95
	0500	Sills, natural cleft, 10″ wide to 6′ long, 1-1/2″ thick	D-11	70	.343	L.F.	9.05	9		18.05	24
	0550	2″ thick		63	.381		10.75	10		20.75	27.50
	0600	Smooth finish, 1-1/2″ thick		70	.343		13.60	9		22.60	29
	0650	2″ thick		63	.381		15.90	10		25.90	33
	0800	Thermal finish, 1-1/2″ thick		70	.343		15.25	9		24.25	31
	0850	2″ thick	↓	63	.381		16.10	10		26.10	33.50
	1000	Stair treads, natural cleft, 12″ wide, 6′ long, 1-1/2″ thick	D-10	115	.348		13.65	9.05	3.40	26.10	32.50
	1050	2″ thick		105	.381		15.25	9.90	3.73	28.88	36
	1100	Smooth finish, 1-1/2″ thick		115	.348		17.85	9.05	3.40	30.30	37.50
	1150	2″ thick		105	.381		18.70	9.90	3.73	32.33	40
	1300	Thermal finish, 1-1/2″ thick		115	.348		19.95	9.05	3.40	32.40	39.50
	1350	2″ thick	↓	105	.381	↓	22.50	9.90	3.73	36.13	44.50
	2000	Coping, finished top & 2 sides, 12″ to 6′									

Important: See the Reference Section for critical supporting data - Reference Nos., Crews, & City Cost Indexes

044 | Stone

	044 200 | Cut Stone	CREW	DAILY OUTPUT	LABOR-HOURS	UNIT	1999 BARE COSTS MAT.	LABOR	EQUIP.	TOTAL	TOTAL INCL O&P	
2100	Natural cleft, 1-1/2″ thick	D-10	115	.348	L.F.	14.50	9.05	3.40	26.95	33.50	204
2150	2″ thick		105	.381		15.85	9.90	3.73	29.48	37	
2200	Smooth finish, 1-1/2″ thick		115	.348		19.30	9.05	3.40	31.75	38.50	
2250	2″ thick		105	.381		20.50	9.90	3.73	34.13	42	
2300	Thermal finish, 1-1/2″ thick		115	.348		21.50	9.05	3.40	33.95	41.50	
2350	2″ thick		105	.381		23	9.90	3.73	36.63	44.50	
	044 500 | Stone Veneer										
0011	**ASHLAR VENEER** 4″ + or - thk, random or random rectangular (R044-500)										504
0150	Sawn face, split joints, low priced stone	D-8	140	.286	S.F.	4.71	7.20		11.91	16.40	
0200	Medium priced stone		130	.308		6.95	7.75		14.70	19.70	
0300	High priced stone		120	.333		9.10	8.40		17.50	23	
0600	Seam face, split joints, medium price stone		125	.320		6	8.05		14.05	19.15	
0700	High price stone		120	.333		10	8.40		18.40	24	
1000	Split or rock face, split joints, medium price stone		125	.320		8	8.05		16.05	21.50	
1100	High price stone		120	.333		10.50	8.40		18.90	24.50	
0011	**LIGHTWEIGHT NATURAL STONE** Lava type										508
0100	Veneer, rubble face, sawed back, irregular shapes	D-10	130	.308	S.F.	5.10	8	3.01	16.11	21.50	
0200	Sawed face and back, irregular shapes	″	130	.308	″	5.25	8	3.01	16.26	21.50	
	044 550 | Marble										
0011	**MARBLE** Ashlar, split face, 4″ + or - thick, random										554
0040	lengths 1′ to 4′ & heights 2″ to 7-1/2″, average	D-8	175	.229	S.F.	13.35	5.75		19.10	23.50	
1000	Facing, polished finish, cut to size, 3/4″ to 7/8″ thick										
1050	Average	D-10	130	.308	S.F.	17.80	8	3.01	28.81	35	
1100	Maximum		130	.308		41	8	3.01	52.01	61	
1300	1-1/4″ thick, average		125	.320		26	8.30	3.13	37.43	45	
1350	Maximum		125	.320		51.50	8.30	3.13	62.93	73	
1500	2″ thick, average		120	.333		30	8.65	3.26	41.91	50	
1550	Maximum		120	.333		51.50	8.65	3.26	63.41	74	
1700	Rubbed finish, cut to size, 4″ thick										
1740	Average	D-10	100	.400	S.F.	30	10.40	3.91	44.31	53.50	
1780	Maximum	″	100	.400	″	57.50	10.40	3.91	71.81	84	
2500	Flooring, polished tiles, 12″ x 12″ x 3/8″ thick										
2510	Thin set, average	D-11	90	.267	S.F.	8.25	7		15.25	19.95	
2600	Maximum		90	.267		30	7		37	44	
2700	Mortar bed, average		65	.369		8.40	9.70		18.10	24.50	
2740	Maximum		65	.369		27.50	9.70		37.20	45	
3210	Stairs, risers, 7/8″ thick x 6″ high	D-10	115	.348	L.F.	11.65	9.05	3.40	24.10	30.50	
3360	Treads, 12″ wide x 1-1/4″ thick	″	115	.348	″	17	9.05	3.40	29.45	36.50	
3500	Thresholds, 3′ long, 7/8″ thick, 4″ to 5″ wide, plain	D-12	24	1.333	Ea.	12.85	33.50		46.35	66	
3550	Beveled	″	24	1.333	″	14.95	33.50		48.45	68.50	
	044 600 | Limestone										
0012	**LIMESTONE,** Cut to size										604
0020	Veneer facing panels										
1400	Sugarcube, textured finish, 4-1/2″ thick, 5′ x 12′	D-10	275	.145	S.F.	19.90	3.77	1.42	25.09	29.50	
1450	5″ thick, 5′ x 14′ panels		275	.145	″	23	3.77	1.42	28.19	33	
2000	Coping, sugarcube finish, top & 2 sides		30	1.333	C.F.	51	34.50	13.05	98.55	124	
2100	Sills, lintels, jambs, trim, stops, sugarcube finish, average		20	2		46	52	19.55	117.55	153	
2150	Detailed		20	2		61	52	19.55	132.55	169	
2300	Steps, extra hard, 14″ wide, 6″ rise		50	.800	L.F.	42	21	7.80	70.80	86.50	
3000	Quoins, plain finish, 6″x12″x12″	D-12	25	1.280	Ea.	101	32		133	161	
3050	6″x16″x24″	″	25	1.280	″	134	32		166	198	

044 650 | Granite

		044 650 | Granite	CREW	DAILY OUTPUT	LABOR-HOURS	UNIT	1999 BARE COSTS MAT.	LABOR	EQUIP.	TOTAL	TOTAL INCL O&P
651	0010	**GRANITE** Cut to size									
	0050	Veneer, polished face, 3/4" to 1-1/2" thick									
	0150	Low price, gray, light gray, etc.	D-10	130	.308	S.F.	18.55	8	3.01	29.56	36
	0180	Medium price, pink, brown, etc.		130	.308		20.50	8	3.01	31.51	38
	0220	High price, red, black, etc.		130	.308		31.50	8	3.01	42.51	50.50
	0300	1-1/2" to 2-1/2" thick, veneer									
	0350	Low price, gray, light gray, etc.	D-10	130	.308	S.F.	20.50	8	3.01	31.51	38
	0500	Medium price, pink, brown, etc.		130	.308		26	8	3.01	37.01	44
	0550	High price, red, black, etc.		130	.308		36	8	3.01	47.01	55
	0700	2-1/2" to 4" thick, veneer									
	0750	Low price, gray, light gray, etc.	D-10	110	.364	S.F.	26	9.45	3.56	39.01	47
	0850	Medium price, pink, brown, etc.		110	.364		31	9.45	3.56	44.01	52.50
	0950	High price, red, black, etc.		110	.364		41	9.45	3.56	54.01	64
	1000	For bush hammered finish, deduct					5%				
	1050	Coarse rubbed finish, deduct					10%				
	1100	Honed finish, deduct					5%				
	1150	Thermal finish, deduct					18%				
	1800	Carving or bas-relief, from templates or plaster molds									
	1850	Minimum	D-10	80	.500	C.F.	118	13	4.89	135.89	155
	1900	Maximum	″	80	.500	″	355	13	4.89	372.89	415
	2000	Intricate or hand finished pieces									
	2010	mouldings, radius cuts, bullnose edges, etc.									
	2050	Add, minimum					30%				
	2100	Add, maximum					300%				
	2450	For radius under 5', add				L.F.	100%				
	2500	Steps, copings, etc., finished on more than one surface									
	2550	Minimum	D-10	50	.800	C.F.	72	21	7.80	100.80	120
	2600	Maximum	″	50	.800	″	108	21	7.80	136.80	160
	2700	Pavers, Belgian block, 8"-13" long, 4"-6" wide, 4"-6" deep	D-11	120	.200	S.F.	18.05	5.25		23.30	28
	2800	Pavers, 4" x 4" x 4" blocks, split face and joints									
	2850	Minimum	D-11	80	.300	S.F.	10.30	7.90		18.20	23.50
	2900	Maximum	″	80	.300	″	20	7.90		27.90	34.50
	3000	Pavers, 4" x 4" x 4", thermal face, sawn joints									
	3050	Minimum	D-11	65	.369	S.F.	18.55	9.70		28.25	35.50
	3100	Maximum	″	65	.369	″	25	9.70		34.70	42
	3500	Curbing, city street type, See Division 025-254									
	3700	Slope, 4-1/2" x 12", split face,									
	3710	sawn top, 2' to 6' lengths	D-10	300	.133	L.F.	7.75	3.46	1.30	12.51	15.30
	3800	Radius curbs, over 5' radius, add					50%				
	3850	Under 5' radius, add					100%				
	4000	Soffits, 2" thick, minimum	D-13	35	1.371	S.F.	31	36	11.20	78.20	102
	4100	Maximum		35	1.371		62	36	11.20	109.20	136
	4200	4" thick, minimum		35	1.371		42	36	11.20	89.20	114

044 700 | Sandstone

		Description	CREW	DAILY OUTPUT	LABOR-HOURS	UNIT	MAT.	LABOR	EQUIP.	TOTAL	TOTAL INCL O&P
704	0011	**SANDSTONE OR BROWNSTONE**									
	0100	Sawed face veneer, 2-1/2" thick, to 2' x 4' panels	D-10	130	.308	S.F.	12.45	8	3.01	23.46	29.50
	0150	4' thick, to 3'-6" x 8'panels		100	.400		12.45	10.40	3.91	26.76	34
	0300	Split face, random sizes		100	.400		8.95	10.40	3.91	23.26	30

044 750 | Slate

		Description	CREW	DAILY OUTPUT	LABOR-HOURS	UNIT	MAT.	LABOR	EQUIP.	TOTAL	TOTAL INCL O&P
754	0010	**SLATE** Pennsylvania, blue gray to gray black; Vermont,									
	0050	Unfading green, mottled green & purple, gray & purple									
	0100	Virginia, blue black									
	0200	Exterior paving, natural cleft, 1" thick									

044 750 | Slate

	044 750 | Slate	CREW	DAILY OUTPUT	LABOR-HOURS	UNIT	1999 BARE COSTS MAT.	LABOR	EQUIP.	TOTAL	TOTAL INCL O&P	
0250	6″ x 6″ Pennsylvania	D-12	100	.320	S.F.	6.35	8.05		14.40	19.50	754
0300	Vermont		100	.320		8.70	8.05		16.75	22	
0350	Virginia		100	.320		9.65	8.05		17.70	23	
0500	24″ x 24″, Pennsylvania		120	.267		8.65	6.70		15.35	19.90	
0550	Vermont		120	.267		11.30	6.70		18	23	
0600	Virginia		120	.267		12.65	6.70		19.35	24.50	
0700	18″ x 30″ Pennsylvania		120	.267		7.55	6.70		14.25	18.70	
0750	Vermont		120	.267		10.55	6.70		17.25	22	
0800	Virginia		120	.267		11.30	6.70		18	23	
1000	Interior flooring, natural cleft, 1/2″ thick										
1100	6″ x 6″ Pennsylvania	D-12	100	.320	S.F.	4.94	8.05		12.99	17.95	
1150	Vermont		100	.320		6.05	8.05		14.10	19.15	
1200	Virginia		100	.320		7.70	8.05		15.75	21	
1300	24″ x 24″ Pennsylvania		120	.267		5.95	6.70		12.65	16.95	
1350	Vermont		120	.267		8.90	6.70		15.60	20	
1400	Virginia		120	.267		9.20	6.70		15.90	20.50	
1500	18″ x 24″ Pennsylvania		120	.267		6.05	6.70		12.75	17.05	
1550	Vermont		120	.267		8.75	6.70		15.45	20	
1600	Virginia		120	.267		9.35	6.70		16.05	20.50	
2000	Facing panels, 1-1/4″ thick, to 4′ x 4′ panels										
2100	Natural cleft finish, Pennsylvania	D-10	180	.222	S.F.	17.60	5.75	2.17	25.52	30.50	
2110	Vermont		180	.222		18.35	5.75	2.17	26.27	31.50	
2120	Virginia		180	.222		20.50	5.75	2.17	28.42	34	
2150	Sand rubbed finish, surface, add					1.91			1.91	2.10	
2200	Honed finish, add					3.77			3.77	4.15	
2500	Ribbon, natural cleft finish, 1″ thick, to 9 S.F.	D-10	80	.500		8.10	13	4.89	25.99	34.50	
2550	Sand rubbed finish		80	.500		8.60	13	4.89	26.49	35	
2600	Honed finish		80	.500		9.60	13	4.89	27.49	36	
2700	1-1/2″ thick		78	.513		10.55	13.30	5	28.85	37.50	
2750	Sand rubbed finish		78	.513		11.75	13.30	5	30.05	39	
2800	Honed finish		78	.513		13	13.30	5	31.30	40.50	
2850	2″ thick		76	.526		12.70	13.65	5.15	31.50	40.50	
2900	Sand rubbed finish		76	.526		13.40	13.65	5.15	32.20	41.50	
2950	Honed finish		76	.526		14.65	13.65	5.15	33.45	43	
3000	Roofing, see division 073-106										
3500	Stair treads, sand finish, 1″ thick x 12″ wide										
3550	Under 3 L.F.	D-10	85	.471	L.F.	13.60	12.20	4.60	30.40	39	
3600	3 L.F. to 6 L.F.	″	120	.333	″	14.60	8.65	3.26	26.51	33	
3700	Ribbon, sand finish, 1″ thick x 12″ wide										
3750	To 6 L.F.	D-10	120	.333	L.F.	9.75	8.65	3.26	21.66	27.50	
4000	Stools or sills, sand finish, 1″ thick, 6″ wide	D-12	160	.200		7.05	5		12.05	15.55	
4100	Honed finish		160	.200		7.85	5		12.85	16.40	
4400	2″ thick, 6″ wide		140	.229		11.35	5.75		17.10	21.50	
4450	Honed finish		140	.229		12.50	5.75		18.25	22.50	
4600	10″ wide		90	.356		16.95	8.90		25.85	32.50	
4650	Honed finish		90	.356		18.50	8.90		27.40	34.50	
4800	For lengths over 3′, add					25%					

r expanded coverage of these items see *Means Concrete & Masonry Cost Data 1999*

045 | Masonry Restoration, Cleaning & Refractories

		045 200 | Masonry Restoration	CREW	DAILY OUTPUT	LABOR-HOURS	UNIT	1999 BARE COSTS MAT.	LABOR	EQUIP.	TOTAL	TOTAL INCL O&P
250	0010	**PATCHING MASONRY** No staging included									
	0500	Concrete patching, includes chipping, cleaning and epoxy									
	0520	Minimum	1 Cefi	65	.123	S.F.	.26	3.22		3.48	5.10
	0540	Average		50	.160		.35	4.18		4.53	6.65
	0580	Maximum		40	.200		.43	5.25		5.68	8.25
270	0010	**POINTING MASONRY**									
	0300	Cut and repoint brick, hard mortar, running bond	1 Bric	60	.133	S.F.	.24	3.68		3.92	5.95
	0320	Common bond		55	.145		.24	4.01		4.25	6.50
	0360	Flemish bond		50	.160		.25	4.42		4.67	7.15
	0400	English bond		45	.178		.25	4.91		5.16	7.95
	0600	Soft old mortar, running bond		80	.100		.25	2.76		3.01	4.56
	0620	Common bond		75	.107		.24	2.94		3.18	4.85
	0640	Flemish bond		70	.114		.25	3.15		3.40	5.20
	0680	English bond		65	.123		.25	3.40		3.65	5.60
	0700	Stonework, hard mortar		120	.067	L.F.	.32	1.84		2.16	3.21
	0720	Soft old mortar		140	.057	"	.32	1.58		1.90	2.80
	1000	Repoint, mask and grout method, running bond		80	.100	S.F.	.32	2.76		3.08	4.64
	1020	Common bond		75	.107		.32	2.94		3.26	4.93
	1040	Flemish bond		70	.114		.32	3.15		3.47	5.25
	1060	English bond		65	.123		.32	3.40		3.72	5.65
	2000	Scrub coat, sand grout on walls, minimum		120	.067		3.15	1.84		4.99	6.30
	2020	Maximum		98	.082		2.25	2.25		4.50	6
		045 650 | Fire Brick									
652	0010	**FIRE BRICK** 9″ x 2-1/2″ x 4-1/2″, low duty, 2000° F	D-1	.60	26.667	M	1,075	655		1,730	2,225
	0050	High duty, 3000° F	"	.60	26.667	"	1,550	655		2,205	2,725
656	0010	**FIREPLACE** For prefabricated fireplace, see div. 103-054									
	0100	Brick fireplace, not incl. foundations or chimneys									
	0110	30″ x 29″ opening, incl. chamber, plain brickwork	D-1	.40	40	Ea.	360	985		1,345	1,925
	0200	Fireplace box only (110 brick)	"	2	8	"	116	197		313	435
	0300	For elaborate brickwork and details, add					35%	35%			
	0400	For hearth, brick & stone, add	D-1	2	8	Ea.	132	197		329	450
	0410	For steel angle, damper, cleanouts, add	"	4	4	"	92	98.50		190.50	254

Division 5 Metals

Estimating Tips

050 Materials, Coatings & Fastenings

- Nuts, bolts, washers, connection angles and plates can add a significant amount to both the tonnage of a structural steel job as well as the estimated cost. As a rule of thumb add 10% to the total weight to account for these accessories.
- Type 2 steel construction, commonly referred to as "simple construction," consists generally of field bolted connections with lateral bracing supplied by other elements of the building, such as masonry walls or x-bracing. The estimator should be aware, however, that shop connections may be accomplished by welding or bolting. The method may be particular to the fabrication shop and may have an impact on the estimated cost.

052 Steel Joists

- In any given project the total weight of open web steel joists is determined by the loads to be supported and the design. However, economies can be realized in minimizing the amount of labor used to place the joists. This is done by maximizing the joist spacing and therefore minimizing the number of joists required to be installed on the job. Certain spacings and locations may be required by the design, but in other cases maximizing the spacing and keeping it as uniform as possible will keep the costs down.

053 Metal Decking

- The takeoff and estimating of metal deck involves more than simply the area of the floor or roof and the type of deck specified or shown on the drawings. Many different sizes and types of openings may exist. Small openings for individual pipes or conduits may be drilled after the floor/roof is installed, but larger openings may require special deck lengths as well as reinforcing or structural support. The estimator should determine who will be supplying this reinforcing. Additionally, some deck terminations are part of the deck package, such as screed angles and pour stops, and others will be part of the steel contract, such as angles attached to structural members and cast-in-place angles and plates. The estimator must ensure that all pieces are accounted for in the complete estimate.

055 Metal Fabrications

- The most economical steel stairs are those that use common materials, standard details and most importantly, a uniform and relatively simple method of field assembly. Commonly available A36 channels and plates are very good choices for the main stringers of the stairs, as are angles and tees for the carrier members. Risers and treads are usually made by specialty shops, and it is most economical to use a typical detail in as many places as possible. The stairs should be pre-assembled and shipped directly to the site. The field connections should be simple and straightforward to be accomplished efficiently and with a minimum of equipment and labor.

Reference Numbers

Reference numbers are shown in bold squares at the beginning of some major classifications. These numbers refer to related items in the Reference Section. The reference information may be an estimating procedure, an alternate pricing method or technical information.

Note: Not all subdivisions listed here necessarily appear in this publication.

050 500 | Metal Fastening

			CREW	DAILY OUTPUT	LABOR-HOURS	UNIT	1999 BARE COSTS MAT.	LABOR	EQUIP.	TOTAL	TOTAL INCL O&P
508	0010	**BOLTS & HEX NUTS** Steel, A307									
	0100	1/4" diameter, 1/2" long				Ea.	.05			.05	.05
	0200	1" long					.06			.06	.06
	0400	3" long					.10			.10	.11
	0600	3/8" diameter, 1" long					.04			.04	.04
	0800	3" long					.09			.09	.09
	1100	1/2" diameter, 1-1/2" long					.22			.22	.24
	1400	6" long					.49			.49	.54
	1600	5/8" diameter, 1-1/2" long					.38			.38	.41
	2000	8" long					1.06			1.06	1.16
	2200	3/4" diameter, 2" long					.66			.66	.73
	2600	10" long					2.07			2.07	2.28
	2800	1" diameter, 3" long					1.61			1.61	1.77
	3000	12" long					4.74			4.74	5.20
	3100	For galvanized, add					75%				
	3200	For stainless, add				↓	350%				
510	0005	**CURB EDGING**									
	0010	Steel angle w/anchors, on forms, 1" x 1", 0.8#/L.F.	E-4	350	.091	L.F.	1.36	2.84	.23	4.43	6.95
	0100	2" x 2" angles, 3.92#/L.F.		330	.097		4.19	3.02	.25	7.46	10.45
	0200	3" x 3" angles, 6.1#/L.F.		300	.107		6.70	3.32	.27	10.29	13.80
	0300	4" x 4" angles, 8.2#/L.F.		275	.116		8.15	3.62	.30	12.07	16
	1000	6" x 4" angles, 12.3#/L.F.		250	.128		11.65	3.98	.33	15.96	20.50
	1050	Steel channels with anchors, on forms, 3" channel, 5#/L.F.		290	.110		5.30	3.43	.28	9.01	12.45
	1100	4" channel, 5.4#/L.F.		270	.119		5.70	3.69	.30	9.69	13.35
	1200	6" channel, 8.2#/L.F.		255	.125		8.85	3.90	.32	13.07	17.25
	1300	8" channel, 11.5#/L.F.		225	.142		11.60	4.42	.37	16.39	21.50
	1400	10" channel, 15.3#/L.F.		180	.178		15.05	5.55	.46	21.06	27
	1500	12" channel, 20.7#/L.F.	↓	140	.229		19.95	7.10	.59	27.64	35.50
	2000	For curved edging, add				↓	35%	10%			
515	0010	**DRILLING** For anchors, up to 4" deep, incl. bit and layout									
	0050	in concrete or brick walls and floors, no anchor									
	0100	Holes, 1/4" diameter	1 Carp	75	.107	Ea.	.08	2.91		2.99	4.67
	0150	For each additional inch of depth, add		430	.019		.02	.51		.53	.82
	0200	3/8" diameter		63	.127		.07	3.47		3.54	5.55
	0250	For each additional inch of depth, add		340	.024		.02	.64		.66	1.03
	0300	1/2" diameter		50	.160		.07	4.37		4.44	6.95
	0350	For each additional inch of depth, add		250	.032		.02	.87		.89	1.39
	0400	5/8" diameter		48	.167		.13	4.55		4.68	7.30
	0450	For each additional inch of depth, add		240	.033		.03	.91		.94	1.47
	0500	3/4" diameter		45	.178		.14	4.85		4.99	7.80
	0550	For each additional inch of depth, add		220	.036		.03	.99		1.02	1.60
	0600	7/8" diameter		43	.186		.17	5.10		5.27	8.20
	0650	For each additional inch of depth, add		210	.038		.04	1.04		1.08	1.69
	0700	1" diameter		40	.200		.20	5.45		5.65	8.80
	0750	For each additional inch of depth, add		190	.042		.05	1.15		1.20	1.87
	0800	1-1/4" diameter		38	.211		.29	5.75		6.04	9.35
	0850	For each additional inch of depth, add		180	.044		.07	1.21		1.28	1.99
	0900	1-1/2" diameter		35	.229		.47	6.25		6.72	10.30
	0950	For each additional inch of depth, add	↓	165	.048	↓	.12	1.32		1.44	2.21
	1000	For ceiling installations, add						40%			
520	0010	**EXPANSION ANCHORS** & shields									
	0100	Bolt anchors for concrete, brick or stone, no layout and drilling									
	0200	Expansion shields, zinc, 1/4" diameter, 1" long, single	1 Carp	90	.089	Ea.	.85	2.43		3.28	4.76
	0300	1-3/8" long, double	↓	85	.094	↓	.97	2.57		3.54	5.10

050 500 | Metal Fastening

	050 500 Metal Fastening	CREW	DAILY OUTPUT	LABOR-HOURS	UNIT	1999 BARE COSTS MAT.	LABOR	EQUIP.	TOTAL	TOTAL INCL O&P	
0600	1/2" diameter, 2-1/2" long, single	1 Carp	80	.100	Ea.	2.40	2.73		5.13	6.95	520
1000	3/4" diameter, 2-3/4" long, single		70	.114		5.10	3.12		8.22	10.50	
1100	4" long, double		65	.123		6.80	3.36		10.16	12.80	
1410	Concrete anchor, w/rod & epoxy cartridge, 1-3/4" diameter x 15" long	E-22	20	1.200		72	34.50		106.50	134	
1415	18" long		17	1.412		86	40.50		126.50	159	
1420	2" diameter x 18" long		16	1.500		110	43		153	189	
1425	24" long		15	1.600		144	46		190	231	
1430	Chemical anchor, w/rod & epoxy cartridge, 3/4" diam. x 9-1/2" long		27	.889		10.45	25.50		35.95	52	
1435	1" diameter x 11-3/4" long		24	1		19.80	28.50		48.30	67.50	
1440	1-1/4" diameter x 14" long		21	1.143		38	33		71	93.50	
1500	Self drilling anchor, snap-off, for 1/4" diameter bolt	1 Carp	26	.308		.93	8.40		9.33	14.20	
1700	1/2" diameter bolt		20	.400		2.11	10.90		13.01	19.50	
1900	3/4" diameter bolt		16	.500		6.45	13.65		20.10	28.50	
3400	1/4" diameter, 3" long		75	.107		.40	2.91		3.31	5	
3600	3/8" diameter, 3" long		70	.114		.88	3.12		4	5.90	
5000	Screw anchors for concrete, masonry,										
5100	stone & tile, no layout or drilling included										
6600	Lead, #6 & #8, 3/4" long				Ea.	.17			.17	.19	
6700	#10 - #14, 1-1/2" long					.26			.26	.29	
6800	#16 & #18, 1-1/2" long					.36			.36	.40	
6900	Plastic, #6 & #8, 3/4" long					.04			.04	.04	
8000	Wedge anchors, not including layout or drilling										
8050	Carbon steel, 1/4" diameter, 1-3/4" long	1 Carp	150	.053	Ea.	.33	1.46		1.79	2.66	
8100	3 1/4" long		145	.055		.50	1.51		2.01	2.92	
8250	1/2" diameter, 2-3/4" long		140	.057		.88	1.56		2.44	3.42	
8300	7" long		130	.062		1.54	1.68		3.22	4.34	
8550	1" diameter, 6" long		100	.080		7.40	2.18		9.58	11.60	
8575	9" long		80	.100		9.60	2.73		12.33	14.85	
8600	12" long		80	.100		10.40	2.73		13.13	15.70	
8750	For type 303 stainless steel, add					350%					
8800	For type 316 stainless steel, add					450%					
0005	**LAG SCREWS**										530
0010	Steel, 1/4" diameter, 2" long	1 Carp	200	.040	Ea.	.05	1.09		1.14	1.78	
0100	3/8" diameter, 3" long		150	.053		.16	1.46		1.62	2.46	
0200	1/2" diameter, 3" long		130	.062		.27	1.68		1.95	2.94	
0300	5/8" diameter, 3" long		120	.067		.53	1.82		2.35	3.45	
0005	**MACHINE SCREWS**										535
0010	Steel, round head, #8 x 1" long				C	1.50			1.50	1.65	
0110	#8 x 2" long					3.39			3.39	3.73	
0200	#10 x 1" long					2.16			2.16	2.37	
0300	#10 x 2" long					4.01			4.01	4.41	
0010	**MACHINERY ANCHORS** Standard, flush mounted,										540
0020	incl. stud w/fiber plug, connecting nut, washer & bolt										
0200	Material only, 1/2" diameter stud & bolt				Ea.	30.50			30.50	33.50	
0300	5/8" diameter					34.50			34.50	38	
0500	3/4" diameter					37.50			37.50	41.50	
0600	7/8" diameter					43.50			43.50	47.50	
0800	1" diameter					47.50			47.50	52.50	
0900	1-1/4" diameter					60.50			60.50	66.50	
0005	**POWDER ACTUATED** Tools & fasteners										550
0010	Stud driver, .22 caliber, buy, minimum				Ea.	287			287	315	
0100	Maximum				"	505			505	555	
0300	Powder charges for above, low velocity				C	15.05			15.05	16.55	

050 500 | Metal Fastening

			CREW	DAILY OUTPUT	LABOR-HOURS	UNIT	1999 BARE COSTS MAT.	LABOR	EQUIP.	TOTAL	TOTAL INCL O&P
550	0400	Standard velocity				C	25			25	27.50
555	0005	**RAILROAD TRACK** R024-520									
	0010	Track bolts				Ea.	1.73			1.73	1.90
	0100	Joint bars				Pr.	26			26	28.50
	0200	Spikes				Ea.	.36			.36	.40
	0300	Tie plates				"	1.50			1.50	1.65
570	0005	**WELD ROD**									
	0010	Steel, type E6011 (all purpose), 1/8" dia, less than 500#				Lb.	1.38			1.38	1.52
	0100	500# to 2,000#					1.23			1.23	1.36
	0200	2,000# to 5,000#					1.14			1.14	1.25
	0300	5/32" diameter, less than 500#					1.32			1.32	1.45
	0310	500# to 2,000#					1.20			1.20	1.31
	0320	2,000# to 5,000#					1.09			1.09	1.20
	0400	3/16" dia, less than 500#					1.32			1.32	1.45
	0500	500# to 2,000#					1.19			1.19	1.30
	0600	2,000# to 5,000#					1.08			1.08	1.19
	0620	Steel, type E6010 (pipe), 1/8" dia, less than 500#					1.47			1.47	1.62
	0630	500# to 2,000#					1.32			1.32	1.45
	0640	2,000# to 5,000#					1.17			1.17	1.29
	0650	Steel, type E7018 (low hydrogen), 1/8" dia, less than 500#					1.32			1.32	1.45
	0660	500# to 2,000#					1.30			1.30	1.43
	0670	2,000# to 5,000#					1.08			1.08	1.19
	0700	Steel, type E7024 (jet weld), 1/8" dia, less than 500#					1.36			1.36	1.50
	0710	500# to 2,000#					1.23			1.23	1.35
	0720	2,000# to 5,000#					1.12			1.12	1.23
	1550	Aluminum, type 4043, 1/8" dia, less than 10#					4.76			4.76	5.25
	1560	10# to 60#					4.54			4.54	5
	1570	Over 60#					4.30			4.30	4.72
	1600	Aluminum, type 5356, 1/8" dia, less than 10#					4.77			4.77	5.25
	1610	10# to 60#					4.54			4.54	4.99
	1620	Over 60#					4.11			4.11	4.52
	1900	Cast iron (cold welding), 1/8" dia, less than 500#					15.35			15.35	16.90
	1910	500# to 1,000#					14.15			14.15	15.60
	1920	Over 1,000#					13.05			13.05	14.35
	2000	Stainless steel, type 308-16, 1/8" dia, less than 500#					5.85			5.85	6.45
	2100	500# to 1000#					5.15			5.15	5.65
	2220	Over 1000#					4.78			4.78	5.25
575	0005	**WELDING STRUCTURAL** R050-520									
	0010	Field welding, 1/8" E6011, cost per welder, no operating engr.	E-14	8	1	Hr.	2.76	32.50	10.30	45.56	74.50
	0200	With 1/2 operating engineer	E-13	8	1.500		2.76	46	10.30	59.06	95
	0300	With 1 operating engineer	E-12	8	2		2.76	60	10.30	73.06	115
	0500	With no operating engineer, 2# weld rod per ton	E-14	8	1	Ton	2.76	32.50	10.30	45.56	74.50
	0600	8# E6011 per ton	"	2	4		11	130	41	182	297
	0800	With one operating engineer per welder, 2# E6011 per ton	E-12	8	2		2.76	60	10.30	73.06	115
	0900	8# E6011 per ton	"	2	8		11	239	41	291	465
	1200	Continuous fillet, stick welding, incl. equipment									
	1300	Single pass, 1/8" thick, 0.1#/L.F.	E-14	240	.033	L.F.	.14	1.09	.34	1.57	2.52
	1400	3/16" thick, 0.2#/L.F.		120	.067		.28	2.17	.69	3.14	5.05
	1500	1/4" thick, 0.3#/L.F.		80	.100		.41	3.26	1.03	4.70	7.55
	1800	3 passes, 3/8" thick, 0.5#/L.F.		48	.167		.69	5.45	1.72	7.86	12.60
	2010	4 passes, 1/2" thick, 0.7#/L.F.		34	.235		.96	7.65	2.42	11.03	17.75
	2600	For all position welding, add, minimum						20%			
	2700	Maximum						300%			

050 | Metal Materials, Coatings & Fastenings

050 500 | Metal Fastening

			CREW	DAILY OUTPUT	LABOR-HOURS	UNIT	1999 BARE COSTS MAT.	LABOR	EQUIP.	TOTAL	TOTAL INCL O&P	
575	2900	For semi-automatic welding, deduct, minimum (R050-520)				L.F.		5%				575
	3000	Maximum						15%				
	4000	Cleaning and welding plates, bars, or rods										
	4010	to existing beams, columns, or trusses	E-14	12	.667	L.F.	.69	21.50	6.85	29.04	48.50	
580	0005	**WIRE**										580
	0010	Barbed wire, galvanized, domestic steel, hi-tensile 15-1/2 ga.				M.L.F.	23.50			23.50	26	
	0200	import, standard, 12-1/2 ga.					31.50			31.50	35	
	0400	Aluminum, 12-1/2 ga.					138			138	152	
	0500	Helical fence topping, stainless steel				C.L.F.	144			144	159	
	0600	Hardware cloth galv., 1/4″ mesh, 23 ga., 2′ wide				C.S.F.	45			45	49.50	
	0700	3′ wide					44			44	48	
	0900	1/2″ mesh, 19 ga., 2′ wide					38.50			38.50	42.50	
	1000	4′ wide					38			38	41.50	
	1200	Chain link fabric, steel, 2″ mesh, 6 ga, galvanized					113			113	124	
	1300	9 ga, galvanized					55			55	60.50	
	1350	Vinyl coated					45			45	49.50	
	1360	Aluminized					76			76	83.50	
	1400	2-1/4″ mesh, 11.5 ga, galvanized					36			36	39.50	
	1600	1-3/4″ mesh (tennis courts), 11.5 ga (core), vinyl coated					49			49	54	
	1700	9 ga, galvanized					64			64	70.50	
	2100	Welded wire fabric, 1″ x 2″, 14 ga.					27.50			27.50	30.50	
	2200	2″ x 4″, 12-1/2 ga.					19.50			19.50	21.50	

051 | Structural Metal Framing

051 100 | Bracing

			CREW	DAILY OUTPUT	LABOR-HOURS	UNIT	1999 BARE COSTS MAT.	LABOR	EQUIP.	TOTAL	TOTAL INCL O&P	
108	0010	**BRACING**										108
	0300	Let-in, "T" shaped, 22 ga. galv. steel, studs at 16″ O.C.	1 Carp	5.80	1.379	C.L.F.	37	37.50		74.50	99.50	
	0400	Studs at 24″ O.C.		6	1.333		37	36.50		73.50	98	
	0500	16 ga. galv. steel straps, studs at 16″ O.C.		6	1.333		48	36.50		84.50	110	
	0600	Studs at 24″ O.C.		6.20	1.290		48	35		83	108	
110	0005	**PIPE SUPPORT FRAMING**										110
	0010	Under 10#/L.F.	E-4	3,900	.008	Lb.	.90	.26	.02	1.18	1.48	
	0200	10.1 to 15#/L.F.		4,300	.007		.78	.23	.02	1.03	1.30	
	0400	15.1 to 20#/L.F.		4,800	.007		.72	.21	.02	.95	1.19	
	0600	Over 20#/L.F.		5,400	.006		.66	.18	.02	.86	1.09	

051 200 | Structural Steel

			CREW	DAILY OUTPUT	LABOR-HOURS	UNIT	MAT.	LABOR	EQUIP.	TOTAL	TOTAL INCL O&P	
204	0010	**ANCHOR BOLTS**										204
	0020	See also divisions 031-110 and 041-504										
	0100	J-type, incl. nut, washer, 1/2″ diameter x 6″ long	2 Carp	70	.229	Ea.	1.11	6.25		7.36	11	
	0110	12″ long		65	.246		1.27	6.70		7.97	11.95	
	0120	18″ long		60	.267		1.73	7.30		9.03	13.35	
	0130	3/4″ diameter x 8″ long		50	.320		1.70	8.75		10.45	15.65	
	0140	12″ long		45	.356		2.11	9.70		11.81	17.55	
	0150	18″ long		40	.400		2.58	10.90		13.48	20	
	0160	1″ diameter x 12″ long		35	.457		3.20	12.50		15.70	23	
	0170	18″ long		30	.533		3.61	14.55		18.16	27	

051 200 | Structural Steel

			CREW	DAILY OUTPUT	LABOR-HOURS	UNIT	1999 BARE COSTS MAT.	LABOR	EQUIP.	TOTAL	TOTAL INCL O&P
204	0180	24″ long	2 Carp	25	.640	Ea.	4.25	17.45		21.70	32
	0190	36″ long		20	.800		5.60	22		27.60	40.50
	0200	1-1/2″ diameter x 18″ long		22	.727		8.95	19.85		28.80	41
	0210	24″ long		16	1		10.15	27.50		37.65	54
	0300	L-type, incl. hex nuts, 3/4″ diameter x 12″ long		45	.356		1.11	9.70		10.81	16.45
	0310	18″ long		40	.400		1.40	10.90		12.30	18.75
	0320	24″ long		35	.457		1.69	12.50		14.19	21.50
	0330	30″ long		30	.533		2.17	14.55		16.72	25.50
	0340	36″ long		25	.640		2.45	17.45		19.90	30
	0350	1″ diameter x 12″ long		35	.457		2.23	12.50		14.73	22
	0360	18″ long		30	.533		2.74	14.55		17.29	26
	0370	24″ long		25	.640		3.29	17.45		20.74	31
	0380	30″ long		23	.696		3.82	19		22.82	34
	0390	36″ long		20	.800		4.34	22		26.34	39.50
	0400	42″ long		18	.889		5.50	24.50		30	44
	0410	48″ long		15	1.067		6.30	29		35.30	53
	0420	1-1/4″ diameter x 18″ long		25	.640		5	17.45		22.45	33
	0430	24″ long		20	.800		5.95	22		27.95	41
	0440	30″ long		20	.800		6.75	22		28.75	42
	0450	36″ long		18	.889		7.65	24.50		32.15	46.50
	0460	42″ long		16	1		8.60	27.50		36.10	52.50
	0470	48″ long		14	1.143		9.65	31		40.65	59.50
	0480	54″ long		12	1.333		10.80	36.50		47.30	69.50
	0490	60″ long		10	1.600		11.65	43.50		55.15	81.50
	0500	1-1/2″ diameter x 18″ long		22	.727		7.95	19.85		27.80	40
	0510	24″ long		19	.842		9	23		32	46
	0520	30″ long		17	.941		10.30	25.50		35.80	52
	0530	36″ long		16	1		11.95	27.50		39.45	56
	0540	42″ long		15	1.067		13.50	29		42.50	61
	0550	48″ long		13	1.231		15.20	33.50		48.70	69.50
	0560	54″ long		11	1.455		16.90	39.50		56.40	81
	0570	60″ long		9	1.778		18.55	48.50		67.05	97
	0580	1-3/4″ diameter x 18″ long		20	.800		12.60	22		34.60	48.50
	0590	24″ long		18	.889		14.45	24.50		38.95	54
	0600	30″ long		17	.941		16.35	25.50		41.85	58.50
	0610	36″ long		16	1		18.50	27.50		46	63.50
	0620	42″ long		14	1.143		20.50	31		51.50	71.50
	0630	48″ long		12	1.333		22.50	36.50		59	82.50
	0640	54″ long		10	1.600		25	43.50		68.50	96
	0650	60″ long		8	2		27	54.50		81.50	116
	0660	2″ diameter x 24″ long		17	.941		18.85	25.50		44.35	61
	0670	30″ long		15	1.067		21	29		50	69
	0680	36″ long		13	1.231		23	33.50		56.50	78.50
	0690	42″ long		11	1.455		25.50	39.50		65	90.50
	0700	48″ long		10	1.600		28.50	43.50		72	99.50
	0710	54″ long		9	1.778		30	48.50		78.50	110
	0720	60″ long		8	2		32.50	54.50		87	122
	0730	66″ long		7	2.286		34.50	62.50		97	136
	0740	72″ long		6	2.667		37.50	73		110.50	157
220	0005	**COLUMNS**									
	0800	Steel, concrete filled, extra strong pipe, 3-1/2″ diameter	E-2	660	.085	L.F.	18.35	2.53	1.53	22.41	26
	0930	6″ diameter		1,200	.047		34.50	1.39	.84	36.73	41.50
	1000	Lightweight units, 3-1/2″ diameter		780	.072		2.17	2.14	1.30	5.61	7.55
	1100	For galvanizing, add				Lb.	.35			.35	.39
	1500	Steel pipe, extra strong, no concrete, 3″ to 5″ diameter	E-2	16,000	.004		.62	.10	.06	.78	.93

Important: See the Reference Section for critical supporting data - Reference Nos., Crews, & City Cost Indexe

051 200 | Structural Steel

		CREW	DAILY OUTPUT	LABOR-HOURS	UNIT	1999 BARE COSTS MAT.	LABOR	EQUIP.	TOTAL	TOTAL INCL O&P	
1600	6" to 12" diameter	E-2	14,000	.004	Lb.	.61	.12	.07	.80	.96	220
3300	Structural tubing, square, A500GrB, 4" to 6" square, light section		11,270	.005		.82	.15	.09	1.06	1.26	
3600	Heavy section		32,000	.002		.78	.05	.03	.86	.98	
5100	Structural tubing, rect, 5" to 6" wide, light section		9,500	.006		.75	.18	.11	1.04	1.26	
5200	Heavy section		31,200	.002		.75	.05	.03	.83	.96	
5300	7" to 10" wide, light section		37,000	.002		.70	.05	.03	.78	.88	
5400	Heavy section	↓	68,000	.001	↓	.70	.02	.01	.73	.83	
5800	Adjustable jack post, 8' maximum height, 2-3/4" diameter				Ea.	21			21	23	
5850	4" diameter				"	34			34	37.50	
6800	W Shape, A36 steel, 2 tier, W8 x 24	E-2	1,080	.052	L.F.	15.85	1.54	.94	18.33	21	
6900	W8 x 48		1,032	.054		31.50	1.62	.98	34.10	39	
7050	W10 x 68		984	.057		45	1.69	1.03	47.72	53.50	
7100	W10 x 112		960	.058		74	1.74	1.05	76.79	85.50	
7200	W12 x 87		984	.057		57.50	1.69	1.03	60.22	67	
7250	W12 x 120		960	.058		79	1.74	1.05	81.79	91	
7350	W14 x 74		984	.057		49	1.69	1.03	51.72	57.50	
7450	W14 x 176	↓	912	.061	↓	116	1.83	1.11	118.94	132	
0010	**LIGHTWEIGHT FRAMING**										230
0400	Angle framing, 4" and larger	E-4	3,000	.011	Lb.	.85	.33	.03	1.21	1.58	
0450	Less than 4" angles		1,800	.018		.91	.55	.05	1.51	2.06	
0600	Channel framing, 8" and larger		3,500	.009		.93	.28	.02	1.23	1.57	
0650	Less than 8" channels	↓	2,000	.016		.94	.50	.04	1.48	1.99	
1000	Continuous slotted channel framing system, minimum	2 Sswk	2,400	.007		1.20	.20		1.40	1.69	
1200	Maximum	"	1,600	.010		2.16	.31		2.47	2.94	
1300	Cross bracing, rods, 3/4" diameter	E-3	700	.034		.96	1.07	.12	2.15	3.16	
1310	7/8" diameter		700	.034		.81	1.07	.12	2	2.99	
1320	1" diameter		700	.034		.81	1.07	.12	2	2.99	
1330	Angle, 5" x 5" x 3/8"		2,800	.009		.84	.27	.03	1.14	1.44	
1350	Hanging lintels, average	↓	850	.028		.90	.88	.10	1.88	2.72	
1380	Roof frames, 3'-0" square, 5' span	E-2	4,200	.013		.81	.40	.24	1.45	1.86	
1400	Tie rod, not upset, 1-1/2" to 4" diameter, with turnbuckle	2 Sswk	800	.020		.90	.61		1.51	2.11	
1420	No turnbuckle		700	.023		.81	.70		1.51	2.17	
1500	Upset, 1-3/4" to 4" diameter, with turnbuckle		800	.020		1.02	.61		1.63	2.24	
1520	No turnbuckle	↓	700	.023	↓	.90	.70		1.60	2.27	
0005	**LINTELS**										232
0010	Plain steel angles, under 500 lb.	1 Bric	550	.015	Lb.	.60	.40		1	1.28	
0100	500 to 1000 lb.		640	.013		.57	.35		.92	1.17	
0200	1,000 to 2,000 lb.		640	.013		.53	.35		.88	1.12	
0300	2,000 to 4,000 lb.	↓	640	.013		.48	.35		.83	1.07	
0500	For built-up angles and plates, add to above					.60			.60	.66	
0700	For engineering, add to above					.15			.15	.16	
0900	For galvanizing, add to above, under 500 lb.					.42			.42	.46	
1000	Over 2,000 lb.				↓	.30			.30	.33	
2000	Steel angles, 3-1/2" x 3", 1/4" thick, 2'-6" long	1 Bric	50	.160	Ea.	8.10	4.42		12.52	15.75	
2100	4'-6" long		45	.178		14.60	4.91		19.51	23.50	
2600	4" x 3-1/2", 1/4" thick, 5'-0" long		40	.200		18.60	5.50		24.10	29	
2700	9'-0" long	↓	35	.229	↓	33.50	6.30		39.80	47	
3500	For precast concrete lintels, see div. 034-802										
0010	**STRUCTURAL STEEL MEMBERS**										250
0020	Common shapes, bolted connections, spans 10'-45', incl erection										
0300	W 8 x 10	E-2	600	.093	L.F.	6.60	2.78	1.69	11.07	14	
0500	x 31		550	.102		20.50	3.03	1.84	25.37	30	
0700	x 22		600	.093		14.50	2.78	1.69	18.97	22.50	
0900	x 49	↓	550	.102	↓	32.50	3.03	1.84	37.37	43	

051 200 | Structural Steel

		051 200 Structural Steel	CREW	DAILY OUTPUT	LABOR-HOURS	UNIT	1999 BARE COSTS MAT.	LABOR	EQUIP.	TOTAL	TOTAL INCL O&P
250	1100	W 12 x 14	E-2	880	.064	L.F.	9.25	1.89	1.15	12.29	14.75
	1700	x 72		640	.087		45.50	2.60	1.58	49.68	56.50
	1900	W 14 x 26		990	.057		17.15	1.68	1.02	19.85	23
	2500	x 120		720	.078		75.50	2.32	1.41	79.23	88.50
	3300	W 18 x 35	E-5	960	.083		23	2.52	1.14	26.66	31
	4100	W 21 x 44		1,064	.075		27.50	2.27	1.03	30.80	35.50
	4900	W 24 x 55		1,110	.072		34.50	2.18	.99	37.67	43
	5900	x 94		1,190	.067		59	2.03	.92	61.95	69.50
	6100	W 30 x 99		1,200	.067		62.50	2.01	.91	65.42	73
	6700	W 33 x 118		1,176	.068		73	2.06	.93	75.99	84.50
	7300	W 36 x 135		1,170	.068		82.50	2.07	.94	85.51	95.50
255	0010	**STRUCTURAL STEEL PROJECTS** Bolted, unless noted otherwise									
	0200	Apartments, nursing homes, etc., 1 to 2 stories	E-5	10.30	7.767	Ton	1,200	235	106	1,541	1,850
	1300	Industrial bldgs., 1 story, beams & girders, steel bearing		12.90	6.202		1,200	187	85	1,472	1,750
	1400	Masonry bearing		10	8		1,200	242	109	1,551	1,875
	2800	Power stations, fossil fuels, minimum	E-6	11	11.636		1,200	350	114	1,664	2,075
	2900	Maximum	"	5.70	22.456		1,800	680	219	2,699	3,425
	5900	Galvanizing structural steel in shop, under 1 ton, add to above					750			750	825
	5950	1 ton to 20 tons					700			700	770
	6000	Over 20 tons					650			650	715
	6100	Cold galvanizing, brush	1 Psst	1,100	.007	S.F.	.06	.19		.25	.43
	6125	Steel surface treatments									
	6170	Wire brush, hand	1 Psst	240	.033	S.F.	.05	.87		.92	1.70
	6180	Power tool	"	600	.013		.05	.35		.40	.72
	6215	Pressure washing, 2800-6000 S.F./day	1 Pord	3,500	.002			.06		.06	.09
	6220	Steam cleaning, 2800-4000 S.F./day		2,400	.003			.08		.08	.13
	6225	Water blasting		3,200	.002			.06		.06	.10
	6230	Brush-off blast	2 Pord	3,800	.004		.07	.10		.17	.23
	6235	Blast (SSPC-6), loose scale, fine pwdr rust, 2.0 #/S.F.		2,500	.006		.13	.16		.29	.39
	6240	Tight mill scale, little/no rust, 3.0 #/S.F.		1,920	.008		.20	.21		.41	.54
	6245	Exist coat blistered/pitted, 4.0 #/S.F.		1,280	.013		.27	.31		.58	.78
	6250	Exist coat badly pitted/nodules, 6.7 #/S.F.		770	.021		.45	.52		.97	1.28
	6255	Near white blast (SSPC-10), loose scale, fine rust, 5.6 #/S.F.		1,025	.016		.38	.39		.77	1
	6260	Tight mill scale, little/no rust, 6.9 #/S.F.		830	.019		.46	.48		.94	1.24
	6265	Exist coat blistered/pitted, 9.0 #/S.F.		640	.025		.60	.62		1.22	1.61
	6270	Exist coat badly pitted/nodules, 11.3 #/S.F.		510	.031		.76	.78		1.54	2.03
	6510	Paints & protective coatings, sprayed									
	6520	Alkyds, primer	2 Psst	3,600	.004	S.F.	.05	.12		.17	.27
	6540	Gloss topcoats		3,200	.005		.04	.13		.17	.29
	6560	Silicone alkyd		3,200	.005		.09	.13		.22	.35
	6610	Epoxy, primer		3,000	.005		.15	.14		.29	.43
	6630	Intermediate or topcoat		2,800	.006		.12	.15		.27	.42
	6650	Enamel coat		2,800	.006		.15	.15		.30	.45
	6810	Latex primer		3,600	.004		.05	.12		.17	.27
	6830	Topcoats		3,200	.005		.05	.13		.18	.30
	6910	Universal primers, one part, phenolic, modified alkyd		2,000	.008		.08	.21		.29	.47
	6940	Two part, epoxy spray		2,000	.008		.13	.21		.34	.53
	7000	Zinc rich primers, self cure, spray, inorganic		1,800	.009		.24	.23		.47	.70
	7010	Epoxy, spray, organic		1,800	.009		.12	.23		.35	.58
	7020	Above grade, simple structures, add						25%			
	7030	Intricate structures, add						50%			
275	0010	**VIBRATION PADS**									
	0300	Laminated synthetic rubber impregnated cotton duck, 1/2" thick	2 Sswk	20	.800	S.F.	50	24.50		74.50	100
	0400	1" thick		20	.800		101	24.50		125.50	156
	0600	Neoprene bearing pads, 1/2" thick		24	.667		18.90	20.50		39.40	58.50

051 | Structural Metal Framing

051 200 | Structural Steel

			CREW	DAILY OUTPUT	LABOR-HOURS	UNIT	1999 BARE COSTS MAT.	LABOR	EQUIP.	TOTAL	TOTAL INCL O&P	
5	0700	1" thick	2 Sswk	20	.800	S.F.	39	24.50		63.50	87.50	275
	0900	Fabric reinforced neoprene, 5000 psi, 1/2" thick		24	.667		8.50	20.50		29	47	
	1000	1" thick		20	.800		17	24.50		41.50	63.50	
	1200	Felt surfaced vinyl pads, cork and sisal, 5/8" thick		24	.667		22	20.50		42.50	62	
	1300	1" thick		20	.800		40	24.50		64.50	89.50	
	1500	Teflon bonded to 10 ga. carbon steel, 1/32" layer		24	.667		36	20.50		56.50	77	
	1600	3/32" layer		24	.667		54	20.50		74.50	97	
	1800	Bonded to 10 ga. stainless steel, 1/32" layer		24	.667		65	20.50		85.50	109	
	1900	3/32" layer	↓	24	.667	↓	83.50	20.50		104	130	
	2100	Circular, encased, rule of thumb				Kip	5.50			5.50	6.05	

051 500 | Steel Wire Rope

			CREW	DAILY OUTPUT	LABOR-HOURS	UNIT	1999 BARE COSTS MAT.	LABOR	EQUIP.	TOTAL	TOTAL INCL O&P	
8	0005	**WIRE ROPE**										508
	0010	6 x 19, bright, fiber core, 5000' rolls, 1/2" diameter				L.F.	.63			.63	.69	
	0050	Steel core					.72			.72	.80	
	0100	Fiber core, 1" diameter					2			2	2.20	
	0150	Steel core					2.20			2.20	2.42	
	0300	6 x 19, galvanized, fiber core, 1/2" diameter					1			1	1.10	
	0350	Steel core					1.12			1.12	1.24	
	0400	Fiber core, 1" diameter					3.03			3.03	3.34	
	0450	Steel core					3.22			3.22	3.54	
	0500	6 x 7, bright, IPS, fiber core, <500 L.F. w/acc., 1/4" diameter	E-17	6,400	.002		.65	.08		.73	.85	
	0510	1/2" diameter		2,100	.008		1.60	.24		1.84	2.20	
	0520	3/4" diameter		960	.017		2.89	.53		3.42	4.15	
	0550	6 x 19, bright, IPS, IWRC, <500 L.F. w/acc., 1/4" diameter		5,760	.003		.99	.09		1.08	1.25	
	0560	1/2" diameter		1,730	.009		1.61	.29		1.90	2.31	
	0570	3/4" diameter		770	.021		2.80	.66		3.46	4.28	
	0580	1" diameter		420	.038		4.75	1.20		5.95	7.40	
	0590	1-1/4" diameter		290	.055		7.90	1.74	.01	9.65	11.90	
	0600	1-1/2" diameter	↓	192	.083		9.70	2.63	.01	12.34	15.50	
	0610	1-3/4" diameter	E-18	240	.167		15.45	5.10	2.14	22.69	28.50	
	0620	2" diameter		160	.250		19.80	7.65	3.20	30.65	39	
	0630	2-1/4" diameter	↓	160	.250		26.50	7.65	3.20	37.35	46	
	0650	6 x 37, bright, IPS, IWRC, <500 L.F. w/acc., 1/4" diameter	E-17	6,400	.002		1.17	.08		1.25	1.43	
	0660	1/2" diameter		1,730	.009		1.96	.29		2.25	2.69	
	0670	3/4" diameter		770	.021		3.14	.66		3.80	4.66	
	0680	1" diameter		430	.037		5.05	1.18		6.23	7.70	
	0690	1-1/4" diameter		290	.055		7.60	1.74	.01	9.35	11.55	
	0700	1-1/2" diameter	↓	190	.084		10.85	2.66	.01	13.52	16.85	
	0710	1-3/4" diameter	E-18	260	.154		17.20	4.70	1.97	23.87	29.50	
	0720	2" diameter		200	.200		22.50	6.10	2.56	31.16	38	
	0730	2-1/4" diameter	↓	160	.250		29.50	7.65	3.20	40.35	49.50	
	0800	6 x 19 & 6 x 37, swaged, 1/2" diameter	E-17	1,220	.013		1.18	.41		1.59	2.06	
	0810	9/16" diameter		1,120	.014		1.35	.45		1.80	2.32	
	0820	5/8" diameter		930	.017		1.63	.54		2.17	2.79	
	0830	3/4" diameter		640	.025		2.06	.79		2.85	3.72	
	0840	7/8" diameter		480	.033		2.59	1.05		3.64	4.78	
	0850	1" diameter		350	.046		3.17	1.44	.01	4.62	6.15	
	0860	1-1/8" diameter		288	.056		3.89	1.76	.01	5.66	7.50	
	0870	1-1/4" diameter		230	.070		4.73	2.20	.01	6.94	9.25	
	0880	1-3/8" diameter	↓	192	.083		5.45	2.63	.01	8.09	10.85	
	0890	1-1/2" diameter	E-18	300	.133	↓	6.60	4.07	1.71	12.38	16.45	

052 | Metal Joists

052 100 | Steel Joists		CREW	DAILY OUTPUT	LABOR-HOURS	UNIT	1999 BARE COSTS MAT.	LABOR	EQUIP.	TOTAL	TOTAL INCL O&P
110 0010	**OPEN WEB JOISTS**, Truckload lots									
0020	K series, horizontal bridging, spans up to 30′, minimum	E-7	15	5.333	Ton	990	161	78.50	1,229.50	1,475
0080	Maximum	↓	9	8.889	↓	1,225	269	131	1,625	1,975
0410	Span 30′ to 50′, minimum		17	4.706		875	142	69	1,086	1,300
0440	Average		17	4.706		900	142	69	1,111	1,325
0460	Maximum		10	8		915	242	118	1,275	1,550
1010	CS series, horizontal bridging									
1020	Spans to 30′, minimum	E-7	15	5.333	Ton	815	161	78.50	1,054.50	1,275
1040	Average	↓	12	6.667	↓	850	201	98	1,149	1,400
1060	Maximum		9	8.889		945	269	131	1,345	1,675
2000	LH series, bolted cross bridging									
2020	Spans to 96′, minimum	E-7	16	5	Ton	930	151	73.50	1,154.50	1,375
2080	Maximum	"	11	7.273	"	1,325	220	107	1,652	1,950
4010	SLH series, bolted cross bridging									
4020	Spans to 200′, minimum	E-7	16	5	Ton	880	151	73.50	1,104.50	1,325
4040	Average	↓	13	6.154	↓	950	186	90.50	1,226.50	1,475
4060	Maximum		11	7.273		1,200	220	107	1,527	1,800

053 | Metal Decking

053 100 | Steel Deck		CREW	DAILY OUTPUT	LABOR-HOURS	UNIT	1999 BARE COSTS MAT.	LABOR	EQUIP.	TOTAL	TOTAL INCL O&P
104 0010	**METAL DECKING** Steel decking									
0200	Cellular units, galvanized, 2″ deep, 20-20 gauge, over 15 squares	E-4	1,460	.022	S.F.	2.81	.68	.06	3.55	4.40
0250	18-20 gauge	↓	1,420	.023	↓	3.27	.70	.06	4.03	4.94
0300	18-18 gauge		1,390	.023		3.69	.72	.06	4.47	5.45
0320	16-18 gauge		1,360	.024		4.16	.73	.06	4.95	6
0340	16-16 gauge		1,330	.024		4.57	.75	.06	5.38	6.50
0400	3″ deep, galvanized, 20-20 gauge		1,375	.023		2.93	.72	.06	3.71	4.62
2100	Open type, galv., 1-1/2″ deep wide rib, 22 gauge, under 50 squares		4,500	.007		.95	.22	.02	1.19	1.47
2400	Over 500 squares		5,100	.006		.72	.20	.02	.94	1.17
2900	18 gauge, under 50 squares		3,800	.008		1.44	.26	.02	1.72	2.08
3000	Over 500 squares		4,300	.007		1.05	.23	.02	1.30	1.60
3050	16 gauge, under 50 squares		3,700	.009		1.81	.27	.02	2.10	2.50
3100	Over 500 squares		4,200	.008		1.36	.24	.02	1.62	1.94
3200	3″ deep, 22 gauge, under 50 squares		3,600	.009		1.55	.28	.02	1.85	2.25
3400	18 gauge, under 50 squares		3,200	.010		2.34	.31	.03	2.68	3.17
3700	4-1/2″ deep, long span roof, over 50 squares, 20 gauge		2,700	.012		2.18	.37	.03	2.58	3.11
4100	6″ deep, long span, 18 gauge		2,000	.016		3.93	.50	.04	4.47	5.30

055 | Metal Fabrications

055 100 | Metal Stairs

		CREW	DAILY OUTPUT	LABOR-HOURS	UNIT	1999 BARE COSTS MAT.	LABOR	EQUIP.	TOTAL	TOTAL INCL O&P	
0010	**STAIR** Steel, safety nosing, steel stringers										104
0020	Grating tread and pipe railing, 3'-6" wide	E-4	35	.914	Riser	83.50	28.50	2.35	114.35	146	
0100	4'-0" wide		30	1.067		92	33	2.74	127.74	165	
0200	Cement fill metal pan, picket rail, 3'-6" wide		35	.914		71.50	28.50	2.35	102.35	133	
0300	4'-0" wide		30	1.067		79	33	2.74	114.74	151	
0350	Wall rail, both sides, 3'-6" wide		53	.604		62	18.80	1.55	82.35	104	
0400	Cast iron tread and pipe rail, 3'-6" wide		35	.914		183	28.50	2.35	213.85	256	
0500	Checkered plate tread, industrial, 3'-6" wide		28	1.143		101	35.50	2.94	139.44	179	
0550	Circular, for tanks, 3'-0" wide		33	.970		87.50	30	2.49	119.99	155	
0600	For isolated stairs, add						100%				
0800	Custom steel stairs, 3'-6" wide, minimum	E-4	35	.914		103	28.50	2.35	133.85	168	
0810	Average		30	1.067		152	33	2.74	187.74	231	
0900	Maximum		20	1.600		222	50	4.11	276.11	340	
1100	For 4' wide stairs, add					10%	5%				
1300	For 5' wide stairs, add					20%	10%				
1500	Landing, steel pan, conventional	E-4	160	.200	S.F.	35.50	6.20	.51	42.21	51.50	
1600	Pre-erected	"	255	.125	"	24	3.90	.32	28.22	33.50	
1700	Pre-erected, steel pan tread, 3'-6" wide, 2 line pipe rail	E-2	87	.644	Riser	229	19.15	11.65	259.80	298	
1810	Spiral aluminum, 5'-0" diameter, stock units	E-4	45	.711		182	22	1.83	205.83	243	
1830	Stock units, 4'-0" diameter, safety treads		50	.640		110	19.90	1.64	131.54	159	
1900	Spiral, cast iron, 4'-0" diameter, ornamental, minimum		45	.711		166	22	1.83	189.83	225	
1920	Maximum		25	1.280		221	40	3.29	264.29	320	
2000	Spiral, steel, industrial checkered plate, 4' diameter		45	.711		166	22	1.83	189.83	225	
2200	Stock units, 6'-0" diameter		40	.800		199	25	2.06	226.06	267	
3900	Industrial ships ladder, 3' W, grating treads, 2 line pipe rail		30	1.067		66	33	2.74	101.74	137	
4000	Aluminum		30	1.067		97.50	33	2.74	133.24	171	
0005	**STAIR TREADS**										108
0010	Aluminum grating, 3' long, 1" x 3/16" bars, 6" wide	1 Sswk	24	.333	Ea.	25	10.20		35.20	46	
0100	12" wide		22	.364		43.50	11.15		54.65	68.50	
0200	1-1/2" x 3/16" bars, 6" wide		24	.333		34.50	10.20		44.70	56.50	
0300	12" wide		22	.364		58.50	11.15		69.65	84.50	
0700	Cast aluminum, abrasive, 3' long x 12" wide, 5/16" thick		15	.533		99.50	16.30		115.80	140	
0800	3/8" thick		15	.533		104	16.30		120.30	144	
0900	1/2" thick		15	.533		118	16.30		134.30	160	
2000	Steel grating, painted, 3' long, 1-1/4" x 3/16" bars, 6" wide		20	.400		19	12.25		31.25	43.50	
2010	12" wide		18	.444		27.50	13.60		41.10	55.50	
2012	Steel grating, 3' long, painted,										
2022	Plain, 3/4" x 1/8", 6" wide	1 Sswk	17	.471	Ea.	14.60	14.40		29	42.50	
2024	12" wide		15	.533		18.75	16.30		35.05	50.50	
2032	1-1/4" x 3/16", 6" wide		17	.471		18.05	14.40		32.45	46.50	
2062	Serrated, 3/4" x 1/8", 6" wide		17	.471		15.15	14.40		29.55	43	
2064	12" wide		15	.533		20	16.30		36.30	52.50	
2072	1-1/4" x 3/16", 6" wide		17	.471		19.45	14.40		33.85	48	
2092	For galvanizing, 3/4" x 1/8", add				S.F.	1.26			1.26	1.39	
2100	Add for abrasive nosing, 3' long, painted				Ea.	12.25			12.25	13.50	
2200	Galvanized					20.50			20.50	22.50	
2500	Expanded steel, 2-1/2" deep, 9" x 3' long, 18 gauge	1 Sswk	20	.400		17.75	12.25		30	42	
2600	14 gauge	"	20	.400		22	12.25		34.25	46.50	

055 150 | Ladders

		CREW	DAILY OUTPUT	LABOR-HOURS	UNIT	MAT.	LABOR	EQUIP.	TOTAL	TOTAL INCL O&P	
0005	**LADDER**										158
0010	Steel, 20" wide, bolted to concrete, with cage	E-4	50	.640	V.L.F.	53	19.90	1.64	74.54	97	
0100	Without cage		85	.376		24.50	11.70	.97	37.17	49.50	
0300	Aluminum, bolted to concrete, with cage		50	.640		100	19.90	1.64	121.54	148	
0400	Without cage		85	.376		55.50	11.70	.97	68.17	83.50	

5 METALS

055 200 | Handrails & Railings

			CREW	DAILY OUTPUT	LABOR-HOURS	UNIT	1999 BARE COSTS MAT.	LABOR	EQUIP.	TOTAL	TOTAL INCL O&P
202	0010	**BUMPER RAILS** For garages, 12 ga. rail, 6″ wide, with steel									
	0020	posts 12′-6″ O.C., Minimum	E-4	190	.168	L.F.	8	5.25	.43	13.68	18.90
	0030	Average		165	.194		10	6.05	.50	16.55	22.50
	0100	Maximum		140	.229		12	7.10	.59	19.69	27
	0300	12″ channel rail, minimum		160	.200		10	6.20	.51	16.71	23
	0400	Maximum		120	.267		15	8.30	.69	23.99	32.50
203	0005	**RAILING, PIPE**									
	0010	Aluminum, 2 rail, satin finish, 1-1/4″ diameter	E-4	160	.200	L.F.	13	6.20	.51	19.71	26.50
	0030	Clear anodized		160	.200		16.10	6.20	.51	22.81	29.50
	0040	Dark anodized		160	.200		18.15	6.20	.51	24.86	32
	0080	1-1/2″ diameter, satin finish		160	.200		15.55	6.20	.51	22.26	29
	0090	Clear anodized		160	.200		17.35	6.20	.51	24.06	31
	0100	Dark anodized		160	.200		19.25	6.20	.51	25.96	33
	0140	Aluminum, 3 rail, 1-1/4″ diam., satin finish		137	.234		19.90	7.25	.60	27.75	36
	0150	Clear anodized		137	.234		25	7.25	.60	32.85	41.50
	0160	Dark anodized		137	.234		27.50	7.25	.60	35.35	44.50
	0200	1-1/2″ diameter, satin finish		137	.234		24	7.25	.60	31.85	40.50
	0210	Clear anodized		137	.234		27	7.25	.60	34.85	44
	0220	Dark anodized		137	.234		29.50	7.25	.60	37.35	46.50
	0500	Steel, 2 rail, on stairs, primed, 1-1/4″ diameter		160	.200		9.35	6.20	.51	16.06	22.50
	0520	1-1/2″ diameter		160	.200		10.30	6.20	.51	17.01	23.50
	0540	Galvanized, 1-1/4″ diameter		160	.200		13	6.20	.51	19.71	26.50
	0560	1-1/2″ diameter		160	.200		14.55	6.20	.51	21.26	28
	0580	Steel, 3 rail, primed, 1-1/4″ diameter		137	.234		13.95	7.25	.60	21.80	29.50
	0600	1-1/2″ diameter		137	.234		14.80	7.25	.60	22.65	30.50
	0620	Galvanized, 1-1/4″ diameter		137	.234		19.60	7.25	.60	27.45	35.50
	0640	1-1/2″ diameter		137	.234		21.50	7.25	.60	29.35	38
	0700	Stainless steel, 2 rail, 1-1/4″ diam. #4 finish		137	.234		32	7.25	.60	39.85	49
	0720	High polish		137	.234		51.50	7.25	.60	59.35	70.50
	0740	Mirror polish		137	.234		64.50	7.25	.60	72.35	85
	0760	Stainless steel, 3 rail, 1-1/2″ diam., #4 finish		120	.267		48	8.30	.69	56.99	69
	0770	High polish		120	.267		79.50	8.30	.69	88.49	103
	0780	Mirror finish		120	.267		97	8.30	.69	105.99	123
	0900	Wall rail, alum. pipe, 1-1/4″ diam., satin finish		213	.150		7.45	4.67	.39	12.51	17.15
	0905	Clear anodized		213	.150		9.05	4.67	.39	14.11	19
	0910	Dark anodized		213	.150		11	4.67	.39	16.06	21
	0915	1-1/2″ diameter, satin finish		213	.150		8.25	4.67	.39	13.31	18.05
	0920	Clear anodized		213	.150		10.35	4.67	.39	15.41	20.50
	0925	Dark anodized		213	.150		12.80	4.67	.39	17.86	23
	0930	Steel pipe, 1-1/4″ diameter, primed		213	.150		5.70	4.67	.39	10.76	15.25
	0935	Galvanized		213	.150		8.25	4.67	.39	13.31	18.05
	0940	1-1/2″ diameter		176	.182		5.85	5.65	.47	11.97	17.25
	0945	Galvanized		213	.150		8.30	4.67	.39	13.36	18.10
	0955	Stainless steel pipe, 1-1/2″ diam., #4 finish		107	.299		25.50	9.30	.77	35.57	46
	0960	High polish		107	.299		52	9.30	.77	62.07	75
	0965	Mirror polish		107	.299		61	9.30	.77	71.07	85.50
206	0005	**RAILINGS, INDUSTRIAL** Welded									
	0010	2 rail, 3′-6″ high, 1-1/2″ pipe	E-4	255	.125	L.F.	13.70	3.90	.32	17.92	22.50
	0100	2″ angle rail	″	255	.125		12.50	3.90	.32	16.72	21.50
	0200	For 4″ high kick plate, 10 gauge, add					2.86			2.86	3.15
	0300	1/4″ thick, add					3.67			3.67	4.04
	0500	For curved rails, add					30%	30%			
208	0005	**RAILINGS, ORNAMENTAL**									
	0010	Aluminum, bronze or stainless, minimum	1 Sswk	24	.333	L.F.	97	10.20		107.20	126
	0100	Maximum		9	.889		545	27		572	650
	0200	Aluminum ornamental rail, minimum		15	.533		44	16.30		60.30	78.50

055 200 | Handrails & Railings

			CREW	DAILY OUTPUT	LABOR-HOURS	UNIT	1999 BARE COSTS MAT.	LABOR	EQUIP.	TOTAL	TOTAL INCL O&P
208	0300	Maximum	1 Sswk	8	1	L.F.	191	30.50		221.50	266
	0400	Hand-forged wrought iron, minimum		12	.667		95	20.50		115.50	142
	0500	Maximum		8	1		405	30.50		435.50	500
	0600	Composite metal and wood or glass, minimum		6	1.333		184	41		225	277
	0700	Maximum	↓	5	1.600	↓	480	49		529	615

055 300 | Gratings & Floor Plates

			CREW	DAILY OUTPUT	LABOR-HOURS	UNIT	MAT.	LABOR	EQUIP.	TOTAL	TOTAL INCL O&P
302	0005	CHECKERED PLATE									
	0010	1/4″ & 3/8″, 2000 to 5000 S.F., bolted	E-4	2,900	.011	Lb.	.59	.34	.03	.96	1.31
	0100	Welded		4,400	.007	″	.55	.23	.02	.80	1.04
	0300	Pit or trench cover and frame, 1/4″ plate, 2′ to 3′ wide	↓	100	.320	S.F.	15.50	9.95	.82	26.27	36
	0400	For galvanizing, add				Lb.	.34			.34	.37
	0500	Platforms, 1/4″ plate, no handrails included, rectangular	E-4	4,200	.008		.93	.24	.02	1.19	1.47
	0600	Circular	″	2,500	.013	↓	1.30	.40	.03	1.73	2.20

			CREW	DAILY OUTPUT	LABOR-HOURS	UNIT	MAT.	LABOR	EQUIP.	TOTAL	TOTAL INCL O&P
304	0010	FLOOR GRATING, ALUMINUM									
	0110	Bearing bars @ 1-3/16″ O.C., cross bars @ 4″ O.C.,									
	0112	1″ x 1/8″ bar	E-4	850	.038	S.F.	9.75	1.17	.10	11.02	13
	0114	Over 300 S.F.		1,000	.032		9.60	1	.08	10.68	12.45
	0122	1-1/4″ x 3/16″ bar		750	.043		14.70	1.33	.11	16.14	18.70
	0124	Over 300 S.F.		1,000	.032		15.65	1	.08	16.73	19.15
	0132	1-1/2″ x 1/8″ bar		700	.046		12.45	1.42	.12	13.99	16.45
	0134	Over 300 S.F.		1,000	.032		10.90	1	.08	11.98	13.90
	0136	1-3/4″ x 3/16″ bar		500	.064		17.90	1.99	.16	20.05	23.50
	0138	Over 300 S.F.		1,000	.032		18.30	1	.08	19.38	22
	0146	2-1/4″ x 3/16″, up to 75 S.F.		600	.053		26.50	1.66	.14	28.30	32.50
	0148	Over 300 S.F.		1,000	.032		19.80	1	.08	20.88	24
	0162	Cross bars @ 2″ O.C., 1″ x 1/8″, up to 75 S.F.		600	.053		17.35	1.66	.14	19.15	22
	0164	Over 300 S.F.		1,000	.032		12.40	1	.08	13.48	15.50
	0172	1-1/4″ x 3/16″, up to 75 S.F.		600	.053		22	1.66	.14	23.80	27
	0174	Over 300 S.F.		1,000	.032		15.90	1	.08	16.98	19.40
	0182	1-1/2″ x 1/8″, up to 75 S.F.		600	.053		17.95	1.66	.14	19.75	23
	0184	Over 300 S.F.		1,000	.032		13.55	1	.08	14.63	16.80
	0186	1-3/4″ x 3/16″, up to 75 S.F.		600	.053		26.50	1.66	.14	28.30	32.50
	0188	Over 300 S.F.	↓	1,000	.032	↓	19.95	1	.08	21.03	24
	0200	For straight cuts, add				L.F.	2.31			2.31	2.54
	0212	Close mesh, 1″ x 1/8″, up to 75 S.F.	E-4	520	.062	S.F.	23	1.91	.16	25.07	29
	0214	Over 300 S.F.		920	.035		16.80	1.08	.09	17.97	20.50
	0222	1-1/4″ x 3/16″, up to 75 S.F.		520	.062		29.50	1.91	.16	31.57	36
	0224	Over 300 S.F.		920	.035		21	1.08	.09	22.17	25
	0232	1-1/2″ x 1/8″, up to 75 S.F.		520	.062		26.50	1.91	.16	28.57	32.50
	0234	Over 300 S.F.	↓	920	.035	↓	19.05	1.08	.09	20.22	23
	0600	For aluminum checkered plate nosings, add				L.F.	3.38			3.38	3.72
	0700	For straight toe plate, add					6.25			6.25	6.90
	0800	For curved toe plate, add					7.70			7.70	8.45
	1000	For cast aluminum abrasive nosings, add				↓	4.99			4.99	5.50
	1200	Expanded aluminum, .65# per S.F., on grade	E-4	1,050	.030	S.F.	3.58	.95	.08	4.61	5.75
	1400	Extruded I bars are 10% less than 3/16″ bars									
	1600	Heavy duty, all extruded panels, 3/4″ deep, 1.7 # per S.F.	E-4	1,100	.029	S.F.	8.70	.90	.07	9.67	11.35
	1700	1-1/4″ deep, 2.5# per S.F.		1,000	.032		12.80	1	.08	13.88	16
	1800	1-3/4″ deep, 4.8# per S.F.		925	.035		17.35	1.08	.09	18.52	21
	1900	2-1/4″ deep, 6.1# per S.F.	↓	875	.037	↓	19.65	1.14	.09	20.88	23.50
	2100	For safety serrated surface, add					15%				
	2400	Close spaced aluminum grating with vinyl tread inserts	E-4	255	.125	S.F.	33.50	3.90	.32	37.72	44.50
	2600	For bottom drainage pan, add	″	510	.063	″	13.80	1.95	.16	15.91	18.95

055 300 | Gratings & Floor Plates

		055 300 Gratings & Floor Plates	CREW	DAILY OUTPUT	LABOR-HOURS	UNIT	1999 BARE COSTS MAT.	LABOR	EQUIP.	TOTAL	TOTAL INCL O&P
308	0010	**FLOOR GRATING, FIBERGLASS**									
	0100	Reinforced polyester, fire retardant, 1" x 4" grid, 1" thick	E-4	510	.063	S.F.	13.75	1.95	.16	15.86	18.90
	0200	1-1/2" thick		500	.064		16.85	1.99	.16	19	22.50
	0300	With grit surface, 1" x 4" grid, 1-1/2" thick		500	.064		17.75	1.99	.16	19.90	23.50
310	0010	**FLOOR GRATING PLANKS**									
	0020	Aluminum, 9-1/2" wide, 14 ga., 2" rib	E-4	950	.034	L.F.	11.20	1.05	.09	12.34	14.30
	0200	Galvanized steel, 9-1/2" wide, 14 ga., 2-1/2" rib		950	.034		7	1.05	.09	8.14	9.70
	0300	4" rib		950	.034		8.25	1.05	.09	9.39	11.05
	0500	12 gauge, 2-1/2" rib		950	.034		8.60	1.05	.09	9.74	11.50
	0600	3" rib		950	.034		9.65	1.05	.09	10.79	12.65
	0800	Stainless steel, type 304, 16 ga., 2" rib		950	.034		18.25	1.05	.09	19.39	22
	0900	Type 316		950	.034		22	1.05	.09	23.14	26.50
312	0010	**FLOOR GRATING, STEEL**									
	0050	Labor for installing, on grade	E-4	845	.038	S.F.		1.18	.10	1.28	2.27
	0100	Elevated		460	.070	"		2.16	.18	2.34	4.17
	0300	Platforms, to 12' high, rectangular		3,150	.010	Lb.	1.07	.32	.03	1.42	1.79
	0400	Circular		2,300	.014	"	1.18	.43	.04	1.65	2.13
	0410	Painted bearing bars @ 1-3/16"									
	0412	Cross bars @ 4" O.C., 3/4" x 1/8" bar, up to 75 S.F.	E-2	500	.112	S.F.	8.40	3.33	2.02	13.75	17.35
	0414	Over 300 S.F.		750	.075		6.15	2.22	1.35	9.72	12.20
	0422	1-1/4" x 3/16", up to 75 S.F.		400	.140		12.15	4.17	2.53	18.85	23.50
	0424	Over 300 S.F.		600	.093		8.35	2.78	1.69	12.82	15.90
	0432	1-1/2" x 1/8", up to 75 S.F.		400	.140		11.20	4.17	2.53	17.90	22.50
	0434	Over 300 S.F.		600	.093		7.75	2.78	1.69	12.22	15.30
	0436	1-3/4" x 3/16", up to 75 S.F.		400	.140		15.05	4.17	2.53	21.75	26.50
	0438	Over 300 S.F.		600	.093		10.50	2.78	1.69	14.97	18.30
	0452	2-1/4" x 3/16", up to 75 S.F.		300	.187		17.50	5.55	3.37	26.42	32.50
	0454	Over 300 S.F.		450	.124		12.15	3.70	2.25	18.10	22.50
	0462	Cross bars @ 2" O.C., 3/4" x 1/8", up to 75 S.F.		500	.112		9.90	3.33	2.02	15.25	19
	0464	Over 300 S.F.		750	.075		7	2.22	1.35	10.57	13.10
	0472	1-1/4" x 3/16", up to 75 S.F.		400	.140		15.80	4.17	2.53	22.50	27.50
	0474	Over 300 S.F.		600	.093		11.10	2.78	1.69	15.57	18.95
	0482	1-1/2" x 1/8", up to 75 S.F.		400	.140		13.55	4.17	2.53	20.25	25
	0484	Over 300 S.F.		600	.093		9.50	2.78	1.69	13.97	17.20
	0486	1-3/4" x 3/16", up to 75 S.F.		400	.140		19.35	4.17	2.53	26.05	31.50
	0488	Over 300 S.F.		600	.093		13.70	2.78	1.69	18.17	22
	0502	2-1/4" x 3/16", up to 75 S.F.		300	.187		24	5.55	3.37	32.92	40
	0504	Over 300 S.F.		450	.124		16.85	3.70	2.25	22.80	27.50
	0601	Painted bearing bars @ 15/16" O.C., cross bars @ 4" O.C.,									
	0612	Up to 75 S.F., 3/4" x 1/8" bars	E-4	850	.038	S.F.	8.05	1.17	.10	9.32	11.10
	0622	1-1/4" x 3/16" bars		600	.053		14.50	1.66	.14	16.30	19.15
	0632	1-1/2" x 1/8" bars		550	.058		12.50	1.81	.15	14.46	17.25
	0636	1-3/4" x 3/16" bars		450	.071		16	2.21	.18	18.39	22
	0652	2-1/4" x 3/16" bars	E-2	300	.187		22.50	5.55	3.37	31.42	38.50
	0662	Cross bars @ 2" O.C., up to 75 S.F., 3/4" x 1/8"		500	.112		8.55	3.33	2.02	13.90	17.50
	0672	1-1/4" x 3/16" bars		400	.140		14.95	4.17	2.53	21.65	26.50
	0682	1-1/2" x 1/8" bars		400	.140		12.80	4.17	2.53	19.50	24
	0686	1-3/4" x 3/16" bars		300	.187		19.20	5.55	3.37	28.12	34.50
	0690	For galvanized grating, add					25%				
	0800	For straight cuts, add				L.F.	3.74			3.74	4.11
	0900	For curved cuts, add					4.70			4.70	5.15
	1000	For straight banding, add					3.59			3.59	3.95
	1100	For curved banding, add					4.70			4.70	5.15
	1200	For checkered plate nosings, add					4.04			4.04	4.44
	1300	For straight toe or kick plate, add					7.25			7.25	8
	1400	For curved toe or kick plate, add					8.10			8.10	8.90

Important: See the Reference Section for critical supporting data - Reference Nos., Crews, & City Cost Indexes

055 300 | Gratings & Floor Plates

			CREW	DAILY OUTPUT	LABOR-HOURS	UNIT	1999 BARE COSTS MAT.	LABOR	EQUIP.	TOTAL	TOTAL INCL O&P
312	1500	For abrasive nosings, add				L.F.	5.90			5.90	6.50
	1600	For safety serrated surface, minimum, add					15%				
	1700	Maximum, add					25%				
	2000	Stainless steel gratings, close spaced, on grade	E-4	340	.094	S.F.	101	2.93	.24	104.17	117
	2100	Standard grating, 3.3# per S.F.		700	.046		3.29	1.42	.12	4.83	6.35
	2200	4.5# per S.F.		650	.049		3.72	1.53	.13	5.38	7.05
	2400	Expanded steel grating, on grade, 3.0# per S.F.		900	.036		2.49	1.11	.09	3.69	4.87
	2500	3.14# per S.F.		900	.036		2.55	1.11	.09	3.75	4.93
	2600	4.0# per S.F.		850	.038		3.18	1.17	.10	4.45	5.75
	2650	4.27# per S.F.		850	.038		3.44	1.17	.10	4.71	6.05
	2700	5.0# per S.F.		800	.040		4.14	1.24	.10	5.48	6.95
	2800	6.25# per S.F.		750	.043		4.99	1.33	.11	6.43	8.05
	2900	7.0# per S.F.	↓	700	.046		5.85	1.42	.12	7.39	9.15
	3100	For flattened expanded steel grating, add					8%				
	3300	For installation above grade, add				↓		15%			
314	0005	GRATING FRAME									
	0010	Aluminum, for gratings 1" to 1-1/2" deep	1 Sswk	70	.114	L.F.	6.20	3.50		9.70	13.25
	0100	For each corner, add				Ea.	4.90			4.90	5.40
370	0010	TRENCH COVER									
	0020	Cast iron grating with bar stops and angle frame, to 18" wide	1 Sswk	20	.400	L.F.	41	12.25		53.25	67.50
	0100	Frame only (both sides of trench), 1" grating		45	.178		9	5.45		14.45	19.85
	0150	2" grating	↓	35	.229	↓	14	7		21	28
	0200	Aluminum, stock units, including frames and									
	0210	3/8" plain cover plate, 4" opening	E-4	205	.156	L.F.	29.50	4.85	.40	34.75	42
	0300	6" opening		185	.173		36.50	5.40	.44	42.34	50.50
	0400	10" opening		170	.188		50	5.85	.48	56.33	66.50
	0500	16" opening	↓	155	.206		69	6.40	.53	75.93	88
	0700	Add per inch for additional widths to 24"					2.73			2.73	3
	0900	For custom fabrication, add					50%				
	1100	For 1/4" plain cover plate, deduct					12%				
	1500	For cover recessed for tile, 1/4" thick, deduct					12%				
	1600	3/8" thick, add					5%				
	1800	For checkered plate cover, 1/4" thick, deduct					12%				
	1900	3/8" thick, add					2%				
	2100	For slotted or round holes in cover, 1/4" thick, add					3%				
	2200	3/8" thick, add					4%				
	2300	For abrasive cover, add				↓	12%				

055 400 | Castings

			CREW	DAILY OUTPUT	LABOR-HOURS	UNIT	MAT.	LABOR	EQUIP.	TOTAL	TOTAL INCL O&P
404	0010	CONSTRUCTION CASTINGS									
	0020	Manhole covers and frames see Division 027-152									
	0100	Column bases, cast iron, 16" x 16", approx. 65 lbs.	E-4	46	.696	Ea.	79	21.50	1.79	102.29	128
	0200	32" x 32", approx. 256 lbs.		23	1.391	"	295	43.50	3.58	342.08	410
	0600	Miscellaneous C.I. castings, light sections, less than 150 lbs		3,200	.010	Lb.	1.21	.31	.03	1.55	1.93
	1100	Heavy sections, more than 150 lb		4,200	.008		.60	.24	.02	.86	1.11
	1300	Special low volume items	↓	3,200	.010		2.12	.31	.03	2.46	2.93
	1500	For ductile iron, add				↓	100%				

055 500 | Metal Specialties

			CREW	DAILY OUTPUT	LABOR-HOURS	UNIT	MAT.	LABOR	EQUIP.	TOTAL	TOTAL INCL O&P
504	0005	LAMP POSTS									
	0010	Aluminum, 7' high, stock units, post only	1 Carp	16	.500	Ea.	65	13.65		78.65	93
	0100	Mild steel, plain	"	16	.500	"	39	13.65		52.65	64.50

055 | Metal Fabrications

055 500 | Metal Specialties

			CREW	DAILY OUTPUT	LABOR-HOURS	UNIT	1999 BARE COSTS MAT.	LABOR	EQUIP.	TOTAL	TOTAL INCL O&P
512	0005	**ALLOY STEEL CHAIN**									
	0010	Self-colored, cut lengths, w/accessories, 1/4"	E-17	4	4	C.L.F.	330	126	.52	456.52	595
	0020	3/8"		2	8		415	253	1.04	669.04	920
	0030	1/2"		1.20	13.333		650	420	1.73	1,071.73	1,475
	0040	5/8"		.72	22.222		1,350	700	2.89	2,052.89	2,750
	0050	3/4"		.48	33.333		1,900	1,050	4.33	2,954.33	4,025
	0060	7/8"		.40	40		2,925	1,275	5.20	4,205.20	5,550
	0070	1"		.35	45.714		4,325	1,450	5.95	5,780.95	7,400
	0080	1-1/4"	▼	.24	66.667	▼	7,225	2,100	8.65	9,333.65	11,800
	0110	Clevis slip hook, 1/4"				Ea.	8.30			8.30	9.15
	0120	3/8"					12.25			12.25	13.50
	0130	1/2"					21.50			21.50	23.50
	0140	5/8"					39.50			39.50	43
	0150	3/4"					81			81	89
	0160	Eye/sling hook w/ hammerlock coupling, 7/8"					185			185	203
	0170	1"					305			305	335
	0180	1-1/4"				▼	495			495	545

058 | Expansion Control

058 100 | Exp. Cover Assemblies

			CREW	DAILY OUTPUT	LABOR-HOURS	UNIT	1999 BARE COSTS MAT.	LABOR	EQUIP.	TOTAL	TOTAL INCL O&P
104	0010	**EXPANSION JOINT ASSEMBLIES** Custom units									
	0200	Floor cover assemblies, 1" space, aluminum	1 Sswk	38	.211	L.F.	13.20	6.45		19.65	26.50
	0300	Bronze		38	.211		25.50	6.45		31.95	40
	2000	Roof closures, aluminum, flat roof, low profile, 1" space		57	.140		24	4.29		28.29	34.50
	2300	Roof to wall, low profile, 1" space	▼	57	.140	▼	13.30	4.29		17.59	22.50

059 | Hydraulic Structures

059 100 | Hydraulic Structures

			CREW	DAILY OUTPUT	LABOR-HOURS	UNIT	1999 BARE COSTS MAT.	LABOR	EQUIP.	TOTAL	TOTAL INCL O&P
104	0010	**SLUICE GATES** Cast iron, AWWA C501									
	0100	Heavy duty, self contained w/crank oper. gate, 18" x 18"	L-5A	1.70	18.824	Ea.	4,275	580	320	5,175	6,075
	0110	24" x 24"		1.20	26.667		5,300	820	455	6,575	7,775
	0120	30" x 30"		1	32		6,200	985	550	7,735	9,150
	0130	36" x 36"		.90	35.556		7,250	1,100	610	8,960	10,500
	0140	42" x 42"		.80	40		9,050	1,225	685	10,960	12,900
	0150	48" x 48"		.50	64		9,800	1,975	1,100	12,875	15,500
	0160	54" x 54"		.40	80		13,600	2,475	1,375	17,450	20,800
	0170	60" x 60"		.30	106		15,900	3,275	1,825	21,000	25,300
	0180	66" x 66"		.30	106		19,100	3,275	1,825	24,200	28,900
	0190	72" x 72"		.20	160		21,200	4,925	2,750	28,875	35,000
	0200	78" x 78"	▼	.20	160	▼	24,500	4,925	2,750	32,175	38,600

Important: See the Reference Section for critical supporting data - Reference Nos., Crews, & City Cost Indexe

059 100 | Hydraulic Structures		CREW	DAILY OUTPUT	LABOR-HOURS	UNIT	1999 BARE COSTS MAT.	LABOR	EQUIP.	TOTAL	TOTAL INCL O&P	
4	0210 84″ x 84″	L-5A	.10	320	Ea.	28,300	9,850	5,475	43,625	54,500	104
	0220 90″ x 90″	E-20	.30	213		32,100	6,375	2,325	40,800	49,100	
	0230 96″ x 96″		.30	213		36,000	6,375	2,325	44,700	53,500	
	0240 108″ x 108″		.20	320		43,700	9,550	3,475	56,725	68,500	
	0250 120″ x 120″		.10	640		51,000	19,100	6,975	77,075	98,000	
	0260 132″ x 132″	↓	.10	640	↓	90,500	19,100	6,975	116,575	141,000	
8	0010 **SLIDE GATES**										108
	0100 Steel, self contained incl. anchor bolts and grout, 12″ x 12″	L-5A	4.60	6.957	Ea.	1,875	214	119	2,208	2,550	
	0110 18″ x 18″		4	8		2,150	246	137	2,533	2,950	
	0120 24″ x 24″		3.50	9.143		2,300	282	157	2,739	3,200	
	0130 30″ x 30″		2.80	11.429		2,600	350	196	3,146	3,675	
	0140 36″ x 36″		2.30	13.913		2,875	430	238	3,543	4,175	
	0150 42″ x 42″		1.70	18.824		3,225	580	320	4,125	4,925	
	0160 48″ x 48″		1.20	26.667		3,600	820	455	4,875	5,925	
	0170 54″ x 54″		.90	35.556		4,500	1,100	610	6,210	7,550	
	0180 60″ x 60″		.55	58.182		5,100	1,800	995	7,895	9,850	
	0190 72″ x 72″	↓	.36	88.889	↓	6,975	2,725	1,525	11,225	14,200	
2	0010 **CANAL GATES** Cast iron body, fabricated frame										112
	0100 12″ diameter	L-5A	4.60	6.957	Ea.	425	214	119	758	970	
	0110 18″ diameter		4	8		700	246	137	1,083	1,350	
	0120 24″ diameter		3.50	9.143		1,000	282	157	1,439	1,775	
	0130 30″ diameter		2.80	11.429		1,625	350	196	2,171	2,625	
	0140 36″ diameter		2.30	13.913		2,000	430	238	2,668	3,225	
	0150 42″ diameter		1.70	18.824		3,525	580	320	4,425	5,250	
	0160 48″ diameter		1.20	26.667		4,450	820	455	5,725	6,850	
	0170 54″ diameter		.90	35.556		6,550	1,100	610	8,260	9,800	
	0180 60″ diameter		.50	64		7,975	1,975	1,100	11,050	13,500	
	0190 66″ diameter		.50	64		9,650	1,975	1,100	12,725	15,300	
	0200 72″ diameter	↓	.40	80	↓	13,600	2,475	1,375	17,450	20,700	
6	0010 **FLAP GATES**										116
	0100 Aluminum, 18″ diameter	L-5A	5	6.400	Ea.	1,675	197	110	1,982	2,300	
	0110 24″ diameter		4	8		2,100	246	137	2,483	2,900	
	0120 30″ diameter		3.50	9.143		2,500	282	157	2,939	3,425	
	0130 36″ diameter		2.80	11.429		3,125	350	196	3,671	4,250	
	0140 42″ diameter		2.30	13.913		4,175	430	238	4,843	5,625	
	0150 48″ diameter		1.70	18.824		5,125	580	320	6,025	7,000	
	0160 54″ diameter		1.20	26.667		7,100	820	455	8,375	9,750	
	0170 60″ diameter		.80	40		8,900	1,225	685	10,810	12,700	
	0180 66″ diameter		.50	64		11,200	1,975	1,100	14,275	17,000	
	0190 72″ diameter	↓	.40	80	↓	11,800	2,475	1,375	15,650	18,800	
0	0010 **KNIFE GATES**										120
	0100 Incl. handwheel operator for hub, 6″ diameter	Q-23	7.70	3.117	Ea.	680	99.50	67	846.50	970	
	0110 8″ diameter		7.20	3.333		950	106	71.50	1,127.50	1,300	
	0120 10″ diameter		4.80	5		1,225	159	107	1,491	1,700	
	0130 12″ diameter		3.60	6.667		1,750	212	143	2,105	2,400	
	0140 14″ diameter		3.40	7.059		2,425	225	151	2,801	3,150	
	0150 16″ diameter		3.20	7.500		3,325	239	161	3,725	4,175	
	0160 18″ diameter		3	8		4,125	255	172	4,552	5,125	
	0170 20″ diameter		2.70	8.889		4,750	283	191	5,224	5,875	
	0180 24″ diameter		2.40	10		5,750	320	214	6,284	7,050	
	0190 30″ diameter		1.80	13.333		10,000	425	286	10,711	12,000	
	0200 36″ diameter	↓	1.20	20	↓	13,300	635	430	14,365	16,100	

or information about Means Estimating Seminars, see yellow pages 11 and 12 in back of book

Division Notes

		CREW	DAILY OUTPUT	LABOR-HOURS	UNIT	1999 BARE COSTS MAT.	LABOR	EQUIP.	TOTAL	TOTAL INCL O&P

Division 6 Wood & Plastics

Estimating Tips

060 Fasteners & Adhesives

- Common to any wood framed structure are the accessory connector items such as screws, nails, adhesives, hangers, connector plates, straps, angles and holdowns. For typical wood framed buildings, such as residential projects, the aggregate total for these items can be significant, especially in areas where seismic loading is a concern. For floor and wall framing, nail quantities can be figured on a "pounds per thousand board feet basis", with 10 to 25 lbs. per MBF the range. Holdowns, hangers and other connectors should be taken off by the piece.

061 Rough Carpentry

- Lumber is a traded commodity and therefore sensitive to supply and demand in the marketplace. Even in "budgetary" estimating of wood framed projects, it is advisable to call local suppliers for the latest market pricing.
- Common quantity units for wood framed projects are "thousand board feet" (MBF). A board foot is a volume of wood, 1″ x 1′ x 1′, or 144 cubic inches. Board foot quantities are generally calculated using nominal material dimensions—dressed sizes are ignored. Board foot per lineal foot of any stick of lumber can be calculated by dividing the nominal cross sectional area by 12. As an example, 2,000 lineal feet of 2 x 12 equates to 4 MBF by dividing the nominal area, 2 x 12, by 12, which equals 2, and multiplying by 2,000 to give 4,000 board feet. This simple rule applies to all nominal dimensioned lumber.
- Waste is an issue of concern at the quantity takeoff for any area of construction. Framing lumber is sold in even foot lengths, i.e., 10′, 12′, 14′, 16′, and depending on spans, wall heights and the grade of lumber, waste is inevitable. A rule of thumb for lumber waste is 5% to 10% depending on material quality and the complexity of the framing.
- Wood in various forms and shapes is used in many projects, even where the main structural framing is steel, concrete or masonry. Plywood as a back-up partition material and 2x boards used as blocking and cant strips around roof edges are two common examples. The estimator should ensure that the costs of all wood materials are included in the final estimate.

062 Finish Carpentry

- It is necessary to consider the grade of workmanship when estimating labor costs for erecting millwork and interior finish. In practice, there are three grades: premium, custom and economy. The Means daily output for base and case moldings is in the range of 200 to 250 L.F. per carpenter per day. This is appropriate for most average custom grade projects. For premium projects an adjustment to productivity of 25% to 50% should be made depending on the complexity of the job.

Reference Numbers

Reference numbers are shown in bold squares at the beginning of some major classifications. These numbers refer to related items in the Reference Section. The reference information may be an estimating procedure, an alternate pricing method or technical information.

Note: Not all subdivisions listed here necessarily appear in this publication.

	060 500 | Fasteners & Adhesives	CREW	DAILY OUTPUT	LABOR-HOURS	UNIT	1999 BARE COSTS MAT.	LABOR	EQUIP.	TOTAL	TOTAL INCL O&P
504 0010	**NAILS** Prices of material only, based on 50# box purchase, copper, plain				Lb.	4.10			4.10	4.51
0400	Stainless steel, plain					5.40			5.40	5.95
0500	Box, 3d to 20d, bright					1.13			1.13	1.24
0520	Galvanized					1.13			1.13	1.24
0600	Common, 3d to 60d, plain					.77			.77	.85
0700	Galvanized					.99			.99	1.09
0800	Aluminum					3.31			3.31	3.64
1000	Annular or spiral thread, 4d to 60d, plain					.66			.66	.73
1200	Galvanized					.88			.88	.97
1400	Drywall nails, plain					.77			.77	.85
1600	Galvanized					1.10			1.10	1.21
1800	Finish nails, 4d to 10d, plain					.90			.90	.99
2000	Galvanized					1.04			1.04	1.14
2100	Aluminum					4.84			4.84	5.30
2300	Flooring nails, hardened steel, 2d to 10d, plain					1.22			1.22	1.34
2400	Galvanized					1.34			1.34	1.47
2500	Gypsum lath nails, 1-1/8″, 13 ga. flathead, blued					1.33			1.33	1.46
2600	Masonry nails, hardened steel, 3/4″ to 3″ long, plain					1.61			1.61	1.77
2700	Galvanized					1.44			1.44	1.58
2900	Roofing nails, threaded, galvanized					1.21			1.21	1.34
3100	Aluminum					4.70			4.70	5.15
3300	Compressed lead head, threaded, galvanized					1.44			1.44	1.58
3600	Siding nails, plain shank, galvanized					1.33			1.33	1.46
3800	Aluminum					4			4	4.40
5000	Add to prices above for cement coating					.07			.07	.08
5200	Zinc or tin plating					.12			.12	.13
5500	Vinyl coated sinkers, 8d to 16d				↓	.55			.55	.61
506 0010	**NAILS** mat. only, for pneumatic tools, framing, per carton of 5000, 2″				Ea.	36.50			36.50	40
0100	2-3/8″					41.50			41.50	45.50
0200	Per carton of 4000, 3″					37			37	40.50
0300	3-1/4″					39			39	43
0400	Per carton of 5000, 2-3/8″, galv.					56.50			56.50	62
0500	Per carton of 4000, 3″, galv.					63.50			63.50	70
0600	3-1/4″, galv.					60			60	66
0700	Roofing, per carton of 7200, 1″					36			36	39.50
0800	1-1/4″					36.50			36.50	40.50
0900	1-1/2″					42			42	46
1000	1-3/4″				↓	47			47	51.50
508 0010	**SHEET METAL SCREWS** Steel, standard, #8 x 3/4″, plain				C	2.70			2.70	2.97
0100	Galvanized					3.33			3.33	3.66
0300	#10 x 1″, plain					3.65			3.65	4.02
0400	Galvanized					4.26			4.26	4.69
0600	With washers, #14 x 1″, plain					10.85			10.85	11.95
0700	Galvanized					12.05			12.05	13.30
0900	#14 x 2″, plain					15.80			15.80	17.40
1000	Galvanized					17.80			17.80	19.55
1500	Self-drilling, with washers, (pinch point) #8 x 3/4″, plain					4.68			4.68	5.15
1600	Galvanized					6.30			6.30	6.90
1800	#10 x 3/4″, plain					6.25			6.25	6.85
1900	Galvanized					7.15			7.15	7.85
3000	Stainless steel w/aluminum or neoprene washers, #14 x 1″, plain					16.45			16.45	18.10
3100	#14 x 2″, plain				↓	24.50			24.50	27
512 0010	**TIMBER CONNECTORS** Add up cost of each part for total									
0020	cost of connection									

060 500 Fasteners & Adhesives		CREW	DAILY OUTPUT	LABOR-HOURS	UNIT	1999 BARE COSTS MAT.	LABOR	EQUIP.	TOTAL	TOTAL INCL O&P
0100	Connector plates, steel, with bolts, straight	2 Carp	75	.213	Ea.	16	5.80		21.80	27
0110	Tee	"	50	.320		24	8.75		32.75	40
0200	Bolts, machine, sq. hd. with nut & washer, 1/2" diameter, 4" long	1 Carp	140	.057		.68	1.56		2.24	3.19
0300	7-1/2" long		130	.062		.94	1.68		2.62	3.67
0500	3/4" diameter, 7-1/2" long		130	.062		1.62	1.68		3.30	4.42
0600	15" long		95	.084	↓	2.13	2.30		4.43	5.95
0800	Drilling bolt holes in timber, 1/2" diameter		450	.018	Inch		.49		.49	.76
0900	1" diameter		350	.023	"		.62		.62	.98
1100	Framing anchors, 2 or 3 dimensional, 10 gauge, no nails incl.		175	.046	Ea.	.39	1.25		1.64	2.39
1250	Holdowns, 3 gauge base, 10 gauge body		8	1		13.45	27.50		40.95	58
1300	Joist and beam hangers, 18 ga. galv., for 2" x 4" joist		175	.046		.48	1.25		1.73	2.48
1400	2" x 6" to 2" x 10" joist		165	.048		.56	1.32		1.88	2.70
1600	16 ga. galv., 3" x 6" to 3" x 10" joist		160	.050		2.24	1.36		3.60	4.61
1700	3" x 10" to 3" x 14" joist		160	.050		2.60	1.36		3.96	5
2000	Two-2" x 6" to two-2" x 10" joists		150	.053		2.15	1.46		3.61	4.66
2100	Two-2" x 10" to two-2" x 14" joists		150	.053		2.15	1.46		3.61	4.66
2300	3/16" thick, 6" x 8" joist		145	.055		4.71	1.51		6.22	7.55
2500	6" x 12" joist		135	.059		6.70	1.62		8.32	9.90
2700	1/4" thick, 6" x 14" joist		130	.062		8.30	1.68		9.98	11.75
2800	Joist anchors, 1/4" x 1-1/4" x 18"	↓	140	.057		3.23	1.56		4.79	6
2900	Plywood clips, extruded aluminum H clip, for 3/4" panels					.11			.11	.12
3000	Galvanized 18 ga. back-up clip					.10			.10	.11
3200	Post framing, 16 ga. galv. for 4" x 4" base, 2 piece	1 Carp	130	.062		4.65	1.68		6.33	7.75
3300	Cap		130	.062		2.29	1.68		3.97	5.15
3500	Rafter anchors, 18 ga. galv., 1-1/2" wide, 5-1/4" long		145	.055		.37	1.51		1.88	2.78
3600	10-3/4" long		145	.055		.76	1.51		2.27	3.21
3800	Shear plates, 2-5/8" diameter		120	.067		1.38	1.82		3.20	4.38
3900	4" diameter		115	.070		3.20	1.90		5.10	6.50
4000	Sill anchors, embedded in concrete or block, 18-5/8" long		115	.070		.96	1.90		2.86	4.05
4100	Spike grids, 4" x 4", flat or curved		120	.067		.40	1.82		2.22	3.29
4400	Split rings, 2-1/2" diameter		120	.067		1.15	1.82		2.97	4.13
4500	4" diameter		110	.073		1.77	1.99		3.76	5.05
4700	Strap ties, 16 ga., 1-3/8" wide, 12" long		180	.044		.91	1.21		2.12	2.91
4800	24" long		160	.050		1.42	1.36		2.78	3.71
5000	Toothed rings, 2-5/8" or 4" diameter		90	.089	↓	.97	2.43		3.40	4.89
5200	Truss plates, nailed, 20 gauge, up to 32' span	↓	17	.471	Truss	7	12.85		19.85	27.50
5400	Washers, 2" x 2" x 1/8"				Ea.	.22			.22	.24
5500	3" x 3" x 3/16"				"	.57			.57	.63
0010	**WOOD SCREWS** #8, 1" long, steel				C	3.46			3.46	3.81
0100	Brass					16.70			16.70	18.35
0200	#8, 2" long, steel					6.55			6.55	7.20
0300	Brass					32.50			32.50	35.50
0400	#10, 1" long, steel					4.22			4.22	4.64
0500	Brass					22.50			22.50	24.50
0600	#10, 2" long, steel					7.50			7.50	8.25
0700	Brass					39.50			39.50	43.50
0800	#10, 3" long, steel					13.75			13.75	15.10
1000	#12, 2" long, steel					10.05			10.05	11.05
1100	Brass					50			50	55
1500	#12, 3" long, steel					15.90			15.90	17.50
2000	#12, 4" long, steel				↓	28.50			28.50	31

	061 100 | Wood Framing		CREW	DAILY OUTPUT	LABOR-HOURS	UNIT	1999 BARE COSTS MAT.	LABOR	EQUIP.	TOTAL	TOTAL INCL O&P
102	0010	**BLOCKING**									
	0020	Miscellaneous, to wood construction	1 Carp	.20	40	M.B.F.	540	1,100		1,640	2,325
	0200	To steel construction	″	.18	44.444	″	540	1,225		1,765	2,500
	1950	Miscellaneous, to wood construction									
	2000	2″ x 4″	1 Carp	250	.032	L.F.	.36	.87		1.23	1.77
	2005	Pneumatic nailed		305	.026		.36	.72		1.08	1.53
	2150	2″ x 10″		178	.045		1.01	1.23		2.24	3.04
	2155	Pneumatic nailed	↓	217	.037	↓	1.01	1.01		2.02	2.69
	2300	To steel construction									
	2320	2″ x 4″	1 Carp	208	.038	L.F.	.36	1.05		1.41	2.05
104	0010	**BRACING** Let-in, with 1″ x 6″ boards, studs @ 16″ O.C.		1.50	5.333	C.L.F.	52	146		198	287
	0200	Studs @ 24″ O.C.	↓	2.30	3.478	″	52	95		147	207
106	0010	**BRIDGING** Wood, for joists 16″ O.C., 1″ x 3″	1 Carp	1.30	6.154	C.Pr.	34.50	168		202.50	300
	0015	Pneumatic nailed		1.70	4.706		34.50	128		162.50	240
	0100	2″ x 3″ bridging		1.30	6.154		37	168		205	305
	0105	Pneumatic nailed		1.70	4.706		37	128		165	243
	0300	Steel, galvanized, 18 ga., for 2″ x 10″ joists at 12″ O.C.		1.30	6.154		85	168		253	360
	0400	24″ O.C.		1.40	5.714		220	156		376	485
	0600	For 2″ x 14″ joists at 16″ O.C.		1.30	6.154		168	168		336	450
	0700	24″ O.C.		1.40	5.714		220	156		376	485
	0900	Compression type, 16″ O.C., 2″ x 8″ joists		2	4		115	109		224	299
	1000	2″ x 12″ joists	↓	2	4	↓	126	109		235	310
108	0011	**FRAMING, SITE STRUCTURES**									
	2870	Redwood joists, 2″ x 6″	2 Carp	1.50	10.667	M.B.F.	4,000	291		4,291	4,850
	2880	2″ x 8″		1.75	9.143		4,000	250		4,250	4,800
	2890	2″ x 10″		1.75	9.143		4,000	250		4,250	4,800
	4000	Platform framing, 2″ x 4″		.70	22.857		540	625		1,165	1,575
	4100	2″ x 6″		.76	21.053		540	575		1,115	1,500
	4200	Pressure treated 6″ x 6″		.80	20		1,325	545		1,870	2,325
	4220	8″ x 8″		.60	26.667		1,400	730		2,130	2,700
	4500	Cedar post, columns & girts, 4″ x 4″		.52	30.769		1,550	840		2,390	3,025
	4510	4″ x 6″		.55	29.091		1,550	795		2,345	2,950
	4520	6″ x 6″		.65	24.615		1,750	670		2,420	2,975
	4530	8″ x 8″		.70	22.857		2,175	625		2,800	3,350
	4540	Redwood post, columns & girts, 4″ x 4″		.52	30.769		4,000	840		4,840	5,725
	4550	4″ x 6″		.55	29.091		4,000	795		4,795	5,650
	4560	6″ x 6″		.65	24.615		4,000	670		4,670	5,450
	4570	8″ x 8″	↓	.70	22.857	↓	4,000	625		4,625	5,375
110	0010	**FRAMING, BEAMS & GIRDERS** R061-010									
	1000	Single, 2″ x 6″	2 Carp	700	.023	L.F.	.54	.62		1.16	1.57
	1005	Pneumatic nailed R061-030		812	.020		.54	.54		1.08	1.44
	1040	2″ x 10″		600	.027		1.01	.73		1.74	2.26
	1045	Pneumatic nailed	↓	696	.023		1.01	.63		1.64	2.10
	1180	4″ x 8″	F-3	1,000	.040		3.07	1.11	.44	4.62	5.60
	2000	Double, 2″ x 6″	2 Carp	625	.026		1.08	.70		1.78	2.29
	2005	Pneumatic nailed		725	.022		1.08	.60		1.68	2.14
	2040	2″ x 10″		550	.029		2.02	.79		2.81	3.48
	2045	Pneumatic nailed		638	.025		2.02	.68		2.70	3.31
	3000	Triple, 2″ x 6″		550	.029		1.62	.79		2.41	3.03
	3005	Pneumatic nailed		638	.025		1.62	.68		2.30	2.86
	3040	2″ x 10″		500	.032		3.03	.87		3.90	4.71
	3045	Pneumatic nailed	↓	580	.028	↓	3.03	.75		3.78	4.52
114	0010	**FRAMING, JOISTS**									
	2000	Joists, 2″ x 4″	2 Carp	1,250	.013	L.F.	.36	.35		.71	.95
	2005	Pneumatic nailed		1,438	.011		.36	.30		.66	.88
	2150	2″ x 8″	↓	1,100	.015	↓	.73	.40		1.13	1.42

061 100 | Wood Framing

	Line	Description	Crew	Daily Output	Labor-Hours	Unit	1999 Bare Costs Mat.	Labor	Equip.	Total	Total Incl O&P
114	2155	Pneumatic nailed	2 Carp	1,265	.013	L.F.	.73	.35		1.08	1.34
	2200	2" x 10"		900	.018		1.01	.49		1.50	1.87
	2205	Pneumatic nailed		1,035	.015		1.01	.42		1.43	1.77
116	0010	**FRAMING, MISCELLANEOUS**									
	2000	Firestops, 2" x 4"	2 Carp	780	.021	L.F.	.36	.56		.92	1.28
	2005	Pneumatic nailed		952	.017		.36	.46		.82	1.12
	5000	Nailers, treated, wood construction, 2" x 4"		800	.020		.64	.55		1.19	1.57
	5005	Pneumatic nailed		960	.017		.64	.45		1.09	1.43
	5200	Steel construction, 2" x 4"		750	.021		.64	.58		1.22	1.63
	7000	Rough bucks, treated, for doors or windows, 2" x 6"		400	.040		.96	1.09		2.05	2.78
	7005	Pneumatic nailed		480	.033		.96	.91		1.87	2.49
	7100	2" x 8"		380	.042		1.14	1.15		2.29	3.06
	7105	Pneumatic nailed		456	.035		1.14	.96		2.10	2.76
	8000	Stair stringers, 2" x 10"		130	.123		1.01	3.36		4.37	6.40
	8100	2" x 12"		130	.123		1.29	3.36		4.65	6.70
118	0010	**FRAMING, COLUMNS**									
	0100	4" x 4"	2 Carp	390	.041	L.F.	.95	1.12		2.07	2.81
	0150	4" x 6"		275	.058		2.31	1.59		3.90	5.05
	0200	4" x 8"		220	.073		3.07	1.99		5.06	6.50
	0250	6" x 6"		215	.074		5.05	2.03		7.08	8.75
	0300	6" x 8"		175	.091		4.98	2.50		7.48	9.45
	0350	6" x 10"		150	.107		7.05	2.91		9.96	12.35
120	0010	**FRAMING, ROOFS**									
	2000	Fascia boards, 2" x 8"	2 Carp	225	.071	L.F.	.73	1.94		2.67	3.85
	5000	Rafters, to 4 in 12 pitch, 2" x 6", ordinary		1,000	.016		.54	.44		.98	1.28
	5020	On steep roofs		800	.020		.54	.55		1.09	1.45
	5060	2" x 8", ordinary		950	.017		.73	.46		1.19	1.52
	5080	On steep roofs		750	.021		.73	.58		1.31	1.72
	5800	Ridge board, #2 or better, 1" x 6"		600	.027		.83	.73		1.56	2.07
	5820	1" x 8"		550	.029		.96	.79		1.75	2.31
122	0010	**FRAMING, SILLS**									
	2000	Ledgers, nailed, 2" x 4"	2 Carp	755	.021	L.F.	.36	.58		.94	1.31
	2050	2" x 6"		600	.027		.54	.73		1.27	1.74
	2100	Bolted, not including bolts, 3" x 6"		325	.049		1.52	1.34		2.86	3.78
	2600	Mud sills, redwood, construction grade, 2" x 4"		895	.018		2.67	.49		3.16	3.70
	2620	2" x 6"		780	.021		4	.56		4.56	5.30
	4000	Sills, 2" x 4"		600	.027		.36	.73		1.09	1.55
	4050	2" x 6"		550	.029		.54	.79		1.33	1.84
	4200	Treated, 2" x 4"		550	.029		.64	.79		1.43	1.96
	4220	2" x 6"		500	.032		.96	.87		1.83	2.43
124	0010	**FRAMING, SLEEPERS**									
	0100	On concrete, treated, 1" x 2"	2 Carp	2,350	.007	L.F.	.17	.19		.36	.47
	0150	1" x 3"		2,000	.008		.25	.22		.47	.62
	0200	2" x 4"		1,500	.011		.64	.29		.93	1.17
	0250	2" x 6"		1,300	.012		.96	.34		1.30	1.59
126	0010	**FRAMING, SOFFITS & CANOPIES**									
	1000	Canopy or soffit framing , 1" x 4"	2 Carp	900	.018	L.F.	.55	.49		1.04	1.37
	1040	1" x 8"		750	.021		.96	.58		1.54	1.98
	1100	2" x 4"		620	.026		.36	.70		1.06	1.51
	1140	2" x 8"		500	.032		.73	.87		1.60	2.17
	1220	3" x 6"		400	.040		1.52	1.09		2.61	3.39
127	0010	**FRAMING, TREATED LUMBER**									
	0020	Water-borne salt, C.C.A., A.C.A., wet, .40 P.C.F. retention									

6 WOOD & PLASTICS

or expanded coverage of these items see *Means Interior Cost Data 1999*

061 100 | Wood Framing

		061 100 | Wood Framing	CREW	DAILY OUTPUT	LABOR-HOURS	UNIT	1999 BARE COSTS MAT.	LABOR	EQUIP.	TOTAL	TOTAL INCL O&P
127	0100	2″ x 4″				M.B.F.	965			965	1,050
	0110	2″ x 6″					965			965	1,050
	0120	2″ x 8″					850			850	935
	0130	2″ x 10″					955			955	1,050
	0140	2″ x 12″					1,100			1,100	1,200
	0200	4″ x 4″					1,325			1,325	1,475
	0210	4″ x 6″					1,275			1,275	1,400
	0220	4″ x 8″					1,275			1,275	1,400
	0250	Add for .60 P.C.F. retention					40%				
	0260	Add for 2.5 P.C.F. retention					200%				
	0270	Add for K.D.A.T.					20%				
128	0010	**FRAMING, WALLS** R061-010									
	2000	Headers over openings, 2″ x 6″	2 Carp	360	.044	L.F.	.54	1.21		1.75	2.50
	2005	2″ x 6″, pneumatic nailed R061-030		432	.037		.54	1.01		1.55	2.18
	5000	Plates, untreated, 2″ x 3″		850	.019		.24	.51		.75	1.08
	5005	2″ x 3″, pneumatic nailed		1,020	.016		.24	.43		.67	.94
	5020	2″ x 4″		800	.020		.36	.55		.91	1.26
	5025	2″ x 4″, pneumatic nailed		960	.017		.36	.45		.81	1.12
	5060	Treated, 2″ x 3″		850	.019		.48	.51		.99	1.34
	5065	2″ x 3″, treated, pneumatic nailed		1,020	.016		.48	.43		.91	1.20
	5080	2″ x 4″		800	.020		.64	.55		1.19	1.57
	5085	2″ x 4″, treated, pneumatic nailed		960	.017		.64	.45		1.09	1.43
	5120	Studs, 8′ high wall, 2″ x 3″		1,200	.013		.24	.36		.60	.84
	5125	2″ x 3″, pneumatic nailed		1,440	.011		.24	.30		.54	.75
	5140	2″ x 4″		1,100	.015		.36	.40		.76	1.02
	5146	2″ x 4″, pneumatic nailed		1,320	.012		.36	.33		.69	.92
130	0010	**FURRING** Wood strips, 1″ x 2″, on walls, on wood	1 Carp	550	.015	L.F.	.18	.40		.58	.82
	0015	On wood, pneumatic nailed		710	.011		.18	.31		.49	.68
	0300	On masonry		495	.016		.18	.44		.62	.89
	0400	On concrete		260	.031		.18	.84		1.02	1.52
	0600	1″ x 3″, on walls, on wood		550	.015		.23	.40		.63	.87
	0605	On wood, pneumatic nailed		710	.011		.23	.31		.54	.73
	0800	On concrete		260	.031		.23	.84		1.07	1.57
	0900	On masonry		320	.025		.23	.68		.91	1.32
138	0010	**PARTITIONS** Wood stud with single bottom plate and									
	0020	double top plate, no waste, std. & better lumber									
	0180	2″ x 4″ studs, 8′ high, studs 12″ O.C.	2 Carp	80	.200	L.F.	3.96	5.45		9.41	12.95
	0185	12″ O.C., pneumatic nailed		96	.167		3.96	4.55		8.51	11.50
	0200	16″ O.C.		100	.160		3.24	4.37		7.61	10.40
	0205	16″ O.C., pneumatic nailed		120	.133		3.24	3.64		6.88	9.30
	0300	24″ O.C.		125	.128		2.52	3.49		6.01	8.25
	0305	24″ O.C., pneumatic nailed		150	.107		2.52	2.91		5.43	7.35
	0580	12′ high, studs 12″ O.C.		65	.246		5.40	6.70		12.10	16.50
	0585	12″ O.C., pneumatic nailed		78	.205		5.40	5.60		11	14.75
	0780	2″ x 6″ studs, 8′ high, studs 12″ O.C.		70	.229		5.95	6.25		12.20	16.30
	0785	12″ O.C., pneumatic nailed		84	.190		5.95	5.20		11.15	14.70
	0900	24″ O.C.		115	.139		3.77	3.80		7.57	10.15
	0905	24″ O.C., pneumatic nailed		138	.116		3.77	3.17		6.94	9.15

061 150 | Sheathing

		061 150 | Sheathing	CREW	DAILY OUTPUT	LABOR-HOURS	UNIT	MAT.	LABOR	EQUIP.	TOTAL	TOTAL INCL O&P
154	0010	**SHEATHING** Plywood on roof, CDX									
	0030	5/16″ thick	2 Carp	1,600	.010	S.F.	.29	.27		.56	.75
	0035	Pneumatic nailed		1,952	.008		.29	.22		.51	.67
	0050	3/8″ thick		1,525	.010		.32	.29		.61	.80

061 150 | Sheathing			CREW	DAILY OUTPUT	LABOR-HOURS	UNIT	1999 BARE COSTS MAT.	LABOR	EQUIP.	TOTAL	TOTAL INCL O&P	
154	0055	Pneumatic nailed	2 Carp	1,860	.009	S.F.	.32	.23		.55	.72	154
	0100	1/2" thick		1,400	.011		.41	.31		.72	.94	
	0105	Pneumatic nailed		1,708	.009		.41	.26		.67	.85	
	0500	Plywood on walls with exterior CDX, 3/8" thick		1,200	.013		.32	.36		.68	.92	
	0505	Pneumatic nailed		1,488	.011		.32	.29		.61	.81	
	0600	1/2" thick		1,125	.014		.41	.39		.80	1.06	
	0605	Pneumatic nailed		1,395	.011		.41	.31		.72	.94	
	1400	With boards, on roof 1" x 6" boards, laid horizontal		725	.022		.99	.60		1.59	2.04	
	1500	Laid diagonal	↓	650	.025		.99	.67		1.66	2.15	
	2000	For steep roofs, add						40%				
	2400	Boards on walls, 1" x 6" boards, laid regular	2 Carp	650	.025		.99	.67		1.66	2.15	
	2500	Laid diagonal		585	.027		.99	.75		1.74	2.26	
	2850	Gypsum, weatherproof, 1/2" thick		1,125	.014		.21	.39		.60	.84	
	3000	Wood fiber, regular, no vapor barrier, 1/2" thick		1,200	.013		.38	.36		.74	.99	
	3100	5/8" thick		1,200	.013		.50	.36		.86	1.12	
	3800	Asphalt impregnated, 25/32" thick		1,200	.013		.34	.36		.70	.94	
	3850	Intermediate, 1/2" thick	↓	1,200	.013	↓	.28	.36		.64	.88	
061 160 | Subfloor												
164	0010	**SUBFLOOR** Plywood, CDX, 1/2" thick R061-020	2 Carp	1,500	.011	SF Flr.	.41	.29		.70	.91	164
	0015	Pneumatic nailed		1,860	.009		.41	.23		.64	.82	
	0100	5/8" thick		1,350	.012		.49	.32		.81	1.05	
	0105	Pneumatic nailed		1,674	.010		.49	.26		.75	.95	
	0200	3/4" thick		1,250	.013		.58	.35		.93	1.19	
	0205	Pneumatic nailed		1,550	.010		.58	.28		.86	1.08	
	0500	With boards, 1" x 10" S4S, laid regular		1,100	.015		.96	.40		1.36	1.68	
	0600	Laid diagonal		900	.018		.97	.49		1.46	1.83	
	1100	Wood fiber, T&G, 2' x 8' planks, 1" thick	↓	1,000	.016	↓	1.19	.44		1.63	2	
168	0010	**UNDERLAYMENT** Plywood, underlayment grade, 3/8" thick R061-020	2 Carp	1,500	.011	SF Flr.	.66	.29		.95	1.19	168
	0015	Pneumatic nailed		1,860	.009		.66	.23		.89	1.10	
	0500	Particle board, 3/8" thick		1,500	.011		.38	.29		.67	.88	
	0505	Pneumatic nailed		1,860	.009		.38	.23		.61	.79	
	0600	1/2" thick		1,450	.011		.40	.30		.70	.91	
	0605	Pneumatic nailed	↓	1,798	.009	↓	.40	.24		.64	.82	
061 250 | Wood Decking												
258	0010	**ROOF DECKS**										258
	0400	Cedar planks, 3" thick	2 Carp	320	.050	S.F.	5.65	1.36		7.01	8.40	
	0500	4" thick		250	.064		7.65	1.75		9.40	11.15	
	0700	Douglas fir, 3" thick		320	.050		2.12	1.36		3.48	4.48	
	0800	4" thick		250	.064		2.84	1.75		4.59	5.85	
	1000	Hemlock, 3" thick		320	.050		2.12	1.36		3.48	4.48	
	1100	4" thick		250	.064		2.83	1.75		4.58	5.85	
	1300	Western white spruce, 3" thick		320	.050		2.05	1.36		3.41	4.41	
	1400	4" thick	↓	250	.064	↓	2.73	1.75		4.48	5.75	
061 800 | Glued-Laminated Const												
804	0010	**LAMINATED FRAMING** Not including decking										804
	0200	Straight roof beams, 20' clear span, beams 8' O.C.	F-3	2,560	.016	SF Flr.	1.38	.43	.17	1.98	2.39	
	0500	40' clear span, beams 8' O.C.	"	3,200	.013		2.66	.35	.14	3.15	3.62	
	0800	60' clear span, beams 8' O.C.	F-4	2,880	.017		4.56	.45	.28	5.29	6	
	1700	Radial arches, 60' clear span, frames 8' O.C.		1,920	.025		6	.68	.41	7.09	8.10	
	2000	100' clear span, frames 8' O.C.	↓	1,600	.030	↓	6.20	.81	.50	7.51	8.65	

or expanded coverage of these items see *Means Interior Cost Data 1999*

061 | Rough Carpentry

061 800 | Glued-Laminated Const

			CREW	DAILY OUTPUT	LABOR-HOURS	UNIT	1999 BARE COSTS MAT.	LABOR	EQUIP.	TOTAL	TOTAL INCL O&P
804	2300	120′ clear span, frames 8′ O.C.	F-4	1,440	.033	SF Flr.	8.25	.90	.55	9.70	11.10
	2600	Bowstring trusses, 20′ O.C., 40′ clear span	F-3	2,400	.017		3.72	.46	.18	4.36	5
	2700	60′ clear span	F-4	3,600	.013		3.34	.36	.22	3.92	4.47
	2800	100′ clear span	″	4,000	.012		4.73	.33	.20	5.26	5.95
	4300	Alternate pricing method: (use nominal footage of									
	4310	components). Straight beams, camber less than 6″	F-3	3.50	11.429	M.B.F.	2,050	315	127	2,492	2,900
	4600	Curved members, radius over 32′		2.50	16		2,275	445	177	2,897	3,400
	4700	Radius 10′ to 32′		3	13.333		2,250	370	148	2,768	3,225
	6000	Laminated veneer members, southern pine or western species									
	6050	1-3/4″ wide x 5-1/2″ deep	2 Carp	480	.033	L.F.	3.51	.91		4.42	5.30
	6100	9-1/2″ deep		480	.033		3.68	.91		4.59	5.50
	6150	14″ deep		450	.036		5.25	.97		6.22	7.35
	6200	18″ deep		450	.036		6.80	.97		7.77	9.05
	6300	Parallel strand members, southern pine or western species									
	6350	1-3/4″ wide x 9-1/4″ deep	2 Carp	480	.033	L.F.	4.21	.91		5.12	6.05
	6400	11-1/4″ deep		450	.036		5.15	.97		6.12	7.25
	6450	14″ deep		400	.040		6.15	1.09		7.24	8.45
	6500	3-1/2″ wide x 9-1/4″ deep		480	.033		8.20	.91		9.11	10.45
	6550	11-1/4″ deep		450	.036		10.10	.97		11.07	12.65
	6600	14″ deep		400	.040		12.05	1.09		13.14	14.95
	6650	7″ wide x 9-1/4″ deep		450	.036		16.15	.97		17.12	19.30
	6700	11-1/4″ deep		420	.038		19.95	1.04		20.99	23.50
	6750	14″ deep		400	.040		24	1.09		25.09	27.50

061 900 | Wood Trusses

			CREW	DAILY OUTPUT	LABOR-HOURS	UNIT	MAT.	LABOR	EQUIP.	TOTAL	TOTAL INCL O&P
908	0010	ROOF TRUSSES (R061 -100)									
	0020	For timber connectors, see div. 060-512									
	5000	Common wood, 2″ x 4″ metal plate connected, 24″ O.C., 4/12 slope									
	5050	20′ span	F-6	62	.645	Ea.	36	16.35	7.15	59.50	73
	5200	28′ span		53	.755		50.50	19.15	8.35	78	94.50
	5220	Gable end		49	.816		58.50	20.50	9.05	88.05	107
	5300	36′ span		46	.870		85	22	9.65	116.65	139

062 | Finish Carpentry

062 200 | Millwork Moldings

			CREW	DAILY OUTPUT	LABOR-HOURS	UNIT	1999 BARE COSTS MAT.	LABOR	EQUIP.	TOTAL	TOTAL INCL O&P
208	0010	MOLDINGS, BASE									
	0500	Base, stock pine, 9/16″ x 3-1/2″	1 Carp	240	.033	L.F.	1.16	.91		2.07	2.71
212	0010	MOLDINGS, CASINGS									
	0090	Apron, stock pine, 5/8″ x 2″	1 Carp	250	.032	L.F.	.95	.87		1.82	2.42
	0110	5/8″ x 3-1/2″	″	220	.036	″	1.40	.99		2.39	3.10
216	0010	MOLDINGS, CEILINGS									
	1200	Cornice molding, stock pine, 9/16″ x 1-3/4″	1 Carp	330	.024	L.F.	.62	.66		1.28	1.72
220	0010	MOLDINGS, EXTERIOR									
	1500	Cornice, boards, pine, 1″ x 2″	1 Carp	330	.024	L.F.	.27	.66		.93	1.33
	1600	1″ x 4″		250	.032		.56	.87		1.43	1.98
	3000	Corner board, sterling pine, 1″ x 4″		200	.040		.46	1.09		1.55	2.23

062 | Finish Carpentry

062 200 | Millwork Moldings

	062 200 Millwork Moldings	CREW	DAILY OUTPUT	LABOR-HOURS	UNIT	1999 BARE COSTS MAT.	LABOR	EQUIP.	TOTAL	TOTAL INCL O&P	
3100	1″ x 6″	1 Carp	200	.040	L.F.	.70	1.09		1.79	2.49	220
3350	Fascia, sterling pine, 1″ x 6″		250	.032		.70	.87		1.57	2.14	
3400	Trim, exterior, sterling pine, back band		250	.032		.55	.87		1.42	1.98	
3500	Casing		250	.032		.58	.87		1.45	2.01	
4100	Verge board, sterling pine, 1″ x 4″		200	.040		.32	1.09		1.41	2.07	
4200	1″ x 6″		200	.040		.70	1.09		1.79	2.49	
0010	**MOLDINGS, TRIM**										224
3300	Half round, stock pine, 1/4″ x 1/2″	1 Carp	270	.030	L.F.	.19	.81		1	1.48	
0010	**MOLDINGS, WINDOW AND DOOR**										228
3150	Door trim set, 1 head and 2 sides, pine, 2-1/2 wide	1 Carp	5.90	1.356	Opng.	12.45	37		49.45	71.50	
3200	Glass beads, stock pine, 1/4″ x 11/16″		285	.028	L.F.	.29	.77		1.06	1.53	
5300	Threshold, oak, 3′ long, inside, 5/8″ x 3-5/8″		32	.250	Ea.	6.10	6.80		12.90	17.45	
5400	Outside, 1-1/2″ x 7-5/8″		16	.500	″	24.50	13.65		38.15	48.50	
5900	Window trim sets, including casings, header, stops,										
5910	stool and apron, 2-1/2″ wide, minimum	1 Carp	13	.615	Opng.	15	16.80		31.80	43	
5950	Average	″	10	.800	″	26	22		48	63	

062 300 | Shelving

	062 300 Shelving	CREW	DAILY OUTPUT	LABOR-HOURS	UNIT	MAT.	LABOR	EQUIP.	TOTAL	TOTAL INCL O&P	
0010	**SHELVING** Pine, clear grade, no edge band, 1″ x 8″	1 Carp	115	.070	L.F.	1.71	1.90		3.61	4.87	304
0100	1″ x 10″		110	.073		2.31	1.99		4.30	5.65	
0200	1″ x 12″		105	.076		4.18	2.08		6.26	7.85	

062 500 | Prefin. Wood Paneling

	062 500 Prefin. Wood Paneling	CREW	DAILY OUTPUT	LABOR-HOURS	UNIT	MAT.	LABOR	EQUIP.	TOTAL	TOTAL INCL O&P	
0010	**PANELING, PLYWOOD**										504
2400	Plywood, prefinished, 1/4″ thick, 4′ x 8′ sheets										
2410	with vertical grooves. Birch faced, minimum	2 Carp	500	.032	S.F.	.72	.87		1.59	2.16	
2430	Maximum		350	.046		1.49	1.25		2.74	3.60	
4600	Plywood, "A" face, birch, V.C., 1/2″ thick, natural		450	.036		1.55	.97		2.52	3.23	
4700	Select		450	.036		1.70	.97		2.67	3.40	
4900	Veneer core, 3/4″ thick, natural		320	.050		1.65	1.36		3.01	3.97	
5200	Lumber core, 3/4″ thick, natural		320	.050		2.47	1.36		3.83	4.87	

062 550 | Prefin. Hardboard Panel

	062 550 Prefin. Hardboard Panel	CREW	DAILY OUTPUT	LABOR-HOURS	UNIT	MAT.	LABOR	EQUIP.	TOTAL	TOTAL INCL O&P	
0010	**PANELING, HARDBOARD**										554
0050	Not incl. furring or trim, hardboard, tempered, 1/8″ thick	2 Carp	500	.032	S.F.	.33	.87		1.20	1.73	
0100	1/4″ thick		500	.032		.37	.87		1.24	1.78	
0300	Tempered pegboard, 1/8″ thick		500	.032		.33	.87		1.20	1.73	
0400	1/4″ thick		500	.032		.41	.87		1.28	1.82	
0600	Untempered hardboard, natural finish, 1/8″ thick		500	.032		.28	.87		1.15	1.68	
2100	Moldings for hardboard, wood or aluminum, minimum		500	.032	L.F.	.32	.87		1.19	1.72	

062 600 | Board Paneling

	062 600 Board Paneling	CREW	DAILY OUTPUT	LABOR-HOURS	UNIT	MAT.	LABOR	EQUIP.	TOTAL	TOTAL INCL O&P	
0010	**PANELING, BOARDS**										604
6400	Wood board paneling, 3/4″ thick, knotty pine	2 Carp	300	.053	S.F.	1.24	1.46		2.70	3.65	
6500	Rough sawn cedar		300	.053		1.60	1.46		3.06	4.05	
6700	Redwood, clear, 1″ x 4″ boards		300	.053		3.75	1.46		5.21	6.40	

062 700 | Misc. Finish Carpentry

	062 700 Misc. Finish Carpentry	CREW	DAILY OUTPUT	LABOR-HOURS	UNIT	MAT.	LABOR	EQUIP.	TOTAL	TOTAL INCL O&P	
0010	**SHUTTERS, EXTERIOR** Aluminum, louvered, 1′-4″ wide, 3′-0″ long	1 Carp	10	.800	Pr.	29.50	22		51.50	67	760
0400	6′-8″ long		9	.889		48	24.50		72.50	90.50	
1000	Pine, louvered, primed, each 1′-2″ wide, 3′-3″ long		10	.800		41	22		63	79.50	
1001	Pine, louvered, primed, each 1′-2″ wide, 3′-3″ long		20	.400	Ea.	24	10.90		34.90	43.50	
1500	Each 1′-6″ wide, 3′-3″ long		10	.800	Pr.	45	22		67	84	

or expanded coverage of these items see *Means Interior Cost Data 1999*

062 | Finish Carpentry

062 700 | Misc. Finish Carpentry

			CREW	DAILY OUTPUT	LABOR-HOURS	UNIT	1999 BARE COSTS MAT.	LABOR	EQUIP.	TOTAL	TOTAL INCL O&P
775	0010	**SOFFITS** Wood fiber, no vapor barrier, 15/32" thick	2 Carp	525	.030	S.F.	.73	.83		1.56	2.11
	0100	5/8" thick	↓	525	.030	"	.79	.83		1.62	2.18
	1000	Exterior AC plywood, 1/4" thick		420	.038	S.F.	.50	1.04		1.54	2.19
	1100	1/2" thick	↓	420	.038	"	.66	1.04		1.70	2.37

063 | Wood Treatment

063 100 | Preservative Treatment

			CREW	DAILY OUTPUT	LABOR-HOURS	UNIT	1999 BARE COSTS MAT.	LABOR	EQUIP.	TOTAL	TOTAL INCL O&P
102	0011	**LUMBER TREATMENT**									
	0400	Fire retardant, wet				M.B.F.	271			271	298
	0500	KDAT					255			255	281
	0700	Salt treated, water borne, .40 lb. retention					122			122	134
	0800	Oil borne, 8 lb. retention					143			143	157
	1000	Kiln dried lumber, 1" & 2" thick, softwoods					81.50			81.50	90
	1100	Hardwoods					87			87	95.50
	1500	For small size 1" stock, add				↓	11			11	12.10
	1700	For full size rough lumber, add					20%				
104	0010	**PLYWOOD TREATMENT** Fire retardant, 1/4" thick				M.S.F.	204			204	224
	0030	3/8" thick					224			224	246
	0050	1/2" thick					240			240	264
	0070	5/8" thick					255			255	281
	0100	3/4" thick					280			280	310
	0200	For KDAT, add					61			61	67
	0500	Salt treated water borne, .25 lb., wet, 1/4" thick					112			112	123
	0530	3/8" thick					117			117	129
	0550	1/2" thick					122			122	134
	0570	5/8" thick					133			133	146
	0600	3/4" thick					138			138	152
	0800	For KDAT add					61			61	67
	0900	For .40 lb., per C.F. retention, add					51			51	56
	1000	For certification stamp, add				↓	30			30	33

064 | Architectural Woodwork

064 300 | Stairwork & Handrails

			CREW	DAILY OUTPUT	LABOR-HOURS	UNIT	1999 BARE COSTS MAT.	LABOR	EQUIP.	TOTAL	TOTAL INCL O&P
306	0011	**STAIRS, PREFABRICATED**									
	0100	Box stairs, prefabricated, 3'-0" wide									
	0110	Oak treads, no handrails, 2' high	2 Carp	5	3.200	Flight	216	87.50		303.50	375
	0200	4' high		4	4		430	109		539	645
	0400	8' high		3	5.333		760	146		906	1,075
	0600	With pine treads for carpet, 2' high		5	3.200		86	87.50		173.50	232
	0900	8' high	↓	3	5.333	↓	266	146		412	520
	1700	Basement stairs, prefabricated, soft wood,									

	064 300 | Stairwork & Handrails	CREW	DAILY OUTPUT	LABOR-HOURS	UNIT	1999 BARE COSTS MAT.	LABOR	EQUIP.	TOTAL	TOTAL INCL O&P	
1710	open risers, 3′ wide, 8′ high	2 Carp	4	4	Flight	555	109		664	785	306
1900	Open stairs, prefabricated prefinished poplar, metal stringers,										
1910	treads 3′-6″ wide, no railings										
2000	3′ high	2 Carp	5	3.200	Flight	224	87.50		311.50	385	
2100	4′ high		4	4		485	109		594	705	
2200	6′ high		3.50	4.571		555	125		680	810	
2300	8′ high		3	5.333		730	146		876	1,025	
4000	Residential, wood, oak treads, prefabricated		1.50	10.667		930	291		1,221	1,475	
4200	Built in place	↓	.44	36.364	↓	1,325	995		2,320	3,000	

	064 400 | Misc. Ornamental Items	CREW	DAILY OUTPUT	LABOR-HOURS	UNIT	MAT.	LABOR	EQUIP.	TOTAL	TOTAL INCL O&P	
0010	**COLUMNS** For base plates, see division 055-404										402
4000	Rough sawn cedar posts, 4″ x 4″	2 Carp	250	.064	V.L.F.	2.43	1.75		4.18	5.40	
4100	4″ x 6″		235	.068		3.62	1.86		5.48	6.90	
4200	6″ x 6″		220	.073		5.45	1.99		7.44	9.10	
4300	8″ x 8″	↓	200	.080	↓	5.60	2.18		7.78	9.60	
0010	**MILLWORK, HIGH DENSITY POLYMER**										410
0100	Base, 9/16″ x 3-3/16″	1 Carp	230	.035	L.F.	1.20	.95		2.15	2.81	
0200	Casing, fluted, 5/8″ x 3-1/4″		215	.037		1.20	1.02		2.22	2.92	
0300	Chair rail, 9/16″ x 2-1/4″		260	.031		.61	.84		1.45	1.99	
0400	5/8″ x 3-1/8″		230	.035		1.15	.95		2.10	2.76	
0500	Corner, inside, 1/2″ x 1-1/8″		220	.036		.60	.99		1.59	2.22	
0600	Cove, 13/16″ x 3-3/4″		260	.031		1.20	.84		2.04	2.64	
0700	Crown, 3/4″ x 3-13/16″		260	.031		1.20	.84		2.04	2.64	
0800	Half round, 15/16″ x 2″	↓	240	.033	↓	.66	.91		1.57	2.16	

or information about Means Estimating Seminars, see yellow pages 11 and 12 in back of book

For expanded coverage of these items see *Means Interior Cost Data 1999*

Division Notes

	CREW	DAILY OUTPUT	LABOR-HOURS	UNIT	1999 BARE COSTS MAT.	LABOR	EQUIP.	TOTAL	TOTAL INCL O&P

Division 7
Thermal & Moisture Protection

Estimating Tips

071 Waterproofing & Dampproofing

Be sure of the job specifications before pricing this subdivision. The difference in cost between waterproofing and dampproofing can be great. Waterproofing will hold back standing water. Dampproofing prevents the transmission of water vapor. Also included in this section are vapor retarding membranes.

072 Insulation & Fireproofing

Insulation and fireproofing products are measured by area, thickness, volume or R value. Specifications may only give what the specific R value should be in a certain situation. The estimator may need to choose the type of insulation to meet that R value.

073 Shingles & Roofing Tiles
074 Preformed Roofing & Siding

- Many roofing and siding products are bought and sold by the square. One square is equal to an area that measures 100 square feet.

This simple change in unit of measure could create a large error if the estimator is not observant. Accessories and fasteners necessary for a complete installation must be figured into any calculations for both material and labor.

075 Membrane Roofing
076 Flashing & Sheet Metal
077 Roof Specialties & Accessories
078 Skylights

- The items in these subdivisions compose a roofing system. No one component completes the installation and all must be estimated. Built-up or single ply membrane roofing systems are made up of many products and installation trades. Wood blocking at roof perimeters or penetrations, parapet coverings, reglets, roof drains, gutters, downspouts, sheet metal flashing, skylights, smoke vents or roof hatches all need to be considered along with the roofing material. Several different installation trades will need to work together on the roofing system. Inherent difficulties in the scheduling and coordination of various trades must be accounted for when estimating labor costs.

079 Joint Sealers

- To complete the weather-tight shell the sealants and caulkings must be estimated. Where different materials meet—at expansion joints, at flashing penetrations, and at hundreds of other locations throughout a construction project—they provide another line of defense against water penetration. Often, an entire system is based on the proper location and placement of caulking or sealants. The detail drawings that are included as part of a set of architectural plans, show typical locations for these materials. When caulking or sealants are shown at typical locations, this means the estimator must include them for all the locations where this detail is applicable. Be careful to keep different types of sealants separate, and remember to consider backer rods and primers if necessary.

Reference Numbers

Reference numbers are shown in bold squares at the beginning of some major classifications. These numbers refer to related items in the Reference Section. The reference information may be an estimating procedure, an alternate pricing method or technical information.

Note: Not all subdivisions listed here necessarily appear in this publication.

071 100 | Sheet Waterproofing

			CREW	DAILY OUTPUT	LABOR-HOURS	UNIT	1999 BARE COSTS MAT.	LABOR	EQUIP.	TOTAL	TOTAL INCL O&P
102	0010	ELASTOMERIC WATERPROOFING									
	0050	Acrylic rubber, fluid applied, 20 mils thick	3 Rofc	1,000	.024	S.F.	1.52	.58		2.10	2.67
	0060	50 mil, reinforced, stucco texture	"	600	.040		2.46	.96		3.42	4.37
	0090	EPDM, plain, 45 mils thick	2 Rofc	580	.028		.74	.66		1.40	1.95
	0100	60 mils thick		570	.028		1.08	.68		1.76	2.35
	0300	Nylon reinforced sheets, 45 mils thick		580	.028		1.04	.66		1.70	2.28
	0400	60 mils thick		570	.028		1.44	.68		2.12	2.74
	0600	Vulcanizing splicing tape for above, 2" wide				C.L.F.	33.50			33.50	36.50
	0700	4" wide				"	67			67	73.50
	0900	Adhesive, bonding, 60 SF per gal				Gal.	13.60			13.60	14.95
	1000	Splicing, 75 SF per gal				"	21			21	23
	1200	Neoprene sheets, plain, 45 mils thick	2 Rofc	580	.028	S.F.	1.09	.66		1.75	2.34
	1300	60 mils thick		570	.028		1.82	.68		2.50	3.17
	1500	Nylon reinforced, 45 mils thick		580	.028		1.26	.66		1.92	2.53
	1600	60 mils thick		570	.028		1.28	.68		1.96	2.57
	1800	120 mils thick		500	.032		2.54	.77		3.31	4.12
	1900	Adhesive, splicing, 150 S.F. per gal. per coat				Gal.	18.15			18.15	19.95
	2100	Fiberglass reinforced, fluid applied, 1/8" thick	2 Rofc	500	.032	S.F.	1.50	.77		2.27	2.98
	2200	Polyethylene and rubberized asphalt sheets, 1/8" thick		550	.029		.54	.70		1.24	1.80
	2210	Asphaltic hardboard protection board, 1/8" thick		500	.032		.30	.77		1.07	1.66
	2220	Asphaltic hardboard protection board, 1/4" thick		450	.036		.52	.86		1.38	2.05
	2400	Polyvinyl chloride sheets, plain, 10 mils thick		580	.028		.14	.66		.80	1.29
	2500	20 mils thick		570	.028		.22	.68		.90	1.40
	2700	30 mils thick		560	.029		.30	.69		.99	1.52
	3000	Adhesives, trowel grade, 40-100 SF per gal				Gal.	20			20	22
	3100	Brush grade, 100-250 SF per gal.				"	20			20	22
	3300	Bitumen modified polyurethane, fluid applied, 55 mils thick	2 Rofc	665	.024	S.F.	.65	.58		1.23	1.71
	3600	Vinyl plastic, sprayed on, 25 to 40 mils thick	"	475	.034	"	.98	.81		1.79	2.48
104	0010	**MEMBRANE WATERPROOFING** On slabs, 1 ply, felt	G-1	3,000	.019	S.F.	.13	.42	.14	.69	1.02
	0100	Glass fiber fabric		2,100	.027		.14	.60	.19	.93	1.40
	0300	2 ply, felt		2,500	.022		.25	.51	.16	.92	1.33
	0400	Glass fiber fabric		1,650	.034		.31	.77	.25	1.33	1.94
	0600	3 ply, felt		2,100	.027		.38	.60	.19	1.17	1.67
	0700	Glass fiber fabric		1,550	.036		.41	.82	.26	1.49	2.15
	0900	For installation on walls, add						15%			
	1000	For adhered 1/4" EPS protection board, add	2 Rofc	3,500	.005		.14	.11		.25	.34
	1050	3/8" thick, add		3,500	.005		.16	.11		.27	.37
	1060	1/2" thick, add		3,500	.005		.18	.11		.29	.39
	1070	Fiberglass fabric, black, 20/10 mesh		116	.138	Sq.	9.95	3.32		13.27	16.65
	1080	White, 20/10 mesh		116	.138	"	10.25	3.32		13.57	17
	1100	1/16" urethane, troweled		200	.080	S.F.	.69	1.93		2.62	4.08
	1200	Roller applied, 2 coats		120	.133	"	.60	3.21		3.81	6.20

071 300 | Bentonite Waterproofing

			CREW	DAILY OUTPUT	LABOR-HOURS	UNIT	MAT.	LABOR	EQUIP.	TOTAL	TOTAL INCL O&P
301	0010	**BENTONITE**, Panels, 4' x 4', 3/16" thick	1 Rofc	625	.013	S.F.	.70	.31		1.01	1.30
	0100	Rolls, 3/8" thick, with geotextile fabric both sides	"	550	.015	"	.84	.35		1.19	1.52
	0300	Granular bentonite, 50 lb. bags (.625 C.F.)				Bag	11.65			11.65	12.85
	0400	3/8" thick, troweled on	1 Rofc	475	.017	S.F.	.58	.41		.99	1.34
	0500	Drain board, expanded polystyrene, binder encapsulated, 1-1/2" thick	1 Rohe	1,600	.005		.40	.09		.49	.60
	0510	2" thick		1,600	.005		.50	.09		.59	.71
	0520	3" thick		1,600	.005		.69	.09		.78	.92
	0530	4" thick		1,600	.005		.88	.09		.97	1.13
	0600	With filter fabric, 1-1/2" thick		1,600	.005		.54	.09		.63	.75
	0625	2" thick		1,600	.005		.63	.09		.72	.85
	0650	3" thick		1,600	.005		.82	.09		.91	1.06
	0675	4" thick		1,600	.005		1.01	.09		1.10	1.27

Important: See the Reference Section for critical supporting data - Reference Nos., Crews, & City Cost Indexe

071 | Waterproofing & Dampproofing

				DAILY	LABOR-		1999 BARE COSTS				TOTAL	
			CREW	OUTPUT	HOURS	UNIT	MAT.	LABOR	EQUIP.	TOTAL	INCL O&P	
		071 300 | Bentonite Waterproofing										
301	0700	Vapor retarder, see polyethelene, 071-922										301
		071 450 | Cement. Waterproofing										
452	0010	**CEMENTITIOUS WATERPROOFING** One coat cement base										452
	0020	1/8″ application, sprayed on	G-2	1,000	.024	S.F.	1.50	.55	.24	2.29	2.76	
	0030	2 coat, cementitious/metallic slurry, troweled, 1/4″ thick	1 Cefi	2.48	3.226	C.S.F.	35	84.50		119.50	165	
	0040	3 coat, 3/8″ thick		1.84	4.348		52.50	114		166.50	228	
	0050	4 coat, 1/2″ thick		1.20	6.667		70	174		244	335	
		071 600 | Bitum. Dampproofing										
602	0010	**BITUMINOUS ASPHALT COATING** For foundation										602
	0030	Brushed on, below grade, 1 coat	1 Rofc	665	.012	S.F.	.06	.29		.35	.57	
	0100	2 coat		500	.016		.10	.39		.49	.77	
	0300	Sprayed on, below grade, 1 coat, 25.6 S.F./gal.		830	.010		.07	.23		.30	.47	
	0400	2 coat, 20.5 S.F./gal.		500	.016		.14	.39		.53	.82	
	0500	Asphalt coating, with fibers				Gal.	3.40			3.40	3.74	
	0600	Troweled on, asphalt with fibers, 1/16″ thick	1 Rofc	500	.016	S.F.	.15	.39		.54	.83	
	0700	1/8″ thick		400	.020		.26	.48		.74	1.12	
	1000	1/2″ thick		350	.023		.85	.55		1.40	1.89	
		071 750 | Water Repellent Coat										
754	0010	**RUBBER COATING** Water base liquid, roller applied	2 Rofc	7,000	.002	S.F.	.55	.06		.61	.70	754
	0200	Silicone or stearate, sprayed on CMU, 1 coat	1 Rofc	4,000	.002		.26	.05		.31	.36	
	0300	2 coats	″	3,000	.003		.51	.06		.57	.67	
		071 800 | Cementitious Dampprfng										
802	0010	**CEMENT PARGING** 2 coats, 1/2″ thick, regular P.C. R071-010	D-1	250	.064	S.F.	.15	1.57		1.72	2.61	802
	0100	Waterproofed Portland cement	″	250	.064	″	.17	1.57		1.74	2.64	
		071 920 | Vapor Retarders										
922	0010	**BUILDING PAPER** Aluminum and kraft laminated, foil 1 side	1 Carp	37	.216	Sq.	3.55	5.90		9.45	13.20	922
	0100	Foil 2 sides		37	.216		5.70	5.90		11.60	15.55	
	0300	Asphalt, two ply, 30#, for subfloors		19	.421		10.90	11.50		22.40	30	
	0400	Asphalt felt sheathing paper, 15#		37	.216		2.68	5.90		8.58	12.25	
	0450	Housewrap, exterior, spun bonded polypropylene										
	0470	Small roll	1 Carp	3,800	.002	S.F.	.10	.06		.16	.20	
	0480	Large roll	″	4,000	.002	″	.09	.05		.14	.19	
	0500	Material only, 3′ x 111.1′ roll				Ea.	33			33	36	
	0520	9′ x 111.1′ roll				″	93			93	102	
	0600	Polyethylene vapor barrier, standard, .002″ thick	1 Carp	37	.216	Sq.	1.15	5.90		7.05	10.55	
	0700	.004″ thick		37	.216		2.32	5.90		8.22	11.85	
	0900	.006″ thick		37	.216		2.86	5.90		8.76	12.45	
	1200	.010″ thick		37	.216		5.45	5.90		11.35	15.30	
	1300	Clear reinforced, fire retardant, .008″ thick		37	.216		8.40	5.90		14.30	18.55	
	1350	Cross laminated type, .003″ thick		37	.216		6.50	5.90		12.40	16.45	
	1400	.004″ thick		37	.216		7.25	5.90		13.15	17.30	
	1500	Red rosin paper, 5 sq rolls, 4 lb per square		37	.216		1.55	5.90		7.45	11	
	1600	5 lbs. per square		37	.216		2	5.90		7.90	11.50	
	1800	Reinf. waterproof, .002″ polyethylene backing, 1 side		37	.216		4.92	5.90		10.82	14.70	
	1900	2 sides		37	.216		6.50	5.90		12.40	16.45	
	2100	Roof deck vapor barrier, class 1 metal decks	1 Rofc	37	.216		8.45	5.20		13.65	18.20	
	2200	For all other decks	″	37	.216		5.75	5.20		10.95	15.25	

071 | Waterproofing & Dampproofing

071 920 | Vapor Retarders

		071 920 Vapor Retarders	CREW	DAILY OUTPUT	LABOR-HOURS	UNIT	1999 BARE COSTS MAT.	LABOR	EQUIP.	TOTAL	TOTAL INCL O&P	
922	2400	Waterproofed kraft with sisal or fiberglass fibers, minimum	1 Carp	37	.216	Sq.	5.30	5.90		11.20	15.15	922
	2500	Maximum	"	37	.216	↓	13.25	5.90		19.15	24	
	3000	Building wrap, spunbonded polyethylene	2 Carp	8,000	.002	S.F.	.10	.05		.15	.20	
	9950	Minimum labor/equipment charge	1 Carp	4	2	Job		54.50		54.50	86	

072 | Insulation & Fireproofing

072 100 | Building Insulation

		072 100 Building Insulation	CREW	DAILY OUTPUT	LABOR-HOURS	UNIT	1999 BARE COSTS MAT.	LABOR	EQUIP.	TOTAL	TOTAL INCL O&P	
108	0010	**MASONRY INSULATION** Vermiculite or perlite, poured										108
	0100	In cores of concrete block, 4" thick wall, .115 CF/SF	D-1	4,800	.003	S.F.	.17	.08		.25	.31	
	0200	6" thick wall, .175 CF/SF		3,000	.005		.25	.13		.38	.48	
	0300	8" thick wall, .258 CF/SF		2,400	.007		.37	.16		.53	.66	
	0400	10" thick wall, .340 CF/SF		1,850	.009		.49	.21		.70	.87	
	0500	12" thick wall, .422 CF/SF	↓	1,200	.013	↓	.61	.33		.94	1.18	
	0550	For sand fill, deduct from above					70%					
	0600	Poured cavity wall, vermiculite or perlite, water repellant	D-1	250	.064	C.F.	1.45	1.57		3.02	4.05	
	0700	Foamed in place, urethane in 2-5/8" cavity	G-2	1,035	.023	S.F.	.38	.53	.23	1.14	1.49	
	0800	For each 1" added thickness, add	"	2,372	.010	"	.12	.23	.10	.45	.60	
109	0600	**PERIMETER INSULATION**, polystyrene, expanded, 1" thick, R4	1 Carp	680	.012	S.F.	.18	.32		.50	.71	109
	0700	2" thick, R8	"	675	.012	"	.33	.32		.65	.88	
116	0010	**WALL INSULATION, RIGID**										116
	0040	Fiberglass, 1.5#/CF, unfaced, 1" thick, R4.1	1 Carp	1,000	.008	S.F.	.23	.22		.45	.59	
	0060	1-1/2" thick, R6.2		1,000	.008		.30	.22		.52	.67	
	0080	2" thick, R8.3		1,000	.008		.37	.22		.59	.74	
	0120	3" thick, R12.4		800	.010		.44	.27		.71	.91	
	0370	3#/CF, unfaced, 1" thick, R4.3		1,000	.008		.29	.22		.51	.66	
	0390	1-1/2" thick, R6.5		1,000	.008		.56	.22		.78	.96	
	0400	2" thick, R8.7		890	.009		.69	.25		.94	1.15	
	0420	2-1/2" thick, R10.9		800	.010		.86	.27		1.13	1.38	
	0440	3" thick, R13		800	.010		1.02	.27		1.29	1.55	
	0520	Foil faced, 1" thick, R4.3		1,000	.008		.68	.22		.90	1.09	
	0540	1-1/2" thick, R6.5		1,000	.008		.91	.22		1.13	1.34	
	0560	2" thick, R8.7		890	.009		1.13	.25		1.38	1.63	
	0580	2-1/2" thick, R10.9		800	.010		1.34	.27		1.61	1.90	
	0600	3" thick, R13		800	.010		1.46	.27		1.73	2.04	
	0670	6#/CF, unfaced, 1" thick, R4.3		1,000	.008		.65	.22		.87	1.05	
	0690	1-1/2" thick, R6.5		890	.009		1.01	.25		1.26	1.50	
	0700	2" thick, R8.7		800	.010		1.41	.27		1.68	1.98	
	0721	2-1/2" thick, R10.9		800	.010		1.55	.27		1.82	2.13	
	0741	3" thick, R13		730	.011		1.86	.30		2.16	2.52	
	0821	Foil faced, 1" thick, R4.3		1,000	.008		.92	.22		1.14	1.35	
	0840	1-1/2" thick, R6.5		890	.009		1.32	.25		1.57	1.84	
	0850	2" thick, R8.7		800	.010		1.72	.27		1.99	2.32	
	0880	2-1/2" thick, R10.9		800	.010		2.07	.27		2.34	2.71	
	0900	3" thick, R13		730	.011		2.47	.30		2.77	3.19	
	1500	Foamglass, 1-1/2" thick, R3.9		800	.010		1.45	.27		1.72	2.03	
	1550	3" thick, R9	↓	730	.011	↓	2.71	.30		3.01	3.45	
	1600	Isocyanurate, 4' x 8' sheet, foil faced, both sides										
	1610	1/2" thick, R3.9	1 Carp	800	.010	S.F.	.28	.27		.55	.74	
	1620	5/8" thick, R4.5	↓	800	.010	↓	.29	.27		.56	.75	

072 | Insulation & Fireproofing

072 100 | Building Insulation

		CREW	DAILY OUTPUT	LABOR-HOURS	UNIT	1999 BARE COSTS MAT.	LABOR	EQUIP.	TOTAL	TOTAL INCL O&P
1630	3/4" thick, R5.4	1 Carp	800	.010	S.F.	.30	.27		.57	.76
1640	1" thick, R7.2		800	.010		.33	.27		.60	.80
1650	1-1/2" thick, R10.8		730	.011		.37	.30		.67	.88
1660	2" thick, R14.4		730	.011		.47	.30		.77	.98
1670	3" thick, R21.6		730	.011		1.10	.30		1.40	1.68
1680	4" thick, R28.8		730	.011		1.36	.30		1.66	1.97
1700	Perlite, 1" thick, R2.77		800	.010		.25	.27		.52	.71
1750	2" thick, R5.55		730	.011		.51	.30		.81	1.03
1900	Extruded polystyrene, 25 PSI compressive strength, 1" thick, R5		800	.010		.31	.27		.58	.77
1940	2" thick R10		730	.011		.61	.30		.91	1.14
1960	3" thick, R15		730	.011		.90	.30		1.20	1.47
2100	Expanded polystyrene, 1" thick, R3.85		800	.010		.15	.27		.42	.59
2120	2" thick, R7.69		730	.011		.32	.30		.62	.82
2140	3" thick, R11.49	↓	730	.011	↓	.52	.30		.82	1.04

072 500 | Fireproofing

		CREW	DAILY OUTPUT	LABOR-HOURS	UNIT	MAT.	LABOR	EQUIP.	TOTAL	TOTAL INCL O&P
0010	**SPRAYED** Mineral fiber or cementitious for fireproofing,									
0050	not incl tamping or canvas protection									
0100	1" thick, on flat plate steel	G-2	3,000	.008	S.F.	.42	.18	.08	.68	.83
0400	Beams		1,500	.016		.42	.37	.16	.95	1.20
0500	Corrugated or fluted decks		1,250	.019		.63	.44	.19	1.26	1.58
0700	Columns, 1-1/8" thick	↓	1,100	.022	↓	.47	.50	.21	1.18	1.53

073 | Shingles & Roofing Tiles

073 100 | Shingles

		CREW	DAILY OUTPUT	LABOR-HOURS	UNIT	1999 BARE COSTS MAT.	LABOR	EQUIP.	TOTAL	TOTAL INCL O&P
0010	**FIBER CEMENT** shingles, 16" x 9.35", 500 lb per square	1 Carp	4	2	Sq.	244	54.50		298.50	355
0110	Starters, 16" x 9.35"		3	2.667	C.L.F.	83	73		156	206
0120	Hip & ridge, 4.75" x 14"		1	8	"	600	218		818	1,000
0200	Shakes, 16" x 9.35", 550 lb per square		2.20	3.636	Sq.	221	99.50		320.50	400
0300	Hip & ridge, 4.75 x 14"		1	8	C.L.F.	600	218		818	1,000
0400	Hexagonal, 16" x 16"		3	2.667	Sq.	165	73		238	297
0500	Square, 16" x 16"	↓	3	2.667		148	73		221	278
2000	For steep roofs (7/12 pitch or greater), add				↓		50%			
0010	**ASPHALT SHINGLES**									
0100	Standard strip shingles									
0150	Inorganic, class A, 210-235 lb/sq	1 Rofc	5.50	1.455	Sq.	26	35		61	89
0155	Pneumatic nailed		7	1.143		26	27.50		53.50	76
0200	Organic, class C, 235-240 lb/sq		5	1.600		34.50	38.50		73	105
0205	Pneumatic nailed	↓	6.25	1.280	↓	34.50	31		65.50	91
0250	Standard, laminated multi-layered shingles									
0300	Class A, 240-260 lb/sq	1 Rofc	4.50	1.778	Sq.	34	43		77	112
0305	Pneumatic nailed		5.63	1.422		34	34.50		68.50	96.50
0350	Class C, 260-300 lb/square, 4 bundles/square		4	2		49.50	48		97.50	138
0355	Pneumatic nailed	↓	5	1.600	↓	49.50	38.50		88	121
0400	Premium, laminated multi-layered shingles									
0450	Class A, 260-300 lb, 4 bundles/sq	1 Rofc	3.50	2.286	Sq.	43	55		98	143
0455	Pneumatic nailed		4.37	1.831		43	44		87	124
0500	Class C, 300-385 lb/square, 5 bundles/square		3	2.667		65.50	64.50		130	183
0505	Pneumatic nailed	↓	3.75	2.133	↓	65.50	51.50		117	161

116 554 103 104

073 | Shingles & Roofing Tiles

	073 100 | Shingles		CREW	DAILY OUTPUT	LABOR-HOURS	UNIT	1999 BARE COSTS MAT.	LABOR	EQUIP.	TOTAL	TOTAL INCL O&P
104	0800	#15 felt underlayment	1 Rofc	64	.125	Sq.	2.68	3.01		5.69	8.15
	0825	#30 felt underlayment		58	.138		5.45	3.32		8.77	11.70
	0850	Self adhering polyethylene and rubberized asphalt underlayment		22	.364	↓	39.50	8.75		48.25	58.50
	0900	Ridge shingles		330	.024	L.F.	.72	.58		1.30	1.80
	0905	Pneumatic nailed	↓	412.50	.019	"	.72	.47		1.19	1.59
	1000	For steep roofs (7 to 12 pitch or greater), add						50%			
106	0010	**SLATE**, Buckingham, Virginia, black									
	0100	3/16" - 1/4" thick	1 Rots	1.75	4.571	Sq.	540	111		651	785
	073 200 | Roofing Tile										
201	0010	**ALUMINUM** Tiles with accessories, .032" thick, mission tile	1 Carp	2.50	3.200	Sq.	350	87.50		437.50	520
	0200	Spanish tiles	"	3	2.667	"	350	73		423	500
202	0010	**CLAY TILE** ASTM C1167, GR 1, severe weathering, acces. incl.									
	0200	Lanai tile or Classic tile, 158 pc per sq	1 Rots	1.65	4.848	Sq.	490	117		607	740
	0300	Americana, 158 pc per sq, most colors		1.65	4.848		490	117		607	740
	0350	Green, gray or brown		1.65	4.848		490	117		607	740
	0400	Blue		1.65	4.848		490	117		607	740
	0600	Spanish tile, 171 pc per sq, red		1.80	4.444		305	108		413	525
	0800	Blend		1.80	4.444		420	108		528	645
	0900	Glazed white		1.80	4.444		500	108		608	735
	1100	Mission tile, 192 pc per sq, machine scored finish, red		1.15	6.957		635	168		803	985
	1700	French tile, 133 pc per sq, smooth finish, red		1.35	5.926		575	143		718	875
	1750	Blue or green		1.35	5.926		685	143		828	1,000
	1800	Norman black 317 pc per sq		1	8		805	194		999	1,225
	2200	Williamsburg tile, 158 pc per sq, aged cedar		1.35	5.926		490	143		633	785
	2250	Gray or green		1.35	5.926	↓	490	143		633	785
	2350	Ridge shingles, clay tile	↓	200	.040	L.F.	8.65	.97		9.62	11.15
204	0010	**CONCRETE TILE** Including installation of accessories									
	0020	Corrugated, 13" x 16-1/2", 90 per sq, 950 lb per sq									
	0050	Earthtone colors, nailed to wood deck	1 Rots	1.35	5.926	Sq.	103	143		246	360
	0150	Custom blues		1.35	5.926		114	143		257	375
	0200	Custom greens		1.35	5.926		114	143		257	375
	0250	Premium colors	↓	1.35	5.926	↓	153	143		296	415
	0500	Shakes, 13" x 16-1/2", 90 per sq, 950 lb per sq									
	0600	All colors, nailed to wood deck	1 Rots	1.50	5.333	Sq.	185	129		314	425
	1500	Accessory pieces, ridge & hip, 10" x 16-1/2", 8 lbs. each				Ea.	2.25			2.25	2.48
	1700	Rake, 6-1/2" x 16-3/4", 9 lbs. each					2.25			2.25	2.48
	1800	Mansard hip, 10" x 16-1/2", 9.2 lbs. each					2.25			2.25	2.48
	1900	Hip starter, 10" x 16-1/2", 10.5 lbs. each					9.50			9.50	10.45
	2000	3 or 4 way apex, 10" each side, 11.5 lbs. each				↓	10.25			10.25	11.30

074 | Preformed Roofing & Siding

	074 100 | Preformed Panels		CREW	DAILY OUTPUT	LABOR-HOURS	UNIT	1999 BARE COSTS MAT.	LABOR	EQUIP.	TOTAL	TOTAL INCL O&P
101	0010	**ALUMINUM ROOFING** Corrugated or ribbed, .0155" thick, natural	G-3	1,200	.027	S.F.	.61	.71		1.32	1.78
	0300	Painted		1,200	.027		.87	.71		1.58	2.07
	0400	Corrugated, .018" thick, on steel frame, natural finish		1,200	.027		.78	.71		1.49	1.97
	0600	Painted	↓	1,200	.027	↓	.97	.71		1.68	2.18

074 100 | Preformed Panels

		Description	Crew	Daily Output	Labor-Hours	Unit	1999 Bare Costs Mat.	Labor	Equip.	Total	Total Incl O&P	
101	0700	Corrugated, on steel frame, natural, .024" thick	G-3	1,200	.027	S.F.	1.14	.71		1.85	2.36	101
	0800	Painted, .024" thick		1,200	.027		1.37	.71		2.08	2.62	
	0900	.032" thick, natural		1,200	.027		1.23	.71		1.94	2.46	
	1200	painted		1,200	.027		1.61	.71		2.32	2.88	
	1300	V-Beam, on steel frame construction, .032" thick, natural		1,200	.027		1.39	.71		2.10	2.64	
	1500	Painted		1,200	.027		1.78	.71		2.49	3.07	
	1600	.040" thick, natural		1,200	.027		1.73	.71		2.44	3.01	
	1800	Painted		1,200	.027		2.15	.71		2.86	3.47	
	1900	.050" thick, natural		1,200	.027		2.10	.71		2.81	3.42	
	2100	Painted		1,200	.027		2.58	.71		3.29	3.94	
	2200	For roofing on wood frame, deduct		4,600	.007	↓	.05	.19		.24	.35	
	2400	Ridge cap, .032" thick, natural	↓	800	.040	L.F.	1.64	1.06		2.70	3.46	
104	0010	**FIBERGLASS** Corrugated panels, roofing, 8 oz per SF	G-3	1,000	.032	S.F.	2.17	.85		3.02	3.72	104
	0100	12 oz per SF		1,000	.032	"	2.87	.85		3.72	4.49	
	0300	Corrugated siding, 6 oz per SF		880	.036	S.F.	1.88	.97		2.85	3.58	
	0400	8 oz per SF		880	.036		2.17	.97		3.14	3.90	
	0500	Fire retardant		880	.036		3	.97		3.97	4.81	
	0600	12 oz. siding, textured		880	.036		2.70	.97		3.67	4.48	
	0700	Fire retardant		880	.036		4.02	.97		4.99	5.95	
	0900	Flat panels, 6 oz per SF, clear or colors		880	.036		1.68	.97		2.65	3.36	
	1100	Fire retardant, class A		880	.036		2.91	.97		3.88	4.71	
	1300	8 oz per SF, clear or colors		880	.036		2.17	.97		3.14	3.90	
	1700	Sandwich panels, fiberglass, 1-9/16" thick, panels to 20 SF		180	.178		21	4.73		25.73	30.50	
	1900	As above, but 2-3/4" thick, panels to 100 SF	↓	265	.121	↓	15.60	3.21		18.81	22	
107	0010	**STEEL ROOFING** on steel frame, corrugated or ribbed, 30 ga galv	G-3	1,100	.029	S.F.	.75	.77		1.52	2.04	107
	0100	28 ga		1,050	.030		.79	.81		1.60	2.14	
	0300	26 ga		1,000	.032		.86	.85		1.71	2.28	
	0400	24 ga		950	.034		1.02	.90		1.92	2.52	
	0600	Colored, 28 ga		1,050	.030		1.05	.81		1.86	2.42	
	0700	26 ga		1,000	.032	↓	1.12	.85		1.97	2.56	
	1200	Ridge, galvanized, 10" wide		800	.040	L.F.	1.68	1.06		2.74	3.51	
	1210	20" wide	↓	750	.043	"	2.95	1.13		4.08	5	

074 600 | Cladding/Siding

		Description	Crew	Daily Output	Labor-Hours	Unit	1999 Bare Costs Mat.	Labor	Equip.	Total	Total Incl O&P	
602	0010	**ALUMINUM SIDING** .019" thick, on steel construction, natural	G-3	775	.041	S.F.	.65	1.10		1.75	2.42	602
	0100	Painted		775	.041		.84	1.10		1.94	2.63	
	0400	Farm type, .021" thick on steel frame, natural		775	.041		.75	1.10		1.85	2.54	
	0600	Painted		775	.041		.92	1.10		2.02	2.72	
	0700	Industrial type, corrugated, on steel, .024" thick, mill		775	.041		.95	1.10		2.05	2.76	
	0900	Painted		775	.041		1.15	1.10		2.25	2.98	
	1000	.032" thick, mill		775	.041		1.30	1.10		2.40	3.14	
	1200	Painted		775	.041		1.65	1.10		2.75	3.52	
	2750	.050" thick, natural		775	.041		1.84	1.10		2.94	3.73	
	3300	For siding on wood frame, deduct from above	↓	2,800	.011	↓	.06	.30		.36	.54	
	3400	Screw fasteners, aluminum, self tapping, neoprene washer, 1"				M	123			123	135	
	3600	Stitch screws, self tapping, with neoprene washer, 5/8"				"	92.50			92.50	101	
	3630	Flashing, sidewall, .032" thick	G-3	800	.040	L.F.	1.75	1.06		2.81	3.59	
	3650	End wall, .040" thick		800	.040		2.05	1.06		3.11	3.92	
	3670	Closure strips, corrugated, .032" thick		800	.040		.52	1.06		1.58	2.23	
	3680	Ribbed, 4" or 8", .032" thick		800	.040		.52	1.06		1.58	2.23	
	3690	V-beam, .040" thick	↓	800	.040	↓	.71	1.06		1.77	2.44	
	3800	Horizontal, colored clapboard, 8" wide, plain	2 Carp	515	.031	S.F.	1.22	.85		2.07	2.67	
	3900	Insulated	"	515	.031	"	1.49	.85		2.34	2.97	
606	0010	**STEEL SIDING**, Beveled, vinyl coated, 8" wide, including fasteners	1 Carp	265	.030	S.F.	1.12	.82		1.94	2.53	606
	0050	10" wide	"	275	.029	↓	1.19	.79		1.98	2.56	

074 | Preformed Roofing & Siding

074 600 | Cladding/Siding

			CREW	DAILY OUTPUT	LABOR-HOURS	UNIT	1999 BARE COSTS MAT.	LABOR	EQUIP.	TOTAL	TOTAL INCL O&P
606	0080	Galv, corrugated or ribbed, on steel frame, 30 gauge	G-3	800	.040	S.F.	.75	1.06		1.81	2.49
	0100	28 gauge		795	.040		.79	1.07		1.86	2.54
	0300	26 gauge		790	.041		.86	1.08		1.94	2.63
	0400	24 gauge		785	.041		1.03	1.08		2.11	2.82
	0600	22 gauge		770	.042		1.17	1.11		2.28	3.02
	0700	Colored, corrugated/ribbed, on steel frame, 10 yr fnsh, 28 ga.		800	.040		1.08	1.06		2.14	2.85
	0900	26 gauge		795	.040		.95	1.07		2.02	2.72
	1000	24 gauge		790	.041		1.09	1.08		2.17	2.88
	1020	20 gauge	↓	785	.041	↓	1.41	1.08		2.49	3.24
607	0010	**VINYL SIDING** Solid PVC panels, 8″ to 10″ wide, plain	1 Carp	255	.031	S.F.	.61	.86		1.47	2.02
	0100	with 3/8″ insulation		255	.031		.75	.86		1.61	2.18
	0200	Soffit and fascia		205	.039	↓	1.33	1.07		2.40	3.14
	0300	Window and door trim moldings		185	.043	L.F.	.31	1.18		1.49	2.20
	0500	Corner posts, outside corner		205	.039		1.04	1.07		2.11	2.82
	0600	Inside corner	↓	205	.039	↓	.58	1.07		1.65	2.32
609	0010	**WOOD SIDING, BOARDS**									
	3200	Wood, cedar bevel, A grade, 1/2″ x 6″	1 Carp	250	.032	S.F.	2.02	.87		2.89	3.59
	3300	1/2″ x 8″		275	.029		1.66	.79		2.45	3.08
	3500	3/4″ x 10″, clear grade		300	.027		2.16	.73		2.89	3.52
	3600	"B" grade		300	.027		3.22	.73		3.95	4.69
	3800	Cedar, rough sawn, 1″ x 4″, A grade, natural		240	.033		3.01	.91		3.92	4.74
	4100	1″ x 12″, board & batten, #3 & Btr., natural		260	.031		2.08	.84		2.92	3.61
	4700	Redwood, clear, beveled, vertical grain, 1/2″ x 4″		200	.040		3.22	1.09		4.31	5.25
	4750	1/2″ x 6″		225	.036		2.70	.97		3.67	4.50
	4800	1/2″ x 8″		250	.032		2.19	.87		3.06	3.78
	5000	3/4″ x 10″		300	.027		3.57	.73		4.30	5.10
	5400	White pine, rough sawn, 1″ x 8″, natural	↓	275	.029	↓	.69	.79		1.48	2.01
611	0010	**WOOD PRODUCT SIDING**									
	0030	Lap siding, hardboard, 7/16″ x 8″, primed									
	0050	Wood grain texture finish	2 Carp	650	.025	S.F.	.95	.67		1.62	2.11
	0700	Particle board, overlaid, 3/8″ thick		750	.021		.63	.58		1.21	1.61
	0900	Plywood, medium density overlaid, 3/8″ thick		750	.021		1.08	.58		1.66	2.11
	1000	1/2″ thick		700	.023		1.26	.62		1.88	2.37
	1100	3/4″ thick		650	.025		1.47	.67		2.14	2.68
	1600	Texture 1-11, cedar, 5/8″ thick, natural		675	.024		1.50	.65		2.15	2.67
	1900	Texture 1-11, fir, 5/8″ thick, natural		675	.024		1.01	.65		1.66	2.13
	2050	Texture 1-11, S.Y.P., 5/8″ thick, natural		675	.024		.83	.65		1.48	1.93
	2200	Rough sawn cedar, 3/8″ thick, natural		675	.024		1.14	.65		1.79	2.27
	2500	Rough sawn fir, 3/8″ thick, natural	↓	675	.024	↓	.61	.65		1.26	1.69

075 | Membrane Roofing

075 100 | Built-Up Roofing

			CREW	DAILY OUTPUT	LABOR-HOURS	UNIT	1999 BARE COSTS MAT.	LABOR	EQUIP.	TOTAL	TOTAL INCL O&P
101	0010	**ASPHALT** Coated felt, #30, 2 sq per roll, not mopped	1 Rofc	58	.138	Sq.	5.45	3.32		8.77	11.70
	0200	#15, 4 sq per roll, plain or perforated, not mopped		58	.138		2.68	3.32		6	8.65
	0300	Roll roofing, smooth, #65		15	.533		12.10	12.85		24.95	35.50
	0500	#90		15	.533		12.65	12.85		25.50	36
	0520	Mineralized		15	.533		15.50	12.85		28.35	39
	0540	D.C. (Double coverage), 19″ selvage edge	↓	10	.800	↓	30.50	19.30		49.80	66.50

075 | Membrane Roofing

075 100 | Built-Up Roofing

	No.	Description	CREW	DAILY OUTPUT	LABOR-HOURS	UNIT	MAT.	LABOR	EQUIP.	TOTAL	TOTAL INCL O&P
							1999 BARE COSTS				
101	0580	Adhesive (lap cement)				Gal.	3.68			3.68	4.05
	0600	Steep, flat or dead level asphalt, 10 ton lots, bulk				Ton	230			230	253
	0800	Packaged				"	255			255	281
102	0010	**BUILT-UP ROOFING**									
	0120	Asphalt flood coat with gravel/slag surfacing, not including									
	0140	Insulation, flashing or wood nailers									
	0200	Asphalt base sheet, 3 plies #15 asphalt felt, mopped	G-1	22	2.545	Sq.	35.50	57.50	18.60	111.60	159
	0350	On nailable decks		21	2.667		39.50	60.50	19.45	119.45	169
	0500	4 plies #15 asphalt felt, mopped		20	2.800		50	63.50	20.50	134	187
	0550	On nailable decks		19	2.947		46	67	21.50	134.50	189
	0700	Coated glass base sheet, 2 plies glass (type IV), mopped		22	2.545		38.50	57.50	18.60	114.60	163
	0850	3 plies glass, mopped		20	2.800		46	63.50	20.50	130	182
	0950	On nailable decks	↓	19	2.947	↓	44	67	21.50	132.50	187
103	0010	**CANTS** 4" x 4", treated timber, cut diagonally	1 Rofc	325	.025	L.F.	.80	.59		1.39	1.90
	0100	Foamglass		325	.025		1.92	.59		2.51	3.13
	0300	Mineral or fiber, trapezoidal, 1"x 4" x 48"		325	.025		.17	.59		.76	1.21
	0400	1-1/2" x 5-5/8" x 48"	↓	325	.025	↓	.29	.59		.88	1.34

075 600 | Roof Maint. & Repairs

	No.	Description	CREW	DAILY OUTPUT	LABOR-HOURS	UNIT	MAT.	LABOR	EQUIP.	TOTAL	TOTAL INCL O&P
604	0010	**ROOF COATINGS** Asphalt				Gal.	2.71			2.71	2.98
	0200	Asphalt base, fibered aluminum coating					7.65			7.65	8.40
	0300	Asphalt primer, 5 gallon				↓	3.25			3.25	3.58
	0600	Coal tar pitch, 200 lb. barrels				Ton	540			540	590
	0700	Tar roof cement, 5 gal. lots				Gal.	5.55			5.55	6.10
	0800	Glass fibered roof & patching cement, 5 gallon				"	3.45			3.45	3.80
	0900	Reinforcing glass membrane, 450 S.F./roll				Ea.	45.50			45.50	50
	1000	Neoprene roof coating, 5 gal, 2 gal/sq				Gal.	20.50			20.50	22.50
	1100	Roof patch & flashing cement, 5 gallon					18.95			18.95	21
	1200	Roof resaturant, glass fibered, 3 gal/sq					7.30			7.30	8.05
	1300	Mineral rubber, 3 gal/sq				↓	4.71			4.71	5.20

076 | Flashing & Sheet Metal

076 200 | Sheet Mtl Flash & Trim

	No.	Description	CREW	DAILY OUTPUT	LABOR-HOURS	UNIT	MAT.	LABOR	EQUIP.	TOTAL	TOTAL INCL O&P
							1999 BARE COSTS				
201	0010	**DOWNSPOUTS** Aluminum 2" x 3", .020" thick, embossed	1 Shee	190	.042	L.F.	.71	1.34		2.05	2.86
	0100	Enameled		190	.042		.68	1.34		2.02	2.83
	0300	Enameled, .024" thick, 2" x 3"		180	.044		1.09	1.41		2.50	3.39
	0400	3" x 4"		140	.057		1.41	1.81		3.22	4.37
	0600	Round, corrugated aluminum, 3" diameter, .020" thick		190	.042		.85	1.34		2.19	3.02
	0700	4" diameter, .025" thick		140	.057		1.29	1.81		3.10	4.24
	4800	Steel, galvanized, round, corrugated, 2" or 3" diam, 28 ga		190	.042		.68	1.34		2.02	2.83
	4900	4" diameter, 28 gauge		145	.055		.90	1.75		2.65	3.71
	5100	5" diameter, 28 gauge		130	.062		1.23	1.95		3.18	4.38
	5400	6" diameter, 28 gauge		105	.076		1.60	2.42		4.02	5.50
	5700	Rectangular, corrugated, 28 gauge, 2" x 3"		190	.042		.53	1.34		1.87	2.66
	5800	3" x 4"		145	.055		1.48	1.75		3.23	4.35
	6000	Rectangular, plain, 28 gauge, galvanized, 2" x 3"		190	.042		.79	1.34		2.13	2.95
	6100	3" x 4"	↓	145	.055	↓	1.21	1.75		2.96	4.05

076 | Flashing & Sheet Metal

		076 200 | Sheet Mtl Flash & Trim	CREW	DAILY OUTPUT	LABOR-HOURS	UNIT	1999 BARE COSTS MAT.	LABOR	EQUIP.	TOTAL	TOTAL INCL O&P
202	0010	**DRIP EDGE**, aluminum, .016″ thick, 5″ wide, mill finish	1 Carp	400	.020	L.F.	.20	.55		.75	1.08
	0100	White finish		400	.020		.22	.55		.77	1.10
	0200	8″ wide, mill finish		400	.020		.30	.55		.85	1.19
	0300	Ice belt, 28″ wide, mill finish		100	.080		3.42	2.18		5.60	7.20
	0400	Galvanized, 5″ wide		400	.020		.22	.55		.77	1.10
	0500	8″ wide, mill finish	↓	400	.020	↓	.33	.55		.88	1.22
204	0010	**FLASHING** Aluminum, mill finish, .013″ thick	1 Shee	145	.055	S.F.	.34	1.75		2.09	3.09
	0030	.016″ thick		145	.055		.50	1.75		2.25	3.27
	0060	.019″ thick		145	.055		.79	1.75		2.54	3.59
	0100	.032″ thick		145	.055		.86	1.75		2.61	3.66
	0200	.040″ thick		145	.055		1.44	1.75		3.19	4.30
	0300	.050″ thick	↓	145	.055		1.80	1.75		3.55	4.70
	0400	Painted finish, add				↓	.24			.24	.26
	1300	Asphalt flashing cement, 5 gallon				Gal.	3.34			3.34	3.67
	4900	Fabric, asphalt-saturated cotton, specification grade	1 Rofc	35	.229	S.Y.	1.93	5.50		7.43	11.60
	5000	Utility grade		35	.229		1.22	5.50		6.72	10.85
	5200	Open-mesh fabric, saturated, 40 oz per S.Y.		35	.229		1.35	5.50		6.85	11
	5300	Close-mesh fabric, saturated, 17 oz per S.Y.		35	.229		1.42	5.50		6.92	11.05
	5500	Fiberglass, resin-coated		35	.229		1.16	5.50		6.66	10.80
	5600	Asphalt-coated, 40 oz per S.Y.	↓	35	.229	↓	7.90	5.50		13.40	18.20
205	0010	**GUTTERS** Aluminum, stock units, 5″ box, .027″ thick, plain	1 Shee	120	.067	L.F.	.87	2.12		2.99	4.25
	0100	Enameled		120	.067		1.07	2.12		3.19	4.47
	0600	5″ x 6″ combination fascia & gutter, .032″ thick, enameled		60	.133		3.45	4.23		7.68	10.35
	2400	Steel, galv, half round or box, 28 ga, 5″ wide, plain		120	.067		.90	2.12		3.02	4.29
	2500	Enameled		120	.067		.94	2.12		3.06	4.33
	2700	26 ga, stock, 5″ wide		120	.067		.95	2.12		3.07	4.34
	2800	6″ wide	↓	120	.067		1.24	2.12		3.36	4.65
	3000	Vinyl, O.G., 4″ wide	1 Carp	110	.073		.85	1.99		2.84	4.06
	3100	5″ wide		110	.073		1	1.99		2.99	4.22
	3300	Wood, clear treated cedar, fir or hemlock, 3″ x 4″		100	.080		6.30	2.18		8.48	10.35
	3400	4″ x 5″	↓	100	.080	↓	7.30	2.18		9.48	11.45

077 | Roof Specialties & Accessories

		077 100 | Prefab Roof Specialties	CREW	DAILY OUTPUT	LABOR-HOURS	UNIT	1999 BARE COSTS MAT.	LABOR	EQUIP.	TOTAL	TOTAL INCL O&P
103	0010	**EXPANSION JOINT**									
	0300	Butyl or neoprene center with foam insulation, metal flanges									
	0400	Aluminum, .032″ thick for openings to 2-1/2″	1 Rofc	165	.048	L.F.	7.40	1.17		8.57	10.15
	0600	For joint openings to 3-1/2″		165	.048		8.65	1.17		9.82	11.50
	0610	For joint openings to 5″		165	.048		10.40	1.17		11.57	13.45
	0620	For joint openings to 8″		165	.048		16.40	1.17		17.57	20
	0700	Copper, 16 oz. for openings to 2-1/2″		165	.048		10.40	1.17		11.57	13.45
	0900	For joint openings to 3-1/2″		165	.048		11.95	1.17		13.12	15.15
	0910	For joint openings to 5″		165	.048		14	1.17		15.17	17.40
	0920	For joint openings to 8″		165	.048		21	1.17		22.17	25
	1000	Galvanized steel, 26 ga. for openings to 2-1/2″		165	.048		6.25	1.17		7.42	8.90
	1200	For joint openings to 3-1/2″		165	.048		7.40	1.17		8.57	10.15
	1210	For joint openings to 5″		165	.048		9.25	1.17		10.42	12.20
	1220	For joint openings to 8″	↓	165	.048	↓	15.75	1.17		16.92	19.35

077 | Roof Specialties & Accessories

	077 100 | Prefab Roof Specialties	CREW	DAILY OUTPUT	LABOR-HOURS	UNIT	1999 BARE COSTS MAT.	LABOR	EQUIP.	TOTAL	TOTAL INCL O&P	
103	1810 For joint openings to 5"	1 Rofc	165	.048	L.F.	13.05	1.17		14.22	16.35	103
	1820 For joint openings to 8"		165	.048		19.85	1.17		21.02	24	
	1900 Neoprene, double-seal type with thick center, 4-1/2" wide		125	.064		8.90	1.54		10.44	12.45	
	1950 Polyethylene bellows, with galv steel flat flanges		100	.080		3.45	1.93		5.38	7.10	
	1960 With galvanized angle flanges		100	.080		3.80	1.93		5.73	7.50	
	2000 Roof joint with extruded aluminum cover, 2"	1 Shee	115	.070		24	2.21		26.21	29.50	
	2700 Wall joint, closed cell foam on PVC cover, 9" wide	1 Rofc	125	.064		3.10	1.54		4.64	6.05	
	2800 12" wide	"	115	.070		3.50	1.68		5.18	6.75	
105	0010 **GRAVEL STOP** Aluminum, .050" thick, 4" face height, mill finish	1 Shee	145	.055	L.F.	2.72	1.75		4.47	5.70	105
	0080 Duranodic finish	"	145	.055	"	3.69	1.75		5.44	6.80	

079 | Joint Sealers

	079 204 | Sealants & Caulkings	CREW	DAILY OUTPUT	LABOR-HOURS	UNIT	1999 BARE COSTS MAT.	LABOR	EQUIP.	TOTAL	TOTAL INCL O&P	
204	0010 **CAULKING AND SEALANTS**										204
	0020 Acoustical sealant, elastomeric, cartridges				Ea.	1.95			1.95	2.15	
	0030 Backer rod, polyethylene, 1/4" diameter	1 Bric	4.60	1.739	C.L.F.	1.26	48		49.26	76	
	0090 1" diameter	"	4.60	1.739	"	8.90	48		56.90	84.50	
	0100 Acrylic latex caulk, white										
	0200 11 fl. oz cartridge				Ea.	1.96			1.96	2.16	
	0500 1/4" x 1/2"	1 Bric	248	.032	L.F.	.16	.89		1.05	1.56	
	0600 1/2" x 1/2"		250	.032		.32	.88		1.20	1.72	
	0800 3/4" x 3/4"		230	.035		.72	.96		1.68	2.28	
	0900 3/4" x 1"		200	.040		.96	1.10		2.06	2.78	
	1000 1" x 1"		180	.044		1.20	1.23		2.43	3.23	
	1400 Butyl based, bulk				Gal.	22			22	24	
	1500 Cartridges				"	26.50			26.50	29.50	
	1700 Bulk, in place 1/4" x 1/2", 154 L.F./gal.	1 Bric	230	.035	L.F.	.14	.96		1.10	1.65	
	1800 1/2" x 1/2", 77 L.F./gal.	"	180	.044	"	.29	1.23		1.52	2.22	
	2000 Latex acrylic based, bulk				Gal.	23			23	25.50	
	2100 Cartridges					26			26	28.50	
	2300 Polysulfide compounds, 1 component, bulk					43.50			43.50	47.50	
	2600 1 or 2 component, in place, 1/4" x 1/4", 308 L.F./gal.	1 Bric	145	.055	L.F.	.14	1.52		1.66	2.52	
	2700 1/2" x 1/4", 154 L.F./gal.		135	.059		.28	1.64		1.92	2.85	
	2900 3/4" x 3/8", 68 L.F./gal.		130	.062		.64	1.70		2.34	3.34	
	3000 1" x 1/2", 38 L.F./gal.		130	.062		1.14	1.70		2.84	3.89	
	3200 Polyurethane, 1 or 2 component				Gal.	48.50			48.50	53	
	3500 Bulk, in place, 1/4" x 1/4"	1 Bric	150	.053	L.F.	.16	1.47		1.63	2.46	
	3600 1/2" x 1/4"		145	.055		.31	1.52		1.83	2.72	
	3800 3/4" x 3/8", 68 L.F./gal.		130	.062		.71	1.70		2.41	3.42	
	3900 1" x 1/2"		110	.073		1.26	2.01		3.27	4.50	
	4100 Silicone rubber, bulk				Gal.	34			34	37.50	
	4200 Cartridges				"	40			40	44	
	4400 Neoprene gaskets, closed cell, adhesive, 1/8" x 3/8"	1 Bric	240	.033	L.F.	.20	.92		1.12	1.65	
	4500 1/4" x 3/4"		215	.037		.48	1.03		1.51	2.13	
	4700 1/2" x 1"		200	.040		1.40	1.10		2.50	3.26	
	4800 3/4" x 1-1/2"		165	.048		2.91	1.34		4.25	5.30	
	5500 Resin epoxy coating, 2 component, heavy duty				Gal.	26			26	28.50	
	5800 Tapes, sealant, P.V.C. foam adhesive, 1/16" x 1/4"				C.L.F.	4.60			4.60	5.05	
	5900 1/16" x 1/2"					6.80			6.80	7.50	

	079 204 | Sealants & Caulkings	CREW	DAILY OUTPUT	LABOR-HOURS	UNIT	1999 BARE COSTS MAT.	LABOR	EQUIP.	TOTAL	TOTAL INCL O&P	
204	5950 1/16" x 1"				C.L.F.	11.30			11.30	12.45	204
	6000 1/8" x 1/2"				↓	7.65			7.65	8.40	
	6200 Urethane foam, 2 component, handy pack, 1 C.F.				Ea.	27.50			27.50	30.50	
	6300 50.0 C.F. pack				C.F.	14.05			14.05	15.45	

For information about Means Estimating Seminars, see yellow pages 11 and 12 in back of book

Division 8
Doors & Windows

Estimating Tips

081 Metal Doors & Frames

Most metal doors and frames look alike, but there may be significant differences among them. When estimating these items be sure to choose the line item that most closely compares to the specification or door schedule requirements regarding:

- type of metal
- metal gauge
- door core material
- fire rating
- finish

082 Wood & Plastic Doors

Wood and plastic doors vary considerably in price. The primary determinant is the veneer material. Lauan, birch and oak are the most common veneers. Other variables include the following:

- hollow or solid core
- fire rating
- flush or raised panel
- finish

If the specifications require compliance with AWI (Architectural Woodwork Institute) standards or acoustical standards, the cost of the door may increase substantially. All wood doors are priced pre-mortised for hinges and predrilled for cylindrical locksets.

083 Special Doors

- There are many varieties of special doors, and they are usually priced per each. Add frames, hardware or operators required for a complete installation.

085 Metal Windows

- Most metal windows are delivered preglazed. However, some metal windows are priced without glass. Refer to 088 Glazing for glass pricing. The grade C indicates commercial grade windows, usually ASTM C-35.

086 Wood & Plastic Windows

- All wood windows are priced preglazed. The two glazing options priced are single pane float glass and insulating glass 1/2″ thick. Add the cost of screens and grills if required.

087 Hardware

- Hardware costs add considerably to the cost of a door. The most efficient method to determine the hardware requirements for a project is to review the door schedule. This schedule, in conjunction with the specifications, is all you should need to take off the door hardware.
- Door hinges are priced by the pair, with most doors requiring 1-1/2 pairs per door. The hinge prices do not include installation labor because it is included in door installation. Hinges are classified according to the frequency of use.

088 Glazing

- Different openings require different types of glass. The three most common types are:
    - float
    - tempered
    - insulating
- Most exterior windows are glazed with insulating glass. Entrance doors and window walls, where the glass is less than 18″ from the floor, are generally glazed with tempered glass. Interior windows and some residential windows are glazed with float glass.

089 Glazed Curtain Walls

- Glazed curtain walls consist of the metal tube framing and the glazing material. The cost data in this subdivision is presented for the metal tube framing alone or the composite wall. If your estimate requires a detailed takeoff of the framing, be sure to add the glazing cost.

Reference Numbers

Reference numbers are shown in bold squares at the beginning of some major classifications. These numbers refer to related items in the Reference Section. The reference information may be an estimating procedure, an alternate pricing method or technical information.

Note: Not all subdivisions listed here necessarily appear in this publication.

081 100 | Steel Doors & Frames

		081 100 Steel Doors & Frames	CREW	DAILY OUTPUT	LABOR-HOURS	UNIT	1999 BARE COSTS MAT.	LABOR	EQUIP.	TOTAL	TOTAL INCL O&P
103	0010	COMMERCIAL STEEL DOORS									
	0015	Flush, full panel, hollow core									
	0020	1-3/8" thick, 20 ga., 2'-0"x 6'-8"	2 Carp	20	.800	Ea.	163	22		185	214
	0060	3'-0" x 6'-8"		17	.941		170	25.50		195.50	228
	0100	3'-0" x 7'-0"		17	.941		178	25.50		203.50	237
	0120	For vision lite, add					51			51	56
	0140	For narrow lite, add					60			60	66
	0160	For bottom louver, add					100			100	110
	0320	Half glass, 20 ga., 2'-0" x 6'-8"	2 Carp	20	.800		208	22		230	264
	0360	3'-0" x 6'-8"		17	.941		218	25.50		243.50	281
	0400	3'-0" x 7'-0"		17	.941		221	25.50		246.50	285
	0500	Hollow core, 1-3/4" thick, full panel, 20 ga., 2'-8" x 6'-8"		18	.889		152	24.50		176.50	205
	0520	3'-0" x 6'-8"		17	.941		173	25.50		198.50	232
	0640	3'-0" x 7'-0"		17	.941		178	25.50		203.50	237
	1000	18 ga., 2'-8" x 6'-8"		17	.941		182	25.50		207.50	241
	1020	3'-0" x 6'-8"		16	1		189	27.50		216.50	251
	1120	3'-0" x 7'-0"		17	.941		210	25.50		235.50	272
	1230	Half glass, 20 ga., 2'-8" x 6'-8"		20	.800		246	22		268	305
	1240	3'-0" x 6'-8"		18	.889		250	24.50		274.50	315
	1320	18 ga., 2'-8" x 6'-8"		18	.889		279	24.50		303.50	345
	1340	3'-0" x 6'-8"		17	.941		274	25.50		299.50	340
	1720	Insulated, 1-3/4" thick, full panel, 18 ga., 3'-0" x 6'-8"		15	1.067		217	29		246	285
	1760	3'-0" x 7'-0"		15	1.067		225	29		254	294
	1820	Half glass, 18 ga., 3'-0" x 6'-8"		16	1		280	27.50		307.50	355
	1860	3'-0" x 7'-0"		16	1		286	27.50		313.50	360
110	0010	FIRE DOOR									
	0015	Steel, flush, "B" label, 90 minute									
	0020	Full panel, 20 ga., 2'-0" x 6'-8"	2 Carp	20	.800	Ea.	208	22		230	264
	0060	3'-0" x 6'-8"		17	.941		210	25.50		235.50	272
	0080	3'-0" x 7'-0"		17	.941		259	25.50		284.50	325
	0140	18 ga., 3'-0" x 6'-8"		16	1		230	27.50		257.50	296
	0220	For "A" label, 3 hour, 18 ga., use same price as "B" label									
	0240	For vision lite, add				Ea.	56.50			56.50	62.50
	0520	Flush, "B" label 90 min., composite, 20 ga., 2'-0" x 6'-8"	2 Carp	18	.889		208	24.50		232.50	267
	0580	3'-0" x 7'-0"		16	1		234	27.50		261.50	300
	0640	Flush, "A" label 3 hour, composite, 18 ga., 3'-0" x 6'-8"		15	1.067		250	29		279	320
	0680	3'-0" x 7'-0"		15	1.067		260	29		289	330
118	0010	STEEL FRAMES, KNOCK DOWN									
	0020	18 ga., up to 5-3/4" deep									
	0025	6'-8" high, 3'-0" wide, single	2 Carp	16	1	Ea.	86	27.50		113.50	138
	0040	6'-0" wide, double		14	1.143		96.50	31		127.50	155
	3600	5-3/4" deep, 7'-0" high, 4'-0" wide, single		15	1.067		75	29		104	129
	3640	8'-0" wide, double		12	1.333		109	36.50		145.50	177
	4400	8-3/4" deep, 7'-0" high, 4'-0" wide, single		15	1.067		91	29		120	146
	4440	8'-0" wide, double		12	1.333		124	36.50		160.50	194
	4900	For welded frames, add					29.50			29.50	32.50
	5400	16 ga., "B" label, up to 5-3/4" deep, 7'-0" high, 4'-0" wide, single	2 Carp	15	1.067		90	29		119	145
	6200	8-3/4" deep, 7'-0" high, 4'-0" wide, single	"	15	1.067		102	29		131	158
	6300	For "A" label use same price as "B" label									

Important: See the Reference Section for critical supporting data - Reference Nos., Crews, & City Cost Indexe

082 050 | Wood & Plastic Doors		CREW	DAILY OUTPUT	LABOR-HOURS	UNIT	1999 BARE COSTS MAT.	LABOR	EQUIP.	TOTAL	TOTAL INCL O&P	
0010	WOOD FRAMES										054
0400	Exterior frame, incl. ext. trim, pine, 5/4 x 4-9/16" deep	2 Carp	375	.043	L.F.	3.70	1.16		4.86	5.90	
0440	6-9/16" deep		375	.043		5.40	1.16		6.56	7.80	
1000	Sills, 8/4 x 8" deep, oak, no horns		100	.160		11.30	4.37		15.67	19.25	
1040	3" horns		100	.160		14.50	4.37		18.87	23	
3000	Interior frame, pine, 11/16" x 3-5/8" deep		375	.043		3.57	1.16		4.73	5.75	
3800	Threshold, oak, 5/8" x 3-5/8" deep		200	.080		2.92	2.18		5.10	6.65	
3820	4-5/8" deep		190	.084		3.72	2.30		6.02	7.70	
3840	5-5/8" deep	↓	180	.089	↓	4.47	2.43		6.90	8.75	
0010	WOOD DOOR, ARCHITECTURAL										062
0015	Flush, int., 1-3/4", 7 ply, hollow core,										
0120	Birch face, 2'-0" x 6'-8"	2 Carp	17	.941	Ea.	46	25.50		71.50	91	
0180	3'-0" x 6'-8"	"	17	.941		65	25.50		90.50	112	
0430	For 7'-0" high, add					13			13	14.30	
0480	For prefinishing, clear, add					28.50			28.50	31.50	
1720	H.P. plastic laminate, 2'-0" x 6'-8"	2 Carp	16	1		195	27.50		222.50	258	
1780	3'-0" x 6'-8"		15	1.067		222	29		251	290	
2020	5 ply particle core, lauan face, 2'-6" x 6'-8"		15	1.067		65	29		94	118	
2080	3'-0" x 7'-0"	↓	13	1.231		74	33.50		107.50	135	
2480	For solid wood core, add				↓	28.50			28.50	31.50	
0010	WOOD FIRE DOORS										070
0020	Particle core, 7 face plys, "B" label,										
0040	1 hour, birch face, 1-3/4" x 2'-6" x 6'-8"	2 Carp	14	1.143	Ea.	152	31		183	216	
0080	3'-0" x 6'-8"		13	1.231		157	33.50		190.50	225	
0090	3'-0" x 7'-0"		12	1.333		173	36.50		209.50	248	
0440	M.D. overlay on hardboard, 2'-6" x 6'-8"		15	1.067		147	29		176	207	
0480	3'-0" x 6'-8"		14	1.143		149	31		180	213	
0540	H.P. plastic laminate, 2'-6" x 6'-8"		13	1.231		233	33.50		266.50	310	
0580	3'-0" x 6'-8"		12	1.333		246	36.50		282.50	330	
0590	3'-0" x 7'-0"		11	1.455		253	39.50		292.50	340	
0740	90 minutes, birch face, 1-3/4" x 2'-6" x 6'-8"		14	1.143		197	31		228	266	
0780	3'-0" x 6'-8"		13	1.231		204	33.50		237.50	277	
0790	3'-0" x 7'-0"	↓	12	1.333	↓	211	36.50		247.50	290	
2200	Custom architectural "B" label, flush, 1-3/4" thick, birch,										
2260	3'-0" x 7'-0"	2 Carp	14	1.143	Ea.	340	31		371	425	
0010	WOOD DOORS, RESIDENTIAL										078
1000	Entrance door, colonial, 1-3/4" x 6'-8" x 2'-8" wide	2 Carp	16	1	Ea.	289	27.50		316.50	365	
1300	Flush, birch, solid core, 1-3/4" x 6'-8" x 2'-8" wide	"	16	1	"	66.50	27.50		94	116	
7310	Passage doors, flush, no frame included										
7700	Birch, hollow core, 1-3/8" x 6'-8" x 1'-6" wide	2 Carp	18	.889	Ea.	27	24.50		51.50	67.50	
7740	2'-6" wide	"	18	.889	"	37.50	24.50		62	79	

083 | Special Doors

083 100 | Sliding Doors		CREW	DAILY OUTPUT	LABOR-HOURS	UNIT	1999 BARE COSTS MAT.	LABOR	EQUIP.	TOTAL	TOTAL INCL O&P	
0010	GLASS, SLIDING										102
0012	Vinyl clad, 1" insul. glass, 6'-0" x 6'-10" high	2 Carp	4	4	Opng.	680	109		789	920	

083 | Special Doors

			CREW	DAILY OUTPUT	LABOR-HOURS	UNIT	1999 BARE COSTS MAT.	LABOR	EQUIP.	TOTAL	TOTAL INCL O&P
		083 100 | Sliding Doors									
102	0030	6'-0" x 8'-0" high	2 Carp	4	4	Ea.	1,425	109		1,534	1,750
	0100	8'-0" x 6'-10" high		4	4	Opng.	1,525	109		1,634	1,850
	0500	3 leaf, 9'-0" x 6'-10" high		3	5.333		1,375	146		1,521	1,750
	0600	12'-0" x 6'-10" high		3	5.333		1,725	146		1,871	2,100
104	0010	GLASS, SLIDING									
	0020	Wood, 5/8" tempered insul. glass, 6' wide, premium	2 Carp	4	4	Ea.	1,150	109		1,259	1,450
	0100	Economy		4	4		665	109		774	900
	0150	8' wide, wood, premium		3	5.333		1,050	146		1,196	1,375
	0200	Economy		3	5.333		780	146		926	1,100
	0250	12' wide, wood, premium		2.50	6.400		1,675	175		1,850	2,125
	0300	Economy		2.50	6.400		1,100	175		1,275	1,475
	0350	Aluminum, 5/8" tempered insulated glass, 6' wide									
	0400	Premium	2 Carp	4	4	Ea.	1,200	109		1,309	1,500
	0450	Economy		4	4		470	109		579	685
	0600	12' wide, premium		2.50	6.400		1,950	175		2,125	2,400
	0650	Economy		2.50	6.400		1,250	175		1,425	1,650
		083 300 | Coiling Doors									
302	0010	COUNTER DOORS									
	0020	Manual, incl. frm and hdwe, galv. stl., 4' roll-up, 6' long	2 Carp	2	8	Opng.	605	218		823	1,000
	0300	Galvanized steel, UL label	"	1.80	8.889	"	760	243		1,003	1,225
		083 600 | Sectional Overhead Drs									
604	0010	OVERHEAD, COMMERCIAL Frames not included									
	1000	Stock, sectional, heavy duty, wood, 1-3/4" thick, 8' x 8' high	2 Carp	2	8	Ea.	400	218		618	785
	1200	12' x 12' high		1.50	10.667		850	291		1,141	1,400
	1300	Chain hoist, 14' x 14' high		1.30	12.308		1,475	335		1,810	2,150
	1500	20' x 8' high		1.30	12.270		1,225	335		1,560	1,875
	2300	Fiberglass and aluminum, heavy duty, sectional, 12' x 12' high		1.50	10.667		1,250	291		1,541	1,825
	2600	Steel, 24 ga. sectional, manual, 8' x 8' high		2	8		355	218		573	735
	2800	Chain hoist, 20' x 14' high		.70	22.857		1,575	625		2,200	2,700
	2850	For 1-1/4" rigid insulation and 26 ga. galv.									
	2860	back panel, add				S.F.	1.73			1.73	1.90
	2900	For electric trolley operator, 1/3 H.P., to 12' x 12', add	1 Carp	2	4	Ea.	415	109		524	625
	2950	Over 12' x 12', 1/2 H.P., add	"	1	8	"	440	218		658	830
		083 700 | Hangar Doors									
701	0010	HANGAR DOOR									
	0020	Bi-fold, ovhd., 20 PSF wind load, incl. elec. oper.									
	0100	Without sheeting, 12' high x 40'	2 Sswk	240	.067	S.F.	9.05	2.04		11.09	13.70
	0200	16' high x 60'		230	.070		16.40	2.13		18.53	22
	0300	20' high x 80'		220	.073		18.60	2.23		20.83	24.50
		083 720 | Special Purpose Doors									
721	0010	BULKHEAD CELLAR DOORS									
	0020	Steel, not incl. sides, 44" x 62"	1 Carp	5.50	1.455	Ea.	180	39.50		219.50	261
	0100	52" x 73"		5.10	1.569		200	43		243	288
	0500	With sides and foundation plates, 57" x 45" x 24"		4.70	1.702		235	46.50		281.50	330
	0600	42" x 49" x 51"		4.30	1.860		283	51		334	390
723	0010	FLOOR, COMMERCIAL									
	0020	Aluminum tile, steel frame, one leaf, 2' x 2' opng.	2 Sswk	3.50	4.571	Opng.	340	140		480	630
	0050	3'-6" x 3'-6" opening	"	3.50	4.571	"	615	140		755	930

083 | Special Doors

083 720 | Special Purpose Doors		CREW	DAILY OUTPUT	LABOR-HOURS	UNIT	1999 BARE COSTS MAT.	LABOR	EQUIP.	TOTAL	TOTAL INCL O&P	
0010	KENNEL DOORS										729
0020	2 way, swinging type, 13" x 19" opening	2 Carp	11	1.455	Opng.	195	39.50		234.50	278	
0100	17" x 29" opening	"	11	1.455	"	195	39.50		234.50	278	
0010	**ROLLING SERVICE DOORS** Steel, manual, 20 ga., incl. hardware										732
0200	20' x 10' high	2 Sswk	1	16	Ea.	1,850	490		2,340	2,925	
2000	Class A fire doors, manual, 20 ga., 8' x 8' high	↓	1.40	11.429	↓	895	350		1,245	1,625	
2200	20' x 10' high	↓	.80	20	↓	2,375	610		2,985	3,750	
083 750 | Swing Doors											
0010	GLASS DOOR, SWING										754
0020	Including hardware, 1/2" thick, tempered, 3' x 7' opening	2 Glaz	2	8	Opng.	1,500	213		1,713	1,975	
0100	6' x 7' opening	"	1.40	11.429	"	3,000	305		3,305	3,775	

084 | Entrances & Storefronts

084 100 | Aluminum		CREW	DAILY OUTPUT	LABOR-HOURS	UNIT	1999 BARE COSTS MAT.	LABOR	EQUIP.	TOTAL	TOTAL INCL O&P	
0010	**STOREFRONT SYSTEMS** Aluminum frame, clear 3/8" plate glass,										111
0020	incl. 3' x 7' door with hardware (400 sq. ft. max. wall)										
0500	Wall height to 12' high, commercial grade	2 Glaz	150	.107	S.F.	11.35	2.84		14.19	16.75	
0700	Monumental grade	↓	115	.139	↓	21.50	3.70		25.20	29	
1000	6' x 7' door with hardware, commercial grade	↓	135	.119	↓	11.60	3.15		14.75	17.55	
1200	Monumental grade	↓	100	.160	↓	29.50	4.26		33.76	39	

085 | Metal Windows

085 100 | Steel Windows		CREW	DAILY OUTPUT	LABOR-HOURS	UNIT	1999 BARE COSTS MAT.	LABOR	EQUIP.	TOTAL	TOTAL INCL O&P	
0010	**STEEL SASH** Custom units, glazing and trim not included										102
0100	Casement, 100% vented	2 Sswk	200	.080	S.F.	33.50	2.45		35.95	41.50	
0200	50% vented	↓	200	.080	↓	28.50	2.45		30.95	36	
0300	Fixed	↓	200	.080	↓	19.20	2.45		21.65	25.50	
2000	Industrial security sash, 50% vented	↓	200	.080	↓	38	2.45		40.45	46	
2100	Fixed	↓	200	.080	↓	32	2.45		34.45	39.50	
085 200 | Aluminum Windows											
0010	ALUMINUM SASH										202
0020	Stock, grade C, glaze & trim not incl., casement	2 Sswk	200	.080	S.F.	26	2.45		28.45	33.50	
0100	Fixed casement	"	200	.080	"	9.55	2.45		12	15.05	
0010	**ALUMINUM WINDOWS** Incl. frame and glazing, grade C										204
1000	Stock units, casement, 3'-1" x 3'-2" opening	2 Sswk	10	1.600	Ea.	255	49		304	370	
3000	Single hung, 2' x 3' opening, enameled, standard glazed	↓	10	1.600	↓	123	49		172	225	
3100	Insulating glass	↓	10	1.600	↓	149	49		198	253	

085 | Metal Windows

		085 200 | Aluminum Windows	CREW	DAILY OUTPUT	LABOR-HOURS	UNIT	1999 BARE COSTS MAT.	LABOR	EQUIP.	TOTAL	TOTAL INCL O&P
204	3890	Awning type, 3′ x 3′ opening standard glass	2 Sswk	14	1.143	Ea.	300	35		335	395
	4000	Sliding aluminum, 3′ x 2′ opening, standard glazed	↓	10	1.600	↓	138	49		187	242
	5500	Sliding, with thermal barrier and screen, 6′ x 4′, 2 track	↓	8	2	↓	520	61		581	680
	5700	4 track	↓	8	2	↓	635	61		696	810

		085 700 | Screens	CREW	DAILY OUTPUT	LABOR-HOURS	UNIT	MAT.	LABOR	EQUIP.	TOTAL	TOTAL INCL O&P
701	0010	SCREENS									
	0020	For metal sash, aluminum or bronze mesh, flat screen	2 Sswk	1,200	.013	S.F.	2.91	.41		3.32	3.95
	0800	Security screen, aluminum frame with stainless steel cloth	"	1,200	.013	"	15.70	.41		16.11	18

8 DOORS & WINDOWS

086 | Wood & Plastic Windows

		086 100 | Wood Windows	CREW	DAILY OUTPUT	LABOR-HOURS	UNIT	1999 BARE COSTS MAT.	LABOR	EQUIP.	TOTAL	TOTAL INCL O&P
120	0010	**CASEMENT WINDOW** Including frame, screen, and exterior trim									
	0100	Avg. quality, bldrs. model, 2′-0″ x 3′-0″ H, standard glazed	1 Carp	10	.800	Ea.	158	22		180	208
	0200	2′-0″ x 4′-6″ high, standard glazed	↓	9	.889	↓	205	24.50		229.50	263
	0522	Vinyl clad, premium, insulating glass, 2′-0″ x 3′-0″	↓	10	.800	↓	192	22		214	246
	0525	2′-0″ x 5′-0″	↓	8	1	↓	262	27.50		289.50	330
	8100	Metal clad, deluxe, insulating glass, 2′-0″ x 3′-0″ high	↓	10	.800	↓	163	22		185	214
	8140	2′-0″ x 5′-0″ high	↓	8	1	↓	222	27.50		249.50	288
124	0010	**DOUBLE HUNG** Including frame, screen, and exterior trim									
	0100	Avg. quality, bldrs. model, 2′-0″ x 3′-0″ high, standard glazed	1 Carp	10	.800	Ea.	105	22		127	151
	0300	4′-0″ x 4′-6″ high, standard glazed	↓	8	1	↓	169	27.50		196.50	229
	1000	Vinyl clad, premium, insulating glass, 2′-6″ x 3′-0″	↓	10	.800	↓	237	22		259	296
	1400	3′-0″ x 5′-0″	↓	8	1	↓	335	27.50		362.50	415
	2000	Metal clad, deluxe, insulating glass, 2′-6″ x 3′-0″ high	↓	10	.800	↓	172	22		194	224
	2300	3′-0″ x 4′-6″ high	↓	9	.889	↓	235	24.50		259.50	296
140	0010	**SLIDING WINDOW** Including frame, screen, and exterior trim									
	0100	Average quality, bldrs. model, 3′-0″ x 3′-0″ high, standard glazed	1 Carp	10	.800	Ea.	126	22		148	174
	0200	4′-0″ x 3′-6″ high, standard glazed	↓	9	.889	↓	150	24.50		174.50	203
	1000	Vinyl clad, premium, insulating glass, 3′-0″ x 3′-0″	↓	10	.800	↓	455	22		477	535
	1100	5′-0″ x 4′-0″	↓	9	.889	↓	630	24.50		654.50	730
	2000	Metal clad, deluxe, insulating glass, 3′-0″ x 3′-0″ high	↓	10	.800	↓	256	22		278	315
	2050	4′-0″ x 3′-6″ high	↓	9	.889	↓	315	24.50		339.50	385

087 | Hardware

		087 100 | Finish Hardware	CREW	DAILY OUTPUT	LABOR-HOURS	UNIT	1999 BARE COSTS MAT.	LABOR	EQUIP.	TOTAL	TOTAL INCL O&P
103	0010	**BOLTS, FLUSH**									
	0020	Standard, concealed	1 Carp	7	1.143	Ea.	17.05	31		48.05	68

087 100 | Finish Hardware

	087 100 Finish Hardware	CREW	DAILY OUTPUT	LABOR-HOURS	UNIT	1999 BARE COSTS MAT.	LABOR	EQUIP.	TOTAL	TOTAL INCL O&P	
0800	Automatic fire exit	1 Carp	5	1.600	Ea.	265	43.50		308.50	360	103
1600	For electric release, add	1 Elec	3	2.667		91	85		176	227	
3000	Barrel, brass, 2" long	1 Carp	40	.200		2.10	5.45		7.55	10.90	
3020	4" long		40	.200		2.05	5.45		7.50	10.85	
3060	6" long	↓	40	.200	↓	4.13	5.45		9.58	13.15	
0010	**DEADLOCKS** Mortise, heavy duty, outside key	1 Carp	9	.889	Ea.	109	24.50		133.50	158	108
0020	Double cylinder		9	.889		120	24.50		144.50	171	
1000	Tubular, standard duty, outside key		10	.800		45.50	22		67.50	84.50	
1010	Double cylinder	↓	10	.800	↓	58.50	22		80.50	99	
0010	**DOORSTOPS** Holder and bumper, floor or wall	1 Carp	32	.250	Ea.	14	6.80		20.80	26	110
1300	Wall bumper, 4" diameter, with rubber pad, aluminum		32	.250		6.35	6.80		13.15	17.75	
1600	Door bumper, floor type, aluminum		32	.250		3.64	6.80		10.44	14.75	
1900	Plunger type, door mounted	↓	32	.250	↓	22.50	6.80		29.30	35.50	
0010	**HINGES** Full mortise, avg. freq., steel base, 4-1/2" x 4-1/2", USP				Pr.	17.55			17.55	19.30	116
0080	US10A					25.50			25.50	28	
0400	Brass base, 4-1/2" x 4-1/2", US10					36			36	40	
0480	US10B					48			48	52.50	
0950	Full mortise, high frequency, steel base, 3-1/2" x 3-1/2", US26D					25.50			25.50	28	
1000	4-1/2" x 4-1/2", USP					43.50			43.50	48	
1400	Brass base, 3-1/2" x 3-1/2", US4					38.50			38.50	42.50	
1430	4-1/2" x 4-1/2", US10				↓	39			39	43	
3000	Half surface, half mortise, or full surface, average frequency										
3010	Steel base, 4-1/2" x 4-1/2", USP				Pr.	36.50			36.50	40.50	
3400	Brass base, 4-1/2" x 4-1/2", US10				"	99			99	109	
0010	**LOCKSET** Standard duty, cylindrical, with sectional trim										120
0020	Non-keyed, passage	1 Carp	12	.667	Ea.	36	18.20		54.20	68	
0100	Privacy		12	.667		44	18.20		62.20	77	
0400	Keyed, single cylinder function		10	.800		64.50	22		86.50	105	
0500	Lever handled, keyed, single cylinder function		10	.800		87	22		109	131	
1700	Residential, interior door, minimum		16	.500		12.25	13.65		25.90	35	
1800	Exterior, minimum	↓	14	.571	↓	27.50	15.60		43.10	54.50	
0010	**MORTISE LOCKSET** Comm., wrought knobs & full escutcheon trim										125
0020	Non-keyed, passage, minimum	1 Carp	9	.889	Ea.	135	24.50		159.50	187	
0100	Keyed, office/entrance/apartment, minimum		8	1		165	27.50		192.50	225	
0110	Maximum	↓	7	1.143	↓	283	31		314	360	
2000	Cast knobs and full escutcheon trim										
2010	Non-keyed, passage, minimum	1 Carp	9	.889	Ea.	191	24.50		215.50	248	
2120	Keyed, single cylinder, typical, minimum		8	1		229	27.50		256.50	295	
2130	Maximum	↓	7	1.143	↓	360	31		391	450	
0010	**PANIC DEVICE** For rim locks, single door, exit only	1 Carp	6	1.333	Ea.	320	36.50		356.50	410	127
0020	Outside key and pull		5	1.600		370	43.50		413.50	475	
0200	Bar and vertical rod, exit only		5	1.600		475	43.50		518.50	590	
0600	Touch bar, exit only		6	1.333		380	36.50		416.50	475	
1000	Mortise, bar, exit only		4	2		425	54.50		479.50	550	
1600	Touch bar, exit only		4	2		485	54.50		539.50	620	
2000	Narrow stile, rim mounted, bar, exit only		6	1.333		510	36.50		546.50	620	
2010	Outside key and pull	↓	5	1.600	↓	555	43.50		598.50	680	
0010	**SPECIAL HINGES**										132
1500	Non template, full mortise, average frequency										
1510	Steel base, 4" x 4", USP				Pr.	26.50			26.50	29	
4600	Spring hinge, single acting, 6" flange, steel				Ea.	41			41	45.50	
4700	Brass				"	72.50			72.50	79.50	

087 | Hardware

		087 200 | Operators	CREW	DAILY OUTPUT	LABOR-HOURS	UNIT	1999 BARE COSTS MAT.	LABOR	EQUIP.	TOTAL	TOTAL INCL O&P
206	0010	**DOOR CLOSER** Rack and pinion	1 Carp	6.50	1.231	Ea.	105	33.50		138.50	169
	0020	Adjustable backcheck, 3 way mount, all sizes, regular arm		6	1.333		111	36.50		147.50	180
	4000	Backcheck, overhead concealed, all sizes, regular arm		5.50	1.455		140	39.50		179.50	217
	4040	Concealed arm		5	1.600		150	43.50		193.50	233
	4900	Floor concealed, all sizes, single acting		2.20	3.636		120	99.50		219.50	288
	4940	Double acting	↓	2.20	3.636	↓	155	99.50		254.50	325
		087 300 | Weatherstripping/Seals									
302	0010	**ASTRAGALS** One piece overlapping									
	2000	"L" extrusion, neoprene bulbs	1 Carp	75	.107	L.F.	1.47	2.91		4.38	6.20
	2400	Spring hinged security seal, with cam	"	75	.107	"	4.71	2.91		7.62	9.80
	3000	One piece stile protection									
	3020	Neoprene fabric loop, nail on aluminum strips	1 Carp	60	.133	L.F.	.49	3.64		4.13	6.30
	3600	Spring bronze strip, nail on type		105	.076		2.23	2.08		4.31	5.70
	4000	Extruded aluminum retainer, flush mount, pile insert	↓	105	.076	↓	1.76	2.08		3.84	5.20
	5000	Two piece overlapping astragal, extruded aluminum retainer									
	5010	Pile insert	1 Carp	60	.133	L.F.	2.29	3.64		5.93	8.25
	5500	Interlocking aluminum, 5/8" x 1" neoprene bulb insert	"	45	.178	"	2.84	4.85		7.69	10.75
304	0010	**THRESHOLD** 3' long door saddles, aluminum	1 Carp	48	.167	L.F.	3.35	4.55		7.90	10.85
	0100	Aluminum, 8" wide, 1/2" thick		12	.667	Ea.	28.50	18.20		46.70	60
	0700	Rubber, 1/2" thick, 5-1/2" wide		20	.400		28.50	10.90		39.40	48
	0800	2-3/4" wide	↓	20	.400	↓	13.15	10.90		24.05	31.50
306	0010	**WEATHERSTRIPPING** Window, double hung, 3' x 5', zinc	1 Carp	7.20	1.111	Opng.	10.60	30.50		41.10	59
	0500	As above but heavy duty, zinc		4.60	1.739		12.85	47.50		60.35	88.50
	1000	Doors, wood frame, interlocking, for 3' x 7' door, zinc		3	2.667		11.55	73		84.55	128
	1300	6' x 7' opening, zinc	↓	2	4	↓	12.60	109		121.60	186
	1700	Wood frame, spring type, bronze									
	1800	3' x 7' door	1 Carp	7.60	1.053	Opng.	15.30	28.50		43.80	62
		087 500 | Door/Window Acces.									
501	0010	**AREA WALL** Galvanized steel, 20 ga., 3'-2" wide, 1' deep	1 Sswk	29	.276	Ea.	12.10	8.45		20.55	29
	0100	2' deep		23	.348		21.50	10.65		32.15	43
	0300	16 ga., galv., 3'-2" wide, 1' deep		29	.276		17.05	8.45		25.50	34.50
	0400	3' deep		23	.348		34.50	10.65		45.15	57.50
	0600	Welded grating for above, 15 lbs., painted		45	.178		33	5.45		38.45	46
	0700	Galvanized	↓	45	.178	↓	62.50	5.45		67.95	78.50

088 | Glazing

		088 100 | Glass	CREW	DAILY OUTPUT	LABOR-HOURS	UNIT	1999 BARE COSTS MAT.	LABOR	EQUIP.	TOTAL	TOTAL INCL O&P
118	0010	**FLOAT GLASS** 3/16" thick, clear, plain	2 Glaz	130	.123	S.F.	3.35	3.27		6.62	8.65
	0200	Tempered, clear	"	130	.123	"	4	3.27		7.27	9.40
132	0010	**INSULATING GLASS** 2 lites 1/8" float, 1/2" thk, under 15 S.F.									
	0020	Clear	2 Glaz	95	.168	S.F.	5.80	4.48		10.28	13.20
	0200	2 lites 3/16" float, for 5/8" thk unit, 15 to 30 S.F., clear		90	.178		7	4.73		11.73	14.90
	0400	1" thk, dbl. glazed, 1/4" float, 30-70 S.F., clear	↓	75	.213	↓	9.95	5.65		15.60	19.55
136	0010	**LAMINATED GLASS** Clear float, .03" vinyl, 1/4" thick	2 Glaz	90	.178	S.F.	6.75	4.73		11.48	14.60
	0200	.06" vinyl, 1/2" thick	"	65	.246	"	12.80	6.55		19.35	24
144	0010	**MIRRORS** No frames, wall type, 1/4" plate glass, polished edge									
	0100	Up to 5 S.F.	2 Glaz	125	.128	S.F.	5.50	3.40		8.90	11.25

088 | Glazing

088 100 | Glass			CREW	DAILY OUTPUT	LABOR-HOURS	UNIT	1999 BARE COSTS MAT.	LABOR	EQUIP.	TOTAL	TOTAL INCL O&P	
144	1000	Float glass, up to 10 S.F., 1/8" thick	2 Glaz	160	.100	S.F.	3.36	2.66		6.02	7.75	144
	1100	3/16" thick	↓	150	.107	↓	3.91	2.84		6.75	8.60	

088 400 | Plastic Glazing			CREW	DAILY OUTPUT	LABOR-HOURS	UNIT	MAT.	LABOR	EQUIP.	TOTAL	TOTAL INCL O&P	
404	0010	**PLEXIGLASS ACRYLIC** Clear, masked, 1/8" thick, cut sheets	2 Glaz	170	.094	S.F.	2.76	2.50		5.26	6.85	404
	0200	Full sheets		195	.082		1.44	2.18		3.62	4.90	
	0900	3/8" thick, cut sheets		155	.103		8.95	2.75		11.70	14.05	
	1000	Full sheets	↓	180	.089	↓	4.82	2.36		7.18	8.90	

089 | Glazed Curtain Walls

089 200 | Glazed Curtain Wall			CREW	DAILY OUTPUT	LABOR-HOURS	UNIT	1999 BARE COSTS MAT.	LABOR	EQUIP.	TOTAL	TOTAL INCL O&P	
202	0010	**CURTAIN WALLS** Aluminum, stock, including glazing, minimum	H-1	205	.156	S.F.	18.70	4.46		23.16	28	202
	0150	Average, double glazed	"	180	.178	"	35.50	5.10		40.60	47.50	
204	0010	**TUBE FRAMING** For window walls and store fronts, aluminum, stock										204
	1000	Flush tube frame, mill finish, 1/4" glass, 1-3/4" x 4", open header	2 Glaz	80	.200	L.F.	7.40	5.30		12.70	16.20	
	1050	Open sill		82	.195		6.40	5.20		11.60	14.95	
	1100	Closed back header		83	.193		10.35	5.15		15.50	19.15	
	1200	Vertical mullion, one piece		75	.213		10.95	5.65		16.60	20.50	
	1300	90° or 180° vertical corner post		75	.213		18.40	5.65		24.05	29	
	5000	Flush tube frame, mill fin., thermal brk., 2-1/4"x 4-1/2", open header		74	.216		11.55	5.75		17.30	21.50	
	5050	Open sill		75	.213		10.10	5.65		15.75	19.75	
	5100	Vertical mullion, one piece		69	.232		13.95	6.15		20.10	25	
	5200	90° or 180° vertical corner post	↓	69	.232	↓	13.40	6.15		19.55	24	
206	0010	**WINDOW WALLS** Aluminum, stock, including glazing, minimum	H-2	160	.150	S.F.	24.50	3.73		28.23	32.50	206
	0050	Average		140	.171		29.50	4.27		33.77	39	
	0100	Maximum	↓	110	.218	↓	93	5.45		98.45	110	

For information about Means Estimating Seminars, see yellow pages 11 and 12 in back of book

or expanded coverage of these items see *Means Interior Cost Data 1999*

Division Notes

		CREW	DAILY OUTPUT	LABOR-HOURS	UNIT	1999 BARE COSTS MAT.	LABOR	EQUIP.	TOTAL	TOTAL INCL O&P

Division 9 Finishes

Estimating Tips

General

- Room Finish Schedule: A complete set of plans should contain a room finish schedule. If one is not available, it would be well worth the time and effort to put one together. A room finish schedule should contain the room number, room name (for clarity), floor materials, base materials, wainscot materials, wainscot height, wall materials (for each wall), ceiling materials and special instructions.
- Surplus Finishes: Review the specifications to determine if there is any requirement to provide certain amounts of extra materials for the owner's maintenance department. In some cases the owner may require a substantial amount of materials, especially when it is a special order item or long lead time item.

092 Lath, Plaster & Gypsum Board

- Lath is estimated by the square yard for both gypsum and metal lath, plus usually 5% allowance for waste. Furring, channels and accessories are measured by the linear foot. An extra foot should be allowed for each accessory miter or stop.
- Plaster is also estimated by the square yard. Deductions for openings vary by preference, from zero deduction to 50% of all openings over 2 feet in width. Some estimators deduct a percentage of the total yardage for openings. The estimator should allow one extra square foot for each linear foot of horizontal interior or exterior angle located below the ceiling level. Also, double the areas of small radius work.
- Each room should be measured, perimeter times maximum wall height. Ceiling areas are equal to length times width.
- Drywall accessories, studs, track, and acoustical caulking are all measured by the linear foot. Drywall taping is figured by the square foot. Gypsum wallboard is estimated by the square foot. No material deductions should be made for door or window openings under 32 S.F. Coreboard can be obtained in a 1″ thickness for solid wall and shaft work. Additions should be made to price out the inside or outside corners.
- Different types of partition construction should be listed separately on the quantity sheets. There may be walls with studs of various widths, double studded, and similar or dissimilar surface materials. Shaft work is usually different construction from surrounding partitions requiring separate quantities and pricing of the work.

093 Tile
094 Terrazzo

- Tile and terrazzo areas are taken off on a square foot basis. Trim and base materials are measured by the linear foot. Accent tiles are listed per each. Two basic methods of installation are used. Mud set is approximately 30% more expensive than the thin set. In terrazzo work, be sure to include the linear footage of embedded decorative strips, grounds, machine rubbing and power cleanup.

095 Acoustical Treatment & Wood Flooring

- Acoustical systems fall into several categories. The takeoff of these materials is by the square foot of area with a 5% allowance for waste. Do not forget about scaffolding, if applicable, when estimating these systems.
- Wood flooring is available in strip, parquet, or block configuration. The latter two types are set in adhesives with quantities estimated by the square foot. The laying pattern will influence labor costs and material waste. In addition to the material and labor for laying wood floors, the estimator must make allowances for sanding and finishing these areas unless the flooring is prefinished.

096 Flooring & Carpet

- Most of the various types of flooring are all measured on a square foot basis. Base is measured by the linear foot. If adhesive materials are to be quantified, they are estimated at a specified coverage rate by the gallon depending upon the specified type and the manufacturer's recommendations.
- Sheet flooring is measured by the square yard. Roll widths vary, so consideration should be given to use the most economical width, as waste must be figured into the total quantity. Consider also the installation methods available, direct glue down or stretched.

099 Painting & Wall Coverings

- Painting is one area where bids vary to a greater extent than almost any other section of a project. This arises from the many methods of measuring surfaces to be painted. The estimator should check the plans and specifications carefully to be sure of the required number of coats.
- Protection of adjacent surfaces is not included in painting costs. When considering the method of paint application, an important factor is the amount of protection and masking required. These must be estimated separately and may be the determining factor in choosing the method of application.
- Wall coverings are estimated by the square foot. The area to be covered is measured, length by height of wall above baseboards, to calculate the square footage of each wall. This figure is divided by the number of square feet in the single roll which is being used. Deduct, in full, the areas of openings such as doors and windows. Where a pattern match is required allow 25%-30% waste. One gallon of paste should be sufficient to hang 12 single rolls of light to medium weight paper.

Reference Numbers

Reference numbers are shown in bold squares at the beginning of some major classifications. These numbers refer to related items in the Reference Section. The reference information may be an estimating procedure, an alternate pricing method or technical information.

Note: Not all subdivisions listed here necessarily appear in this publication.

092 | Lath, Plaster & Gypsum Board

		092 100 | Gypsum Plaster	CREW	DAILY OUTPUT	LABOR-HOURS	UNIT	1999 BARE COSTS MAT.	LABOR	EQUIP.	TOTAL	TOTAL INCL O&P	
108	0010	**GYPSUM PLASTER** 80# bag, less than 1 ton (R092-105)				Bag	13			13	14.30	108
	0100	Over 1 ton				″	11.90			11.90	13.10	
	0300	2 coats, no lath included, on walls	J-1	105	.381	S.Y.	3.15	9.15	.40	12.70	17.95	
	0600	2 coats on and incl. 3/8″ gypsum lath on steel, on walls	J-2	97	.495		6.95	12.15	.44	19.54	26.50	
	0900	3 coats, no lath included, on walls	J-1	87	.460		4.32	11.05	.49	15.86	22	
	1200	3 coats on and including painted metal lath, on wood studs	J-2	86	.558		6.35	13.70	.49	20.54	28.50	

		092 150 | Veneer Plaster	CREW	DAILY OUTPUT	LABOR-HOURS	UNIT	MAT.	LABOR	EQUIP.	TOTAL	TOTAL INCL O&P	
154	0010	**THIN COAT** Plaster, 1 coat veneer, not incl. lath	J-1	3,600	.011	S.F.	.10	.27	.01	.38	.53	154
	1000	In 50 lb. bags				Bag	7.75			7.75	8.55	

		092 300 | Aggregate Coatings	CREW	DAILY OUTPUT	LABOR-HOURS	UNIT	MAT.	LABOR	EQUIP.	TOTAL	TOTAL INCL O&P	
304	0010	**STUCCO**, 3 coats 1″ thick, float finish, with mesh, on wood frame	J-2	135	.356	S.Y.	5.40	8.70	.31	14.41	19.60	304
	0100	On masonry construction, no mesh incl.	J-1	200	.200		2.04	4.81	.21	7.06	9.85	
	0300	For trowel finish, add	1 Plas	170	.047			1.21		1.21	1.85	
	1000	Exterior stucco, with bonding agent, 3 coats, on walls, no mesh incl.	J-1	200	.200		3.25	4.81	.21	8.27	11.20	

		092 600 | Gypsum Board Systems	CREW	DAILY OUTPUT	LABOR-HOURS	UNIT	MAT.	LABOR	EQUIP.	TOTAL	TOTAL INCL O&P	
608	0010	**DRYWALL** Gypsum plasterboard, nailed or screwed to studs										608
	0150	3/8″ thick, on walls, standard, no finish included	2 Carp	2,000	.008	S.F.	.16	.22		.38	.52	
	0300	1/2″ thick, on walls, standard, no finish included		2,000	.008		.15	.22		.37	.51	
	0350	Taped and finished		965	.017		.24	.45		.69	.98	
	2000	5/8″ thick, on walls, standard, no finish included		2,000	.008		.24	.22		.46	.60	
	2200	Water resistant, no finish included		2,000	.008		.29	.22		.51	.66	
	4000	Fireproofing, beams or columns, 2 layers, 1/2″ thick, incl finish		330	.048		.57	1.32		1.89	2.71	
	4050	5/8″ thick		300	.053		.60	1.46		2.06	2.95	
620	0010	**PARTITION WALL** Stud wall, 8′ to 12′ high										620
	3200	5/8″, interior, gypsum drywall, standard, taped both sides										
	3400	Installed on and including 2″ x 4″ wood studs, 16″ O.C.	2 Carp	300	.053	S.F.	1.06	1.46		2.52	3.45	
	3800	Metal studs, NLB, 25 ga., 16″ O.C., 3-5/8″ wide		340	.047		.84	1.28		2.12	2.95	
	4800	Water resistant, on 2″ x 4″ wood studs, 16″ O.C.		300	.053		1.17	1.46		2.63	3.57	
	5200	Metal studs, NLB, 25 ga. 16″ O.C., 3-5/8″ wide		340	.047		.95	1.28		2.23	3.07	
	6000	Fire res., 2 layers, 2 hr., on 2″ x 4″ wood studs, 16″ O.C.		205	.078		1.33	2.13		3.46	4.82	
	6400	Metal studs, NLB, 25 ga., 16″ O.C., 3-5/8″ wide		245	.065		1.21	1.78		2.99	4.13	
622	0010	**PLASTER PARTITION WALL**										622
	0400	Stud walls, 3.4 lb. metal lath, 3 coat gypsum plaster, 2 sides										
	0600	2″ x 4″ wood studs, 16″ O.C.	J-2	315	.152	S.F.	2.29	3.74	.13	6.16	8.35	
	0700	2-1/2″ metal studs, 25 ga., 12″ O.C.	″	325	.148	″	2.12	3.62	.13	5.87	8	
	0900	Gypsum lath, 2 coat vermiculite plaster, 2 sides										
	1000	2″ x 4″ wood studs, 16″ O.C.	J-2	355	.135	S.F.	2.55	3.32	.12	5.99	8	
	1200	2-1/2″ metal studs, 25 ga., 12″ O.C.	″	365	.132	″	2.26	3.22	.12	5.60	7.55	

093 | Tile

		093 100 | Ceramic Tile	CREW	DAILY OUTPUT	LABOR-HOURS	UNIT	1999 BARE COSTS MAT.	LABOR	EQUIP.	TOTAL	TOTAL INCL O&P	
102	0010	**CERAMIC TILE**										102
	3000	Floors, natural clay, random or uniform, thin set, color group 1	D-7	183	.087	S.F.	3.55	2.10		5.65	7	
	3100	Color group 2		183	.087		3.82	2.10		5.92	7.30	
	3300	Porcelain type, 1 color, color group 2, 1″ x 1″		183	.087		4.11	2.10		6.21	7.65	

093 | Tile

093 100 | Ceramic Tile

		CREW	DAILY OUTPUT	LABOR-HOURS	UNIT	1999 BARE COSTS MAT.	LABOR	EQUIP.	TOTAL	TOTAL INCL O&P	
3400	2″ x 2″ or 2″ x 1″, thin set	D-7	190	.084	S.F.	4.34	2.03		6.37	7.75	102
5400	Walls, interior, thin set, 4-1/4″ x 4-1/4″ tile		190	.084		2.21	2.03		4.24	5.40	
5700	8-1/2″ x 4-1/4″ tile		190	.084		3.19	2.03		5.22	6.50	
6300	Exterior walls, frostproof, mud set, 4-1/4″ x 4-1/4″		102	.157		4.02	3.77		7.79	10	
6400	1-3/8″ x 1-3/8″	↓	93	.172	↓	3.71	4.14		7.85	10.20	

093 300 | Quarry Tile

		CREW	DAILY OUTPUT	LABOR-HOURS	UNIT	MAT.	LABOR	EQUIP.	TOTAL	TOTAL INCL O&P	
0010	**QUARRY TILE** Base, cove or sanitary, 2″ or 5″ high, mud set										304
0100	1/2″ thick	D-7	110	.145	L.F.	3.51	3.50		7.01	9	
0700	Floors, mud set, 1,000 S.F. lots, red, 4″ x 4″ x 1/2″ thick		120	.133	S.F.	3.67	3.21		6.88	8.80	
0900	6″ x 6″ x 1/2″ thick	↓	140	.114	″	2.65	2.75		5.40	7	

094 | Terrazzo

094 100 | Portland Cem. Terrazzo

		CREW	DAILY OUTPUT	LABOR-HOURS	UNIT	1999 BARE COSTS MAT.	LABOR	EQUIP.	TOTAL	TOTAL INCL O&P	
0010	**TILE OR TERRAZZO BASE** Scratch coat only	1 Mstz	150	.053	S.F.	.30	1.42		1.72	2.42	108
0500	Scratch and brown coat only	″	75	.107	″	.55	2.83		3.38	4.80	

095 | Acoustical Treatment & Wood Flooring

095 100 | Acoustical Ceilings

		CREW	DAILY OUTPUT	LABOR-HOURS	UNIT	1999 BARE COSTS MAT.	LABOR	EQUIP.	TOTAL	TOTAL INCL O&P	
0010	**SUSPENDED CEILINGS, COMPLETE** Including standard										106
0100	suspension system but not incl. 1-1/2″ carrier channels										
0600	Fiberglass ceiling board, 2′ x 4′ x 5/8″, plain faced,	1 Carp	500	.016	S.F.	.99	.44		1.43	1.78	
1800	Tile, Z bar suspension, 5/8″ mineral fiber tile	″	150	.053	″	1.43	1.46		2.89	3.86	

096 | Flooring & Carpet

096 350 | Brick Flooring

		CREW	DAILY OUTPUT	LABOR-HOURS	UNIT	1999 BARE COSTS MAT.	LABOR	EQUIP.	TOTAL	TOTAL INCL O&P	
0010	**FLOORING**										354
0020	Acid proof shales, red, 8″ x 3-3/4″ x 1-1/4″ thick	D-7	.43	37.209	M	695	895		1,590	2,100	
0050	2-1/4″ thick	D-1	.40	40		755	985		1,740	2,350	
0200	Acid proof clay brick, 8″ x 3-3/4″ x 2-1/4″ thick	″	.40	40	↓	755	985		1,740	2,350	
0260	Cast ceramic, pressed, 4″ x 8″ x 1/2″, unglazed	D-7	100	.160	S.F.	4.94	3.85		8.79	11.15	
0270	Glazed		100	.160		6.60	3.85		10.45	12.95	
0280	Hand molded flooring, 4″ x 8″ x 3/4″, unglazed		95	.168		6.55	4.05		10.60	13.20	
0290	Glazed	↓	95	.168	↓	8.20	4.05		12.25	15	

096 | Flooring & Carpet

		096 350 | Brick Flooring	CREW	DAILY OUTPUT	LABOR-HOURS	UNIT	1999 BARE COSTS MAT.	LABOR	EQUIP.	TOTAL	TOTAL INCL O&P
354	0300	8" hexagonal, 3/4" thick, unglazed	D-7	85	.188	S.F.	7.15	4.53		11.68	14.60
	0310	Glazed	↓	85	.188		12.95	4.53		17.48	21
	0400	Heavy duty industrial, cement mortar bed, 2" thick, not incl. brick	D-1	80	.200		.71	4.92		5.63	8.45
	0450	Acid proof joints, 1/4" wide	"	65	.246		1.13	6.05		7.18	10.65
	0500	Pavers, 8" x 4", 1" to 1-1/4" thick, red	D-7	95	.168		2.88	4.05		6.93	9.15
	0510	Ironspot	"	95	.168		4.07	4.05		8.12	10.50
	0540	1-3/8" to 1-3/4" thick, red	D-1	95	.168		2.78	4.14		6.92	9.50
	0560	Ironspot		95	.168		4.02	4.14		8.16	10.85
	0580	2-1/4" thick, red		90	.178		2.83	4.37		7.20	9.90
	0590	Ironspot	↓	90	.178	↓	4.38	4.37		8.75	11.60
	0800	For sidewalks and patios with pavers, see division 025-158									
	0870	For epoxy joints, add	D-1	600	.027	S.F.	2.15	.66		2.81	3.39
	0880	For Furan underlayment, add	"	600	.027		1.78	.66		2.44	2.98
	0890	For waxed surface, steam cleaned, add	D-5	1,000	.008	↓	.15	.22		.37	.51

		096 600 | Resilient Flooring	CREW	DAILY OUTPUT	LABOR-HOURS	UNIT	MAT.	LABOR	EQUIP.	TOTAL	TOTAL INCL O&P
601	0010	**RESILIENT FLOORING**									
	0800	Base, cove, rubber or vinyl, .080" thick									
	1100	Standard colors, 2-1/2" high	1 Tilf	315	.025	L.F.	.39	.68		1.07	1.43
	6700	Synthetic turf, 3/8" thick	"	90	.089	S.F.	3.16	2.37		5.53	7
	6750	Interlocking 2' x 2' squares, 1/2" thick, not									
	6810	cemented, for playgrounds, minimum	1 Tilf	210	.038	S.F.	2.61	1.02		3.63	4.37
	6850	Maximum		190	.042		6.70	1.12		7.82	9
	7000	Vinyl composition tile, 12" x 12", 1/16" thick		500	.016		.68	.43		1.11	1.38
	7350	1/8" thick, marbleized		500	.016		.98	.43		1.41	1.71
	7500	Vinyl tile, 12" x 12", .050" thick, minimum		500	.016		1.54	.43		1.97	2.32
	7550	Maximum	↓	500	.016	↓	3.01	.43		3.44	3.94

		096 780 | Resilient Accessories	CREW	DAILY OUTPUT	LABOR-HOURS	UNIT	MAT.	LABOR	EQUIP.	TOTAL	TOTAL INCL O&P
781	0010	**STAIR TREADS AND RISERS** See index for materials other									
	0100	than rubber and vinyl									
	0300	Rubber, molded tread, 12" wide, 5/16" thick, black	1 Tilf	115	.070	L.F.	6.55	1.85		8.40	10
	1800	Risers, 7" high, 1/8" thick, flat		250	.032		3.26	.85		4.11	4.85
	2100	Vinyl, molded tread, 12" wide, colors, 1/8" thick		115	.070		2.63	1.85		4.48	5.65
	2400	Riser, 7" high, 1/8" thick, coved	↓	175	.046	↓	1.61	1.22		2.83	3.57

099 | Painting & Wall Coverings

		099 100 | Exterior Painting	CREW	DAILY OUTPUT	LABOR-HOURS	UNIT	1999 BARE COSTS MAT.	LABOR	EQUIP.	TOTAL	TOTAL INCL O&P
106	0010	**SIDING** Exterior, Alkyd (oil base)									
	0800	Paint 2 coats, brushwork	2 Pord	1,300	.012	S.F.	.10	.31		.41	.58
	1200	Stucco, rough, oil base, paint 2 coats, brushwork		1,300	.012		.10	.31		.41	.58
	1400	Roller		1,625	.010		.10	.25		.35	.49
	1800	Texture 1-11 or clapboard, oil base, primer coat, brushwork		1,300	.012		.08	.31		.39	.55
	2400	Paint 2 coats, brushwork	↓	810	.020		.14	.49		.63	.90
	8000	For latex paint, deduct				↓	10%				

099 200 | Interior Painting

	099 200 Interior Painting	CREW	DAILY OUTPUT	LABOR-HOURS	UNIT	1999 BARE COSTS MAT.	LABOR	EQUIP.	TOTAL	TOTAL INCL O&P	
0010	**COATINGS & PAINTS** In 5 gallon lots										208
0050	For 100 gallons or more, deduct					10%				10%	
0100	Paint, Exterior alkyd (oil base)										
0200	Flat				Gal.	21			21	23	
0300	Gloss					19.50			19.50	21.50	
0400	Primer					19.25			19.25	21	
0500	Latex (water base)										
0600	Acrylic stain				Gal.	16.75			16.75	18.45	
0700	Gloss enamel					22			22	24	
0800	Flat					17			17	18.70	
0900	Primer					18			18	19.80	
1000	Semi-gloss					21			21	23	
2400	Masonry, Exterior										
2500	Alkali resistant primer				Gal.	19			19	21	
2600	Block filler, epoxy					20			20	22	
2700	Latex					9.75			9.75	10.75	
2800	Latex, flat					16.75			16.75	18.45	
2900	Semi-gloss					16.75			16.75	18.45	
4000	Metal										
4100	Galvanizing paint				Gal.	24.50			24.50	26.50	
4200	High heat					32			32	35	
4300	Heat resistant					21.50			21.50	23.50	
4400	Machinery enamel, alkyd					20.50			20.50	22.50	
4500	Metal pretreatment (polyvinyl butyral)					23			23	25.50	
4600	Rust inhibitor, ferrous metal					21.50			21.50	23.50	
4700	Zinc chromate					19.50			19.50	21.50	
4800	Zinc rich primer					47.50			47.50	52.50	
5500	Coatings										
5600	Heavy duty										
5700	Acrylic urethane				Gal.	50.50			50.50	55.50	
5800	Chlorinated rubber					29.50			29.50	32.50	
5900	Coal tar epoxy					22			22	24	
6000	Polyamide epoxy, finish					31			31	34	
6100	Primer					31			31	34	
6200	Silicone alkyd					31.50			31.50	34.50	
6300	2 component solvent based acrylic epoxy					32.50			32.50	36	
6400	Polyester epoxy					41			41	45	
6500	Vinyl					26.50			26.50	29	
6600	Special/Miscellaneous										
6700	Aluminum				Gal.	22.50			22.50	25	
6900	Dry fall out, flat					12			12	13.20	
7000	Fire retardant, intumescent					36			36	39.50	
7100	Linseed oil					10			10	11	
7200	Shellac					18			18	19.80	
7300	Swimming pool, epoxy or urethane base					36			36	39.50	
7400	Rubber base					25			25	27	
7500	Texture paint					11			11	12.10	
7600	Turpentine					10.50			10.50	11.55	
7700	Water repellent, 5% silicone					17.50			17.50	19.25	
0010	**MISCELLANEOUS PAINTING** R099-100										220
2400	Floors, conc./wood, oil base, primer/sealer coat, brushwork	2 Pord	1,950	.008	S.F.	.05	.20		.25	.36	
2450	Roller	"	5,200	.003		.05	.08		.13	.18	
3800	Grilles, per side, oil base, primer coat, brushwork	1 Pord	520	.015		.08	.38		.46	.68	
3900	Spray	"	1,140	.007		.10	.17		.27	.39	
5000	Pipe, to 4" diameter, primer or sealer coat, oil base, brushwork	2 Pord	1,250	.013	L.F.	.05	.32		.37	.55	

or expanded coverage of these items see *Means Interior Cost Data 1999*

099 200 | Interior Painting

			CREW	DAILY OUTPUT	LABOR-HOURS	UNIT	1999 BARE COSTS MAT.	LABOR	EQUIP.	TOTAL	TOTAL INCL O&P
220	5350	Paint 2 coats, brushwork (R099-100)	2 Pord	775	.021	L.F.	.09	.51		.60	.89
	6300	To 16" diameter, primer or sealer coat, brushwork		310	.052		.20	1.29		1.49	2.19
	6500	Paint 2 coats, brushwork		195	.082		.37	2.04		2.41	3.54
	8900	Trusses and wood frames, primer coat, oil base, brushwork	1 Pord	800	.010	S.F.	.04	.25		.29	.43
	9220	Paint 2 coats, brushwork		500	.016		.09	.40		.49	.71
	9260	Stain, brushwork, wipe off		600	.013		.04	.33		.37	.55
	9280	Varnish, 3 coats, brushwork		275	.029		.16	.72		.88	1.28
224	0010	**WALLS AND CEILINGS** (R099-100)									
	0100	Concrete, dry wall or plaster, oil base, primer or sealer coat									
	0200	Smooth finish, brushwork	1 Pord	1,150	.007	S.F.	.04	.17		.21	.32
	0800	Paint 2 coats, smooth finish, brushwork		680	.012		.07	.29		.36	.53
	0880	Spray		1,625	.005		.06	.12		.18	.26

For information about Means Estimating Seminars, see yellow pages 11 and 12 in back of book

Division 10 Specialties

'stimating Tips

eneral

The items in this division are usually priced per square foot or each.

Many items in Division 10 require some type of support system that is not usually furnished with the item. Examples of these systems include blocking for the attachment of grab bars and support angles for ceiling hung toilet partitions. The required blocking or supports must be added to the estimate in the appropriate division.

Some items in Division 10, such as lockers, may require assembly before installation. Verify the amount of assembly required. Assembly can often exceed installation time.

101 Visual Display Boards, Compartments & Cubicles

- Toilet partitions are priced by the stall. A stall consists of a side wall, pilaster and door with hardware. Toilet tissue holders and grab bars are extra.

106 Partitions & Storage Shelving

- The required acoustical rating of a folding partition can have a significant impact on costs. Verify the sound transmission coefficient rating of the panel priced to the specification requirements.

Reference Numbers

Reference numbers are shown in bold squares at the beginning of some major classifications. These numbers refer to related items in the Reference Section. The reference information may be an estimating procedure, an alternate pricing method or technical information.

Note: Not all subdivisions listed here necessarily appear in this publication.

102 | Louvers, Corner Protection & Access Flooring

102 600 | Wall & Corner Guards

			CREW	DAILY OUTPUT	LABOR-HOURS	UNIT	1999 BARE COSTS MAT.	LABOR	EQUIP.	TOTAL	TOTAL INCL O&P
604	0010	**CORNER GUARDS** Steel angle w/anchors, 1" x 1" x 1/4", 1.5#/L.F.	2 Carp	160	.100	L.F.	2.77	2.73		5.50	7.35
	0100	2" x 2" x 1/4" angles, 3.2#/L.F.		150	.107		6.50	2.91		9.41	11.75
	0200	3" x 3" x 5/16" angles, 6.1#/L.F.		140	.114		8.45	3.12		11.57	14.20
	0300	4" x 4" x 5/16" angles, 8.2#/L.F.	↓	120	.133		12.45	3.64		16.09	19.40
	0350	For angles drilled and anchored to masonry, add					15%	120%			
	0370	Drilled and anchored to concrete, add				↓	20%	170%			
	0400	For galvanized angles, add					35%				
	0450	For stainless steel angles, add				L.F.	100%				
	0500	Steel door track/wheel guards, 4' - 0" high	E-4	22	1.455	Ea.	129	45	3.74	177.74	229
	0800	Pipe bumper for truck doors, 8' long, 6" diameter, filled		20	1.600		179	50	4.11	233.11	293
	0900	8" diameter	↓	20	1.600	↓	275	50	4.11	329.11	400

103 | Fireplaces, Exterior Specialties & Flagpoles

103 520 | Ground Set Flagpoles

			CREW	DAILY OUTPUT	LABOR-HOURS	UNIT	1999 BARE COSTS MAT.	LABOR	EQUIP.	TOTAL	TOTAL INCL O&P
524	0010	**FLAGPOLE**									
	0050	Not including base or foundation									
	0100	Aluminum, tapered, ground set 20' high	K-1	2	8	Ea.	590	196	90	876	1,050
	0200	25' high		1.70	9.412		710	231	106	1,047	1,250
	0300	30' high		1.50	10.667		920	262	120	1,302	1,525
	0400	35' high		1.40	11.429		1,275	280	128	1,683	1,975
	0500	40' high		1.20	13.333		1,675	325	150	2,150	2,525
	0600	50' high		1	16		2,625	390	180	3,195	3,675
	0700	60' high		.90	17.778		4,125	435	199	4,759	5,450
	0800	70' high		.80	20		5,625	490	224	6,339	7,200
	1100	Counterbalanced, internal halyard, 20' high		1.80	8.889		1,875	218	99.50	2,192.50	2,500
	1200	30' high		1.50	10.667		2,425	262	120	2,807	3,200
	1300	40' high		1.30	12.308		3,275	300	138	3,713	4,225
	1400	50' high		1	16		5,775	390	180	6,345	7,150
	2820	Aluminum, electronically operated, 30' high		1.40	11.429		3,800	280	128	4,208	4,750
	2840	35' high		1.30	12.308		3,975	300	138	4,413	5,000
	2860	39' high		1.10	14.545		4,500	355	163	5,018	5,675
	2880	45' high		1	16		5,625	390	180	6,195	6,975
	2900	50' high		.90	17.778		6,175	435	199	6,809	7,700
	3000	Fiberglass, tapered, ground set, 23' high		2	8		835	196	90	1,121	1,325
	3100	29"-7" high		1.50	10.667		1,225	262	120	1,607	1,850
	3200	36'-1" high		1.40	11.429		1,475	280	128	1,883	2,200
	3300	39'-5" high		1.20	13.333		1,575	325	150	2,050	2,425
	3400	49'-2" high		1	16		4,000	390	180	4,570	5,200
	3500	59' high	↓	.90	17.778	↓	5,125	435	199	5,759	6,525
	4300	Concrete, direct imbedded installation									
	4400	Internal halyard, 20' high	K-1	2.50	6.400	Ea.	1,200	157	72	1,429	1,650
	4500	25' high		2.50	6.400		1,350	157	72	1,579	1,825
	4600	30' high		2.30	6.957		1,575	171	78	1,824	2,075
	4700	40' high		2.10	7.619		1,900	187	85.50	2,172.50	2,450
	4800	50' high		1.90	8.421		3,400	206	94.50	3,700.50	4,175
	5000	60' high		1.80	8.889		5,300	218	99.50	5,617.50	6,300
	5100	70' high		1.60	10		6,375	245	112	6,732	7,500
	5200	80' high	↓	1.40	11.429	↓	8,275	280	128	8,683	9,675

103 | Fireplaces, Exterior Specialties & Flagpoles

103 520 | Ground Set Flagpoles

			CREW	DAILY OUTPUT	LABOR-HOURS	UNIT	1999 BARE COSTS MAT.	LABOR	EQUIP.	TOTAL	TOTAL INCL O&P	
524	5300	90′ high	K-1	1.20	13.333	Ea.	10,900	325	150	11,375	12,700	524
	5500	100′ high	↓	1	16	↓	11,300	390	180	11,870	13,200	
	6400	Wood poles, tapered, clear vertical grain fir with tilting										
	6410	base, not incl. foundation, 4″ butt, 25′ high	K-1	1.90	8.421	Ea.	465	206	94.50	765.50	940	
	6800	6″ butt, 30′ high	″	1.30	12.308	″	580	300	138	1,018	1,250	
	7300	Foundations for flagpoles, including										
	7400	excavation and concrete, to 35′ high poles	C-1	10	3.200	Ea.	390	82.50		472.50	555	
	7600	40′ to 50′ high		3.50	9.143		705	236		941	1,150	
	7700	Over 60′ high	↓	2	16	↓	790	415		1,205	1,525	

103 540 | Wall-Mounted Flagpoles

			CREW	DAILY OUTPUT	LABOR-HOURS	UNIT	1999 BARE COSTS MAT.	LABOR	EQUIP.	TOTAL	TOTAL INCL O&P	
544	0010	**FLAGPOLE** Structure mounted										544
	0100	Fiberglass, vertical wall set, 19′-8″ long	K-1	1.50	10.667	Ea.	1,000	262	120	1,382	1,625	
	0200	23′ long		1.40	11.429		960	280	128	1,368	1,625	
	0300	26′-3″ long		1.30	12.308		1,425	300	138	1,863	2,200	
	0800	19′-8″ long outrigger		1.30	12.308		1,125	300	138	1,563	1,875	
	1300	Aluminum, vertical wall set, tapered, with base, 20′ high		1.20	13.333		725	325	150	1,200	1,475	
	1400	29′-6″ high		1	16		1,550	390	180	2,120	2,525	
	2400	Outrigger poles with base, 12′ long		1.30	12.308		1,100	300	138	1,538	1,825	
	2500	14′ long	↓	1	16	↓	1,125	390	180	1,695	2,025	

104 | Identifying & Pedestrian Control Devices

104 100 | Directories

			CREW	DAILY OUTPUT	LABOR-HOURS	UNIT	1999 BARE COSTS MAT.	LABOR	EQUIP.	TOTAL	TOTAL INCL O&P	
104	0010	**DIRECTORY BOARDS**										104
	0900	Outdoor, weatherproof, black plastic, 36″ x 24″	2 Carp	2	8	Ea.	300	218		518	675	
	1000	36″ x 36″	″	1.50	10.667	″	635	291		926	1,150	

104 300 | Signs

			CREW	DAILY OUTPUT	LABOR-HOURS	UNIT	1999 BARE COSTS MAT.	LABOR	EQUIP.	TOTAL	TOTAL INCL O&P	
304	0010	**SIGNS** Letters, 2″ high, 3/8″ deep, cast bronze	1 Carp	24	.333	Ea.	15.35	9.10		24.45	31	304
	0140	1/2″ deep, cast aluminum		18	.444		15.35	12.15		27.50	36	
	0160	Cast bronze		32	.250		21.50	6.80		28.30	35	
	0300	6″ high, 5/8″ deep, cast aluminum		24	.333		17.15	9.10		26.25	33	
	0400	Cast bronze		24	.333		42.50	9.10		51.60	61	
	0600	8″ high, 3/4″ deep, cast aluminum		14	.571		27.50	15.60		43.10	55	
	0700	Cast bronze		20	.400		64	10.90		74.90	87.50	
	0900	10″ high, 1″ deep, cast aluminum		18	.444		37.50	12.15		49.65	60	
	1000	Bronze		18	.444		71	12.15		83.15	97	
	1200	12″ high, 1-1/4″ deep, cast aluminum		12	.667		45	18.20		63.20	78	
	1500	Cast bronze		18	.444		103	12.15		115.15	132	
	1600	14″ high, 2-5/16″ deep, cast aluminum		12	.667		57	18.20		75.20	91.50	
	1800	Fabricated stainless steel, 6″ high, 2″ deep		20	.400		89.50	10.90		100.40	116	
	1900	12″ high, 3″ deep		18	.444		106	12.15		118.15	136	
	2100	18″ high, 3″ deep		12	.667		160	18.20		178.20	205	
	2200	24″ high, 4″ deep		10	.800		223	22		245	281	
	2700	Acrylic, on high density foam,		20	.400		17.80	10.90		28.70	37	
	2800	12″ high, 2″ deep	↓	18	.444	↓	39.50	12.15		51.65	62.50	

104 | Identifying & Pedestrian Control Devices

		104 300 | Signs	CREW	DAILY OUTPUT	LABOR-HOURS	UNIT	1999 BARE COSTS MAT.	LABOR	EQUIP.	TOTAL	TOTAL INCL O&P	
304	3900	Plaques, custom, 20" x 30", for up to 450 letters, cast aluminum	2 Carp	4	4	Ea.	740	109		849	980	304
	4000	Cast bronze		4	4		980	109		1,089	1,250	
	4200	30" x 36", up to 900 letters cast aluminum		3	5.333		1,475	146		1,621	1,850	
	4300	Cast bronze		3	5.333		1,925	146		2,071	2,325	
	4500	36" x 48", for up to 1300 letters, cast bronze		2	8		2,900	218		3,118	3,550	
	4800	Signs, reflective alum. street signs, dbl. face, 2-way, w/bracket		30	.533		39	14.55		53.55	66	
	4900	4-way		30	.533		69.50	14.55		84.05	99.50	
	5100	Exit signs, 24 ga. alum., 14" x 12" surface mounted	1 Carp	30	.267		12.70	7.30		20	25.50	
	5200	10" x 7"		20	.400		7.75	10.90		18.65	26	
	5400	Bracket mounted, double face, 12" x 10"		30	.267		22	7.30		29.30	36	
	5500	Sticky back, stock decals, 14" x 10"	1 Clab	50	.160		4.65	3.43		8.08	10.50	
	6000	Interior elec., wall mount, fiberglass panels, 2 lamps, 6"	1 Elec	8	1		47	32		79	99	
	6100	8"	"	8	1		39.50	32		71.50	91	
	6400	Replacement sign faces, 6" or 8"	1 Clab	50	.160		21	3.43		24.43	28.50	

		104 560 | Turnstiles	CREW	DAILY OUTPUT	LABOR-HOURS	UNIT	MAT.	LABOR	EQUIP.	TOTAL	TOTAL INCL O&P	
561	0010	**TURNSTILES** One way, 4 arm, 46" diameter, economy, manual	2 Carp	5	3.200	Ea.	249	87.50		336.50	410	561
	0100	Electric		1.20	13.333		835	365		1,200	1,500	
	0420	Three arm, 24" opening, light duty, manual		2	8		585	218		803	990	
	0450	Heavy duty		1.50	10.667		600	291		891	1,125	
	0460	Manual, with registering & controls, light duty		2	8		680	218		898	1,100	
	0470	Heavy duty		1.50	10.667		1,200	291		1,491	1,775	

105 | Lockers, Protective Covers & Postal Specialties

		105 200 | Fire Protect. Specialties	CREW	DAILY OUTPUT	LABOR-HOURS	UNIT	1999 BARE COSTS MAT.	LABOR	EQUIP.	TOTAL	TOTAL INCL O&P	
225	0010	**FIRE EXTINGUISHERS**										225
	0120	CO2, portable with swivel horn, 5 lb.				Ea.	101			101	111	
	0140	With hose and "H" horn, 10 lb.					150			150	165	
	0160	15 lb.					172			172	189	
	0360	Wheeled type, cart mounted, 50 lb.					1,200			1,200	1,325	
	1000	Dry chemical, pressurized										
	1040	Standard type, portable, painted, 2-1/2 lb.				Ea.	25			25	27.50	
	1060	5 lb.					40			40	44	
	1080	10 lb.					65			65	71.50	
	1100	20 lb.					90			90	99	
	1120	30 lb.					145			145	160	
	1300	Standard type, wheeled, 150 lb.					1,400			1,400	1,550	
	2000	ABC all purpose type, portable, 2-1/2 lb.					25			25	27.50	
	2060	5 lb.					40			40	44	
	2080	9-1/2 lb.					60			60	66	
	2100	20 lb.					85			85	93.50	
	2300	Wheeled, 45 lb.					650			650	715	
	2360	150 lb.					1,450			1,450	1,600	
	3000	Dry chemical, outside cartridge to -65°F, painted, 9 lb.					200			200	220	
	3060	26 lb.					250			250	275	
	5000	Pressurized water, 2-1/2 gallon, stainless steel					75			75	82.50	
	5060	With anti-freeze					110			110	121	
	6000	Soda & acid, 2-1/2 gallon, stainless steel					75			75	82.50	
	9400	Installation of extinguishers, 12 or more, on wood	1 Carp	30	.267			7.30		7.30	11.45	

105 | Lockers, Protective Covers & Postal Specialties

105 200 | Fire Protect. Specialties

		CREW	DAILY OUTPUT	LABOR-HOURS	UNIT	1999 BARE COSTS MAT.	LABOR	EQUIP.	TOTAL	TOTAL INCL O&P	
225	9420	On masonry or concrete	1 Carp	15	.533	Ea.		14.55		14.55	23

105 380 | Canopies

			CREW	DAILY OUTPUT	LABOR-HOURS	UNIT	MAT.	LABOR	EQUIP.	TOTAL	TOTAL INCL O&P
384	0010	**CANOPIES** Wall hung, .032", aluminum, prefinished, 8' x 10'	K-2	1.30	18.462	Ea.	1,250	525	138	1,913	2,450
	0300	8' x 20'		1.10	21.818		2,500	620	163	3,283	4,000
	0500	10' x 10'		1.30	18.462		1,400	525	138	2,063	2,625
	0700	10' x 20'		1.10	21.818		2,600	620	163	3,383	4,100
	1000	12' x 20'		1	24		3,200	680	180	4,060	4,925
	1360	12' x 30'		.80	30		4,700	850	224	5,774	6,925
	1700	12' x 40'	↓	.60	40		5,000	1,125	299	6,424	7,825
	1900	For free standing units, add				↓	20%	10%			
	4700	Canvas awnings, including canvas, frame & lettering									
	5000	Minimum	2 Carp	100	.160	S.F.	36	4.37		40.37	46.50
	5300	Average		90	.178		43.50	4.85		48.35	55.50
	5500	Maximum	↓	80	.200	↓	88.50	5.45		93.95	106
	7000	Carport, baked vinyl finish, .032", 20' x 10', no foundations, min.	K-2	4	6	Car	2,250	170	45	2,465	2,825
	7250	Maximum		2	12	"	5,350	340	90	5,780	6,575
	7500	Walkway cover, to 12' wide, stl., vinyl finish, .032",no fndtns., min.		250	.096	S.F.	13.20	2.72	.72	16.64	20
	7750	Maximum	↓	200	.120	"	14.35	3.40	.90	18.65	22.50

SPECIALTIES 10

106 | Partitions & Storage Shelving

106 100 | Folding Gates

			CREW	DAILY OUTPUT	LABOR-HOURS	UNIT	1999 BARE COSTS MAT.	LABOR	EQUIP.	TOTAL	TOTAL INCL O&P
101	0010	**SECURITY GATES** For roll up type, see division 083-302									
	0300	Scissors type folding gate, ptd. steel, single, 6' high, 5-1/2'wide	2 Sswk	4	4	Opng.	112	122		234	345
	0400	7-1/2' wide		4	4		155	122		277	395
	0600	Double gate, 7-1/2' high, 8' wide		2.50	6.400		170	196		366	545
	0750	14' wide	↓	2	8	↓	310	245		555	790

107 | Telephone Specialties

107 550 | Telephone Enclosures

			CREW	DAILY OUTPUT	LABOR-HOURS	UNIT	1999 BARE COSTS MAT.	LABOR	EQUIP.	TOTAL	TOTAL INCL O&P
551	0010	**TELEPHONE ENCLOSURE**									
	0600	Booth type, painted steel, indoor or outdoor, minimum	2 Carp	1.50	10.667	Ea.	2,975	291		3,266	3,725
	0700	Maximum (stainless steel)		1.50	10.667		9,850	291		10,141	11,300
	1300	Outdoor, acoustical, on post		3	5.333		1,250	146		1,396	1,600
	1400	Phone carousel, pedestal mounted with dividers		.60	26.667		4,800	730		5,530	6,425
	1900	Outdoor, drive-up type, wall mounted		4	4		775	109		884	1,025
	2000	Post mounted, stainless steel posts	↓	3	5.333	↓	1,200	146		1,346	1,550

108 800 | Scales

			CREW	DAILY OUTPUT	LABOR-HOURS	UNIT	1999 BARE COSTS MAT.	LABOR	EQUIP.	TOTAL	TOTAL INCL O&P
801	0010	**SCALES** Built-in floor scale, not incl. foundations									
	0100	Dial type, 5 ton capacity, 8′ x 6′ platform	3 Carp	.50	48	Ea.	4,350	1,300		5,650	6,825
	0400	10 ton capacity, steel platform, 8′ x 6′ platform	"	.40	60	"	7,050	1,650		8,700	10,300
	0700	Truck scales, incl. steel weigh bridge,									
	0800	not including foundation, pits									
	0900	Dial type, mech., 24′ x 10′ platform, 20 ton cap.	3 Carp	.30	80	Ea.	8,300	2,175		10,475	12,600
	1000	30 ton capacity		.20	120		9,375	3,275		12,650	15,500
	1100	50 ton capacity, 50′ x 10′ platform		.14	171		14,200	4,675		18,875	23,000
	1200	70′ x 10′ platform		.12	200		17,700	5,450		23,150	28,000
	1400	60 ton capacity, 60′ x 10′ platform		.13	184		16,200	5,050		21,250	25,700
	1500	70′ x 10′ platform		.10	240		18,400	6,550		24,950	30,500
	1550	Digital, electronic, 100 ton capacity, steel deck 12′ x 10′ platform		.20	120		8,650	3,275		11,925	14,700
	1600	40′ x 10′ platform		.14	171		13,700	4,675		18,375	22,500
	1640	60′ x 10′ platform		.13	184		23,200	5,050		28,250	33,400
	1680	70′ x 10′ platform	↓	.12	200		27,300	5,450		32,750	38,700
	2000	For standard automatic printing device, add					1,900			1,900	2,100
	2100	For remote reading electronic system, add					2,000			2,000	2,200
	2300	Concrete foundation pits for above, 8′ x 6′, 5 C.Y. required	C-1	.50	64		710	1,650		2,360	3,375
	2400	14′ x 6′ platform, 10 C.Y. required		.35	91.429		1,000	2,375		3,375	4,825
	2600	50′ x 10′ platform, 30 C.Y. required		.25	128		1,400	3,300		4,700	6,750
	2700	70′ x 10′ platform, 40 C.Y. required	↓	.15	213		2,350	5,525		7,875	11,300
	2750	Crane scales, dial, 1 ton capacity					805			805	890
	2780	5 ton capacity					960			960	1,050
	2800	Digital, 1 ton capacity					1,800			1,800	1,975
	2850	10 ton capacity					3,425			3,425	3,775
	3800	Portable, beam type, capacity 1000#, platform 18″ x 24″					490			490	540
	3900	Dial type, capacity 2000#, platform 24″ x 24″					850			850	935
	4000	Digital type, capacity 1000#, platform 24″ x 30″					1,450			1,450	1,600
	4100	Portable contractor truck scales, 50 ton cap., 40′ x 10′ platform					25,300			25,300	27,800
	4200	60′ x 10′ platform				↓	28,300			28,300	31,100

For information about Means Estimating Seminars, see yellow pages 11 and 12 in back of book

Division 11 Equipment

Estimating Tips

General

- The items in this division are usually priced per square foot or each. Many of these items are purchased by the owner for installation by the contractor. Check the specifications for responsibilities, and include time for receiving, storage, installation and mechanical and electrical hook-ups in the appropriate divisions.
- Many items in Division 11 require some type of support system that is not usually furnished with the item. Examples of these systems include blocking for the attachment of casework and support angles for ceiling hung projection screens. The required blocking or supports must be added to the estimate in the appropriate division.
- Some items in Division 11 may require assembly or electrical hook-ups. Verify the amount of assembly required or the need for a hard electrical connection and add the appropriate costs.

Reference Numbers

Reference numbers are shown in bold squares at the beginning of some major classifications. These numbers refer to related items in the Reference Section. The reference information may be an estimating procedure, an alternate pricing method or technical information.

Note: Not all subdivisions listed here necessarily appear in this publication.

111 400 | Service Station Equip

			CREW	DAILY OUTPUT	LABOR-HOURS	UNIT	1999 BARE COSTS MAT.	LABOR	EQUIP.	TOTAL	TOTAL INCL O&P
401	0010	**AUTOMOTIVE**									
	0030	Compressors, electric, 1-1/2 H.P., standard controls	L-4	1.50	16	Ea.	320	410		730	995
	1100	Product dispenser with vapor recovery for 6 nozzles, installed, not									
	1110	including piping to storage tanks				Ea.	15,000			15,000	16,500
	2810	Hydraulic lifts, above ground, 2 post, clear floor, 6000 lb cap	L-4	2.67	8.989		4,700	231		4,931	5,550
	2815	9000 lb capacity		2.29	10.480		11,200	269		11,469	12,700
	2820	15,000 lb capacity		2	12		25,900	310		26,210	29,000
	2825	30,000 lb capacity		1.60	15		27,900	385		28,285	31,200
	2830	4 post, ramp style, 25,000 lb capacity		2	12		10,900	310		11,210	12,500
	2835	35,000 lb capacity		1	24		51,000	615		51,615	57,000
	2840	50,000 lb capacity		1	24		57,000	615		57,615	63,500
	2845	75,000 lb capacity		1	24		66,000	615		66,615	73,500
	2850	For drive thru tracks, add, minimum					700			700	770
	2855	Maximum					1,200			1,200	1,325
	2860	Ramp extensions, 3'(set of 2)					575			575	635
	2865	Rolling jack platform					2,000			2,000	2,200
	2870	Elec/hyd jacking beam					5,350			5,350	5,875
	2880	Scissor lift, portable, 6000 lb capacity					5,250			5,250	5,775

111 500 | Parking Control Equip

			CREW	DAILY OUTPUT	LABOR-HOURS	UNIT	MAT.	LABOR	EQUIP.	TOTAL	TOTAL INCL O&P
501	0010	**PARKING EQUIPMENT**									
	5000	Barrier gate with programmable controller	2 Elec	3	5.333	Ea.	2,500	170		2,670	3,000
	5020	Industrial	"	3	5.333		3,500	170		3,670	4,100
	5100	Card reader	1 Elec	2	4		2,000	128		2,128	2,400
	5120	Proximity with customer display	2 Elec	1	16		5,000	510		5,510	6,275
	5200	Cashier booth, average	B-22	1	30		10,000	760	204	10,964	12,400
	5300	Collector station, pay on foot	2 Elec	.20	80		110,000	2,550		112,550	125,000
	5320	Credit card only		.50	32		45,000	1,025		46,025	51,000
	5500	Exit verifier		1	16		22,500	510		23,010	25,600
	5600	Fee computer	1 Elec	1.50	5.333		13,300	170		13,470	15,000
	5700	Full sign, 4" letters	"	2	4		900	128		1,028	1,175
	5800	Inductive loop	2 Elec	4	4		500	128		628	740
	5900	Ticket spitter with time/date stamp, standard		2	8		6,000	255		6,255	6,975
	5920	Mag stripe encoding		2	8		16,000	255		16,255	18,000
	5950	Vehicle detector, microprocessor based	1 Elec	3	2.667		400	85		485	565
	6000	Parking control software, minimum		.50	16		18,000	510		18,510	20,600
	6020	Maximum		.20	40		75,000	1,275		76,275	84,500

111 600 | Loading Dock Equipment

			CREW	DAILY OUTPUT	LABOR-HOURS	UNIT	MAT.	LABOR	EQUIP.	TOTAL	TOTAL INCL O&P
601	0010	**LOADING DOCK**									
	0020	Bumpers, rubber blocks 4-1/2" thk, 10" H, 14" long	1 Carp	26	.308	Ea.	37	8.40		45.40	54
	0300	36" long		17	.471		68	12.85		80.85	95
	0500	12" high, 14" long		25	.320		61	8.75		69.75	81
	0600	36" long		15	.533		76	14.55		90.55	107
	0800	Rubber blocks 6" thick, 10" high, 14" long		22	.364		57.50	9.95		67.45	78.50
	0900	36" long		13	.615		93	16.80		109.80	130
	0920	Extruded rubber bumpers, T section, 22" x 22" x 3" thick		41	.195		41.50	5.35		46.85	54
	0940	Molded rubber bumpers, 24" x 12" x 3" thick		20	.400		39	10.90		49.90	60
603	0010	**DOCK BUMPERS** Bolts not included									
	0020	2" x 6" to 4" x 8", average	1 Carp	.30	26.667	M.B.F.	885	730		1,615	2,125

111 700 | Waste Handling Equip

			CREW	DAILY OUTPUT	LABOR-HOURS	UNIT	MAT.	LABOR	EQUIP.	TOTAL	TOTAL INCL O&P
701	0010	**WASTE HANDLING**									
	0020	Compactors, 115 volt, 250#/hr., chute fed	L-4	1	24	Ea.	8,225	615		8,840	10,000

Important: See the Reference Section for critical supporting data - Reference Nos., Crews, & City Cost Indexes

111 | Mercantile, Commercial & Detention Equipment

111 700 | Waste Handling Equip		CREW	DAILY OUTPUT	LABOR-HOURS	UNIT	1999 BARE COSTS MAT.	LABOR	EQUIP.	TOTAL	TOTAL INCL O&P
701 1400	For handling hazardous waste materials, 55 gallon drum packer, std.				Ea.	11,700			11,700	12,900
1410	55 gallon drum packer w/HEPA filter					14,900			14,900	16,400
1420	55 gallon drum packer w/charcoal & HEPA filter					19,000			19,000	20,900
1430	All of the above made explosion proof, add				↓	9,075			9,075	9,975
4750	Large municipal incinerators, incl. stack, minimum	Q-3	.25	128	Ton/day	12,900	3,975		16,875	20,200
4850	Maximum	"	.10	320	"	34,300	9,950		44,250	53,000
6000	Transfer station compactor, with power unit									
6050	and pedestal, not including pit, 50 ton per hour				Ea.	121,000			121,000	133,000

114 | Food Service, Residential, Darkroom, Athletic Equipment

114 800 | Athletic/Recreational		CREW	DAILY OUTPUT	LABOR-HOURS	UNIT	1999 BARE COSTS MAT.	LABOR	EQUIP.	TOTAL	TOTAL INCL O&P
805 0010	SCHOOL EQUIPMENT									
0200	For exterior equipment, see division 028									
7000	Scoreboards, baseball, minimum	R-3	1.30	15.385	Ea.	2,300	485	104	2,889	3,400
7200	Maximum		.05	400		16,800	12,600	2,725	32,125	40,500
7300	Football, minimum		.86	23.256		5,675	735	158	6,568	7,500
7400	Maximum	↓	.20	100	↓	52,500	3,150	680	56,330	63,500

Division Notes

	CREW	DAILY OUTPUT	LABOR-HOURS	UNIT	1999 BARE COSTS MAT.	LABOR	EQUIP.	TOTAL	TOTAL INCL O&P

Division 12 Furnishings

'stimating Tips

eneral

The items in this division are usually priced per square foot or each. Most of these items are purchased by the owner and placed by the supplier. Do not assume the items in Division 12 will be purchased and installed by the supplier. Check the specifications for responsibilities and include receiving, storage, installation and mechanical and electrical hook-ups in the appropriate divisions.

- Some items in this division require some type of support system that is not usually furnished with the item. Examples of these systems include blocking for the attachment of casework and heavy drapery rods. The required blocking must be added to the estimate in the appropriate division.

Reference Numbers

Reference numbers are shown in bold squares at the beginning of some major classifications. These numbers refer to related items in the Reference Section. The reference information may be an estimating procedure, an alternate pricing method or technical information.

Note: Not all subdivisions listed here necessarily appear in this publication.

126 | Furniture & Accessories

126 100 | Landscape Partitions

		126 100 Landscape Partitions	CREW	DAILY OUTPUT	LABOR-HOURS	UNIT	1999 BARE COSTS MAT.	LABOR	EQUIP.	TOTAL	TOTAL INCL O&P	
107	0010	**POSTS**										107
	0020	Portable for pedestrian traffic control, standard, minimum				Ea.	87.50			87.50	96	
	0100	Maximum					134			134	147	
	0300	Deluxe posts, minimum					148			148	163	
	0400	Maximum					285			285	315	
	0600	Ropes for above posts, plastic covered, 1-1/2" diameter				L.F.	8.10			8.10	8.95	
	0700	Chain core				"	8.25			8.25	9.05	

126 900 | Floor Mats & Frames

		126 900 Floor Mats & Frames	CREW	DAILY OUTPUT	LABOR-HOURS	UNIT	1999 BARE COSTS MAT.	LABOR	EQUIP.	TOTAL	TOTAL INCL O&P	
901	0010	**FLOOR MATS**										901
	0020	Recessed, in-laid black rubber, 3/8" thick, solid	1 Clab	155	.052	S.F.	19.05	1.11		20.16	22.50	
	0500	Link mats, including nosings, aluminum, 3/8" thick		155	.052		16.55	1.11		17.66	19.95	
	0550	Black rubber with galvanized tie rods		155	.052		13	1.11		14.11	16.05	
	0600	Steel, galvanized, 3/8" thick		155	.052		6.25	1.11		7.36	8.65	
	0650	Vinyl, in colors		155	.052		16.85	1.11		17.96	20.50	
	0750	Add for nosings, rubber				L.F.	6.05			6.05	6.65	
	0850	Recess frames for above mats, aluminum	1 Carp	100	.080		3.74	2.18		5.92	7.55	
	0870	Bronze	"	100	.080		7.90	2.18		10.08	12.15	
	0900	Skate lock tile, 24" x 24" x 1/2" thick, rubber, black	1 Clab	125	.064	S.F.	12.70	1.37		14.07	16.10	
	0950	Color		125	.064	"	14.60	1.37		15.97	18.25	
	1000	12" x 24" border, black		75	.107	L.F.	21	2.29		23.29	26.50	
	1100	Color		75	.107	"	25.50	2.29		27.79	31.50	
	1150	12" x 12" outside corner, black		100	.080	S.F.	20.50	1.72		22.22	25	
	1200	Color		100	.080		23.50	1.72		25.22	28.50	
	1500	Duckboard, aluminum slats		155	.052		17.45	1.11		18.56	21	
	1700	Hardwood strips on rubber base, to 54" wide		155	.052		12.25	1.11		13.36	15.25	
	1800	Assembled with brass rods and vinyl spacers, to 48" wide		155	.052		16.55	1.11		17.66	19.95	
	1850	Tire fabric, 3/4" thick		155	.052		10.05	1.11		11.16	12.80	
	1900	Vinyl, 36" wide, in colors, hollow top & bottoms		155	.052		4.16	1.11		5.27	6.30	
	1950	Solid top & bottom members		155	.052		7.60	1.11		8.71	10.10	

128 | Interior Plants & Planters

128 100 | Interior Plants

		128 100 Interior Plants	CREW	DAILY OUTPUT	LABOR-HOURS	UNIT	1999 BARE COSTS MAT.	LABOR	EQUIP.	TOTAL	TOTAL INCL O&P	
105	0010	**PLANTS** Permanent only, weighted										105
	0020	Preserved or polyester leaf, natural wood										
	0030	Or molded trunk. For pots see division 121-750										
	0040	For fill see 028-310										
	0100	Plants, acuba, 5' high				Ea.	136			136	150	
	0200	Apidistra, 4' high					128			128	141	
	0300	Beech, variegated, 5' high					125			125	138	
	0400	Birds nest fern, 6' high					225			225	248	
	0500	Croton, 4' high					50			50	55	
	0600	Diffenbachia, 3' high					50			50	55	
	0700	Ficus Benjamina, 3' high					60			60	66	
	0720	6' high					120			120	132	
	0760	Nitida, 3' high					135			135	149	
	0780	6' high					270			270	297	

128 100 | Interior Plants

		Description	CREW	DAILY OUTPUT	LABOR-HOURS	UNIT	1999 BARE COSTS MAT.	LABOR	EQUIP.	TOTAL	TOTAL INCL O&P
105	0800	Helicona				Ea.	125			125	138
	1200	Palm green date fan, 5′ high					130			130	143
	1220	7′ high					140			140	154
	1280	Green chamdora date, 5′ high					115			115	127
	1300	7′ high					180			180	198
	1400	Green giant for 7′ high					280			280	310
	1600	Rubber plant, 5′ high					95			95	105
	1800	Schefflera, 3′ high					65			65	71.50
	1820	4′ high					80			80	88
	1840	5′ high					125			125	138
	1860	6′ high					170			170	187
	1880	7′ high					260			260	286
	1900	Spathiphyllum, 3′ high				▼	90			90	99
	4000	Trees, polyester or preserved, with									
	4020	Natural trunks									
	4100	Acuba, 10′ high				Ea.	710			710	780
	4120	12′ high					995			995	1,100
	4140	14′ high					1,400			1,400	1,550
	4400	Bamboo, 10′ high					310			310	340
	4420	12′ high					330			330	365
	4800	Beech, 10′ high					950			950	1,050
	4820	12′ high					1,100			1,100	1,200
	4840	14′ high					1,500			1,500	1,650
	5000	Birch, 10′ high					950			950	1,050
	5020	12′ high					1,100			1,100	1,200
	5040	14′ high					1,500			1,500	1,650
	5060	16′ high					1,200			1,200	1,300
	5080	18′ high					1,325			1,325	1,450
	5500	Ficus Benjamina, 10′ high					700			700	770
	5520	12′ high					900			900	990
	5540	14′ high					1,000			1,000	1,100
	5560	16′ high					1,400			1,400	1,550
	5580	18′ high					2,100			2,100	2,300
	6000	Magnolia, 10′ high					1,025			1,025	1,125
	6020	12′ high					1,450			1,450	1,600
	6040	14′ high					1,975			1,975	2,175
	6500	Maple, 10′ high					100			100	110
	6520	12′ high					1,300			1,300	1,425
	6540	14′ high					1,600			1,600	1,750
	7000	Palms, chamadora, 10′ high					200			200	220
	7020	12′ high					255			255	281
	7040	Date, 10′ high					275			275	305
	7060	12′ high				▼	315			315	345

128 150 | Planters

		Description	CREW	DAILY OUTPUT	LABOR-HOURS	UNIT	MAT.	LABOR	EQUIP.	TOTAL	TOTAL INCL O&P
155	0010	PLANTERS									
	1000	Fiberglass, hanging, 12″ diameter, 7″ high				Ea.	38			38	41.50
	1100	15″ diameter, 7″ high					59.50			59.50	65.50
	1200	36″ diameter, 8″ high					133			133	147
	1500	Rectangular, 48″ long, 16″ high x 15″ wide					310			310	340
	1550	16″ high x 24″ wide					425			425	465
	1600	24″ high x 24″ wide					515			515	570
	1650	60″ long, 30″ high, 28″ wide					660			660	725
	1700	72″ long, 16″ high, 15″ wide					655			655	720
	1750	21″ high, 24″ wide				▼	750			750	825

For expanded coverage of these items see *Means Interior Cost Data 1999*

	128 150 | Planters	CREW	DAILY OUTPUT	LABOR-HOURS	UNIT	1999 BARE COSTS MAT.	LABOR	EQUIP.	TOTAL	TOTAL INCL O&P
155 1800	30″ high, 24″ wide				Ea.	890			890	975
2000	Round, 12″ diameter, 13″ high					66			66	72.50
2050	25″ high					78.50			78.50	86.50
2150	14″ diameter, 15″ high					73			73	80
2200	16″ diameter, 16″ high					83			83	91.50
2250	18″ diameter, 19″ high					99			99	109
2300	23″ high					129			129	142
2350	20″ diameter, 16″ high					105			105	116
2400	18″ high					107			107	118
2450	21″ high					127			127	139
2500	22″ diameter, 10″ high					157			157	172
2550	24″ diameter, 16″ high					187			187	206
2600	19″ high					218			218	239
2650	25″ high					289			289	320
2700	36″ high					300			300	330
2750	48″ high					315			315	350
2800	30″ diameter, 16″ high					220			220	242
2850	18″ high					252			252	277
2900	21″ high					255			255	280
3000	24″ high					281			281	310
3350	27″ high					305			305	335
3400	36″ diameter, 16″ high					320			320	350
3450	18″ high					350			350	385
3500	21″ high					445			445	490
3550	24″ high					380			380	420
3600	27″ high					410			410	455
3650	30″ high					460			460	505
3700	48″ diameter, 16″ high					490			490	540
3750	21″ high					525			525	580
3800	24″ high					595			595	655
3850	27″ high					635			635	695
3900	30″ high					715			715	785
3950	36″ high					915			915	1,000
4000	60″ diameter, 16″ high					730			730	805
4100	21″ high					825			825	905
4150	27″ high					990			990	1,100
4200	30″ high					910			910	1,000
4250	33″ high					940			940	1,025
4300	36″ high					980			980	1,075
4400	39″ high					1,050			1,050	1,150
5000	Square, 10″ side, 20″ high					91			91	100
5100	14″ side, 15″ high					92.50			92.50	102
5200	18″ side, 19″ high					151			151	166
5300	20″ side, 16″ high					180			180	198
5320	18″ high					185			185	204
5340	21″ high					192			192	211
5400	24″ side, 16″ high					232			232	255
5420	21″ high					269			269	296
5440	25″ high					305			305	335
5460	30″ side, 16″ high					310			310	340
5480	24″ high					410			410	450
5490	27″ high					525			525	575
5500	Round, 36″ diameter, 16″ high					470			470	515
5510	18″ high					435			435	475
5520	21″ high					525			525	575
5530	24″ high				↓	575			575	630

128 150 | Planters		CREW	DAILY OUTPUT	LABOR-HOURS	UNIT	1999 BARE COSTS MAT.	LABOR	EQUIP.	TOTAL	TOTAL INCL O&P
5540	27" high				Ea.	665			665	735
5550	30" high					590			590	650
5800	48" diameter, 16" high					680			680	745
5820	21" high					705			705	775
5840	24" high					835			835	920
5860	27" high					890			890	975
5880	30" high					975			975	1,075
5900	60" diameter, 16" high					915			915	1,000
5920	21" high					1,025			1,025	1,125
5940	27" high					1,100			1,100	1,225
5960	30" high					1,275			1,275	1,400
5980	36" high					1,475			1,475	1,625
6000	Metal bowl, 32" diameter, 8" high, minimum					310			310	340
6050	Maximum					465			465	510
6100	Rectangle, 30" long x 12" wide, 6" high, minimum					257			257	283
6200	Maximum					340			340	375
6300	36" long 12" wide, 6" high, minimum					320			320	350
6400	Maximum					415			415	455
6500	Square, 15" side, minimum					320			320	350
6600	Maximum					410			410	455
6700	20" side, minimum					505			505	555
6800	Maximum					600			600	660
6900	Round, 6" diameter x 6" high, minimum					23.50			23.50	26
7000	Maximum					28			28	31
7100	8" diameter x 8" high, minimum					31.50			31.50	34.50
7200	Maximum					33.50			33.50	37
7300	10" diameter x 11" high, minimum					62			62	68.50
7400	Maximum					72			72	79.50
7420	12" diameter x 13" high, minimum					70			70	77
7440	Maximum					92.50			92.50	102
7500	14" diameter x 15" high, minimum					92			92	101
7550	Maximum					112			112	124
7580	16" diameter x 17" high, minimum					101			101	111
7600	Maximum					138			138	151
7620	18" diameter x 19" high, minimum					122			122	135
7640	Maximum					157			157	173
7680	22" diameter x 20" high, minimum					157			157	173
7700	Maximum					191			191	210
7750	24" diameter x 21" high, minimum					207			207	228
7800	Maximum					249			249	274
7850	31" diameter x 18" high, minimum					620			620	685
7900	Maximum					960			960	1,050
7950	38" diameter x 24" high, minimum					760			760	835
8000	Maximum					1,150			1,150	1,250
8050	48" diameter x 24" high, minimum					1,075			1,075	1,175
8150	Maximum				↓	1,625			1,625	1,775
8500	Plastic laminate faced, fiberglass liner, square									
8520	14" sq., 15" high				Ea.	237			237	261
8540	24" sq., 16" high					300			300	330
8580	36" sq., 21" high					425			425	465
8600	Rectangle 36" long, 12" wide, 10" high					257			257	282
8650	48" long, 12" wide, 10" high					286			286	315
8700	48" long, 12" wide, 24" high				↓	425			425	465
8750	Wood, fiberglass liner, square									
8780	14" square, 15" high, minimum				Ea.	171			171	188
8800	Maximum				↓	210			210	232

or expanded coverage of these items see *Means Interior Cost Data 1999*

	128 150 | Planters	CREW	DAILY OUTPUT	LABOR-HOURS	UNIT	1999 BARE COSTS MAT.	LABOR	EQUIP.	TOTAL	TOTAL INCL O&P
155	8820 24″ sq., 16″ high, minimum				Ea.	210			210	232
	8840 Maximum					279			279	305
	8860 36″ sq., 21″ high, minimum					271			271	298
	8880 Maximum					405			405	450
	9000 Rectangle, 36″ long x 12″ wide, 10″ high, minimum					188			188	207
	9050 Maximum					245			245	270
	9100 48″ long x 12″ wide, 10″ high, minimum					204			204	224
	9120 Maximum					257			257	282
	9200 48″ long x 12″ wide, 24″ high, minimum					245			245	270
	9300 Maximum					360			360	395
	9400 Plastic cylinder, molded, 10″ diameter, 10″ high					6.95			6.95	7.65
	9500 11″ diameter, 11″ high					22			22	24
	9600 13″ diameter, 12″ high					29			29	32
	9700 16″ diameter, 14″ high				▼	30			30	33

Division 13 Special Construction

[E]stimating Tips

[G]eneral

The items and systems in this division are usually estimated, purchased, supplied and installed as a unit by one or more subcontractors. The estimator must ensure that all parties are operating from the same set of specifications and assumptions and that all necessary items are estimated and will be provided. Many times the complex items and systems are covered but the more common ones such as excavation or a crane are overlooked for the very reason that everyone assumes nobody could miss them. The estimator should be the central focus and be able to ensure that all systems are complete.

Another area where problems can develop in this division is at the interface between systems. The estimator must ensure, for instance, that anchor bolts, nuts and washers are estimated and included for the air-supported structures and pre-engineered buildings to be bolted to their foundations. Utility supply is a common area where essential items or pieces of equipment can be missed or overlooked due to the fact that each subcontractor may feel it is the others' responsibility. The estimator should also be aware of certain items which may be supplied as part of a package but installed by others, and ensure that the installing contractor's estimate includes the cost of installation. Conversely, the estimator must also ensure that items are not costed by two different subcontractors, resulting in an inflated overall estimate.

131 Pre-Engineered Structures, Aquatic Facilities & Ice Rinks

- The foundations and floor slab, as well as rough mechanical and electrical, should be estimated, as this work is required for the assembly and erection of the structure. Generally, as noted in the book, the pre-engineered building comes as a shell and additional features must be included by the estimator. Here again, the estimator must have a clear understanding of the scope of each portion of the work and all the necessary interfaces.

132 Tanks, Tank Covers, Filtration Equipment

- The prices in this subdivision for above and below ground storage tanks do not include foundations or hold-down slabs. The estimator should refer to Divisions 2 and 3 for foundation system pricing. In addition to the foundations, required tank accessories such as tank gauges, leak detection devices, and additional manholes and piping must be added to the tank prices.

Reference Numbers

Reference numbers are shown in bold squares at the beginning of some major classifications. These numbers refer to related items in the Reference Section. The reference information may be an estimating procedure, an alternate pricing method or technical information.

Note: Not all subdivisions listed here necessarily appear in this publication.

130 100 | Air Supported Structures

			CREW	DAILY OUTPUT	LABOR-HOURS	UNIT	1999 BARE COSTS MAT.	LABOR	EQUIP.	TOTAL	TOTAL INCL O&P
111	0010	**AIR SUPPORTED STRUCTURES**									
	0020	Site preparation, incl. anchor placement and utilities	B-11B	2,000	.008	SF Flr.	.66	.20	.52	1.38	1.61
	0030	For concrete curb, see division 033-130									
	0050	Warehouse, polyester/vinyl fabric, 28 oz., over 10 yr. life, welded									
	0060	Seams, tension cables, primary & auxiliary inflation system,									
	0070	airlock, personnel doors and liner									
	0100	5000 S.F.	4 Clab	5,000	.006	SF Flr.	15.40	.14		15.54	17.15
	0250	12,000 S.F.	″	6,000	.005		10.20	.11		10.31	11.40
	0400	24,000 S.F.	8 Clab	12,000	.005		7.80	.11		7.91	8.80
	0500	50,000 S.F.	″	12,500	.005	↓	6.20	.11		6.31	6.95
	0700	12 oz. reinforced vinyl fabric, 5 yr. life, sewn seams,									
	0710	accordian door, including liner									
	0750	3000 S.F.	4 Clab	3,000	.011	SF Flr.	6.40	.23		6.63	7.40
	0800	12,000 S.F.	″	6,000	.005		5.10	.11		5.21	5.80
	0850	24,000 S.F.	8 Clab	12,000	.005		4.15	.11		4.26	4.75
	0950	Deduct for single layer					.58			.58	.64
	1000	Add for welded seams					.90			.90	.99
	1050	Add for double layer, welded seams included				↓	1.85			1.85	2.04
	1250	Tedlar/vinyl fabric, 28 oz., with liner, over 10 yr. life,									
	1260	incl. overhead and personnel doors									
	1300	3000 S.F.	4 Clab	3,000	.011	SF Flr.	13.50	.23		13.73	15.20
	1450	12,000 S.F.	″	6,000	.005		9.45	.11		9.56	10.60
	1550	24,000 S.F.	8 Clab	12,000	.005		7.40	.11		7.51	8.35
	1700	Deduct for single layer				↓	1.29			1.29	1.42
	2250	Greenhouse/shelter, woven polyethylene with liner, 2 yr. life,									
	2260	sewn seams, including doors									
	2300	3000 S.F.	4 Clab	3,000	.011	SF Flr.	5.90	.23		6.13	6.85
	2350	12,000 S.F.	″	6,000	.005		3.09	.11		3.20	3.58
	2450	24,000 S.F.	8 Clab	12,000	.005		2.69	.11		2.80	3.14
	2550	Deduct for single layer				↓	.56			.56	.62
	2600	Tennis/gymnasium, polyester/vinyl fabric, 28 oz., over 10 yr. life,									
	2610	including thermal liner, heat and lights									
	2650	7200 S.F.	4 Clab	6,000	.005	SF Flr.	13	.11		13.11	14.50
	2750	13,000 S.F.	″	6,500	.005		9.50	.11		9.61	10.60
	2850	Over 24,000 S.F.	8 Clab	12,000	.005		8.50	.11		8.61	9.55
	2860	For low temperature conditions, add				↓	.53			.53	.58
	2870	For average shipping charges, add				Total	2,000			2,000	2,200
	2900	Thermal liner, translucent reinforced vinyl				SF Flr.	.75			.75	.83
	2950	Metalized mylar fabric and mesh, double liner				″	1.30			1.30	1.43
	3050	Stadium/convention center, teflon coated fiberglass, heavy weight,									
	3060	over 20 yr. life, incl. thermal liner and heating system									
	3100	Minimum	9 Clab	26,000	.003	SF Flr.	35	.06		35.06	38.50
	3110	Maximum	″	19,000	.004	″	40	.08		40.08	44
	3400	Doors, air lock, 15′ long, 10′ x 10′	2 Carp	.80	20	Ea.	12,500	545		13,045	14,700
	3600	15′ x 15′	″	.50	32		18,500	875		19,375	21,800
	3700	For each added 5′ length, add					2,000			2,000	2,200
	3900	Revolving personnel door, 6′ diameter, 6′-6″ high	2 Carp	.80	20	↓	8,800	545		9,345	10,500
	4200	Double wall, self supporting, shell only, minimum				SF Flr.					21
	4300	Maximum				″					39
115	0010	**AIR SUPPORTED STORAGE TANK COVERS** Vinyl polyester									
	0100	scrim, double layer, with hardware, blower, standby & controls									
	0200	Round, 75′ diameter	B-2	4,500	.009	S.F.	5.60	.19		5.79	6.45
	0300	100′ diameter		5,000	.008		5	.17		5.17	5.80
	0400	150′ diameter		5,000	.008		3.55	.17		3.72	4.19
	0500	Rectangular, 20′ x 20′	↓	4,500	.009	↓	12	.19		12.19	13.50

130 | Special Construction

130 100 | Air Supported Structures

			CREW	DAILY OUTPUT	LABOR-HOURS	UNIT	1999 BARE COSTS MAT.	LABOR	EQUIP.	TOTAL	TOTAL INCL O&P
115	0600	30′ x 40′	B-2	4,500	.009	S.F.	12	.19		12.19	13.50
	0700	50′ x 60′	↓	4,500	.009		12	.19		12.19	13.50
	0800	For single wall construction, deduct, minimum					.27			.27	.30
	0900	Maximum					1.06			1.06	1.17
	1000	For maximum resistance to atmosphere or cold, add				↓	.33			.33	.36
	1100	For average shipping charges, add				Total	1,050			1,050	1,150

131 | Pre-Eng. Structures, Aquatic Facilities and Ice Rinks

131 200 | Pre-Eng. Structures

			CREW	DAILY OUTPUT	LABOR-HOURS	UNIT	1999 BARE COSTS MAT.	LABOR	EQUIP.	TOTAL	TOTAL INCL O&P
202	0010	**DOMES**									
	1500	Bulk storage, shell only, dual radius hemispher. arch, steel									
	1600	framing, corrugated steel covering, 150′ diameter	E-2	550	.102	SF Flr.	27	3.03	1.84	31.87	37
	1700	400′ diameter	″	720	.078		22	2.32	1.41	25.73	29.50
	1800	Wood framing, wood decking, to 400′ diameter	F-4	400	.120	↓	20	3.25	1.98	25.23	29
	1900	Radial framed wood (2″ x 6″), 1/2″ thick									
	2000	plywood, asphalt shingles, 50′ diameter	F-3	2,000	.020	SF Flr.	25	.55	.22	25.77	28.50
	2100	60′ diameter		1,900	.021		19	.58	.23	19.81	22
	2200	72′ diameter		1,800	.022		17	.62	.25	17.87	19.95
	2300	116′ diameter		1,730	.023		15	.64	.26	15.90	17.80
	2400	150′ diameter	↓	1,500	.027	↓	13	.74	.30	14.04	15.80
203	0010	**GEODESIC DOME** Shell only, interlocking plywood panels R131-310									
	0400	30′ diameter	F-5	1.60	20	Ea.	9,950	555		10,505	11,800
	0500	34′ diameter		1.14	28.070		12,200	780		12,980	14,600
	0600	39′ diameter	↓	1	32		14,200	890		15,090	17,000
	0700	45′ diameter	F-3	1.13	35.556		17,100	985	395	18,480	20,800
	0750	55′ diameter		1	40		25,500	1,100	445	27,045	30,200
	0800	60′ diameter		1	40		33,200	1,100	445	34,745	38,700
	0850	65′ diameter	↓	.80	50	↓	39,400	1,375	555	41,330	46,100
	1100	Aluminum panel, with 6″ insulation									
	1200	100′ diameter				SF Flr.					30
	1300	500′ diameter				″					25
	1600	Aluminum framed, plexiglass closure panels									
	1700	40′ diameter				SF Flr.					80
	1800	200′ diameter				″					70
	2100	Aluminum framed, aluminum closure panels									
	2200	40′ diameter				SF Flr.					50
	2300	100′ diameter									30
	2400	200′ diameter									25
	2500	For VRP faced bonded fiberglass insulation, add				↓					10
	2700	Aluminum framed, fiberglass sandwich panel closure									
	2800	6′ diameter	2 Carp	150	.107	SF Flr.	33	2.91		35.91	41
	2900	28′ diameter	″	350	.046	″	25	1.25		26.25	29.50
204	0010	**GARAGE COSTS**									
	0301	Prefab shell, stock, wood, single car, minimum	2 Carp	1	16	Total	2,550	435		2,985	3,475
	0350	Maximum		.67	23.881		5,575	650		6,225	7,150
	0400	Two car, minimum		.67	23.881		3,650	650		4,300	5,025
	0450	Maximum	↓	.50	32	↓	7,525	875		8,400	9,650

		131 200 | Pre-Eng. Structures	CREW	DAILY OUTPUT	LABOR-HOURS	UNIT	1999 BARE COSTS MAT.	LABOR	EQUIP.	TOTAL	TOTAL INCL O&P	
205	0010	**SILOS** Concrete stave industrial, not incl. foundations, conical or										205
	0100	sloping bottoms, 12′ diameter, 35′ high	D-8	.11	363	Ea.	15,000	9,175		24,175	30,700	
	0200	16′ diameter, 45′ high		.08	500		20,000	12,600		32,600	41,600	
	0400	25′ diameter, 75′ high		.05	800		49,000	20,200		69,200	85,500	
	0500	Steel, factory fab., 30,000 gallon cap., painted, minimum	L-5	1	56		10,900	1,725	550	13,175	15,700	
	0700	Maximum	"	.50	112		16,900	3,450	1,100	21,450	26,000	
207	0010	**TENSION STRUCTURES** Rigid steel/alum. frame, vyl. coated polyester										207
	0100	fabric shell, 60′ clear span, not incl. foundations or floors										
	0200	6,000 S.F.	B-41	1,000	.044	SF Flr.	10.20	.98	.18	11.36	12.95	
	0300	12,000 S.F.		1,100	.040		8.65	.89	.16	9.70	11.10	
	0400	80′ clear span, 20,800 S.F.		1,220	.036		9.35	.80	.14	10.29	11.70	
	0410	100′ clear span, 10,000 S.F.	L-5	2,175	.026		9.90	.79	.25	10.94	12.60	
	0430	26,000 S.F.		2,300	.024		8.85	.75	.24	9.84	11.35	
	0450	36,000 S.F.		2,500	.022		8.50	.69	.22	9.41	10.80	
	0460	120′ clear span, 24,000 S.F.		3,000	.019		11.40	.57	.18	12.15	13.75	
	0470	150′ clear span, 30,000 S.F.		6,000	.009		10.45	.29	.09	10.83	12.10	
	0480	200′ clear span, 40,000 S.F.	E-6	8,000	.016		14.35	.48	.16	14.99	16.85	
	0500	For roll-up door, 12′ x 14′, add	L-2	1	16	Ea.	3,600	385		3,985	4,550	
	0600	For personnel doors, add, minimum				SF Flr.	5%					
	0700	Add, maximum					15%					
	0800	For site work, simple foundation, etc., add, minimum								1.25	1.95	
	0900	Add, maximum								2.75	3.05	

		131 210 | Pre-Engineered Buildings	CREW	DAILY OUTPUT	LABOR-HOURS	UNIT	MAT.	LABOR	EQUIP.	TOTAL	TOTAL INCL O&P	
211	0010	**HANGARS** Prefabricated steel T hangars, Galv. steel roof &										211
	0100	walls, incl. electric bi-folding doors, 4 or more units,										
	0110	not including floors or foundations, minimum	E-2	1,275	.044	SF Flr.	8.85	1.31	.79	10.95	12.90	
	0130	Maximum	"	1,063	.053	"	9.50	1.57	.95	12.02	14.25	
	1200	Alternate pricing method:										
	1300	Galv. roof and walls, electric bi-folding doors, minimum	E-2	1.06	52.830	Plane	10,600	1,575	955	13,130	15,400	
	1500	Maximum	"	.91	61.538	"	11,700	1,825	1,100	14,625	17,300	
212	0010	**PRE-ENGINEERED STEEL BUILDINGS** R131-210										212
	0100	Clear span rigid frame, 26 ga. colored roofing and siding										
	0200	30′ to 40′ wide, 10′ eave height	E-2	1,800	.031	SF Flr.	4.36	.93	.56	5.85	7.05	
	0300	14′ eave height		1,800	.031		4.42	.93	.56	5.91	7.10	
	0400	16′ eave height		1,400	.040		4.77	1.19	.72	6.68	8.15	
	0500	20′ eave height		1,000	.056		5.20	1.67	1.01	7.88	9.75	
	0600	24′ eave height		1,000	.056		6.30	1.67	1.01	8.98	11	
	0700	50′ to 100′ wide, 10′ eave height		1,800	.031		4.05	.93	.56	5.54	6.70	
	0800	14′ eave height		1,800	.031		4.04	.93	.56	5.53	6.70	
	0900	16′ eave height		1,400	.040		4.35	1.19	.72	6.26	7.70	
	1000	20′ eave height		1,000	.056		4.61	1.67	1.01	7.29	9.10	
	1100	24′ eave height		1,000	.056		5.25	1.67	1.01	7.93	9.85	
	1200	Clear span tapered beam frame, 26 ga. colored roofing and siding										
	1300	30′ wide, 10′ eave height	E-2	1,800	.031	SF Flr.	4.77	.93	.56	6.26	7.50	
	1400	14′ eave height		1,800	.031		5.40	.93	.56	6.89	8.15	
	1500	16′ eave height		1,400	.040		5.85	1.19	.72	7.76	9.30	
	1600	20′ eave height		1,000	.056		6.55	1.67	1.01	9.23	11.25	
	1700	40′ wide, 10′ eave height		1,800	.031		4.39	.93	.56	5.88	7.05	
	1800	14′ eave height		1,800	.031		4.78	.93	.56	6.27	7.50	
	1900	16′ eave height		1,400	.040		5	1.19	.72	6.91	8.40	
	2000	20′ eave height		1,000	.056		5.60	1.67	1.01	8.28	10.20	
	2100	50′ to 80′ wide, 10′ eave height		1,800	.031		4.42	.93	.56	5.91	7.10	
	2200	14′ eave height		1,800	.031		4.50	.93	.56	5.99	7.20	
	2300	16′ eave height		1,400	.040		4.69	1.19	.72	6.60	8.05	

Important: See the Reference Section for critical supporting data - Reference Nos., Crews, & City Cost Indexe

131 210 | Pre-Engineered Buildings

		CREW	DAILY OUTPUT	LABOR-HOURS	UNIT	1999 BARE COSTS MAT.	LABOR	EQUIP.	TOTAL	TOTAL INCL O&P
2400	20′ eave height R131-210	E-2	1,000	.056	SF Flr.	5.55	1.67	1.01	8.23	10.20
2500	Single post 2-span frame, 26 ga. colored roofing and siding									
2600	80′ wide, 14′ eave height	E-2	1,800	.031	SF Flr.	3.71	.93	.56	5.20	6.30
2700	16′ eave height		1,400	.040		3.96	1.19	.72	5.87	7.25
2800	20′ eave height		1,000	.056		4.25	1.67	1.01	6.93	8.70
2900	24′ eave height		1,000	.056		4.65	1.67	1.01	7.33	9.15
3000	100′ wide, 14′ eave height		1,800	.031		3.45	.93	.56	4.94	6.05
3100	16′ eave height		1,400	.040		3.83	1.19	.72	5.74	7.10
3200	20′ eave height		1,000	.056		4.23	1.67	1.01	6.91	8.70
3300	24′ eave height		1,000	.056		4.45	1.67	1.01	7.13	8.95
3400	120′ wide, 14′ eave height		1,800	.031		3.51	.93	.56	5	6.10
3500	16′ eave height		1,400	.040		3.64	1.19	.72	5.55	6.90
3600	20′ eave height		1,000	.056		3.89	1.67	1.01	6.57	8.30
3700	24′ eave height	↓	1,000	.056	↓	4.32	1.67	1.01	7	8.80
3800	Double post 3-span frame, 26 ga. colored roofing and siding									
3900	150′ wide, 14′ eave height	E-2	1,800	.031	SF Flr.	3.28	.93	.56	4.77	5.85
4000	16′ eave height		1,400	.040		3.24	1.19	.72	5.15	6.45
4100	20′ eave height		1,000	.056		3.42	1.67	1.01	6.10	7.80
4200	24′ eave height	↓	1,000	.056	↓	4.33	1.67	1.01	7.01	8.80
4300	Triple post 4-span frame, 26 ga. colored roofing and siding									
4400	160′ wide, 14′ eave height	E-2	1,800	.031	SF Flr.	3.02	.93	.56	4.51	5.55
4500	16′ eave height		1,400	.040		3.14	1.19	.72	5.05	6.35
4600	20′ eave height		1,000	.056		3.42	1.67	1.01	6.10	7.80
4700	24′ eave height		1,000	.056		4.16	1.67	1.01	6.84	8.60
4800	200′ wide, 14′ eave height		1,800	.031		3.04	.93	.56	4.53	5.60
4900	16′ eave height		1,400	.040		3.21	1.19	.72	5.12	6.40
5000	20′ eave height		1,000	.056		3.42	1.67	1.01	6.10	7.80
5100	24′ eave height	↓	1,000	.056	↓	3.74	1.67	1.01	6.42	8.15

212

131 220 | Metal Building Systems

		CREW	DAILY OUTPUT	LABOR-HOURS	UNIT	MAT.	LABOR	EQUIP.	TOTAL	TOTAL INCL O&P
0010	**SHELTERS** Aluminum frame, acrylic glazing, 3′ x 9′ x 8′ high	2 Sswk	1.14	14.035	Ea.	2,550	430		2,980	3,575
0100	9′ x 12′ x 8′ high	″	.73	21.918	″	4,250	670		4,920	5,900

221

131 230 | Greenhouses

		CREW	DAILY OUTPUT	LABOR-HOURS	UNIT	MAT.	LABOR	EQUIP.	TOTAL	TOTAL INCL O&P
0010	**GREENHOUSE** Shell only, stock units, not incl. 2′ stub walls,									
0020	foundation, floors, heat or compartments									
0300	Residential type, free standing, 8′-6″ long x 7′-6″ wide	2 Carp	59	.271	SF Flr.	35	7.40		42.40	50
0400	10′-6″ wide		85	.188		27	5.15		32.15	37.50
0600	13′-6″ wide		108	.148		24	4.04		28.04	33
0700	17′-0″ wide		160	.100		27	2.73		29.73	34
0900	Lean-to type, 3′-10″ wide		34	.471		31	12.85		43.85	54
1000	6′-10″ wide	↓	58	.276	↓	24	7.55		31.55	38.50
1100	Wall mounted, to existing window, 3′ x 3′	1 Carp	4	2	Ea.	335	54.50		389.50	455
1120	4′ x 5′	″	3	2.667	″	500	73		573	665
1200	Deluxe quality, free standing, 7′-6″ wide	2 Carp	55	.291	SF Flr.	69	7.95		76.95	88.50
1220	10′-6″ wide		81	.198		64	5.40		69.40	79
1240	13′-6″ wide		104	.154		60	4.20		64.20	72.50
1260	17′-0″ wide		150	.107		51	2.91		53.91	60.50
1400	Lean-to type, 3′-10″ wide		31	.516		80	14.10		94.10	110
1420	6′-10″ wide		55	.291		75	7.95		82.95	95
1440	8′-0″ wide	↓	97	.165	↓	70	4.50		74.50	84
1500	Commercial, custom, truss frame, incl. equip., plumbing, elec.,									
1510	benches and controls, under 2,000 S.F., minimum				SF Flr.					30
1550	Maximum									38
1700	Over 5,000 S.F., minimum									23
1750	Maximum				↓					30

231

131 230 | Greenhouses

		131 230 Greenhouses	CREW	DAILY OUTPUT	LABOR-HOURS	UNIT	1999 BARE COSTS MAT.	LABOR	EQUIP.	TOTAL	TOTAL INCL O&P
231	2000	Institutional, custom, rigid frame, including compartments and									
	2010	multi-controls, under 500 S.F., minimum				SF Flr.					77
	2050	Maximum									99
	2150	Over 2,000 S.F., minimum									39
	2200	Maximum									65
	2400	Concealed rigid frame, under 500 S.F., minimum									90
	2450	Maximum									110
	2550	Over 2,000 S.F., minimum									68
	2600	Maximum									79
	2800	Lean-to type, under 500 S.F., minimum									80
	2850	Maximum									120
	3000	Over 2,000 S.F., minimum									44
	3050	Maximum				↓					73
	3600	For 1/4" clear plate glass, add				SF Surf	1.40			1.40	1.54
	3700	For 1/4" tempered glass, add				"	3.15			3.15	3.47
	3900	For cooling, add, minimum				SF Flr.	2.10			2.10	2.31
	4000	Maximum					5.20			5.20	5.70
	4200	For heaters, 13.6 MBH, add					4			4	4.40
	4300	60 MBH, add				↓	1.50			1.50	1.65
	4500	For benches, 2' x 3'-6", add				SF Hor.	17.90			17.90	19.70
	4600	3' x 10', add				S.F.	9.70			9.70	10.65
	4800	For controls, add, minimum				Total	1,800			1,800	1,975
	4900	Maximum				"	10,700			10,700	11,800
	5100	For humidification equipment, add				M.C.F.	4.60			4.60	5.05
	5200	For vinyl shading, add				S.F.	.97			.97	1.07
	6000	Geodesic hemisphere, 1/8" plexiglass glazing									
	6050	8' diameter	2 Carp	2	8	Ea.	2,025	218		2,243	2,575
	6150	24' diameter		.35	45.714		10,400	1,250		11,650	13,400
	6250	48' diameter	↓	.20	80	↓	28,200	2,175		30,375	34,400

131 240 | Portable Buildings

		131 240 Portable Buildings	CREW	DAILY OUTPUT	LABOR-HOURS	UNIT	1999 BARE COSTS MAT.	LABOR	EQUIP.	TOTAL	TOTAL INCL O&P
242	0010	**COMFORT STATIONS** Prefab., stock, w/doors, windows & fixt.									
	0100	Not incl. interior finish or electrical									
	0300	Mobile, on steel frame, minimum				S.F.	37.50			37.50	41
	0350	Maximum					63.50			63.50	70
	0400	Permanent, including concrete slab, minimum	B-12J	50	.320		138	8.55	12.75	159.30	179
	0500	Maximum	"	43	.372	↓	200	9.95	14.80	224.75	251
	0600	Alternate pricing method, mobile, minimum				Fixture	1,725			1,725	1,900
	0650	Maximum					2,575			2,575	2,825
	0700	Permanent, minimum	B-12J	.70	22.857		9,900	610	910	11,420	12,800
	0750	Maximum	"	.50	32	↓	16,500	855	1,275	18,630	20,900
244	0010	**GARDEN HOUSE** Prefab wood, no floors or foundations									
	0100	32 to 200 S.F., minimum	2 Carp	200	.080	SF Flr.	13.45	2.18		15.63	18.25
	0300	Maximum	"	48	.333	"	24.50	9.10		33.60	41.50
245	0010	**KIOSKS** Round, 5' diameter, 8' high, 1/4" fiberglass wall				Ea.	5,500			5,500	6,050
	0100	1" insulated double wall, fiberglass					6,250			6,250	6,875
	0500	Rectangular, 5' x 9', 7'-6" high, 1/4" fiberglass wall					8,000			8,000	8,800
	0600	1" insulated double wall, fiberglass				↓	9,500			9,500	10,500
247	0010	**PORTABLE BOOTHS** Prefab aluminum with doors, windows, ext. roof									
	0100	lights wiring & insulation, 15 S.F. building, O.D., painted, minimum				S.F.	255			255	281
	0300	30 S.F. building, minimum					175			175	193
	0400	50 S.F. building, minimum					130			130	143
	0600	80 S.F. building, minimum					105			105	116
	0700	100 S.F. building, minimum				↓	97			97	107

131 240 | Portable Buildings		CREW	DAILY OUTPUT	LABOR-HOURS	UNIT	1999 BARE COSTS MAT.	LABOR	EQUIP.	TOTAL	TOTAL INCL O&P	
0900	Acoustical booth, 27 Db @ 1,000 Hz, 15 S.F. floor				Ea.	2,875			2,875	3,175	247
1000	7′ x 7′-6″, including light & ventilation					5,900			5,900	6,500	
1200	Ticket booth, galv. steel, not incl. foundations., 4′ x 4′					4,425			4,425	4,875	
1300	4′ x 6′					5,175			5,175	5,700	
131 250 | Grandstands/Bleachers											
0010	**GRANDSTANDS** Permanent, municipal, including foundation										251
0050	Steel understructure w/aluminum closed deck, minimum				Seat					130	
0100	Maximum									225	
0300	Steel, minimum									25	
0400	Maximum									80	
0600	Aluminum, extruded, stock design, minimum									70	
0700	Maximum									115	
0900	Composite, steel, wood and plastic, stock design, minimum									30	
1000	Maximum									90	
131 520 | Swimming Pools											
0010	**SWIMMING POOL ENCLOSURE** Translucent, free standing, R131-520										521
0020	not including foundations, heat or light										
0200	Economy, minimum	2 Carp	200	.080	SF Hor.	10	2.18		12.18	14.45	
0300	Maximum		100	.160		21	4.37		25.37	30	
0400	Deluxe, minimum		100	.160		23	4.37		27.37	32.50	
0600	Maximum		70	.229		250	6.25		256.25	285	
0700	For motorized roof, 40% opening, solid roof, add					3			3	3.30	
0800	Skylight type roof, add					4.50			4.50	4.95	
0900	Air-inflated, including blowers and heaters, minimum									3.50	
1000	Maximum									6.70	
0010	**SWIMMING POOL EQUIPMENT** Diving stand, stainless steel, 3 meter	2 Carp	.40	40	Ea.	4,725	1,100		5,825	6,900	523
0300	1 meter		2.70	5.926		2,650	162		2,812	3,175	
0600	Diving boards, 16′ long, aluminum		2.70	5.926		1,250	162		1,412	1,625	
0700	Fiberglass		2.70	5.926		900	162		1,062	1,250	
0900	Filter system, sand or diatomite type, incl. pump, 6,000 gal./hr.	2 Plum	1.80	8.889	Total	1,000	290		1,290	1,550	
1020	Add for chlorination system, 800 S.F. pool		3	5.333	Ea.	270	174		444	560	
1040	5,000 S.F. pool		3	5.333	″	560	174		734	880	
1100	Gutter system, stainless steel, with grating, stock,										
1110	contains supply and drainage system	E-1	20	1.200	L.F.	160	36	4.12	200.12	244	
1120	Integral gutter and 5′ high wall system, stainless steel	″	10	2.400	″	240	72.50	8.25	320.75	400	
1200	Ladders, heavy duty, stainless steel, 2 tread	2 Carp	7	2.286	Ea.	251	62.50		313.50	375	
1500	4 tread	″	6	2.667		340	73		413	490	
1900	Portable					2,000			2,000	2,200	
2100	Lights, underwater, 12 volt, with transformer, 300 watt	1 Elec	.40	20		150	640		790	1,125	
2200	110 volt, 500 watt, standard		.40	20		140	640		780	1,100	
2400	Low water cutoff type		.40	20		90	640		730	1,050	
2800	Heaters, see division 155-150										
3000	Pool covers, reinforced vinyl, material only				S.F.	.35			.35	.39	
3050	Automatic, electric									9.65	
3100	Vinyl water tube, material only, minimum					.25			.25	.28	
3200	Maximum					.45			.45	.50	
3250	Sealed air bubble polyethylene solar blanket					.17			.17	.19	
3300	Slides, tubular, fiberglass, aluminum handrails & ladder, 5′-0″, straight	2 Carp	1.60	10	Ea.	1,025	273		1,298	1,550	
3320	8′-0″, curved		3	5.333		9,600	146		9,746	10,800	
3400	10′-0″, curved		1	16		12,000	435		12,435	13,900	
3420	12′-0″, straight with platform		1.20	13.333		950	365		1,315	1,625	
0010	**SWIMMING POOLS** Residential in-ground, vinyl lined, concrete sides										525
0020	Sides including equipment, sand bottom	B-52	300	.187	SF Surf	10.40	4.71	1.44	16.55	20.50	

131 520 | Swimming Pools

		131 520 Swimming Pools	CREW	DAILY OUTPUT	LABOR-HOURS	UNIT	1999 BARE COSTS MAT.	LABOR	EQUIP.	TOTAL	TOTAL INCL O&P
525	0100	Metal or polystyrene sides	B-14	410	.117	SF Surf	8.70	2.66	.53	11.89	14.30
	0200	Add for vermiculite bottom				↓	.67			.67	.74
	0500	Gunite bottom and sides, white plaster finish									
	0600	12′ x 30′ pool	B-52	145	.386	SF Surf	17.25	9.75	2.98	29.98	37.50
	0720	16′ x 32′ pool	↓	155	.361	↓	15.50	9.10	2.79	27.39	34.50
	0750	20′ x 40′ pool	↓	250	.224	↓	13.85	5.65	1.73	21.23	26
	0810	Concrete bottom and sides, tile finish									
	0820	12′ x 30′ pool	B-52	80	.700	SF Surf	17.40	17.65	5.40	40.45	52.50
	0830	16′ x 32′ pool	↓	95	.589	↓	14.40	14.85	4.55	33.80	44.50
	0840	20′ x 40′ pool	↓	130	.431	↓	11.45	10.85	3.33	25.63	33.50
	1100	Motel, gunite with plaster finish, incl. medium									
	1150	capacity filtration & chlorination	B-52	115	.487	SF Surf	21	12.30	3.76	37.06	47
	1200	Municipal, gunite with plaster finish, incl. high									
	1250	capacity filtration & chlorination	B-52	100	.560	SF Surf	27.50	14.10	4.32	45.92	57
	1350	Add for formed gutters				L.F.	47.50			47.50	52.50
	1360	Add for stainless steel gutters				"	141			141	155
	1600	For water heating system, see division 155-150									
	1700	Filtration and deck equipment only, as % of total				Total				20%	20%
	1800	Deck equipment, rule of thumb, 20′ x 40′ pool				SF Pool					1.30
	1900	5000 S.F. pool				"					1.90
	3000	Painting pools, preparation + 3 coats, 20′ x 40′ pool, epoxy	2 Pord	.33	48.485	Total	595	1,200		1,795	2,500
	3100	Rubber base paint, 18 gallons	"	.33	48.485	↓	450	1,200		1,650	2,350
	3500	42′ x 82′ pool, 75 gallons, epoxy paint	3 Pord	.14	171	↓	2,500	4,275		6,775	9,275
	3600	Rubber base paint	"	.14	171	↓	1,925	4,275		6,200	8,650

131 600 | Ice Rinks

		131 600 Ice Rinks	CREW	DAILY OUTPUT	LABOR-HOURS	UNIT	MAT.	LABOR	EQUIP.	TOTAL	TOTAL INCL O&P
601	0010	**ICE SKATING** Equipment incl. refrigeration and plumbing, not									
	0020	including building or slab, 85′ x 200′ rink									
	0300	55° system, 5 mos., 100 ton				Total					325,000
	0700	90° system, 12 mos., 135 ton				"					350,000
	1000	Dasher boards, 1/2″ H.D. polyethylene faced steel frame, 3′ acrylic									
	1020	screen at sides, 5′ acrylic ends, 85′ x 200′	F-5	.06	533	Ea.	115,000	14,800		129,800	150,000
	1100	Fiberglass & aluminum construction, same sides and ends	"	.06	533	↓	125,000	14,800		139,800	161,000
	1200	Subsoil heating system (recycled from compressor), 85′ x 200′	Q-7	.27	118	↓	18,500	3,700		22,200	26,000
	1300	Subsoil insulation, 2 lb. polystyrene with vapor barrier, 85′ x 200′	2 Carp	.14	114	↓	24,000	3,125		27,125	31,300

132 | Tanks, Tank Covers, Filtration Equipment

132 050 | Ground Storage Tanks

		132 050 Ground Storage Tanks	CREW	DAILY OUTPUT	LABOR-HOURS	UNIT	1999 BARE COSTS MAT.	LABOR	EQUIP.	TOTAL	TOTAL INCL O&P
051	0010	**TANKS** Not incl. pipe or pumps, prestress conc., 250,000 gal.				Ea.					275,000
	0100	500,000 gallons				↓					375,000
	0300	1,000,000 gallons				↓					530,000
	0400	2,000,000 gallons				↓					800,000
	0600	4,000,000 gallons				↓					1,270,000
	0700	6,000,000 gallons				↓					1,730,000
	0750	8,000,000 gallons				↓					2,200,000
	0800	10,000,000 gallons				↓					2,650,000
	0900	Steel, ground level, not incl. foundations, 100,000 gallons				↓					96,000
	1000	250,000 gallons				↓					135,000